Springer Series in Information Sciences 29

Editor: Thomas S. Huang

Springer Series in Information Sciences

Editors: Thomas S. Huang Teuvo Kohonen Manfred R. Schroeder
Managing Editor: H. K. V. Lotsch

1 **Content-Addressable Memories**
By T. Kohonen 2nd Edition

2 **Fast Fourier Transform and Convolution Algorithms**
By H. J. Nussbaumer 2nd Edition

3 **Pitch Determination of Speech Signals**
Algorithms and Devices By W. Hess

4 **Pattern Analysis and Understanding**
By H. Niemann 2nd Edition

5 **Image Sequence Analysis**
Editor: T. S. Huang

6 **Picture Engineering**
Editors: King-sun Fu and T. L. Kunii

7 **Number Theory in Science and Communication** With Applications in Cryptography, Physics, Digital Information, Computing, and Self-Similarity By M. R. Schroeder
2nd Edition

8 **Self-Organization and Associative Memory**
By T. Kohonen 3rd Edition

9 **Digital Picture Processing**
An Introduction By L. P. Yaroslavsky

10 **Probability, Statistical Optics, and Data Testing**
A Problem Solving Approach
By B. R. Frieden 2nd Edition

11 **Physical and Biological Processing of Images** Editors: O. J. Braddick and A. C. Sleigh

12 **Multiresolution Image Processing and Analysis** Editor: A. Rosenfeld

13 **VLSI for Pattern Recognition and Image Processing** Editor: King-sun Fu

14 **Mathematics of Kalman-Bucy Filtering**
By P. A. Ruymgaart and T. T. Soong
2nd Edition

15 **Fundamentals of Electronic Imaging Systems**
Some Aspects of Image Processing
By W. F. Schreiber 3rd Edition

16 **Radon and Projection Transform-Based Computer Vision**
Algorithms, A Pipeline Architecture, and Industrial Applications By J. L. C. Sanz, E. B. Hinkle, and A. K. Jain

17 **Kalman Filtering**
with Real-Time Applications
By C. K. Chui and G. Chen 2nd Edition

18 **Linear Systems and Optimal Control**
By C. K. Chui and G. Chen

19 **Harmony: A Psychoacoustical Approach**
By R. Parncutt

20 **Group-Theoretical Methods in Image Understanding**
By Ken-ichi Kanatani

21 **Linear Prediction Theory**
A Mathematical Basis for Adaptive Systems
By P. Strobach

22 **Psychoacoustics** Facts and Models
By E. Zwicker and H. Fastl

23 **Digital Image Restoration**
Editor: A. K. Katsaggelos

24 **Parallel Algorithms in Computational Science**
By D. W. Heermann and A. N. Burkitt

25 **Radar Array Processing**
Editors: S. Haykin, J. Litva, and T. J. Shepherd

26 **Signal Processing and Systems Theory**
Selected Topics
By C. K. Chui and G. Chen

27 **3D Dynamic Scene Analysis**
A Stereo Based Approach
By Z. Zhang and O. Faugeras

28 **Theory of Reconstruction from Image Motion**
By S. Maybank

29 **Motion and Structure from Image Sequences**
By J. Weng, T. S. Huang, and N. Ahuja

Juyang Weng Thomas S. Huang
Narendra Ahuja

Motion and Structure from Image Sequences

With 119 Figures

Springer-Verlag
Berlin Heidelberg New York
London Paris Tokyo
Hong Kong Barcelona
Budapest

Professor Juyang Weng
Professor Thomas S. Huang
Professor Narendra Ahuja

University of Illinois at Urbana-Champaign, Beckman Institute,
405 North Mathews Ave., Urbana, IL 61801, USA

ISBN 3-540-55672-9 Springer-Verlag Berlin Heidelberg New York
ISBN 0-387-55672-9 Springer-Verlag New York Berlin Heidelberg

Library of Congress Cataloging-in-Publication Data. Weng, Juyang, 1957- Motion and structure from image sequences / Juyang Weng, Thomas S. Huang, Narendra Ahuja. p. cm. – (Springer series in information sciences; 29) Includes bibliographical references and index. ISBN 3-540-55672-9 (Berlin). – ISBN 0-387-55672-9 (New York) 1. Image processing. 2. Computer vision. I. Huang, Thomas S., 1936- . II. Ahuja, Narendra, 1950- . III. Title. IV. Series. TA 1632.W46 1993 006.4'2–dc 92-28268

Typesetting: Camera-ready by authors

54/3140-5 4 3 2 1 0 – Printed on acid-free paper

Preface

This book has grown out of our research activities in the area of deriving geometric and motion descriptions of dynamic scenes from visual images. It should prove useful for researchers who are interested in a comprehensive treatment of this topic and practitioners who wish to find algorithms they can use.

The results presented here are mostly very recent, some of them being published here for the first time. A rich collection of practical algorithms is presented, applicable to monocular or stereo views.

Effort has been made to present a balanced treatment of the theoretical and practical issues. Theoretical issues of broad and practical significance have been explored in detail, while other more specific ones are not covered. The topics covered in detail include image matching and motion estimation, error analysis, error estimation, stability, optimality, and experiments with simulated and real image data, with the objective being to present general techniques that have a mathematical foundation and engineering applications.

This book is meant for graduate students and practicing engineers who are interested in computer vision. The organization of the material is intended to facilitate extraction and use of the algorithms without necessitating study of the rest of the material.

Chapter 1 introduces the problem of motion analysis and briefly reviews past work. Chapter 2 is concerned with image matching, a step which is prerequisite for all further discussions in the book. Chapter 3 presents a closed-form solution for motion and structure parameters from two views of a rigid object. In the presence of noise, this solution is required as an initial condition for the optimization approach investigated in Chapter 4. The principles covered in Chapter 4 are used in later chapters whenever optimization is discussed. Chapters 5 and 6 address two separate issues, planar surfaces and line features for motion analysis, each of which can be skipped without loss of continuity. While Chapters 3 to 6 all assume monocular views, Chapter 7 is devoted to the analysis of image pairs acquired using stereo cameras. The techniques of robust statistics introduced in Chapter 7 can also be applied to monocular images. The subject of analyzing long image sequences, instead of only two or three views, is also treated in Chapter 7. Modeling and prediction of motion are covered in Chapter 8. Although both Chapter 7 and Chapter 8 consider long image sequences, they are different. Chapter 7 does not impose any restriction on changes in motion with time, and thus interframe motions can be very different from one to the next. In Chapter 8, however, the motion is assumed to be smooth in the sense that it follows certain specified, although not very restrictive, object dynamics. The method in Chapter 7 is useful when motion is unpredictable and quite irregular, while that in Chapter 8 can be used when motion is known to be smooth and therefore predictable. The following flow chart indicates the interdependence among the material in different chapters.

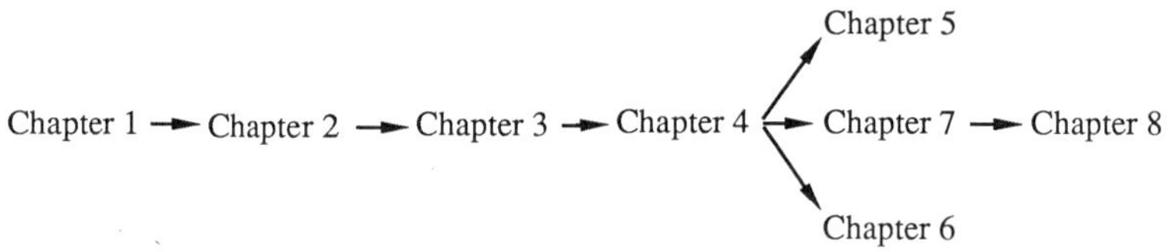

Most of the work presented in this book was done as part of the Ph.D. program of one of the authors (J. W.) at the University of Illinois during 1985-1988, while the other two authors were his research advisors. A small portion of the work was done during 1989-1990 when one of the authors (J. W.) was with Centre de Recherche Informatique de Montréal and Ecole Polytechnique, University of Montreal, Montreal, Canada. Many colleagues have had valuable influence on this book. First and foremost, we are indebted to Roger Y. Tsai and Bob Yen, former Ph.D. students of one of the authors (T. S. H.), for their pioneering work in motion estimation using points and lines, respectively. During 1986, intensive discussions with Xinhua Zhuang significantly helped us in developing the algorithm presented in Chapter 3. Collaborations with J. K. Aggarwal greatly influenced our work on optimization. Discussions with Rama Chellappa made us investigate the inherent stability of motion problems more critically. The conjecture of a sufficient condition for the decomposability of the

E matrix came to our attention during our interaction with Olivier Faugeras. Discussions with John Aloimonos and Minas Spetsakis triggered our investigation on the equivalence between two methods of linear motion estimation. Interaction with colleagues Bill Hoff, Steve Blostein, Homer Chen, Christian Debrunner, Yuncai Liu, Dorothea Blostein, Edward Altman, Toshifumi Tsukiyama, Gary Silverman, Lynn Abbott, Dmitry Goldgof, Paul Cohen, Ning Cui, and Nicola Rebibo had a significant impact on the work presented here.

Much of our work in motion estimation has been supported by the National Science Foundation. We gratefully acknowledge the encouragement and support of Drs. E. Barrett, S. S. Chen, Y. T. Chien, K. Laws, and H. Moraff.

Juyang Weng
Urbana, Illinois *Thomas S. Huang*
December 1991 *Narendra Ahuja*

Contents

1. Introduction

Computer vision is concerned with extracting information about a scene from its images. Analogous to the roles of motion analysis in biological vision, the objectives of computer interpretation of images of a dynamic scene include detecting moving objects, determining their positions and structure, tracking them, predicting their future locations, and even using the motion information for recognition of moving objects. In much of the work on motion analysis so far, the objective has been to estimate the three-dimensional (3-D) motion and structure of the scene such as location, orientation, velocity, trajectory, and shape. This book presents theories, methodologies, techniques, and algorithms developed for extracting this type of information from motion image sequences.

There is a long list of applications that have motivated the strong current interest in motion research. The abilities of motion detection, tracking, and estimation are essential to many automation tasks such as traffic flow monitoring and control, traffic speed monitoring, and tollway billing. Moving target detection and tracking is of great interest to the defense industry. In coding of image sequences for transmission, effective bandwidth reduction can be achieved through the estimation of motion in the image sequence. A large reduction of bandwidth is particularly in demand for video conferencing and videophone systems. Estimation, characterization and understanding of human motion in athletic activities, dancing, pilot training and patient rehabilitation are of interest in several disciplines. The analysis of human heart motion allows diagnosis of heart problems. Velocity estimation for fluid flow is of interest to meteorology and experimental fluid mechanics with applications ranging from weather forecasting to aircraft design. Obstacle detection by a moving vehicle is necessary for collision avoidance and path planning for autonomous navigation.

1.1 Formulation of the Problem

A proper formulation of the motion estimation problem requires the identification of several characteristics of the image acquisition process and the scene, including: the camera model, the number of cameras, the allowed amount of interframe motion, the reference coordinates for motion, and the type of motion. A discussion of these issues is in order here.

There are two basic types of camera projection models: perspective and orthographic. Perspective projection, also called central projection, is a good model of the actual geometry of image formation. Mathematically, perspective projection corresponds to a pin hole camera model. Orthographic projection, on the other hand, neglects the depth differences in the visible scene. The objects are projected to the image plane orthographically and compressed in size using a scale factor (which is assumed to be known or is estimated). This scale factor is related to the distance between the camera and the objects. Orthographic projection is often used as an approximation of perspective projection in order to simplify mathematics. Such an approximation is good only when the object is far away so that the depth variation in the field of view is very small compared to the actual depths. However, since orthographic projection is only an approximation and its applicability is limited, we shall use only perspective projection as camera model in this book.

The number of cameras acquiring images is another variable in formulating the problem. If two image sensors are used and the relative position between the sensors is known, stereo matching (matching between the two images) can be used to determine the depth through triangulation. Although three or more image sensors are useful under some conditions, they basically provide the same information as two sensors. If a single image sensor is used, however, the absolute depth information of the scene is lost. Interestingly, with one image sensor, the motion and structure of the scene can still be determined from two or more image frames, but the magnitude of translation and the depth of the scene can be determined only up to a global scale factor. We will investigate these issues. Mathematically the monocular problem is more complicated than its binocular counterpart. Many methods used for monocular situation can be easily extended to binocular situations.

With regard to the allowed amount of interframe motion, distinction has been made between small interframe motion and large interframe motion. The former is sometimes called short term motion and the latter is called long term motion. An image is a projection of the scene at one time instant. If the interframe motion is small enough, the difference between two images can be used to approximate instantaneous velocity. Then, velocity based mathematical analysis can be used to approximate the small three-dimensional displacement. This approximation sometimes simplifies mathematical treatment, although it introduces a source of error. For the mathematical formulation used in this book, we do not restrict the amount of motion and thus the formulation is applicable to either large or small interframe motion.

Regarding reference coordinate systems, two cases are common: moving observer and stationary observer. The former case is usually called "passive navigation", implying an observer moving in a stationary background. The latter case assumes stationary image sensors. If only kinematics is considered, both cases can be viewed as basically the same. The difference is just the location of the reference coordinate system. In the former case, the coordinate system is on the scene. In the latter case, the coordinate system is on the camera. The objective of both cases is to determine the relative motion between the image sensors and the scene, as well as the geometrical structure of the visible scene. With long image sequences, the motion represented in the above two coordinate systems have different meanings. We will discuss interpretations in these two types of coordinate systems where appropriate. If motion dynamics is considered, as in Chapter 8, the reference coordinate system should be stationary relative to the ground (inertia system) so that object dynamics can be treated effectively.

With respect to the type of motion, it is characterized as rigid or nonrigid. In general, the field of view may contain independently moving objects. The scene has to be segmented into different segments, each segment corresponding to an object. Each object may be rigid or nonrigid. If rigid, then rigid motion analysis can be performed for each segment separately. We will assume rigidity except when discussing image matching (Chapter 2). Motion estimation for nonrigid objects, e.g., human heart or fluid, is beyond the scope of this book.

1.2 Overview of Recent Progress

Among psychologists, motion related visual perception has been an active research area for many years. But motion analysis for computer vision has a short history, mainly the last 15 years. In recent years, we have seen greater theoretical, methodological and experimental advances.

Theoretically, many important open questions have been answered and significant insights have been gained into other remaining open questions. The issue of motion estimation from two views is the back bone of analyzing long image sequences. Although the mathematical solution of motion from noise-free three-dimensional point data is relatively simple, the nontrivial problem is to determine the "best" solution from noisy data. The least-squares solutions independently developed by several researchers are a significant result. Since then, the scalar weighted solution has reduced the error of the least-squares solutions by roughly a factor of two. Later, another factor of two improvement over the scalar weighted solution has been brought about by the matrix-weighted closed-form solution. The solution of motion from monocular views is not a trivial problem, even from noise-free data. The closed-form solution using a linear algorithm was a significant development. Although it requires more data than inherently necessary, a closed-form solution made a direct solution possible and first answered the uniqueness question. Recent results have provided some insight into cases where the linear algorithm fails but there still exists a unique solution. For the case of planar surfaces, the issue has become clear. It is known now that the closed-form solution from the plane-based linear algorithm solves all the cases that are solvable. Line-based closed-form solution has also become available and the uniqueness of the solution has been established, but a complete story on solution using fewer data than required by the available linear algorithm is not yet available.

Methodologically, motion analysis has used a number of approaches. From the point of view of image primitives, we have seen methods that use raw intensities, zero-crossing contours, edges, points, lines, regions, correlations, local energy, and local phase. The outcome of matching using these primitives is given in terms of sparsely located matched pairs of primitives, dense optical flow (velocity), or dense displacement field, depending on which primitives are used and the amount of interframe motion allowed. Some of the methods for computing motion parameters are iterative and some give closed-form solutions. There are also direct methods that do not require complete image matching, although surface type is more restricted. A series of efforts has been devoted to the study of the stability of solutions. There have been algorithm-dependent and algorithm independent error analyses. Along this line, optimization methods have been employed. Some of these methods appear to have reached theoretical lower error bounds under assumed noise distributions. Recently, robust statistical methods have been applied to motion problems in order to make the results insensitive to outliers in the data and to noise distribution. Kalman filter and other related techniques have played an important role in the processing of long image sequences. In the past several years, we have also seen developments in motion analysis of nonrigid objects, such as articulated objects, human heart, fluid, human face, and human body.

Experimentally, it is encouraging to see work with real imagery. Some of the difficulties in performing motion related experiments include accurate camera calibration, precise motion control, acquisition of ground truth represented in the sensor-centered coordinate system, and working with long image sequences.

This book uses an organization which closely follows the paradigm common in motion research. Accordingly, the task of estimating 3-D motion parameters and structure of scene from a long image sequence is divided into three subtasks. The first subtask is to establish feature correspondences between different images at two time instants. Chapter 2 addresses this problem of image matching. The second subtask is to compute the structure of the scene and determine the interframe motion parameters (called two-view motion parameters). This is the subject of Chapters 3 to 6 and part of Chapter 7. The third subtask is to perform motion and structure estimation from a sequence of many views taken at many time instants, which allows the use of repeated observations of the same part of the scene. This is the subject of the last part of Chapter 7. Knowledge about object dynamics and thus temporal coherence of motion can be used to further constrain the estimation process. One such approach is discussed in Chapter 8.

1.3 Bibliographical Notes

Motion analysis in computer vision started in the early 70's. A book edited by Huang in 1975 [1.6] consists of a collection of earlier papers on the subject. Since the first workshop on motion held in Philadelphia in 1979 [1.2], motion has attracted a broad interest. For example, the series of IEEE Workshops on Visual Motion, is devoted to motion related research. There are several books on motion. *The Interpretation of Visual Motion* [1.14] by Ullman presents a study of both the psychological aspects and the computational formulations of motion estimation. *The Measurement of Visual Motion* [1.5] by Hildreth is concerned with the measurement of image plane (or retinal) motion. The edited books on motion analysis include [1.7, 3, 8, 10, 4]. A large number of research papers have appeared in conferences, journals, and books. A partial list of publications on motion can be found in a host of survey articles [1.9, 11-13, 1].

2. Image Matching

Establishing image correspondences is one of the most challenging steps in motion and structure analysis. The algorithms discussed in the following chapters all require the result of image matching as a prerequisite. In this chapter, we first briefly introduce some existing approaches to image matching. Then we describe how image attributes can be used to perform this task. We also address the related important issues such as structural discontinuity and occlusion. The matching algorithm presented in this chapter is directly applicable to general two-view matching. Stereo matching is a special case of general two-view matching in that the motion between two views is known and thus the epipolar constraint can be used. The matching algorithm presented in this chapter can be easily specialized to perform stereo matching using the epipolar constraint.

2.1 Approaches to Establishing Correspondences

In order to determine motion from time-varying image sequences, it is necessary to establish, either explicitly or implicitly, certain type of correspondences between images, i.e., to identify image elements in different images that correspond to the same element of the sensed scene. The matching elements, or tokens, can vary significantly from one approach to another. Although it is possible for one to process many images simultaneously to establish image matches, the underlying basic problem is to establish the correspondences between two images. Existing techniques for image matching roughly fall into two categories: continuous and discrete.

(1) Continuous approaches. Although the objective of the approaches in this category is to determine image velocity field instead of performing explicit matching between images, the computed

velocity field amounts to image matching. Each velocity vector approximates the correspondence between two points in different images. Ideally, one needs the projection of 3-D velocity on the image plane. However, since such a projection is not directly available from visual sensors, an optical flow field (the field representing the apparent motion of the brightness pattern) is used to approximate the actual image plane velocity field. The techniques in this category typically need the assumption that the interframe motion is small and the intensity function is smooth and well-behaved. Optical flow may be computed based on spatiotemporal variation of image intensity function [2.20, 38, 33, 23, 12, 1, 17]. Though the flow computed from intensity variation is not exactly the same as the projection of 3-D velocities [2.20, 37], they are considered to be similar under certain conditions [2.21]. From the assumption that the intensity i of the corresponding points is preserved from image to image, a linear equation in the two components of the velocity vector $(\alpha, \beta) \triangleq (\frac{du}{dt}, \frac{dv}{dt})$ at the image point (u, v) can be derived:

$$\frac{\partial i(u, v, t)}{\partial u}\alpha + \frac{\partial i(u, v, t)}{\partial v}\beta + \frac{\partial i(u, v, t)}{\partial t} = 0.$$

This equation is called *optical flow constraint equation.* This equation alone is not sufficient to determine the two components of the velocity vector (α, β). A variety of smoothness constraints is proposed to solve this underdetermined problem. A typical one is minimizing $\iint \|\nabla\alpha\|^2 + \|\nabla\beta\|^2 du dv$ proposed by Horn and Schunck [2.20]. Since this isotropic smoothness constraint is inappropriate across the image of occluding edges, Nagel and Enkelmann [2.33] introduced some controlled smoothness constraints with the goal of smoothing only along the edge direction at edge points, and smoothing isotropically at points having small spatial gradient. These methods are commonly called (intensity) gradient based methods. Another existing method is based on energy variation in spatiotemporal data (a pile of images with very small interframe disparities) [2.1, 17]. A family of Gabor-energy filters tuned to different spatial orientations and temporal frequencies are applied to this spatiotemporal data. Image velocity (α, β) is determined by minimizing the difference between the measured motion energies (from the response of Gabor-filters) and those predicted for a pattern with a flat power spectrum.

(2) Discrete approaches. The techniques in this category treat the images as samples of the scene taken at discrete times, and select

discrete features as tokens that are to be matched. Points with high intensity variation are often used as the matching tokens [2.36, 4, 10]. Other features used for matching include closed contours of zero crossings of Laplacian-of-Gaussian images to compute velocity field [2.18], edges for stereo matching [2.28, 30, 15, 35, 19], lines for stereo matching [2.3], correlation of intensity patterns [2.32, 13, 2], or some aspects of higher level scene structure [2.22, 16, 26].

With a continuous approach one usually computes optical flow field along a pixel grid. There is no need for explicit feature extraction and matching. Matching is performed through numerical minimization. These approaches can potentially derive dense depth maps. However, they face the following problems: (1) The existing approaches resort to smoothness constraint to make the underdetermined problem solvable. When discontinuities occur in the velocity field, severe errors result. (2) Since the interframe motion is restricted to be small, the magnitudes of flow vectors are also small (usually within a couple of pixels). So, the velocities computed can be easily overridden by pixel level perturbations. Such a flow with very low signal to noise ratio inherently limits the stability of motion and structure analysis. (3) The assumption that the intensity of the same object patch is constant from one image to the next is not strictly true. (4) Well behaved and well textured intensity images are required for the techniques to be applicable.

Discrete approaches allow either small motion or large motion, corresponding to short range process or long range process, respectively [2.8, 9]. Accurate estimation of motion parameters and structure of the scene is possible under a relatively large motion. Discrete approaches do not suffer much from the problem of varying image intensity as continuous approaches, since discrete features are more invariant than intensity values. Moreover, using discrete approaches, the intensity variation needs not be smooth. However, the approaches in this category also have problems: (1) It is difficult to reliably match discrete features between two images. Because the number of motion-invariant attributes associated with the selected features are very limited, there exist a large number of potential candidates for matching and it becomes difficult to select the correct one. (2) Since the features are generally sparse, only sparse depth data can be obtained. This makes it harder to estimate surfaces. Usually surface interpolation has to be performed to give a complete surface from the sparse depth data. However the interpolated surfaces are often quite different from the real surfaces. (3) Features may be detected in one image but not in the other, e.g., due to occlusion. These lead to problems of unmatchable or mismatched features. (4) To make matching possible, usually various smoothness constraints are used, which may be invalid at occlusion and motion boundaries. Because the features are sparse, the lack of spatial contiguity makes the detection of discontinuities and occlusions very difficult.

In this chapter, we present an approach to image matching that associates multiple attributes with the images to obtain an overdetermined system of matching constraints. This accommodates, to varying degrees, image noise and slight variations in image intensity that result from changes in viewing position, lighting, shading and reflection. More importantly, the displacement vectors are determined in this overdetermined system without resorting to smoothness constraint. A multi-resolution multi-grid computational structure is employed to deal with relatively large image disparities caused by large interframe motions. The approach is capable of dealing with uniform non-textured object surfaces that are often present in real world images. We also address the problem of discontinuities in the field of displacement, and occlusion, which has received little attention in the literature. The algorithm presented in this chapter computes the displacement field and occlusion maps along a dense pixel grid, based on two perspective images of a scene. The algorithm has been tested on images of real world scenes that contain significant occlusions and depth discontinuities. To demonstrate the performance of matching, motion parameters and structure of rigid scenes are computed from the obtained matches, using the methods discussed in Chapter 4.

2.2 An Approach to Image Matching

We first introduce some notation. Matrices are denoted by capital italics of English letters (e.g., R) or capital Greek symbols (e.g., Φ), and so are the sets that have a single-letter name. Vectors are denoted by bold fonts either capital or small (e.g., $\mathbf{X}$ and $\mathbf{x}$). A vector is always represented by a column vector unless specified otherwise. For notation conciseness, vector operations such as cross product, $\mathbf{X}\times\mathbf{Y}$, and matrix operations such as matrix multiplication, $R\,\mathbf{x}$, can be appled to 3-dimensional vectors. Matrix operations precede vector operations. $\mathbf{0}$ denotes a zero vector. Unless otherwise stated, for a matrix $A = [a_{ij}]$, $\|A\|$ denotes the Euclidean norm of the matrix, i.e., $\|[a_{ij}]\| = (\sum_{ij} a_{ij}^2)^{1/2}$. Similarly, for a vector $\mathbf{x} = (a, b, c)^t$, $\|\mathbf{x}\| = (a^2 + b^2 + c^2)^{1/2}$. Euclidean space is implied for any k-dimensional (k-D) space.

Formally, two images, I and I′, are defined by two functions: $i: U \to B$ and $i': U \to B$, where U is a subset in 2-D space, and B is a subset of 1-D space for monochrome images (3-D space for color images). U defines the image plane. Functions i and i' map each point in image plane to an intensity value. Without loss of generality, we assume the image intensity is given in a finite normalized image plane

$$U = \{(u, v)^t \mid -h \le u \le h, -w \le v \le w\}$$

with intensity in a range $B = \{i \mid 0 \le i \le b\}$ for monochrome images. ($b = 255$ in the normalized range we used.) A 3-D scene is characterized by a set Π of 3-D points in 3-D Euclidean space. In the context of vision, we are only concerned with the visible surface of the scene. Therefore, we can restrict Π to consists of only the surface points of the scene. The result of object motion can be given by a mapping from Π to Π': $\mathbf{m}: \Pi \to \Pi'$. A motion $\mathbf{m}$ is a one to one and onto mapping (one to one correspondence).

Let the camera be located at the origin of the 3-D coordinate system. The perspective projection of a point $\mathbf{x} = (x, y, z)^t$ in a scene Π is defined by a mapping $\mathbf{p}$ from Π into 2-D space such that

$$\mathbf{p}(\mathbf{x}) = (x/z, y/z)^t \tag{2.2.1}$$

which corresponds to the perspective projection of a pinhole camera with a normalized unit focal length. We assume that for any image point $\mathbf{u}$, there will always be some point in the scene whose projection is $\mathbf{u}$. Namely, the projections of Π and Π' cover entire image plane U: $U \subset \mathbf{p}(\Pi)$, $U \subset \mathbf{p}(\Pi')$.

A point $\mathbf{x} \in \Pi$ is visible if and only if $\mathbf{p}(\mathbf{x}) \in U$ and there is no $\mathbf{y} \in \Pi$ such that $\mathbf{p}(\mathbf{x}) = \mathbf{p}(\mathbf{y})$ and $\|\mathbf{y}\| < \|\mathbf{x}\|$. An occlusion map, O, for image I consists of those image points in U whose corresponding 3-D point $\mathbf{x} \in \Pi$ is visible but the moved point $\mathbf{m}(\mathbf{x})$ is not visible. Similarly, we define O' as the occlusion map for image I′.

An image matching from I to I′ is a mapping $\boldsymbol{\kappa}: U \to U$ such that for any $\mathbf{u} \in U - O$ (the symbol "$-$" denotes set subtraction), $\boldsymbol{\kappa}(\mathbf{u}) = \mathbf{p}(\mathbf{m}(\mathbf{x}))$, where $\mathbf{x}$ is the 3-D point corresponding to $\mathbf{u}$, $\mathbf{p}(\mathbf{x}) = \mathbf{u}$ and $\mathbf{m}$ is the motion of Π. Notice that the mapping from an occluded point $\mathbf{u} \in O$ is arbitrary. Similarly we define κ' as the matching from I′ to I. The displacement field is defined by $\mathbf{d} = \kappa - \mathbf{e}$, where $\mathbf{e}$ is an identity mapping $\mathbf{e}(\mathbf{u}) = \mathbf{u}$ for all $\mathbf{u}$ in 2-D space. Therefore, the matched image point for $\mathbf{u}$ is $\mathbf{u}' = \kappa(\mathbf{u}) = \mathbf{u} + \mathbf{d}(\mathbf{u})$. We will use the term displacement field to refer to the result of image matching.

2.2.1 Image Attributes

How do we distinguish image points so as to find their matches? A direct simple answer to this question is to characterize a point by its intensity. Under certain conditions, e.g., for matte surfaces illuminated by extended light sources, the image intensity value of a scene is in fact quite stable under motion. Fig. 2.16 in Section 2.6 shows a pair of monochrome images of a laboratory scene taken at two different positions. From those two images, it can be seen that the corresponding scene patches have similar intensity values. We call this *intensity similarity criterion.*

However, if matching is based on intensity only, a point can be matched to any point with the same or similar intensity. A 1-D example is shown in Fig. 2.1(a). This problem is more severe in two dimensional images, where a point can be potentially matched to a wide region with the similar intensity.

Even when intensity may vary, certain relationships among intensity values of nearby points may be relatively stable under motion. These relationships provide structural information of the scene. A candidate of such image structural information is sharp transitions of intensity, or edges. In the example shown in Fig. 2.1(b), matching edges reduces uncertainty in matching. Edges are less sensitive to motion since a sharp intensity transition is likely to remain to be a

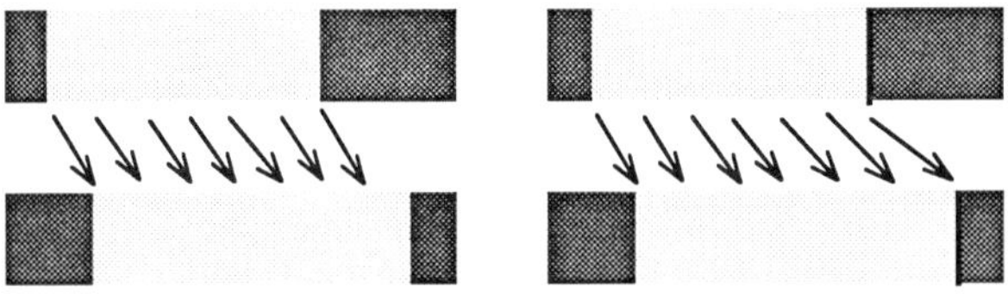

(a) (b)

Fig. 2.1. Matching intensity is not sufficient to determine matching. **(a)** A point can be matched to any point with similar intensity. **(b)** The use of edges reduces the uncertainty in matching.

sharp transition after a limited amount of motion, even though the intensity values are changed. (We will define an edgeness measure later.) The criterion that a given edge should be matched to another edge with similar edgeness measure is called *edgeness similarity criterion.*

The intensity similarity and edgeness similarity together are often not sufficient to determine correct matches. For example, if a closed contour is rotated as shown in Fig. 2.2, the intensity similarity, the edgeness similarity and smoothness are not sufficient to determine the correct match. Even if the variation of displacement field is minimized along the contour so as to minimize matching ambiguity, the resulting displacement field is still not correct [2.18]. In such cases, features of the shape of the local intensity surface are useful in distinguishing otherwise similar looking points. For example, different points on the border of a region may have the same intensity and edgeness values, but the local border shape may very from point to point on the border. A point at a geometrical corner may be early distinguished from others. The cornerness or the curvature of the region border thus can be used as a matching criterion. (We will define a cornerness measure later.) The criterion that a point should be matched to a point with similar cornerness measure is called *cornerness similarity criterion.*

More global attributes of image points such as region shapes computed from larger neighborhoods are in general even more invariant to motion but their use requires care since other effects such as perspective projection and occlusion must be taken into account. The algorithm described in this chapter uses the above local attributes, i.e., the intensity, edgeness and cornerness attributes for matching. The framework of our approach is such that additional attributes (e.g., color) could be easily included.

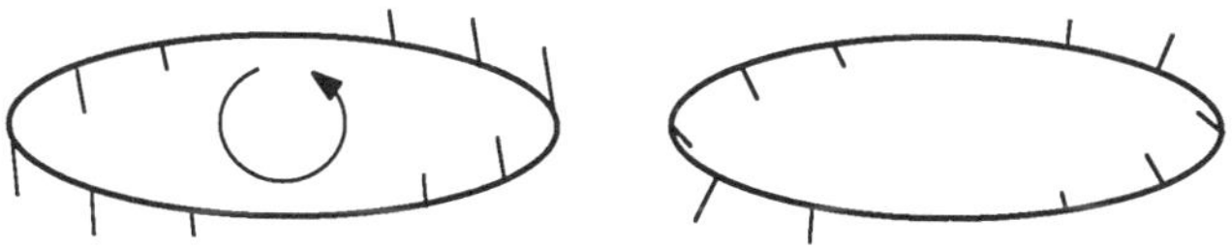

(a) (b)

Fig. 2.2. Intensity and edges are not sufficient to yield a correct match. **(a)** A closed contour is rotated. The thin line segments show true displacement. **(b)** The displacement vectors determined from local edge flow. Even if the variation of displacement field is minimized along the contour, the resulting displacement field is not correct [2.18].

2.2.2 Relationships of the Attributes

The intensity, edgeness and cornerness attributes measure different properties of the local intensity surfaces associated with an image point.

A corner point is isolated — it constitutes a zero dimensional point set. A pair of matched corners gives the displacement vector completely. Edges usually form a contour — a one-dimensional point set. Locally, if a section of edge is matched with another edge, the displacement vector, starting from an edge point, can terminate at any point on the matched edge (see Fig. 2.3). This uncertainty is commonly referred to as the *aperture problem.* Similarly, a point can be matched to any point in a region having the same intensity. This is a *two-dimensional aperture problem.*

Although matched corners completely determine the displacement vector, corners alone are not sufficient to determine the entire displacement field. First, we cannot guarantee that corners are available everywhere in images. Second, clusters of corners are difficult to match without additional support from other attributes. Fig. 2.4 shows one example, where without edge and intensity information, the corners tend to be mismatched. For similar reasons, corners and edges may not suffice without the intensity information (see Fig. 2.5).

Together, intensity, edgeness and cornerness attributes constrain the matching process and generally provide overdetermination to obtain matches. This overdetermination also makes it possible to tolerate small differences in attributes between two images.

2.2.3 Intra-Regional Smoothness and Occlusion

In order to deal with real world images, we must be able to handle regions with uniform intensity (see Fig. 2.16). Regions with uniform

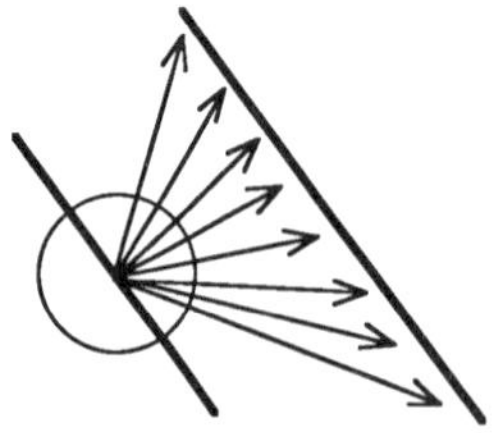

Fig. 2.3. Aperture problem in matching based on edge. An edge point can be matched potentially to any edge point on the matched edge curve.

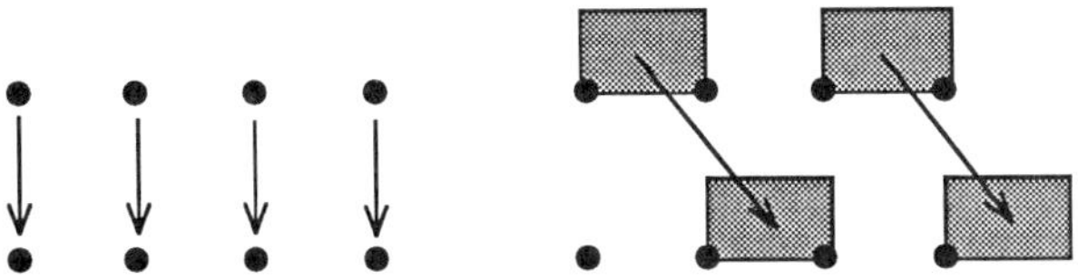

(a) (b)

Fig. 2.4. Without edge and intensity information, corner points tend to be mismatched. **(a)** Without intensity and edge information. **(b)** With intensity and edge information.

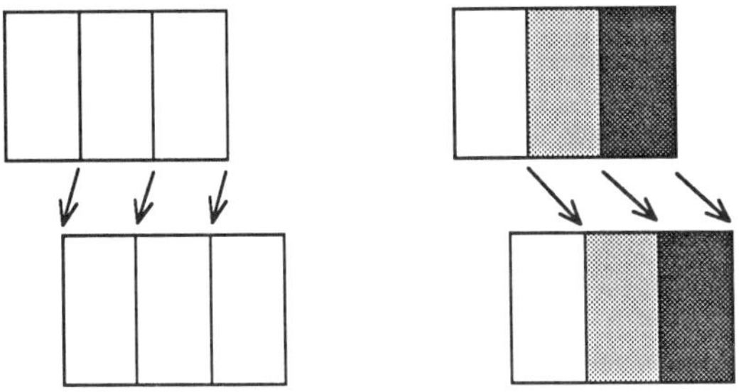

(a) (b)

Fig. 2.5. Without intensity information, edges points tend to be mismatched, although in principle edge contours can be matched globally. Intensity gives information that makes matching more reliable and easier. **(a)** Without intensity information. **(b)** With intensity information.

intensity often result from the same continuous surface. This suggests that a uniform region will have a uniform displacement field. We call this *intra-regional smoothness criterion.* The objective of this criterion is to fill displacement information into those areas where no significant intensity variation occurs. We cannot generally assume smoothness across different regions.

Obviously, this intra-regional smoothness assumption is not always satisfied. Fig. 2.6 shows an example in which image information is not sufficient to correctly infer true physical motion. The *intra-regional smoothness criterion* may be consistent with what we perceive, but may be inconsistent with reality.

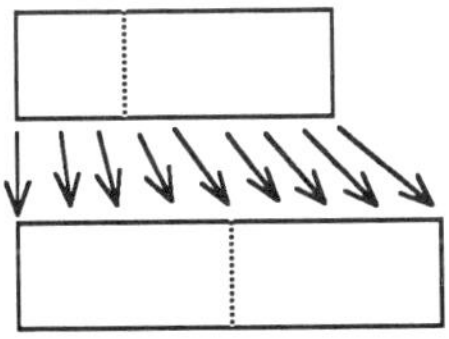

Fig. 2.6. A sheet of white paper on top of another. Neither the boundary between the sheets (*dashed*) nor the texture of the paper is visible. When the top sheet slides along the bottom sheet, the motion is perceived as stretching of the union of both sheets.

To correctly match two images, those scene regions which are occluded in one or the other image must be identified. Occlusion occurs when a part of scene visible in one image is occluded in the other by the scene itself, or a part of the scene near the image boundary moves out of the field of view in the other image. If occluded regions are not detected, they may be incorrectly matched to nearby regions, interfering with the correct matching of these regions. To identify the occluded regions, we define two occlusion maps, occlusion map 1 showing parts of image 1 not visible in image 2, and similarly occlusion map 2 for image 2 (see Fig 2.7, where black areas denote the occluded regions). We first determine the displacement field from image 2 to image 1 (say, based on similarity of attributes), without occlusion information. The objective of this preliminary matching process is to compute occlusion map 1. This matching may ''jam'' the occluded parts of image 2 into parts of image 1 (e.g., see the right-most section in Fig. 2.7). This generally will not affect the computation of occlusion map 1, since the occluded regions of image 1 to be detected may only occur on the opposite side across the ''jammed'' region (in Fig. 2.7, e.g., the occluded region of image 1 is to the left of a ''jammed'' region). Those regions in image 1 that have not been matched (in Fig. 2.7, no arrows pointing to them) are occluded in image 2 and are therefore marked in occlusion map 1 (black in Fig. 2.7). These unmatched patches may also be located at the center of the images, if they are occluded by other parts of the scene. Once the occlusion map 1 is obtained, we then compute the displacement field from image 1 to image 2 except for the occluded regions of image 1. The results of this step determine occlusion map 2 analogously (see Fig. 2.7).

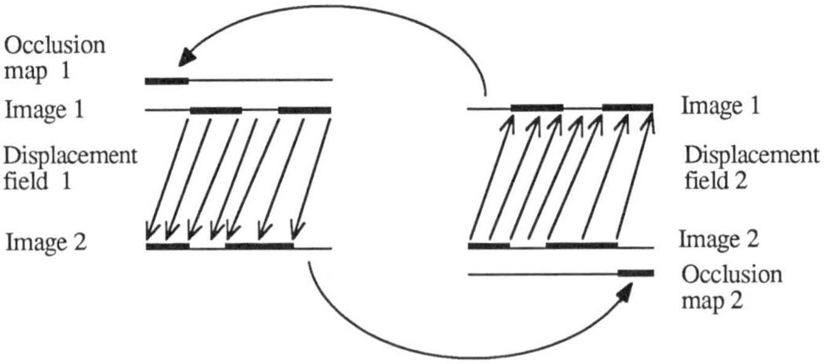

Fig. 2.7. A 1-D illustration of determining occlusion maps (see text). Images are represented by lines as one dimensional images. The displacement fields shown here just illustrate the correspondences between two 1-D images, not as actual 2-D displacement fields.

Formally, for any image point $\mathbf{u} \in U$, there is a scene point $\mathbf{x} \in \Pi$ such that $\mathbf{p}(\mathbf{x}) = \mathbf{u}$. From the definition of $\boldsymbol{\kappa}$ and $\boldsymbol{\kappa}'$, it is clear that $\boldsymbol{\kappa}$ and $\boldsymbol{\kappa}'$ are one to one correspondences from $U - O$ to $U - O'$, and from $U - O'$ to $U - O$, respectively. Therefore, the occlusion map O is given by $O = U - \boldsymbol{\kappa}'(U - O')$, and similarly $O' = U - \boldsymbol{\kappa}(U - O)$. However, this procedure is recursive. Once one occlusion map is determined, the other can also be determined. The procedure outlined in Fig. 2.7 used preliminary $\boldsymbol{\kappa}'$ that is computed without information about O' to determine O. Since regions in O and O' are generally far apart, this preliminary $\boldsymbol{\kappa}'$ may suffice to determine O.

2.2.4 Multi-Resolution Multi-Grid Structure

Large disparities are crucial to general image matching, since matches may be spatially well separated. However, to find such matches requires that we know approximate locations of the matches, since otherwise multiple matches may be found. One solution to this problem is using multi-resolution hierarchy as shown in Fig. 2.8.

At a coarse resolution level, the images to be matched may be blurred to filter out high spatial frequency components. However, blurred intensity image has very few features left. Therefore, instead of blurring the image first and then measuring edgeness and cornerness, we blur the original edgeness and cornerness images (called attribute images here). Since the cornerness measure has a sign (e.g., corner of a dark region over a bright background versus that of a bright region over a dark background), nearby positive and negative corners may be blurred to give almost zero values, which is the same as the result of blurring an area without corners. We therefore separate positive and negative corners into two attribute images. Blurring is done for positive and negative images separately. Such blurred edgeness and cornerness images are not directly related to the blurred intensity images. They are related to the strength and frequency of occurrence of the corresponding features, or to the texture content of the original images. While texture is lost in intensity images at coarse levels, the blurred edgeness and cornerness images retain a representation of texture, which is used for coarse matching. The intraregional smoothness constraint at coarse levels applies to blurred uniform texture regions (with averaged intensity). When the computation proceeds to finer levels, the sharper edgeness and cornerness measures lead to more accurate matching. Therefore, in general the algorithm applys to both textured and non-textured surfaces.

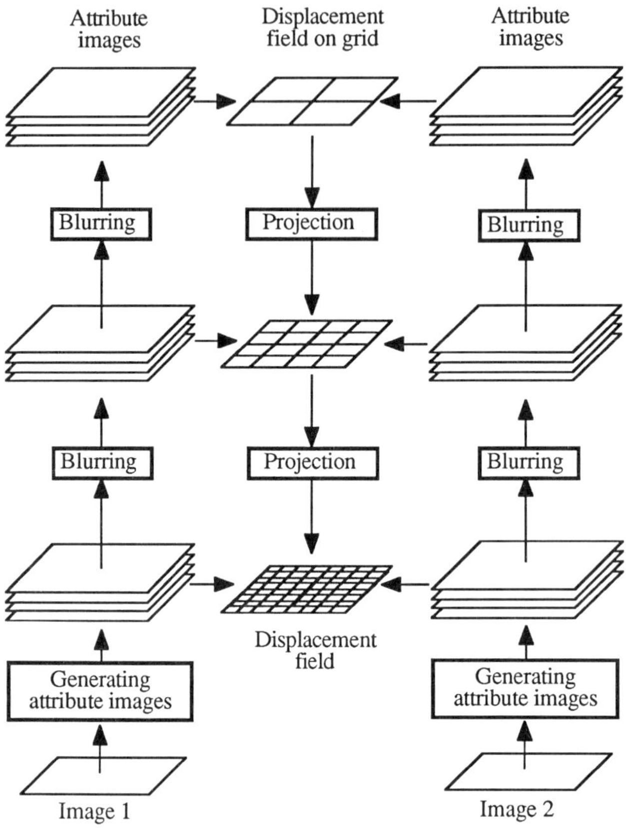

Fig. 2.8. Computational structure and data flow.

With a coarse resolution, the displacement field only needs to be computed along a coarse grid, since the displacement computed at a coarse resolution is not accurate and a low sample rate suffices. A coarse grid also helps to speed up the propagation of results within uniform regions. In the approach described in this chapter, the coarse displacement field is projected to the next finer level (copied to the four corresponding grid points), where it is refined. Such a projection and refinement continues down to finer levels successively until we get the final results at the original resolution. The computational structure and data flow used in this process are shown in Fig. 2.8.

2.2.5 Limitations

It should be noted that our approach is not intended for situations where matching criteria involve image interpretation. For example, the algorithm will not always correctly match to satisfy high-level context sensitive criteria such as illustrated in Fig. 2.9. Another limitation of our approach is that the criteria for similarity of matching attributes may be violated in some situations, e.g., along motion boundaries. For example, corners may not always correspond to a physical point along motion boundaries, as shown in Fig. 2.10, where a gray board is in front of a black board. When the viewing direction changes, the corner points (heavy dots) move, because they do not correspond to physical points fixed on the surface. Since at coarse levels corners are blurred to contribute to texture measure, a small portion of non-physical corners or edges will not cause severe problems at coarse levels. The influence of the non-physical corners is expected to be overcome by other attributes, intra-regional smoothness and occlusion information (For example, in Fig. 2.10, if the occlusion is identified, the points at the non-physical corners will be marked to

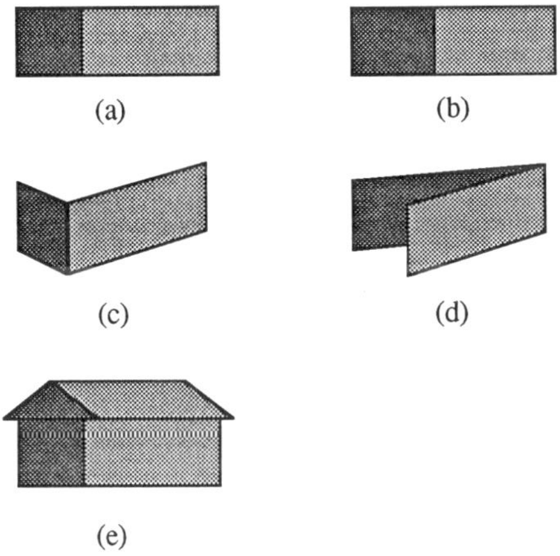

Fig. 2.9. Ambiguity in finding displacement field for textureless surfaces. **(a)** and **(b)** Two views of two textureless surfaces. **(c)** and **(d)** Two different 3-D interpretations. **(e)** With more context, understanding of the scene helps to select the correct interpretation.

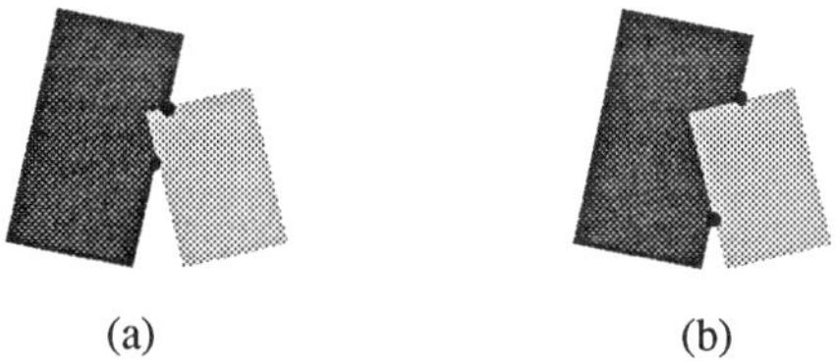

Fig. 2.10. Two image corners along motion boundary that do not correspond to the same physical points in two views **(a)** and **(b)**.

be occluded and no matching will be attempted for these points). At finer levels, the weight for cornerness should be reduced since cornerness is not as reliable as edgeness and intensity and the strength of cornerness at finer levels begins to dominate the influence of intensities. A similar but slower reduction is performed for the edgeness weights relative to the intensity weights for the same reasons.

2.3 Algorithm

The approach presented in Section 2.2 is general. The specific attributes used and details of the computational steps can vary considerably. In this section we will describe an algorithm based on the above approach.

In choosing attributes for matching, we must realize that there exist no exactly invariant attributes between two matched tokens (points, edges, regions etc) under a general setting. Changes in lighting, surface orientation and distance, viewing position, light path in an optical system etc all affect the sensed intensity values of two corresponding image points. The situation is even more complicated in the presence of surface discontinuities and occlusions. So, we can only choose attributes to minimize their sensitivity to the underlying motion. Before addressing the choice of attributes, we will first characterize the motion under which we expect the attributes to locally remain unchanged.

2.3.1 Locally Rigid Motion

In Section 2.2, motion $\mathbf{m}$ is required to yield one to one correspondence between two scenes so that inverse motion exists to define matching $\boldsymbol{\kappa}'$. For the matching algorithm to perform well, motion has to be such that the assumptions of similarity of attribute and intraregional smoothness are approximately true. We will assume that the motion can be locally approximated by piecewise rigid motion although globally the motion is non-rigid and may also involve individually moving objects. In order words, each object does not deform drastically in a small region.

Suppose that the visible scene Π can be partitioned into a finite number of subsets $\Pi = \{\Pi_i\}$ such that each Π_i has continuously differentiable visible surfaces. We assume each surface Π_i undergoes *locally rigid motion* defined as follows.

Definition 2.1. A motion $\mathbf{m}: \Pi \to \Pi'$ is *locally rigid* if for any non-boundary point $\mathbf{x}_0$ on a continuously differentiable surface of Π_i, there is a positive number $\delta > 0$ such that for any $\mathbf{x}$ with $\|\mathbf{x} - \mathbf{x}_0\| < \delta$, $\mathbf{m}(\mathbf{x})$ can be expressed by

$$\mathbf{m}(\mathbf{x}) = \mathbf{m}(\mathbf{x}_0) + R(\mathbf{x}_0)(\mathbf{x} - \mathbf{x}_0) + o(\|\mathbf{x} - \mathbf{x}_0\|)$$

or

$$\mathbf{m}(\mathbf{x}) = R(\mathbf{x}_0)\mathbf{x} + \mathbf{T}(\mathbf{x}_0) + o(\|\mathbf{x} - \mathbf{x}_0\|) \tag{2.3.1}$$

where

$$R(\mathbf{x}_0) = \left.\frac{\partial \mathbf{m}(\mathbf{x})}{\partial \mathbf{x}}\right|_{\mathbf{x} = \mathbf{x}_0}$$

is a rotation matrix and $\mathbf{T}(\mathbf{x}_0) = \mathbf{m}(\mathbf{x}_0) - R(\mathbf{x}_0)\mathbf{x}_0$, and $o(v)$ denotes a term that satisfies $\lim_{v \to 0} o(v)/v = 0$.

Similarly, we define locally rigid motion in the two-dimensional space.

A rigid motion of Π_i is a special case of the locally rigid motion in which $R(\mathbf{x}_0)$ and $\mathbf{T}(\mathbf{x}_0)$ are constant (independent of $\mathbf{x}_0$) and the higher order term $o(\|\mathbf{x} - \mathbf{x}_0\|)$ is exactly equal to zero. In a locally rigid motion, the motion is rigid, in a small neighborhood around a point $\mathbf{x}_0$, if the higher order term $o(\|\mathbf{x} - \mathbf{x}_0\|)$ is neglected. However, the global motion may still be significantly different from rigid motion, because the definition only restricts the infinitesimal behavior of the motion. Since $R(\mathbf{x}_0)$ and $\mathbf{T}(\mathbf{x}_0)$ may vary with $\mathbf{x}_0$, the nonzero term $o(\|\mathbf{x} - \mathbf{x}_0\|)$ allows significant deviation from a rigid motion globally.

2.3.2 3-D Motion and Image Plane Motion

The projection of a 3-D rigid motion onto the image plane is not, in general, a 2-D rigid motion. In the following, we investigate to what degree the image plane displacement that corresponds to a 3-D locally rigid motion can be locally approximated by a rigid image plane motion.

According to (2.3.1), the locally rigid 3-D motion in the neighborhood of a point $\mathbf{x}_0$ can be represented by

$$\mathbf{x}' = R\mathbf{x} + \mathbf{T} + o(\|\mathbf{x} - \mathbf{x}_0\|) \tag{2.3.2}$$

where $\mathbf{x}' = \mathbf{m}(\mathbf{x})$, and R and $\mathbf{T}$ depend on $\mathbf{x}_0$. As defined in (2.2.1), let the perspective projection of $\mathbf{x} = (x, y, z)^t$ and $\mathbf{x}' = (x', y', z')^t$ be $\mathbf{u}$ and $\mathbf{u}'$, respectively. Equation (2.3.2) gives

$$\mathbf{u}' = (z/z')\{R_{11}\mathbf{u} + \mathbf{R}_{12} + \mathbf{T}_1/z + o(\|\mathbf{x}-\mathbf{x}_0\|)/z\} \tag{2.3.3}$$

where R_{11} is the 2×2 upper left submatrix of R, $\mathbf{R}_{12}$ is the 2×1 upper right submatrix (vector) of R, and $\mathbf{T}_1$ consists of the first two components of $\mathbf{T}$. In general, the submatrix R_{11} is not a rotation matrix. However, for a rotation about unit vector $\mathbf{n}$ by a small angle θ, the rotation matrix can be approximated by

$$R \approx \begin{bmatrix} 1 & \gamma & \beta \\ -\gamma & 1 & -\alpha \\ \beta & \alpha & 1 \end{bmatrix}$$

where $(\alpha, \beta, \gamma) = \theta\mathbf{n}$. Therefore,

$$R_{11} \approx \begin{bmatrix} 1 & \gamma \\ -\gamma & 1 \end{bmatrix}$$

which approximates a two-dimensional rotation with an angle γ. In other words, under a small rotation, R_{11} can be approximated by a two-dimensional rotation matrix M. In fact, a more careful analysis leads to the conclusion that R_{11} can be approximated by a two-dimensional rotation matrix M as long as the rotation is small about x- and y-axis, and it can still have a relatively large rotational component about z-axis.

The term z/z' in (2.3.3) represents a scaling. If the change in depth $|z' - z|$ is significantly smaller than the depth itself (i.e., $|z' - z|/z \ll 1$), which is usually the case, then

$$\frac{z}{z'} = \frac{1}{1 + (z' - z)/z} \approx 1.$$

Using the above two approximations, we can rewrite (2.3.3) by

$$\mathbf{u}' \approx M_{11}\mathbf{u} + \mathbf{V} + o(\|\mathbf{x} - \mathbf{x}_0\|)/z$$

where $\mathbf{V} = \mathbf{R}_{12} + \mathbf{T}_1/z$ depends on the depth of the point $\mathbf{x}$. In a small neighborhood of $\mathbf{x}_0$, if the depth difference $|z - z_0|$ is small compared to z_0 (i.e., $|z - z_0|/z_0 \ll 1$), we have

$$z = \frac{z_0}{1 + (z - z_0)/z_0} \approx z_0.$$

Then $\mathbf{V}$ can be approximated by $\mathbf{V}_0 \triangleq \mathbf{R}_{12} + \mathbf{T}_1/z_0$. Equation (2.3.3) is then approximated by

$$\mathbf{u}' \approx M_{11}\mathbf{u} + \mathbf{V}_0 + o(\|\mathbf{x}-\mathbf{x}_0\|)/z_0 \qquad (2.3.4)$$

where $M_{11}$is a rotation matrix depending only on $\mathbf{x}_0$ and $\mathbf{V}_0$ is a vector depending also only on $\mathbf{x}_0$. If we can prove

$$\lim_{\mathbf{u} \to \mathbf{u}_0} \frac{o(\|\mathbf{x} - \mathbf{x}_0\|)}{z_0 \ \|\mathbf{u}-\mathbf{u}_0\|} = 0 \qquad (2.3.5)$$

we can rewrite (2.3.4) by

$$\mathbf{u}' \approx M_{11}\mathbf{u} + \mathbf{V}_0 + o(\|\mathbf{u}-\mathbf{u}_0\|). \qquad (2.3.6)$$

In order to obtain (2.3.5) we assume that in a small neighborhood of $\mathbf{x}_0$, the depth of the surface z can be expressed as a continuously differentiable function of image coordinate vector $\mathbf{u}$:

$$z = f(\mathbf{u}). \qquad (2.3.7)$$

This implies that the viewing line is not tangential to the surface at $\mathbf{x}_0$ so that the surface does not degenerate to a curve locally. Using (2.3.7) the 3-D point $\mathbf{x}$ on the surface can be represented as a function of $\mathbf{u}$:

$$\mathbf{x} = f(\mathbf{u}) \begin{bmatrix} \mathbf{u} \\ 1 \end{bmatrix}.$$

So, $\mathbf{x}$ is continuously differentiable with respect to $\mathbf{u}$. According to the law of the mean, there exists a positive number k such that

$$\|\mathbf{x} - \mathbf{x}_0\| \leq k \ \|\mathbf{u}-\mathbf{u}_0\|.$$

For all $\mathbf{u}$ in a sufficiently small neighborhood of $\mathbf{u}_0$, it follows that

$$\lim_{\mathbf{u} \to \mathbf{u}_0} \frac{o(\|\mathbf{x} - \mathbf{x}_0\|)}{z_0 \|\mathbf{u} - \mathbf{u}_0\|} = \lim_{\mathbf{x} \to \mathbf{x}_0} \frac{o(\|\mathbf{x} - \mathbf{x}_0\|)}{\|\mathbf{x} - \mathbf{x}_0\|} \frac{\|\mathbf{x} - \mathbf{x}_0\|}{z_0 \|\mathbf{u} - \mathbf{u}_0\|}$$

$$\leq \frac{k}{z_0} \lim_{\mathbf{x} \to \mathbf{x}_0} \frac{o(\|\mathbf{x} - \mathbf{x}_0\|)}{\|\mathbf{x} - \mathbf{x}_0\|} = 0.$$

This proves (2.3.5). Equation (2.3.6) concludes that the projection of a locally rigid 3-D motion onto the image plane can be approximated by a locally rigid 2-D image plane motion, in a small neighborhood of any non-boundary point $\mathbf{x}_0$ on the scene surface, provided (1) change in depth due to the motion is small compared to the original depth, (2) rotation does not have large components about x- and y-axis, (3) surface depth variation in the small neighborhood of $\mathbf{x}_0$ is small compared to the depth of $\mathbf{x}_0$, and (4) the surface can be represented by a continuously differentiable function of image coordinates in a neighborhood of $\mathbf{x}_0$. Since those conditions are usually fairly well satisfied within each piece of smooth surface, we expect that the projection of locally rigid motion of piecewise smooth scene surface can be approximately represented by piecewise locally rigid image plane motion.

2.3.3 Motion Insensitive Image Attributes

Even though the image plane motion can be generally approximated by piecewise locally rigid motion, the intensity (or color) of an image is influenced by a series of factors in a complicated way. For example, the measured intensity of an object point varies due to changes in geometry between light sources and objects (motion between object and light sources), in geometry between objects and image sensors (motion between sensors and object), in optical attenuation (e.g., lens peripheral attenuation) and photo-electrical sensitivity of image sensors (e.g., inconsistency in CCD sensor array) [2.7, 24, 37]. The factors related to the structure of image sensors can be controlled in manufacturing process, or the camera can be calibrated in application so that those factors are well compensated for. However, the geometrical relations among light sources, objects and image sensors can not be completely controlled in practice. We must exclude specular or glossy surfaces from consideration, since a slight change in lighting direction or viewing direction will significantly change the apparent brightness of the surface. With time-invariant, extended lighting source, matte surfaces do not drastically change their apparent brightness if slight changes in the geometry among the light source, objects and image sensors occur. Fine micro surface structure of those surfaces constitutes a macro structure that reflects light relatively uniformly in all directions. We will concern only such matte surfaces in this chapter. Thus, we assume that the object surfaces are such that the image intensity of a point varies only slightly between two images.

Let the corresponding image positions of a point in two images be $\mathbf{u}$ and $\mathbf{u}'$, respectively, and the corresponding intensities be $i(\mathbf{u})$ and $i'(\mathbf{u}')$, respectively. The above assumption can be expressed by

$$i'(\mathbf{u}') = i(\mathbf{u}) + s_i(\mathbf{u})$$

where $s_i(\mathbf{u})$ is a small value.

We are ready to introduce the image plane motion invariant attributes we will use. First we need to extend the domain of an image: an intended image i is a mapping from the whole 2-D space into B, the 1-D space. S is the set of all possible infinite images. Let $h = fg$ denote a product of two mappings f and g: $h(\mathbf{u}) = f(g(\mathbf{u}))$. A two-dimensional *image plane rigid motion* is a mapping $m: S \rightarrow S$, such that $m\ i \triangleq i_m$ is given by $i_m(\mathbf{u}) = i(R_2\mathbf{u} + \mathbf{T}_2)$, where R_2 is a 2×2 rotation matrix and $\mathbf{T}_2$ is a two-dimensional vector. Namely, the two-dimensional image plane rigid motion m is a mapping that maps an infinite image to another rotated and translated image.

Definition 2.2. A planar rigid motion invariant (PRMI) operator $g: S \rightarrow S$ is a mapping from S to S such that

$$gmi = mgi \tag{2.3.8}$$

for all $i \in S$ and all possible two-dimensional image plane rigid motion m.

The resulting image of an operator is called attribute image, and the value of the attribute image at a point is called the attribute of the point. The PRMI operators are such that the result of the operator on a rigidly moved image is the same as that obtained by moving the result of the operator on the original image. The attribute defined by a PRMI operator is called a PRMI attribute. The PRMI attribute at the moved point is equal to the attribute of the corresponding original point. In other words, under image plane rigid motion, the matched points have the same value in the PRMI attribute.

Obviously, intensity is a PRMI attribute since g here is an identity mapping and (2.3.8) holds trivially. The edgeness and cornerness defined in the following section are also PRMI attributes.

2.3.4 Edgeness and Cornerness

To get a continuous measure of edgeness, we define *edgeness* as the magnitude of the gradient of intensity. Namely $e = gi = \|\nabla i\|$, where g is the operator that maps i to e:

$$e(\mathbf{u}) = \left\| \frac{\partial i(\mathbf{u})}{\partial \mathbf{u}} \right\|. \tag{2.3.9}$$

Property 2.1. Edgeness defined in (2.3.9) is a type of PRMI attribute.

Proof. Given any intensity image i and any image plane rigid motion m, let $gmi = e'$ where

$$e'(\mathbf{u}) = \left\| \frac{\partial i(R_2\mathbf{u} + \mathbf{T}_2)}{\partial \mathbf{u}} \right\| = \left\| \frac{\partial i(\mathbf{v})}{\partial \mathbf{v}} R_2 \right\|_{\mathbf{v} = R_2\mathbf{u} + \mathbf{T}_2}$$

$$= \left\| \frac{\partial i(\mathbf{v})}{\partial \mathbf{v}} \right\|_{\mathbf{v} = R_2\mathbf{u} + \mathbf{T}_2} = e(\mathbf{v})\big|_{\mathbf{v} = R_2\mathbf{u} + \mathbf{T}_2}. \tag{2.3.10}$$

Equation (2.3.10) holds because R_2 is a rotation matrix. Therefore, we get $e' = me$, which gives $gmi = e' = me = mgi$. □

A few methods for detecting corners can be found in the literature [2.10, 25, 42]. Usually a local polynomial fit to image intensity is performed and then the corners are detected based on the fitted polynomial. But the polynomial fitting process is computationally expensive. Since the pre-processing to be discussed in Subsection 2.3.8 removes image noise to a significant extent, polynomial fitting can be avoided. The cornerness of a point may be defined by instantaneous rate of change in the direction of gradient along an edge curve that passes through the point [2.25, 42]. However, it has been reported that instantaneous change performs poorer than the incremental change, which is measured between the direction of gradients at two points of intersection between a circle, centered at the point under consideration, and the iso-intensity contour passing through the point.

There are some problems with these corner detectors. First, the value of the directional change of the gradient is thresholded to select corners. Therefore, a change of 180° is more likely to be selected than a change of 90°. However, the location of a right angle corner is more reliable than a corner with an acute angle or an obtuse angle. Second, cornerness measure should include a sign to distinguish a corner of a black region in a white background from that of a white region in a black background.

We define the *cornerness* in the following way that does not use computationally expensive polynomial fitting but achieves very good performance on real world images. As we mentioned earlier, we define positive and negative cornerness separately. Roughly speaking, the edgeness at a point **u** measures the change in the direction of the gradient at two nearby points, weighted by the gradient at the point. These two points, $\mathbf{u} + \mathbf{r}_a$ and $\mathbf{u} + \mathbf{r}_b$ (see Fig. 2.11) are located on a circle centered at **u**. The radius of the circle is determined by the

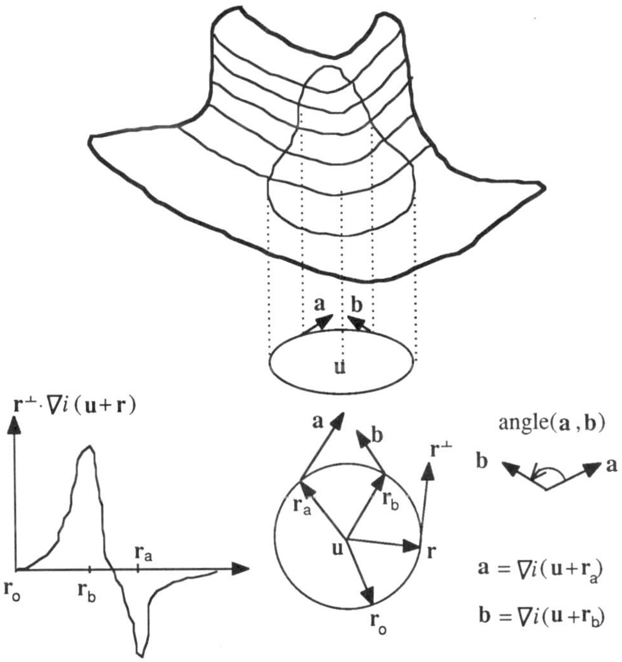

Fig. 2.11. Definition of cornerness. The upper center shows a part of the intensity surface whose height indicates the intensity value, and a circle in the image plane centered at the point **u**. Lower left shows the profile of the directional derivative of the intensity along the circle, which is used to determine the positions $\mathbf{r}_a$ and $\mathbf{r}_b$. Lower center shows a top view of the circle as well as those vectors. Lower right indicates the definition of the function angle(**a**, **b**) as well as the definition of the vectors **a** and **b**.

level of resolution. We choose $\mathbf{r}_a$ and $\mathbf{r}_b$ such that the directional derivative along the circle reaches the minimum and the maximum values, respectively (see Fig. 2.11). Let $\mathbf{a} = \nabla i(\mathbf{u} + \mathbf{r}_a)$, $\mathbf{b} = \nabla i(\mathbf{u} + \mathbf{r}_b)$, and angle($\mathbf{a}$, $\mathbf{b}$) be the angle from $\mathbf{a}$ to $\mathbf{b}$ measured in radians counter-clockwise, ranging from $-\pi$ to π. The closer the angle is to $\pi/2$, the higher the positive cornerness measure should be. In addition, the measure should be weighted by the magnitude of gradient at the point $\mathbf{u}$ since the direction of the gradient in a uniform region is very unreliable. Mathematically, the *positive cornerness* and *negative cornerness* are defined, respectively, by

$$p(\mathbf{u}) = \begin{cases} e(\mathbf{u})(1 - |1 - 2/\pi \text{ angle}(\mathbf{a}, \mathbf{b})|) & 0 \le \text{angle}(\mathbf{a}, \mathbf{b}) \le \pi \\ 0 & \text{otherwise} \end{cases} \tag{2.3.11}$$

and

$$n(\mathbf{u}) = \begin{cases} e(\mathbf{u})(1 - |1 + 2/\pi \text{ angle}(\mathbf{a}, \mathbf{b})|) & -\pi \le \text{angle}(\mathbf{a}, \mathbf{b}) \le 0 \\ 0 & \text{otherwise} \end{cases} \tag{2.3.12}$$

where column vectors $\mathbf{a}$ and $\mathbf{b}$ are intensity gradients at $\mathbf{u} + \mathbf{r}_a$, and $\mathbf{u} + \mathbf{r}_b$, respectively:

$$\mathbf{a}^t = \left.\frac{\partial i(\mathbf{s})}{\partial \mathbf{s}}\right|_{\mathbf{s} = \mathbf{u} + \mathbf{r}_a},$$

$$\mathbf{b}^t = \left.\frac{\partial i(\mathbf{s})}{\partial \mathbf{s}}\right|_{\mathbf{s} = \mathbf{u} + \mathbf{r}_b},$$

where $\mathbf{r}_a$ and $\mathbf{r}_b$ are such that $\|\mathbf{r}_a\| = \|\mathbf{r}_b\| = r$ and

$$\left.\frac{\partial i(\mathbf{v})}{\partial \mathbf{v}}\right|_{\mathbf{v} = \mathbf{u} + \mathbf{r}_a} \cdot \mathbf{r}_a^{\perp} = \min_{\|\mathbf{r}\| = r} \left.\frac{\partial i(\mathbf{v})}{\partial \mathbf{v}}\right|_{\mathbf{v} = \mathbf{u} + \mathbf{r}} \cdot \mathbf{r}^{\perp} \tag{2.3.13}$$

and

$$\left.\frac{\partial i(\mathbf{v})}{\partial \mathbf{v}}\right|_{\mathbf{v} = \mathbf{u} + \mathbf{r}_b} \cdot \mathbf{r}_b^{\perp} = \max_{\|\mathbf{r}\| = r} \left.\frac{\partial i(\mathbf{v})}{\partial \mathbf{v}}\right|_{\mathbf{v} = \mathbf{u} + \mathbf{r}} \cdot \mathbf{r}^{\perp} \tag{2.3.14}$$

where the superscript "$\perp$" denotes the corresponding perpendicular vector: if $\mathbf{r} = (r_u, r_v)^t$, then $\mathbf{r}^{\perp} = (-r_v, r_u)^t$. If $\mathbf{r}_a$ and $\mathbf{r}_b$ that correspond to the minimum in (2.3.13) and the maximum in (2.3.14), respectively, are not unique, we choose those that minimize $p(\mathbf{u})$ in (2.3.11) and (2.3.12), in addition to satisfying (2.3.13) and (2.3.14). The cornerness is defined only if the involved derivatives exist. The value r is a parameter of cornerness, and is directly related to image resolution. In discrete version, r is equal to the pixel size.

Property 2.2. The positive cornerness and negative cornerness defined above are PRMI attributes.

Proof. See Appendix of this Chapter. □

The edgeness and cornerness of a point are defined in a small neighborhood around the point. A 3×3 neighborhood is used in our implementation. Therefore, to the extent that the motion does not exhibit non-rigidity in that small neighborhood centered at a point, the attributes are motion invariant. The relationships between the 3-D motion and image plane motion discussed in Subsection 2.3.2 indicate that a 3-D locally rigid motion results in a locally rigid 2-D image plane motion under some regularity conditions. Although locally rigid motion is not a necessary condition for our approach to image matching discussed here (we need only the validity of the similarity of attributes), the study of locally rigid motion, its relationships to image plane motion as well as the plane motion invariance of the attributes leads to a class of motion with which the conditions about the similarity of the attributes are approximately satisfied.

2.3.5 Smoothness

Smoothness constraints impose similarity of the displacement vectors over a neighborhood. In addition to considering the smoothness of the overall displacement vectors, we separately consider the smoothness of the orientation of these vectors. The reason for emphasizing orientational smoothness is that (1) the orientation of the displacement vectors projected from a coarse level is generally more reliable than their magnitudes, and (2) at a fine level, the gradient perpendicular to the displacement vector can easily lead the displacement vector in a wrong direction if orientational smoothness is not emphasized.

Clearly, smoothness constraint should be enforced only over points whose displacements are related, e.g., over adjacent points from the same surface. To selectively apply the smoothness constraint to two points, we use the similarity of intensities and the similarity of available displacement vector estimates at the two points. We represent the displacement vector field in the vicinity of a point $\mathbf{u}_0$ by a vector $\overline{\mathbf{d}}(\mathbf{u}_0)$. It is intended to approximate the displacement field within the region that $\mathbf{u}_0$ belongs to. In the implementation, $\overline{\mathbf{d}}(\mathbf{u}_0)$ is computed as

$$\overline{\mathbf{d}}(\mathbf{u}_0) = \iint_{0 < \|\mathbf{u} - \mathbf{u}_0\| < r} w(i(\mathbf{u}) - i(\mathbf{u}_0), \mathbf{d}(\mathbf{u}) - \mathbf{d}(\mathbf{u}_0))\, \mathbf{d}(\mathbf{u})\, d\mathbf{u} \tag{2.3.15}$$

where $0 < \|\mathbf{u} - \mathbf{u}_0\| < r$ denotes a region around $\mathbf{u}_0$, and $w(\cdot, \cdot)$ is a weight function to be discussed. In digital implementation, the integration is replaced by a summation over $\mathbf{u}_0$'s 8 neighboring pixels. The weight is a function of intensity difference $i(\mathbf{u}) - i(\mathbf{u}_0)$ and displacement vector difference $\mathbf{d}(\mathbf{u}) - \mathbf{d}(\mathbf{u}_0)$. The objective that $\overline{\mathbf{d}}(\mathbf{u}_0)$ represents the neighboring displacement vectors of the region of $\mathbf{u}_0$ suggests the following requirements on the weight.

(1) The weight is large if the intensity difference is small. We assume that small intensity difference is observed when two neighboring points $\mathbf{u}$ and $\mathbf{u}_0$ belong to the same region, and therefore, their displacement vectors should be similar.

(2) If $\mathbf{u}$ and $\mathbf{u}_0$ have similar intensity but the corresponding displacement vectors are different, the weight should remain to be large. This case occurs when the displacement field is projected from a coarse level to the finer level. Two adjacent points with the same intensity may take quite different initial displacement vectors if they belong to different grid points at the coarse level.

(3) If $\mathbf{u}$ and $\mathbf{u}_0$ have different intensities and their displacement vectors are very different, the weight should be extremely small to suppress the influence of $\mathbf{u}$ on $\overline{\mathbf{d}}(\mathbf{u}_0)$.

Let $\eta_i = |i(\mathbf{u}) - i(\mathbf{u}_0)|$ and $\boldsymbol{\eta}_\mathbf{d} = \mathbf{d}(\mathbf{u}) - \mathbf{d}(\mathbf{u}_0)$. A definition of weight that satisfies the above criteria is as follows:

$$w(\eta_i, \boldsymbol{\eta}_\mathbf{d}) = \frac{c}{\varepsilon + |\eta_i|(1 + \|\boldsymbol{\eta}_\mathbf{d}\|^2)} \tag{2.3.16}$$

where ε is a small positive number to reduce the effects of noise in intensity and prevent denominator from becoming 0, and c is a normalization constant which makes the integration of weights equal to 1:

$$\iint_{0 < \|\mathbf{u} - \mathbf{u}_0\| < r} w(i(\mathbf{u}) - i(\mathbf{u}_0), \mathbf{d}(\mathbf{u}) - \mathbf{d}(\mathbf{u}_0))\, d\mathbf{u} = 1.$$

To ensure that requirement (2) is met, a small scale factor can be applied to the term $\|\eta_\mathbf{d}\|^2$, or alternatively, it can be set to zero which gives a simpler form:

$$w(\eta_i, \eta_\mathbf{d}) = \frac{c}{\varepsilon + |\eta_i|}. \tag{2.3.17}$$

When the displacement field is computed along with an occlusion map, the weights in (2.3.16) and (2.3.17) should be modified if, in (2.3.15), $\mathbf{u}_0$ is not an occluded point but $\mathbf{u}$ is. In this case, the weight corresponding to the occluded point $\mathbf{u}$ should be zero, since $\overline{\mathbf{d}}(\mathbf{u}_0)$ should not take the meaningless $\mathbf{d}(\mathbf{u})$ into account. But if $\mathbf{u}_0$ is an occluded point, the weight needs not be set to zero no matter whether $\mathbf{u}$ is an occluded point or not, since the displacement vector at an occluded point may conveniently take the value of such a $\overline{\mathbf{d}}(\mathbf{u}_0)$.

Thus, the weight is automatically determined based on intensity difference and displacement difference. The smoothness constraint imposes similarity of $\mathbf{d}(\mathbf{u}_0)$ and $\overline{\mathbf{d}}(\mathbf{u}_0)$. The larger the difference in intensity, the more easily the fields for two adjacent regions can differ. If two regions get different displacements after some iterations, the quadratic term $\|\eta_\mathbf{d}\|^2$ results in very small weight to reduce their interactions. On the other hand, the displacement vectors in the same region will be similar since the corresponding weight is large. Since intensity difference is usually much larger than the magnitude of displacement difference, $|\eta_i|$ is not squared in (2.3.16), unlike $\eta_\mathbf{d}$, otherwise the weight will be too sensitive to small changes in intensity. The weights, thus, implicitly take into account discontinuitics and occlusions. The registered value $\overline{\mathbf{d}}(\mathbf{u}_0)$ allows us to perform matching using uniform numerical optimization despite the presence of discontinuities and occlusions. This is discussed in the following.

2.3.6 Minimizing Residuals

Given a displacement vector field, we can define measures of similarity, or residual errors, between the attributes of estimated corresponding points. If we denote the attributes of the second image by adding a prime, the various residuals between $\mathbf{u}$ in the first image and $\mathbf{u}+\mathbf{d}$ in the second image are defined as follows:

(1) Residual of the intensity:

$$r_i(\mathbf{u}, \mathbf{d}) = i'(\mathbf{u} + \mathbf{d}) - i(\mathbf{u}).$$

(2) Residual of the edgeness:

$$r_e(\mathbf{u}, \mathbf{d}) = e'(\mathbf{u} + \mathbf{d}) - e(\mathbf{u}).$$

(3) Residual of the positive cornerness:

$$r_p(\mathbf{u}, \mathbf{d}) = p'(\mathbf{u} + \mathbf{d}) - p(\mathbf{u}).$$

(4) Residual of the negative cornerness:

$$r_n(\mathbf{u}, \mathbf{d}) = n'(\mathbf{u} + \mathbf{d}) - n(\mathbf{u}).$$

(5) Residual of the orientation smoothness:

$$r_o(\mathbf{u}, \mathbf{d}) = \text{cross}\{\mathbf{d}(\mathbf{u}), \bar{\mathbf{d}}(\mathbf{u})\} / \|\bar{\mathbf{d}}(\mathbf{u})\|,$$

where $\text{cross}\{(a, b), (c, d)\} = ac - bd$. It is obvious that $r_o(\mathbf{u}, \mathbf{d}) = 0$ if $\mathbf{d}(\mathbf{u})$ is aligned with $\bar{\mathbf{d}}(\mathbf{u})$.

(6) Residual of the displacement smoothness:

$$r_d(\mathbf{u}, \mathbf{d}) = \|\mathbf{d}(\mathbf{u}) - \bar{\mathbf{d}}(\mathbf{u})\|.$$

As we mentioned, those residuals account for changes of a wide variety of factors. Under the conditions we discussed above, the similarity of attributes approximately holds. So, we determine displacement vector $\mathbf{d}$ such that the weighted sum of squares of residuals is minimized:

$$\min_{\mathbf{d}} \sum_{\mathbf{u}} \{r_i^2(\mathbf{u}, \mathbf{d}) + \lambda_e r_e^2(\mathbf{u}, \mathbf{d}) + \lambda_p r_p^2(\mathbf{u}, \mathbf{d})$$
$$+ \lambda_n r_n^2(\mathbf{u}, \mathbf{d}) + \lambda_o r_o^2(\mathbf{u}, \mathbf{d}) + \lambda_d r_d^2(\mathbf{u}, \mathbf{d})\} \tag{2.3.18}$$

where r_i, r_e, r_p, r_n, r_o and r_d are weighting parameters that are dynamically adjusted at different resolutions. Let

$$\mathbf{r} \triangleq (r_i, r_e, r_p, r_n, r_o, r_d)^t. \tag{2.3.19}$$

With a preliminary estimate of the displacement vector $\mathbf{d}$ (initially $\mathbf{d}$ is a zero vector at the highest level), we need to find increment $\boldsymbol{\delta}_{\mathbf{d}}$. Expanding $\mathbf{r}(\mathbf{u}, \mathbf{d} + \boldsymbol{\delta}_{\mathbf{d}})$ at $\boldsymbol{\delta}_{\mathbf{d}} = \mathbf{0}$, we have (suppressing variable $\mathbf{u}$ for conciseness)

$$\mathbf{r}(\mathbf{d} + \boldsymbol{\delta}_{\mathbf{d}}) = \mathbf{r}(\mathbf{d}) + \frac{\partial \mathbf{r}(\mathbf{d})}{\partial \mathbf{d}} \boldsymbol{\delta}_{\mathbf{d}} + o(\|\boldsymbol{\delta}_{\mathbf{d}}\|) \triangleq \mathbf{r} + J\boldsymbol{\delta}_{\mathbf{d}} + o(\|\boldsymbol{\delta}_{\mathbf{d}}\|) \tag{2.3.20}$$

where

$$J = \frac{\partial \mathbf{r}(\mathbf{d})}{\partial \mathbf{d}} = \begin{bmatrix} \frac{\partial i'}{\partial u} & \frac{\partial i'}{\partial v} \\ \frac{\partial e'}{\partial u} & \frac{\partial e'}{\partial v} \\ \frac{\partial p'}{\partial u} & \frac{\partial p'}{\partial v} \\ \frac{\partial n'}{\partial u} & \frac{\partial n'}{\partial v} \\ -\bar{d}_v / \|\bar{\mathbf{d}}\| & \bar{d}_u / \|\bar{\mathbf{d}}\| \\ 1 & 0 \\ 0 & 1 \end{bmatrix} \tag{2.3.21}$$

where $(\bar{d}_u, \bar{d}_v)^t = \bar{\mathbf{d}}$, and the partial derivative $\frac{\partial i'}{\partial u}$ denotes the partial derivative of $i'(u, v)$ with respect to u at point $\mathbf{u} + \mathbf{d}$, and so on. Let

$$\Lambda = \mathrm{diag}\{1, \lambda_e, \lambda_p, \lambda_n, \lambda_o, \lambda_d\},$$

which is the diagonal matrix with the corresponding diagonal elements. We need to solve for $\boldsymbol{\delta}_{\mathbf{d}}$ such that the sum of squared residuals in (2.3.18) at the point is minimized. Neglecting high order terms and minimizing $\|\Lambda(\mathbf{r} + J\boldsymbol{\delta}_{\mathbf{d}})\|^2$, from (2.3.20) we get the formula for updating $\mathbf{d}$:

$$\boldsymbol{\delta}_{\mathbf{d}} = -(J^t \Lambda^2 J)^{-1} J^t \Lambda^2 \mathbf{r}. \tag{2.3.22}$$

The partial derivatives in the entries of J are computed by a finite difference method in our implementation. Let s denote the distance between two adjacent points on a grid, along which the finite deference of the attributes is be computed, assuming a unit spacing between adjacent pixels. Then s should vary with the resolution. In addition, s should also vary with successive iterations within a resolution level. A large spacing is necessary for a rough displacement estimate when iterations start at each level. As iterations progress, the accuracy of the displacement field increases and s should be reduced to measure local structure more accurately. The mask to compute the finite differences is shown in Fig. 2.12, where spacing s at level l is equal to 2^l for the first one-half number of iterations at level l, and is reduced by a factor of 2 for the second half, except for $l = 0$. At the original resolution ($l = 0$), the spacing is always equal to 1, since no smaller spacing is available on pixel grid.

For each grid point along which the displacement vectors are to be computed at a resolution level (see Fig. 2.8), the displacement vector $\mathbf{d}$ is replaced by $\mathbf{d} + \delta_{\mathbf{d}}$ according to (2.3.22). An iteration consists of such an updating for every grid point. At each resolution level, a fixed number of iterations (e.g. 20) are performed before the displacement field along the grid is projected to the next finer level. The final displacement field is obtained at the original image resolution.

2.3.7 Recursive Blurring

As shown in Fig. 2.8, the original attribute images need to be blurred to obtain higher levels. The original images are first pre-processed by the methods to be discussed in Subsection 2.3.8. Then four pairs of attribute images are generated (intensity, edgeness, positive cornerness and negative cornerness). The attribute images are extended in four directions to provide context for the points that are near the image border. The extension is made by repeating the border row or column. We use recursive blurring to speed up computation. Only integer summations and a few integer divisions are needed to perform such a simple blurring. The blurring for level $l + 1$ is done using the corresponding attribute image at level l: For each pixel at level $l + 1$, the value is equal to the sum of the values of four pixels at level l divided by k ($k = 4$ for the intensity, $k = 3$ for the edgeness and

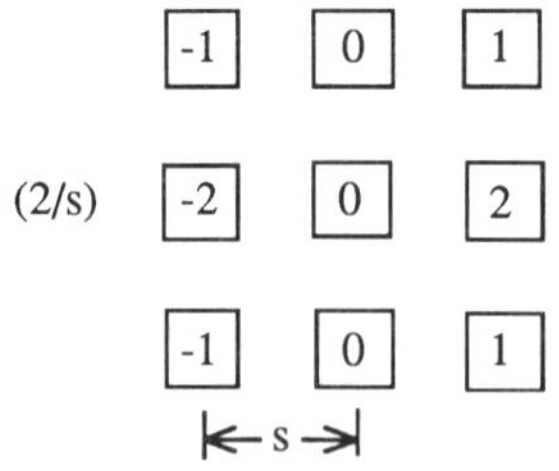

Fig. 2.12. Mask for computing derivatives.

$k = 2$ for the cornerness). The locations of these four pixels are such that each is centered at a quadrant of a square of $a \times a$ (See Fig. 2.13). a is equal to 2^l at level l. Therefore, the blurred intensity image at level l is equal to the average over all pixels in a square of size $a \times a$. To enhance sparse edges and corners, k is smaller than 4 for the edgeness and the cornerness. So, the results can be larger than 255. If this occurs, the resulting value is limited to 255. This multilevel recursive normalization is useful for the algorithm to adapt to different scenes.

2.3.8 Preprocessing and Normalization

The matching algorithm has to cope with images of a wide variety of scenes. The purposes of preprocessing are (1) to normalize the images so that the algorithm can use a set of standard parameters for different scenes and (2) to filter out noise in the images.

The pair of intensity images to be matched is first normalized by a linear function so that the minimum and the maximum intensities are equal to 0 and 255, respectively. (This range from 0 to 255 is to adapt to the representation with 8 bits/pixel, but it can be changed as needed.) Then it is filtered with a small (3×3) low pass filer to suppress gray level noise.

Similarly the edgeness and the cornerness also need to be normalized. The following considerations motivate the normalization. First, small gradients are more susceptible to intensity noise and are not reliable. Second, strong gradients may overwhelm moderate gradients in the edgeness measurement. Third, different scenes have different ranges of gradient magnitude and the algorithm should treat them in a

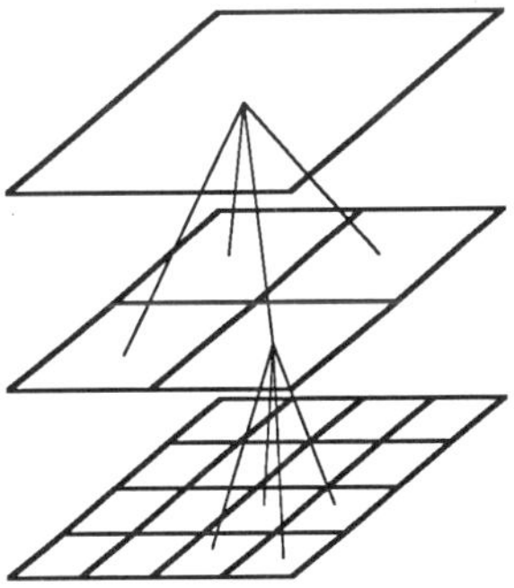

Fig. 2.13. Recursive blurring and limiting (see text).

systematic way. Therefore, we slightly modify the definition of edgeness presented in (2.3.9). Edgeness is the magnitude of the gradient normalized and transformed by a function f shown in Fig. 2.14

$$e(\mathbf{u}) = f(\|\nabla i(\mathbf{u})\|). \tag{2.3.23}$$

The function f maps the magnitude of gradient onto the range [0, 255]. It has two transition points x_0 and x_1. From $x = 0$ to $x = x_0$, $f(x) \approx 0$ to suppress noise. From $x = x_0$ to $x = x_1$, $f(x)$ increases from near 0 to almost 255 gradually and smoothly. The smooth transition interval $[x_0, x_1]$ allows continuous variation of edgeness for the gradient with a moderate magnitude. For $x > x_1$, $f(x) \approx 255$, to limit strong edges and relatively enhance the moderate edges. The values of two transition points x_0 and x_1 are determined automatically through an analysis of the histogram of gradient magnitudes such that the fractions of the pixels in the edgeness images that have values below $f(x_0)$ and above $f(x_1)$ are maintained at predetermined levels.

The edgeness $e(\mathbf{u})$ used in the definition of cornerness, (2.3.11) and (2.3.12), should also used the modified definition (2.3.23) as well. Note that such modified edgeness and cornerness are still PRMI attributes, and all the related proofs still hold if g is replaced by fg.

The preprocessing and normalization steps make the algorithm to perform consistently for a wide variety of images using a set of standard parameters which are selected based on a moderate number of

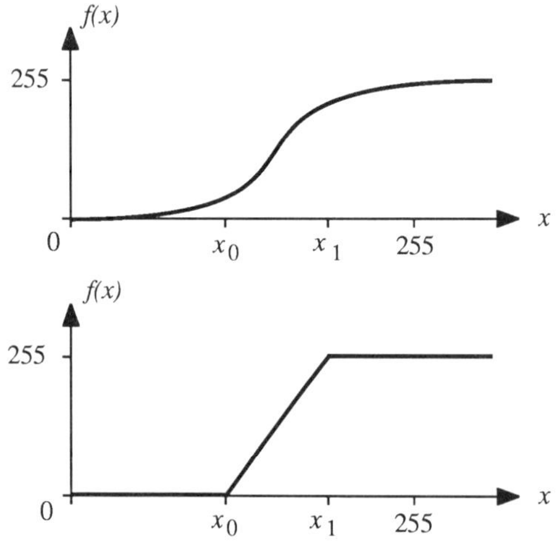

Fig. 2.14. Two normalization functions for the edgeness.

image examples. In the present implementation, the parameters are determined through trials. A set of parameters, e.g, those in (2.3.16) or (2.3.17) and (2.3.19), are determined for each level of resolution. At coarse levels, the edgeness, cornerness and smoothness have relatively large weights. Their weights are reduced gradually down to finer levels, since smoothness constraint should be reduced at finer levels where details of the displacement are obtained, and the cornerness and edgeness measurements at finer levels are more susceptible to noise than the significantly blurred measurements.

2.3.9 Outline of the Matching Algorithm

The steps of the algorithm are outlined in the following.

(1) Filter two images using a 3×3 low pass filter to remove noise (Subsection 2.3.8).

(2) Normalize the pair of images (Subsection 2.3.8).

(3) Generate attribute images: the intensity, edgeness, positive cornerness and negative cornerness (Subsection 2.3.4).

(4) Set the level to the highest *e.g.*, $l = 6$, and set displacement field on the grid to zero.

(5) Blur attribute images to level l (Subsection 2.3.7).

(6) Compute displacement field along the grid (Subsection 2.3.6). Perform a number of iterations (e.g., 20).

(7) If $l = 0$, the procedure returns with resulting displacement field, otherwise go to (8).

(8) Project the displacement field on the grid of level l to the grid of level $l-1$ by copying the vector at each grid point to the four corresponding grid points of level $l-1$. Let $l \leftarrow l-1$ and go to (5).

Suppose we need to determine the displacement field from image 1 to image 2. To obtain occlusion map 1, first compute displacement field from image 2 to image 1 using the above procedure, without occlusion information (assuming image 2 has no occluded region in image 1). The displacement field computed is then used to determine the occlusion map 1 for image 1 as discussed in Subsection 2.2.3. In the implementation, the occlusion maps are filtered by 3×3 median filters to remove single-pixel wide occlusion and noise. Then, the displacement field from image 1 to image 2 is computed using the occlusion map 1 by calling the above procedure, starting at step (4). In step (6), if a point in image 1 is included in occlusion map 1, it is not visible in image 2, and so, the displacement vector from this point cannot be determined. We just copy the vector $\overline{\mathbf{d}}$ to this occluded point. The finally computed displacement field assigns a displacement vector to each pixel in image 1.

2.4 Motion and Structure Computation

Here we briefly introduce the concept of motion and structure computation in order to present the method of refinement based on rigidity in Section 2.5, and the test data in Section 2.6. Mathematically, the displacement field is equivalent to point correspondences between two images. If the scene is rigid, the algorithm presented in Chapter 3 and Chapter 4 can be used to determine the motion between two images as well as the surface of the scene (since we have very dense displacement field). A rigid motion can be represented by a rotation followed by a translation. Under the rigid motion, if a point $\mathbf{x}$ is moved to $\mathbf{x}'$, they are related by

$$\mathbf{x}' = R\mathbf{x} + \mathbf{T} \tag{2.4.1}$$

where R is a rotation matrix and $\mathbf{T}$ is a translational vector. We first solve for motion parameters and 3-D positions of the points using a closed-form solution to be discussed in Chapter 3, assuming the scene is rigid. The results are then improved through the optimization to be discussed in Chapter 4. A large number of point correspondences is desirable for the motion and structure estimation algorithm to combat noise. The depths computed at all pixels give a surface, sampled at every pixel.

The general matching algorithm can be easily modified to solve a stereo matching problem. In stereo case, the "motion" from one camera to the other, relative to the scene, is known through stereo camera calibration. For each image point $\mathbf{u}$ in one image, there is a line in the image, called epipolar line, which goes through the point $\mathbf{u}$ (see Chapter 7). The displacement vector lies in this epipolar line. Letting the direction of this epipolar line be represented by a known 2-D unit vector $\mathbf{e}$, the displacement vector from the point $\mathbf{u}$ can be represented by $\mathbf{d} = d\mathbf{e}$, where d is a scalar indicating the stereo disparity. Therefore, the minimization problem in (2.3.18) then becomes a problem of one variable d at each point. For example, similar to (2.3.20), we need to expand $\mathbf{r}(\mathbf{u}, d + \delta_d)$ at $\delta_d = 0$. If the epipolar line is horizontal or vertical, the scalar d is identical to a component of $\mathbf{d}$. After stereo matching, the depth of each point is determined using triangulation.

2.5 Refinements

The computational structure illustrated in Fig. 2.8 can be extended further in order to improve the computed displacement field. This section discusses two techniques that we have implemented: (1) refinement through bottom-up and top-down computational scheme, and (2) refinement using a rigidity constraint.

2.5.1 Bottom-Up and Top-Down

The data flow characterized by the projections shown in Fig. 2.8 is of a top-down fashion in the sense that the displacement fields are computed from high levels (coarse resolution) down to low levels. At high levels, coarse estimates of the field are computed to cope with large disparities. At low levels, details of the displacement fields are computed. However, at a coarse level, different initial estimates may result in different results. The more accurate the initial estimate is, generally the more accurate the final result will be. Since the result at a lower level is generally more accurate than that at a higher level, the result at a lower level can be used as an initial estimate for a higher level. We may also observe this issue in a slightly different way: The result of a coarse level needs to be verified and refined at low level where more detailed image is available. Such refined local field needs to be propagated to wider areas. One computationally efficient way to do that is to go up to the higher levels where a coarse grid is available. These considerations motivate the bottom-up scheme — the result of a lower level is projected up to a higher level as an initial estimate. The upward projection is done as follows: The initial value of the grid point at a higher level is the average of the values of the corresponding grid points at the lower level. Then another pass of computation is performed from the higher level to the lower level, at which a refined field is obtained. A multiple of such top-down, bottom-up structures can be embedded in the entire algorithm. Fig. 2.15 shows an example of the computational structure that we have implemented, which leads to tangible improvement over a straight top-down scheme for some images.

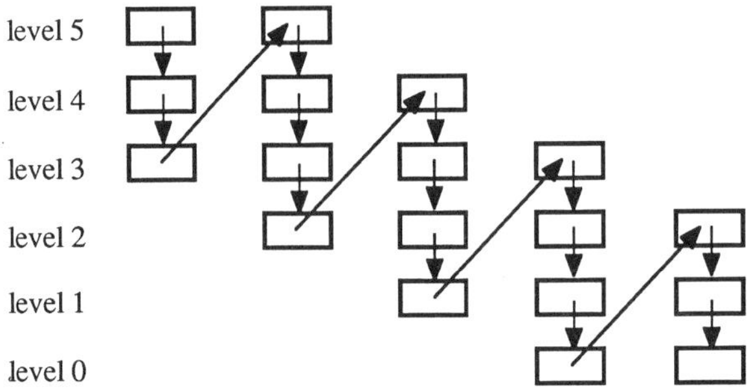

Fig. 2.15. Bottom-up and top-down refinement.

2.5.2 Refinement Based on Rigidity

The matching algorithm applies to a scene which is not necessarily rigid. In the case where the scene can be decomposed into several rigid objects, the displacement field can be used to segment the scene into rigid objects, for each of which motion parameters can be estimated. The segmentation can be done by, e.g., first segmenting the displacement field based on field discontinuity and then grouping the segments according to consistency of motion parameters between the segments. It is beyond the scope of this chapter to discuss such a segmentation. For the case where a camera moves in a static scene, the segmentation is not needed since the entire visible scene is rigid. Without loss of generality, we assume, for the remainder of this section, that the scene is rigid. The refinement discussed in this section can be applied to any rigid portion of the scene in general cases.

Assume $\mathbf{T} \neq \mathbf{0}$. From two images, translation and depths of the 3-D points can only be determined up to an unknown constant factor α, $\alpha = \|\mathbf{T}\|^{-1}$, while rotation matrix R can be determined completely. Letting $\mathring{\mathbf{T}} = \mathbf{T}/\|\mathbf{T}\|$, $\tilde{z} = \alpha z$ and $\tilde{z}' = \alpha z'$, from (2.4.1) we have,

$$\tilde{z}'\mathbf{X}' = \tilde{z}R\mathbf{X} + \mathring{\mathbf{T}}$$

where $\mathbf{X} = \mathbf{x}/z$ and $\mathbf{X}' = \mathbf{x}'/z'$. The motion estimation algorithm to be discussed later estimates rotation matrix R, translation direction $\mathring{\mathbf{T}}$, and scaled depths $\tilde{z}$ and $\tilde{z}'$ for each 3-D point. A dense map of $\tilde{z}\mathbf{X} = \alpha\mathbf{x}$ gives the scaled surface of the scene.

All displacement vectors should satisfy the rigidity condition, i.e., they describe a rigid motion. Due to noise, this is generally not true. Since the motion parameters are estimated from a large number of displacement vectors based on some optimality criterion (Chapter 4), they are generally more reliable than an individual displacement vector. So, each displacement vector can be modified based on the estimated motion parameters. Since we determine displacement on a fixed grid, the starting image coordinates of each displacement vector are fixed. To satisfy the rigidity condition, the end point of every displacement vector is corrected to be the projection of the right-hand side of (2.4.1), which is calculated from the estimated R, $\mathring{\mathbf{T}}$ and $\tilde{z}$. In other words, we move the estimated structure from the viewpoint of the first position to the new position using the estimated motion parameters. Then the projection of the structure at the new position determines the new displacement field. This resulting displacement field exactly satisfies the rigidity constraint.

2.6 Examples

In this section we present a few examples of the experiment with real world scenes. A CCD monochrome video camera with roughly 512×480 pixels was used as an image sensor. The focal length of the camera was calibrated but no correction was made for camera distortion. The camera took two images at different positions for each scene. The number of levels used for experiments was equal to 7. 20 iterations were performed at each level.

First, we present the results for the pair of images shown in Fig. 2.16, which is called the Mac scene. Significant depth discontinuities occur in the scene. Books and manuals lie irregularly on the table. Such a surface is very difficult to estimate accurately from sparse depth data. The largest disparity for this pair of 512×512 images is about 80 pixels. The two edgeness images are shown in Fig. 2.17, and the positive and negative cornerness images are shown in Fig. 2.18 and Fig. 2.19, respectively. The blurred attribute images at the highest level ($l = 6$) and the next finer level ($l = 5$) are shown in Fig. 2.20. It can be seen that the different attribute images of the same image look quite different. Although all the attribute images are derived from the original intensity images, different attribute images characterize different properties of the intensity images. Such a difference among different attribute images is what we need to provide overdetermined system of matching criteria. At the coarsest level, the grid on which displacement is to be computed is very sparse (8×8 at $l = 6$), and it is also the grid on which the finite differences are computed in the first 10 iterations. The samples of the computed displacement field along the grid are presented at different levels, superimposed on the corresponding blurred intensity images of image 1 (they have been extended to provide context for the borders). The computed displacement fields are shown in Fig. 2.21 at level 6, and in Fig. 2.22 at level 5. For levels lower than 5, the grid for displacement field is too dense to display the field clearly. Instead, only samples along a sparse 16 by 16 grid are shown for lower levels. Fig. 2.23 shows the sample of displacement field at level 4. The samples of displacement field at level 1 are shown in Fig. 2.24. Examed by flickering between two images on a Sun workstation, 95 percent of the vectors shown in Fig. 2.24 appear to have no visible errors. Between the top of Macintosh computer in the foreground and the central SUN workstation in the background is a dark region corresponding to the wall. Due to drastic depth change across this region and absence of texture inside the region, this region appears to undergo a deformation. So, the resulting displacement vectors in this region are consistent with the deformation interpretation, which is not physically correct. A correct solution to the displacement field in this region may require greater resolution and brighter lighting in order to pick up the fine texture on the wall. The occlusion map 1 is shown in Fig. 2.25, where black areas indicate that the corresponding areas in image 1 (the first image in Fig. 2.16) are not visible in image 2 (the second image in Fig. 2.16). The occlusion map is to show relatively large occluded regions (more than one pixel wide) instead of thin occlusion boundaries, which can be easily detected by analyzing discontinuities in the constructed depth map.

Fig. 2.16(a)

Fig. 2.16(b)

Fig. 2.16. Two views of a laboratory scene (called the Mac scene). **(a)** Image 1. **(b)** Image 2.

Fig. 2.17(a)

Fig. 2.17(b)

Fig. 2.17. Edgeness images of the Mac scene.

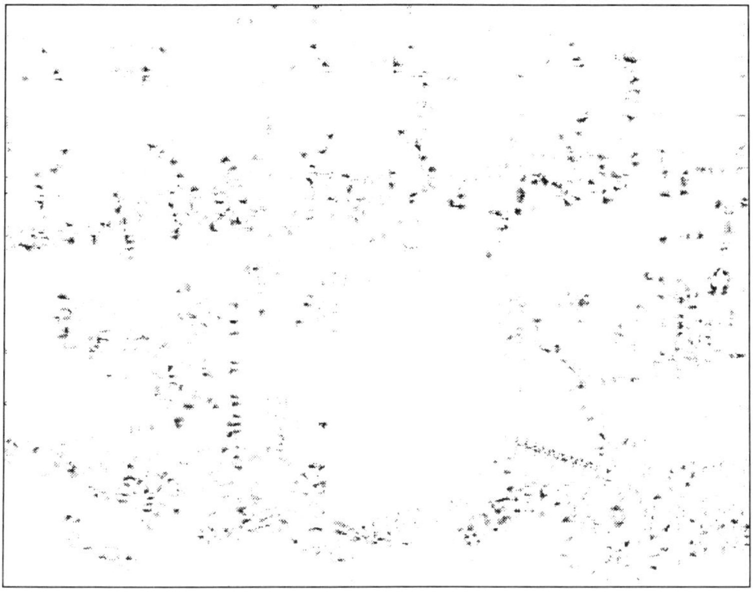

Fig. 2.18(a)

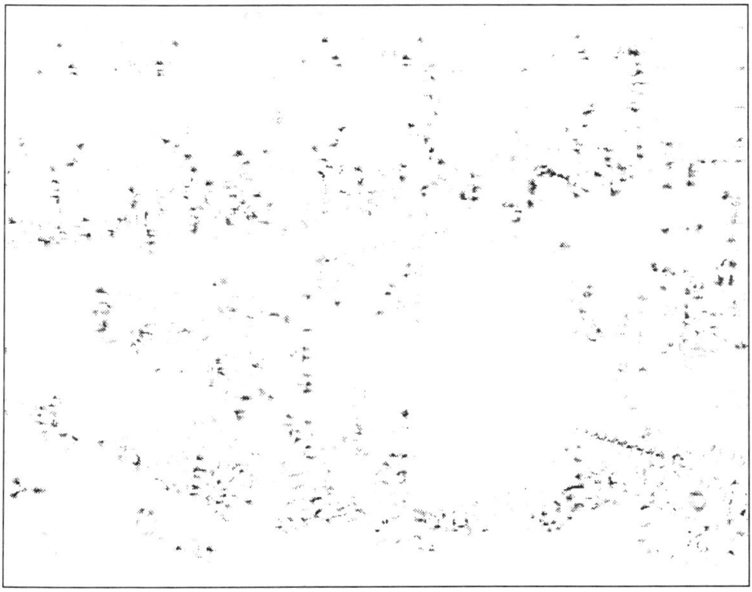

Fig. 2.18(b)

Fig. 2.18. Positive cornerness images of the Mac scene.

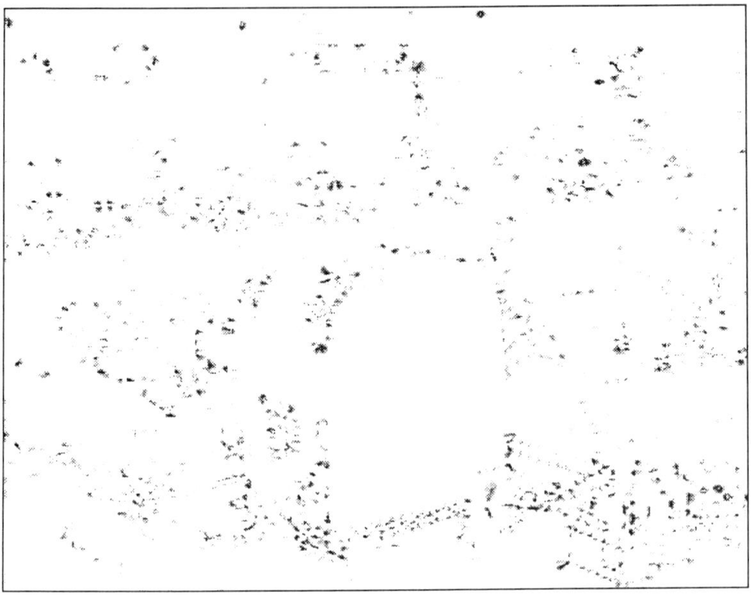

Fig. 2.19(a)

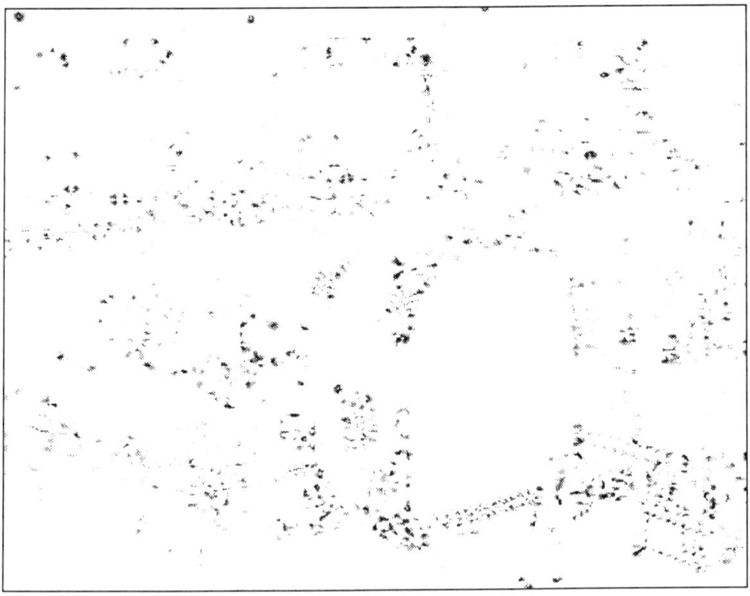

Fig. 2.19(b)

Fig. 2.19. Negative cornerness images of the Mac scene.

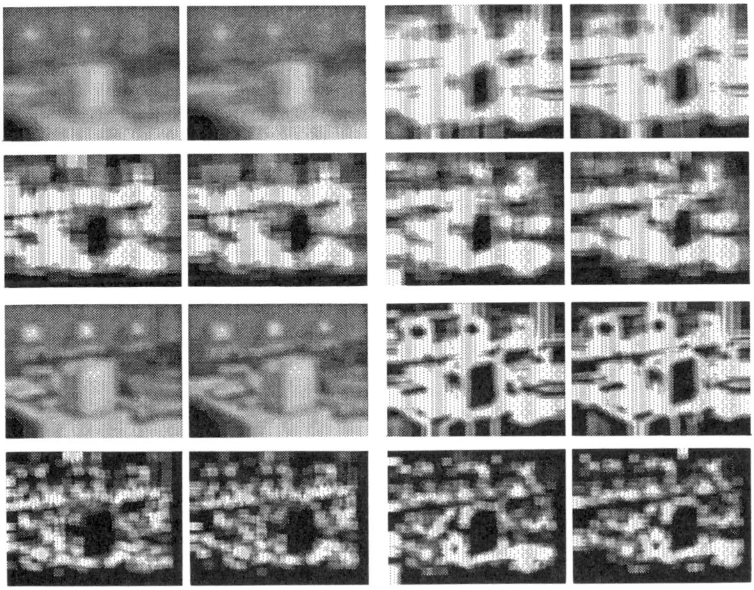

Fig. 2.20. Blurred attribute images. *Upper two rows*: level 6. *In the first row*: left pair — blurred intensity, right pair — blurred edgeness. *In the second row*: left pair — blurred positive cornerness, right pair — blurred negative cornerness. *Lower two rows*: level 5.

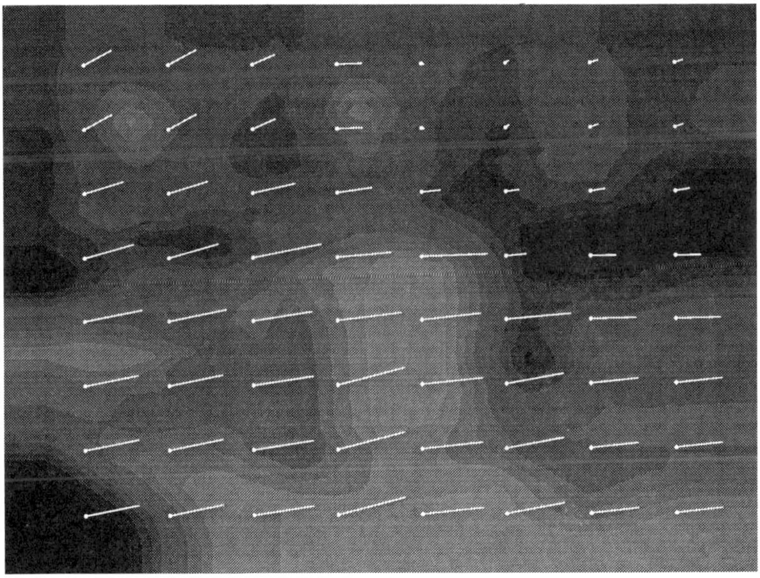

Fig. 2.21. Computed displacement field at level 6 for the Mac scene, superimposed on the blurred extended image 1.

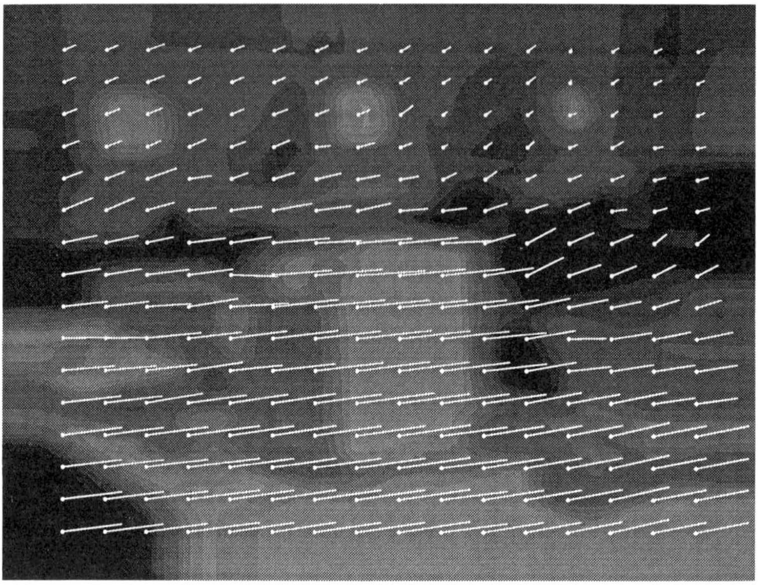

Fig. 2.22. Computed displacement field at level 5 for the Mac scene, superimposed on the blurred extended image 1.

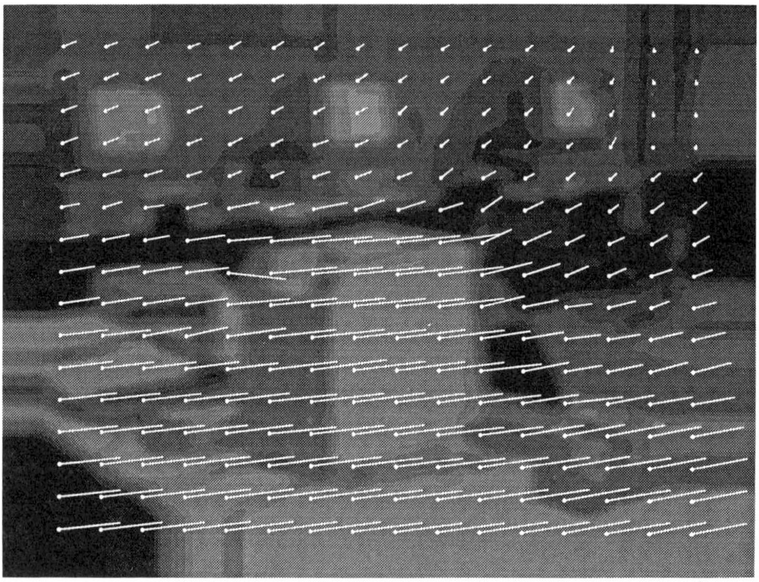

Fig. 2.23. Samples of the computed displacement field at level 4 for the Mac scene, superimposed on the blurred extended image 1.

Fig. 2.24(a)

Fig. 2.24(b)

Fig. 2.24. **(a)** Samples of the computed displacement field at level 1 for the Mac scene, superimposed on the blurred extended image 1. **(b)** The extended image 2.

Fig. 2.25. Computed occlusion map 1 for the Mac scene. Black areas indicate that the corresponding areas in image 1 are not visible in image 2.

Since the scene is rigid in this case, the algorithms presented in Chapter 4, which uses the algorithm in Chapter 3 to compute a preliminary solution, were employed to compute the motion parameters and the 3-D structure of the scene from the computed displacement field. The reconstructed 3-D surface represented by the reciprocal of depth is shown as an intensity image in Fig. 2.26 and is plotted as a wire frame in Fig. 2.27. The resulting surface agrees fairly well with that observed in the real scene. It is worth mentioning that the complicated surfaces on the manuals and books on the front table have been recovered. The estimated parameters of the motion of the scene, relative to the camera, are shown in Table 2.1. The translation direction and rotation axis are represented by three components, (up, right, forward). We can measure the discrepancies between the projection of the recovered 3-D position of the scene points and the actual observed projection. Let us define the *image plane error*, or *image error* for short, as

$$\textit{image error} = \left[\sum_{i=1}^{n} (d_i^2 + d_i'^2)/2n \right]^{1/2} \qquad (2.6.1)$$

where n is the number of points in each image (number of displacement vectors), d_i is the distance between the projection of the computed 3-D point i and its observed projection on image 1, and d_i' is analogous distance for image 2. Remember that the computed 3-D positions of points at two time instances exactly satisfy the rigidity constraint. If the displacement field is not correct that it does not correspond to the motion of a rigid scene, the image error will be large no matter how good the performance of the motion and structure estimation algorithm is. On the other hand, if the errors in the displacement field do not violate the rigidity constraint, the image error can still be small provided the performance of the motion and structure estimation algorithm is good. As shown in Table 2.1, the image error is within half of the pixel width. Thus, the performance of the algorithm for motion and structure estimation is good, and the matching algorithm at least does not make large errors that violate the rigidity constraint. The displacement field could still conceivably make systematic errors so as to depict a rigid scene, different from the real one. But such systematic errors are unlikely except those caused by the existence of multiple interpretations shown in Fig. 2.9 (all those interpretations satisfy the rigidity constraint). A quantitative estimation for the accuracy of the estimated motion parameters is presented in Chapter 4.

Table 2.1. Data and results for the Mac scene

Parameters	x	y	z
Translation	0.016	0.991	0.13
Rotation axis	0.966	0.18	-0.19
Rotation angle		1.6°	
Image error		0.00033	
Pixel width		0.00094	

In order to compare our algorithm with one that uses only intensity gradients, we set the weights for the edgeness and cornerness, λ_e, λ_p and λ_n to zero. The depth from the resulting matching is shown in Fig. 2.28, which can be seen to contain many errors. The result without using occlusion map is shown in Fig. 2.29, from which we can see that severe errors occur around the occluded regions.

Two images of another scene (called the Desk scene) are shown in Fig. 2.30. The sample of the computed displacement field is presented in Fig. 2.31, and the estimated motion parameters are given in Table 2.2. The resulting 3-D surface is shown with value $1/z$ as intensity image in Fig. 2.32.

Table 2.2. Data and results for the Desk scene

Parameters	x	y	z
Translation	-0.045	0.943	-0.328
Rotation axis	-0.87	0.38	0.30
Rotation angle		-1.4°	
Image error		0.00051	
Pixel width		0.00094	

Fig. 2.33 shows one more scene (called the Path scene), and the samples of the computed displacement field is presented in Fig. 2.34. The results of motion estimation are given in Table 2.3.

Table 2.3. Data and results for the Path scene

Parameters	x	y	z
Translation	0.095	-0.057	0.9938
Rotation axis	0.71	0.41	0.58
Rotation angle		0.13°	
Image error		0.00032	
Pixel width		0.00094	

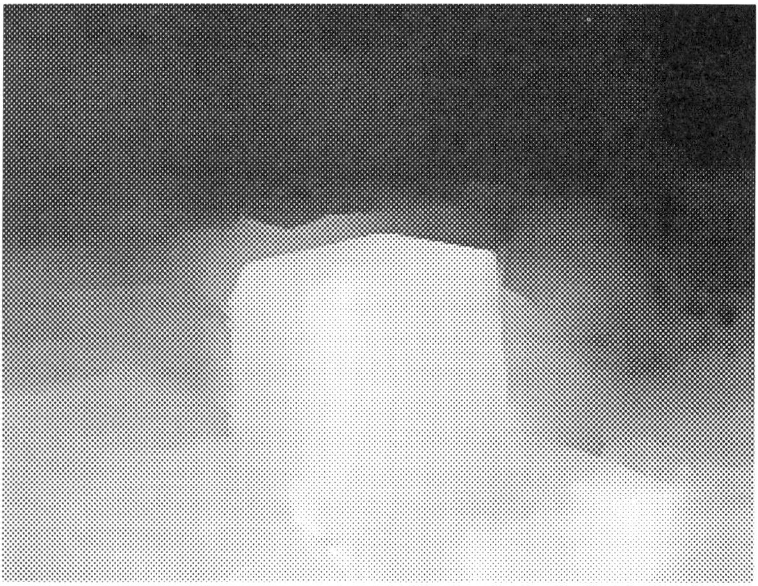

Fig. 2.26. The computed 3-D surface ($1/z$) shown as intensity image for the Mac scene (registered from the viewpoint used for image 1).

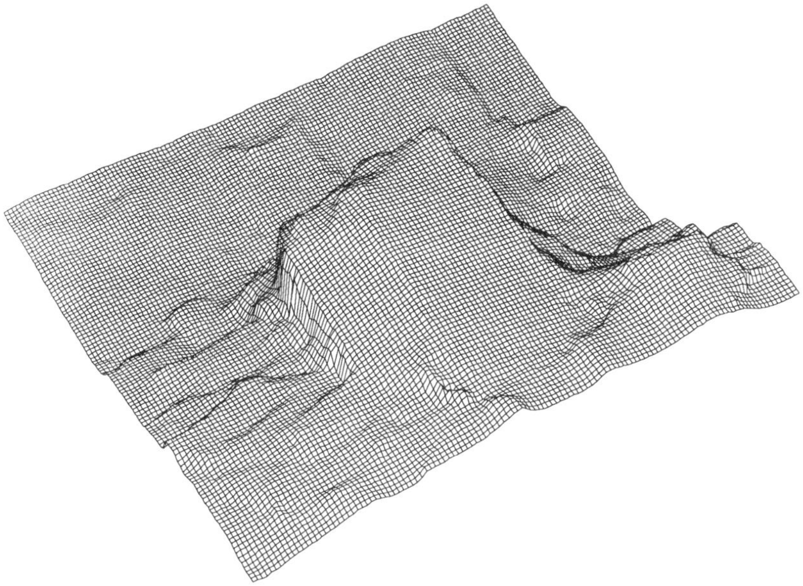

Fig. 2.27. A perspective plot of computed 3-D surface ($1/z$) for the Mac scene (registered at the viewpoint used for image 1).

Fig. 2.28. The performance using only intensity: The computed 3-D surface ($1/z$) shown as intensity image for the Mac scene (registered from the viewpoint used for image 1).

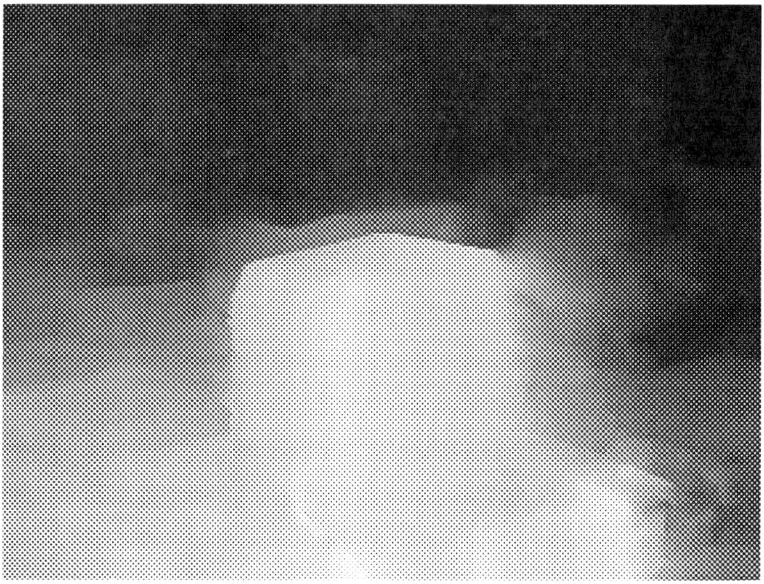

Fig. 2.29. The performance without identifying occlusion regions: The computed 3-D surface ($1/z$) shown as intensity image for the Mac scene (registered from the viewpoint used for image 1).

Fig. 2.30(a)

Fig. 2.30(b)

Fig. 2.30. Two images of the Desk scene. (**a**) Image 1. (**b**) Image 2.

Fig. 2.31(a)

Fig. 2.31(b)

Fig. 2.31. (**a**) The samples of the computed displacement field at level 1 for the Desk scene, superimposed on the blurred extended image 1. (**b**) The extended image 2.

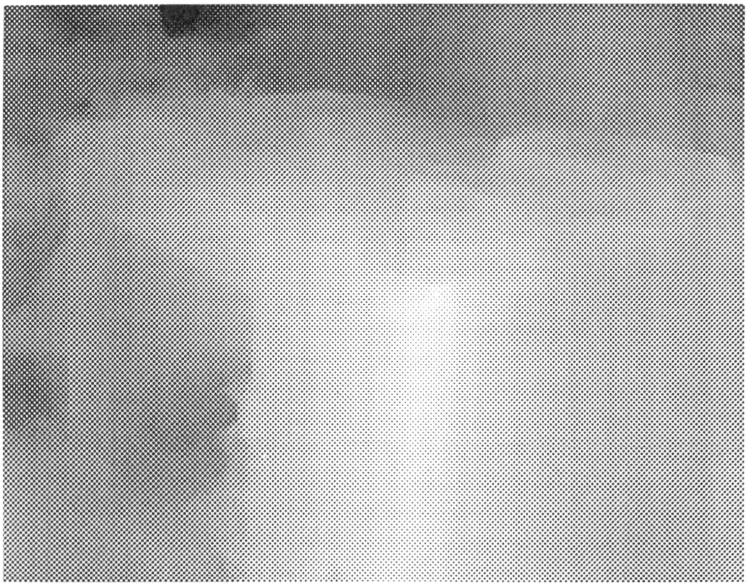

Fig. 2.32. The computed 3-D surface ($1/z$) shown as intensity image for the Desk scene (registered from the viewpoint used for image 1).

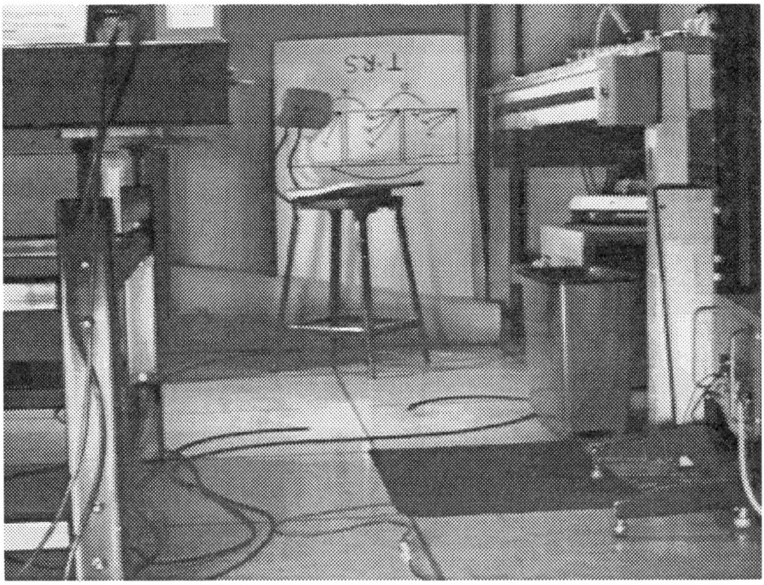

Fig. 2.33(a)

Fig. 2.33(b)

Fig. 2.33. Two images of the Path scene. **(a)** Image 1. **(b)** Image 2.

Fig. 2.34(a)

Fig. 2.34(b)

Fig. 2.34. **(a)** The samples of the computed displacement field at level 1 for the Path scene, superimposed on the blurred extended image 1. **(b)** The extended image 2.

2.7 Summary

We have presented an approach to computing displacement field between two images taken from different view points. The approach employs multiple image attributes to yield an overdetermined system of matching constraints. The continuities and discontinuities in displacement field and occlusion are taken into account to analyze complicated real world scenes.

In the current implementation of the algorithm, intensity, edgeness, and cornerness are used as matching attributes. Those attributes are invariant under image plane rigid motion. The motion is modeled as locally rigid and its relationships with image plane motion motivate the use of image plane motion invariant attributes for matching. Since the edgeness and cornerness attributes are low level attributes defined in a very small neighborhood around a point (specifically 3×3-pixel neighborhood), the attributes are insensitive to those motions that do not exhibit significant deformation in the small neighborhood.

The matching algorithm does not require extensively textured images. From the matches obtained, dense 3-D surface and occlusion maps are computed for real world scenes, assuming the scene is rigid. The discrepancy between the projection of the computed 3-D points and the observed image points (standard image error) is about one half of pixel width.

In order to relate the presented algorithm with others, let us first make some observations about the role of J given in (2.3.21). The top four rows of J determine the matching, and the bottom three rows account for the intra-regional smoothness. At a point of image $i'(\mathbf{u})$ where there are strong transitions of intensity, edgeness and cornerness, or a subset of them, the first four rows of J are relatively strong and determine the optimal $\boldsymbol{\delta}_\mathbf{d}$ to update displacement vector. The bottom three rows are relatively weak and they are used to control the intraregional uniformity of the field in the neighborhood. At a point where the intensity, edgeness and cornerness are flat, the top four rows of J are weak and the three bottom rows play a major role. The displacement is updated such that it is consistent with the neighboring displacement vectors of the same region. The resulting effect is extrapolating across a uniform region. The first four linear equations of

$$\mathbf{r}(\mathbf{d} + \boldsymbol{\delta}_\mathbf{d}) = \mathbf{r}(\mathbf{d}) + \frac{\partial \mathbf{r}(\mathbf{d})}{\partial \mathbf{d}} \boldsymbol{\delta}_\mathbf{d} = 0 \tag{2.7.1}$$

yield four linear equations in term of two components of $\boldsymbol{\delta}_d$, which determine four lines in the space of $\boldsymbol{\delta}_d$. Since the measurements are relatively noisy, those lines are not very reliable and generally do not intersect at a single point. A weighted least squares solution of (2.7.1) determines a point that minimizes the weighted sums of squared residuals.

The existing gradient-based matching methods use only one linear equation based on intensity similarity. Namely, only the first of the four lines is used. This line does not determine a point in the plane (an underdetermined system). Those methods resort to smoothness constraint. However, many incorrect solutions that satisfy the intensity constraint can also be very smooth, and very often can be even smoother than the correct solution. In other words, there is a huge class of solutions that satisfy, numerically, both the linear equation and the smoothness constraint but may be very different from the correct solution. The final solution obtained by those methods can be any one in this class. Therefore, those methods do not give correct solution in general. This partially accounts for the problems encounted by intensity gradient based methods.

In our approach, the system is generally overdetermined, especially at coarse levels, and smoothness is used mainly for filling in uniform regions. Although the available information for matching is just the original intensity images, the matching criteria here are based on not only individual intensity values, but also relationships between those intensity values. Edgeness and cornerness characterize some meaningful local relationships at a point and they are approximately invariant under locally rigid image plane displacement. These attributes provide additional information that is needed to guide the matching. At coarse level, they provide texture content of the original images. More importantly, they lead to a generally overdetermined system based solely on attribute matching. Such overdetermination significantly improves the stability of the solution.

Since the intensity, edgeness and cornerness used in our algorithm are point-based local properties, the algorithm is pixel oriented: simple, uniform and easy to implement on certain parallel computer architectures. This is an advantage over symbolic matching approaches that use high level primitives and provide only sparse matches.

2.8 Bibliographical Notes

Some of the early work on time-varying imagery was largely motivated by image coding. See, for example, Limb & Murphy [2.27], and Netravali & Robbins [2.34]. Image matching with image understanding as its motive has received intensive attention in the last decade. Tremendous activities have been reported in two major areas: stereo matching and motion analysis.

In the area of stereo matching, the representative work of Marr-Poggio-Grimson approach has been reported by Marr & Poggio [2.28], in the book by Marr, *Vision* [2.29], and in the book by Grimson *From Images to Surfaces* [2.14]. The matching primitive is the zero-crossings of the Laplacian-of-Gaussian images. To deal with matching ambiguity, various continuities have been proposed, including the *figural continuity* used by Mayhew & Frisby [2.30], Grimson [2.14], Baker & Binford [2.5], and Ohta & Kanade [2.35], the similarity of nearby disparity employed by Medioni & Nevatia [2.31] and Ayache & Faverjon [2.3], as well as the piecewise continuity utilized by Witkin, Terzopoulos & M. Kass [2.41], and Hoff & Ahuja [2.19].

In the area of motion matching, most works were concentrated on the cases with small image disparities. Fennema & Thompson [2.11] reported their early work on determining image plane velocity. Horn & Schunck [2.20] developed an iterative algorithm for estimating the optical flow. Dreschler & Nagel [2.10] approached the problem by tracking brightness corners in an image sequence. Provided that a dense image sequence is available and the motion is known, tracking feature was performed through spatiotemporal analysis by Baker & Bolles [2.6]. Hildreth [2.18] has developed methods by which zero-crossings of the Laplacian-of-Gaussian images are matched to determine the visual motion. Nagel [2.33] has modified the isotropic smoothness so as to restrict the application of smoothness along the direction across intensity edges. Weng, Ahuja & Huang [2.39, 40] developed an approach to image matching which utilizes multiple blurred image attributes, controls smoothness according to neighboring intensity and displacement, and identifies the occluded regions. Other motion matching methods include those by Anandan & Weiss [2.2], Heeger [2.17], Lim & Binford [2.26], and Fogel [2.12].

Appendix

Property 2.2. The positive cornerness and negative cornerness defined are PRMI attributes.

Proof. Let $p = gi$, where g is the operator that maps i to p, the positive cornerness image. For convenience, denote the moved image mi by i_m: $i_m = mi$, and the edgeness image of i_m by e_m. We need to prove $gmi = mgi$, or equivalently using the above notation, $gi_m = mp$. According to the definition of positive cornerness, we have

$$gi_m(\mathbf{u}) = \begin{cases} e_m(\mathbf{u})(1 - |1 - 2/\pi \,\text{angle}(\mathbf{a}, \mathbf{b})|) & 0 \le \text{angle}(\mathbf{a}, \mathbf{b}) \le \pi \\ 0 & \text{otherwise} \end{cases} \tag{2.A.1}$$

where $\mathbf{a}$ and $\mathbf{b}$ are intensity gradients at $\mathbf{u} + \mathbf{r}_a$, and $\mathbf{u} + \mathbf{r}_b$, respectively:

$$\mathbf{a}^t = \left. \frac{\partial i_m(\mathbf{s})}{\partial \mathbf{s}} \right|_{\mathbf{s} = \mathbf{u} + \mathbf{r}_a}$$

$$\mathbf{b}^t = \left. \frac{\partial i_m(\mathbf{s})}{\partial \mathbf{s}} \right|_{\mathbf{s} = \mathbf{u} + \mathbf{r}_b}$$

where $\|\mathbf{r}_a\| = \|\mathbf{r}_b\| = r$ and $\mathbf{r}_a$ and $\mathbf{r}_b$ are such that

$$\left. \frac{\partial i_m(\mathbf{v})}{\partial \mathbf{v}} \right|_{\mathbf{v} = \mathbf{u} + \mathbf{r}_a} \cdot \mathbf{r}_a^{\perp} = \min_{\|\mathbf{r}\| = r} \left. \frac{\partial i(\mathbf{v})}{\partial \mathbf{v}} \right|_{\mathbf{v} = \mathbf{u} + \mathbf{r}} \cdot \mathbf{r}^{\perp} \tag{2.A.2}$$

and

$$\left. \frac{\partial i_m(\mathbf{v})}{\partial \mathbf{v}} \right|_{\mathbf{v} = \mathbf{u} + \mathbf{r}_b} \cdot \mathbf{r}_b^{\perp} = \max_{\|\mathbf{r}\| = r} \left. \frac{\partial i(\mathbf{v})}{\partial \mathbf{v}} \right|_{\mathbf{v} = \mathbf{u} + \mathbf{r}} \cdot \mathbf{r}^{\perp}. \tag{2.A.3}$$

Since the edgeness is a PRMI attribute, we have

$$e_m(\mathbf{u}) = e(\mathbf{v})|_{\mathbf{v} = R_2\mathbf{u} + \mathbf{T}_2} \tag{2.A.4}$$

where R_2 and $\mathbf{T}_2$ are the image plane motion parameters of m. Let $\mathbf{r}'_a \triangleq R_2\mathbf{r}_a$, $\mathbf{r}'_b \triangleq R_2\mathbf{r}_b$, $\mathbf{b}' \triangleq R_2\mathbf{b}$, and $\mathbf{r}' \triangleq R_2\mathbf{r}$. According to the definition of the function angle, we have

$$\text{angle}(\mathbf{a}, \mathbf{b}) = \text{angle}(\mathbf{a}', \mathbf{b}'). \tag{2.A.5}$$

From (2.A.4) and (2.A.5), we can rewrite (2.A.1) by

$$gi_m(\mathbf{u}) = e(\mathbf{v})\big|_{\mathbf{v} = R_2\mathbf{u} + \mathbf{T}_2}(1 - |1 - 2/\pi \text{ angle}(\mathbf{a}', \mathbf{b}')|) \tag{2.A.6}$$

if $0 \leq \text{angle}(\mathbf{a}', \mathbf{b}') \leq \pi$, and $gi_m(\mathbf{u}) = 0$ otherwise.

Next, we need to derive the relationships among $\mathbf{a}'$, $\mathbf{b}'$ and the image i. Since $i_m(\mathbf{s}) = i(R_2\mathbf{s} + \mathbf{T}_2)$, we have

$$\left.\frac{\partial i_m(\mathbf{s})}{\partial \mathbf{s}}\right|_{\mathbf{s} = \mathbf{u} + \mathbf{r}_a} = \left.\frac{\partial i(\mathbf{v})}{\partial \mathbf{v}} R_2\right|_{\mathbf{v} = R_2(\mathbf{u} + \mathbf{r}_a) + \mathbf{T}_2}$$

$$= \left.\frac{\partial i(\mathbf{v})}{\partial \mathbf{v}} R_2\right|_{\mathbf{v} = R_2\mathbf{u} + \mathbf{T}_2 + \mathbf{r}'_a}, \tag{2.A.7}$$

or,

$$(\mathbf{a}')^t = \left.\frac{\partial i(\mathbf{v})}{\partial \mathbf{v}}\right|_{\mathbf{v} = R_2\mathbf{u} + \mathbf{T}_2 + \mathbf{r}'_a}. \tag{2.A.8}$$

Similarly,

$$(\mathbf{b}')^t = \left.\frac{\partial i(\mathbf{v})}{\partial \mathbf{v}}\right|_{\mathbf{v} = R_2\mathbf{u} + \mathbf{T}_2 + \mathbf{r}'_b}. \tag{2.A.9}$$

From (2.A.7), it follows that

$$\left.\frac{\partial i_m(\mathbf{v})}{\partial \mathbf{v}}\right|_{\mathbf{v} = \mathbf{u} + \mathbf{r}_a} \cdot \mathbf{r}_a^{\perp} = \left.\frac{\partial i(\mathbf{v})}{\partial \mathbf{v}}\right|_{\mathbf{v} - R_2\mathbf{u} + \mathbf{T}_2 + \mathbf{r}'_a} \cdot R_2\mathbf{r}_a^{\perp}$$

$$= \left.\frac{\partial i(\mathbf{v})}{\partial \mathbf{v}}\right|_{\mathbf{v} = R_2\mathbf{u} + \mathbf{T}_2 + \mathbf{r}'_a} \cdot (\mathbf{r}'_a)^{\perp}.$$

Therefore equations (2.A.2) and (2.A.3) lead to

$$\left.\frac{\partial i(\mathbf{v})}{\partial \mathbf{v}}\right|_{\mathbf{v} = R_2\mathbf{u} + \mathbf{T}_2 + \mathbf{r}'_a} \cdot (\mathbf{r}'_a)^{\perp} = \min_{\|\mathbf{r}'\| = r} \left.\frac{\partial i(\mathbf{v})}{\partial \mathbf{v}}\right|_{\mathbf{v} = R_2\mathbf{u} + \mathbf{T}_2 + \mathbf{r}'} \cdot (\mathbf{r}')^{\perp} \tag{2.A.10}$$

and

$$\left.\frac{\partial i(\mathbf{v})}{\partial \mathbf{v}}\right|_{\mathbf{v} = R_2\mathbf{u} + \mathbf{T}_2 + \mathbf{r}_b'} \cdot (\mathbf{r}_b')^{\perp} = \max_{\|\mathbf{r}'\| = r} \left.\frac{\partial i(\mathbf{v})}{\partial \mathbf{v}}\right|_{\mathbf{v} = R_2\mathbf{u} + \mathbf{T}_2 + \mathbf{r}'} \cdot (\mathbf{r}')^{\perp}. \tag{2.A.11}$$

From equations (2.A.6), (2.A.8) to (2.A.11), and the definition of positive cornerness, it follows that

$$gi_m(\mathbf{u}) = p(\mathbf{v})|_{\mathbf{v} = R_2\mathbf{u} + \mathbf{T}_2}$$

or, in terms of operator: $gi_m = mp$. An analogous proof leads to the corresponding conclusion for negative cornerness. □

3. Two-View Analysis

The displacement field computed by the method discussed in the previous chapter provides point correspondences for motion and structure analysis. This chapter deals with estimating motion and structure of the scene from point correspondences between two perspective views. First, an algorithm is presented that computes a closed-form solution for motion and structure parameters. The algorithm exploits redundancy in the data to obtain more reliable estimates in the presence of noise. Then, an approach is introduced to estimating the errors in the computed solution. Specifically, standard deviation of the error is estimated in terms of the variance of the errors in the coordinates of the image points. The estimated errors indicate the reliability of the solution, as well as any degeneracy or near degeneracy that causes the failure of the motion estimation algorithm. The presented approach to error estimation is applicable to a wide variety of problems that involve least-squares optimization or pseudo-inverse. Finally, the relationships between errors and other parameters are analyzed.

3.1 Some Basic Issues

Suppose that some image points arise from a rigid object in the scene. Our objective is to determine the motion (displacement) of the object relative to the camera, as well as the structure of the scene (depths of the points from the camera).

First of all, we need a closed-form solution to the problem. It is always possible to write some equations that relate the unknowns to the image coordinates of the observed points. However, since those equations are generally nonlinear in the unknowns and thus there exist problems of local extrema, directly applying an iterative numerical method to solve those equations often ends up with a wrong solution.

On the other hand, a brute-force exhaustive search through the space of unknown parameters is computationally very expensive. If a closed-form solution is available, a non-iterative algorithm can be designed to compute the solution directly and quickly. Based on the closed-form solution, one can also investigate such important problems as the uniqueness of the solution, the properties of the solutions, and conditions that guarantee a unique solution.

If an algorithm only requires solving linear equations, we call it a linear algorithm. A linear algorithm is often available for a closed-form solution, since the steps to compute the closed-form solution usually require no iterative computations. The requirement of the eigenvalues and eigenvectors of symmetrical matrices, up to size 4×4, is also allowed by our linear algorithms. Since, theoretically, non-iterative methods exist for computing eigenvalues and eigenvectors of up to 4×4 symmetrical matrices. Practical numerical algorithms for eigenvalues and eigenvectors are reliable and fast for low order symmetrical matrices, although some of these algorithms are iterative.

Stability of the solution in the presence of noise is also a very important issue. The observed data are always corrupted by noise or errors. An unstable algorithm is sensitive to the noise in the data. The formulation of the closed-form solution and the design of the associated linear algorithm both affect the stability of the solution. Error analysis and error estimation are two types of studies related to the stability of solution. The former concerns the way noise in observation affects the accuracy of the solution under different situations, while the latter deals with estimating errors in the solution. The error analysis may provide guidelines to avoid unstable designs and identify unstable situations. The error estimation is important since a solution is hardly useful if one has no idea how accurate the solution is. As we will see in this chapter, error in the solution to our problem varies widely under different configurations, which makes error estimation very crucial.

In this chapter, we first present a closed-form solution and the associated linear algorithm. The algorithm exploits redundancy in the available data to improve accuracy of the solution. Using this algorithm as an example, we introduce a method of estimating errors in the computed solutions. The errors are estimated in terms of the variance of errors in the image coordinates of image points. Then, we investigate how different parameters affect the accuracy of the solution.

3.2 An Algorithm

We now present an algorithm whose objective is to reliably estimate the parameters of the relative motion between a camera and a rigid scene, and the structure of the scene points, from the correspondences between points in the image plane. We first present an overview of the algorithm. Intermediate motion parameters are introduced which are called "essential parameters" by Tsai and Huang [3.21]. The essential parameters are elements of a 3×3 matrix E, defined in terms of motion parameters. A set of equations is established that relates image coordinates of the feature points and the elements of matrix E. Since those equations are linear and homogeneous in the elements of E, the essential parameter matrix E can be determined up to a scale factor. Then we solve for motion parameters from the essential parameters. Finally the relative depth (depth scaled by the magnitude of translation) of each point is determined from motion parameters and the observed projection of the point.

The essential parameter matrix E has 8 degrees of freedom (the norm of E is pre-determined). Each point correspondence gives one linear equation for E. This is why we need at least 8 point correspondences to solve for E. The relative motion between a camera and a rigid scene has 6 degrees of freedom (3 for rotation and 3 for translation). As we will see soon, the magnitude of the translation cannot be determined from monocular images. Therefore the motion parameters to be determined have 5 degrees of freedom. Determining unknowns with 5 degrees of freedom from the matrix E with 8 degrees of freedom involves overdetermination, which is exploited in the following algorithm to combat noise. Determining relative depth of each point also involves overdetermination. A least-squares solution is obtained for relative depths. Finally, because of the noise in the observed image coordinates of the feature points, the estimated structure may not be consistent with the model of a rigid motion between two images. The 3-D structure is corrected to satisfy the rigidity constraint.

We first define a mapping $[\cdot]_\times$ from a 3-dimensional vector to a 3×3 matrix:

$$[(x_1, x_2, x_3)^t]_\times = \begin{bmatrix} 0 & -x_3 & x_2 \\ x_3 & 0 & -x_1 \\ -x_2 & x_1 & 0 \end{bmatrix}. \tag{3.2.1}$$

Using this mapping, we can express cross operation of two vectors by the matrix multiplication of a 3×3 matrix and a column matrix:

$$\mathbf{X} \times \mathbf{Y} = [\mathbf{X}]_\times \mathbf{Y}. \tag{3.2.2}$$

3.2.1 Problem Statement

Let the coordinate system be fixed on the camera with the origin coinciding with the projection center of the camera, and the z-axis coinciding with the optical axis and pointing toward the scene (Fig. 3.1). Since we are only interested in the ratio of image coordinates to the focal length and one can always measure the image coordinates in the unit of focal length, we assume, without loss of generality, that the focal length is unity. We call such a camera model *normalized* camera model. Thus, in the normalized camera model, the image plane is located at $z = 1$. Visible objects are always located in front of the camera, *i.e.*, $z > 0$. Notice that $0 < z < 1$ may occur since the camera model is normalized.

Consider a point P on the object which is visible at two time instants. The following notation is used for the spatial vectors and the image vectors.

$\mathbf{x} = (x, y, z)^t$ spatial vector of P at time t_1;

$\mathbf{x}' = (x', y', z')^t$ spatial vector of P at time t_2;

$\mathbf{X} = (u, v, 1)^t = (\frac{x}{z}, \frac{y}{z}, 1)^t$ image vector of P at time t_1;

$\mathbf{X}' = (u', v', 1)^t = (\frac{x'}{z'}, \frac{y'}{z'}, 1)^t$ image vector of P at time t_2;

where (u, v) and (u', v') are the image coordinates of the point. Therefore, the spatial vector and image vector are related by

$$\mathbf{x} = z\mathbf{X}, \qquad \mathbf{x}' = z'\mathbf{X}'.$$

Fig. 3.1 shows the geometry and the camera model of the setup.

From the figure we can see that the image of a point determines nothing but the projection line, the line that passes through the point and the projection center. The direction of this projection line is all that we get, and the position of the image plane is immaterial. That is why we can normalized the focal length to unity. It is obvious that the model in Fig. 3.1 is not meant to describe optical path in a conventional camera. But rather, it is a simple geometrical model that is mathematically equivalent to an ideal pin hole camera. A conventional camera can be calibrated so that every point in the actual image plane can be transformed to a point in the image plane of this normalized model.

Let R and $\mathbf{T}$ be the rotation matrix and the translational vector, respectively. The spatial points at the two time instants are related by

$$\mathbf{x}' = R\,\mathbf{x} + \mathbf{T}$$

or for image vectors:

$$z'\mathbf{X}' = zR\,\mathbf{X} + \mathbf{T}. \tag{3.2.3}$$

If $\|\mathbf{T}\| \neq 0$, from (3.2.3) we get

$$\frac{z'}{\|\mathbf{T}\|}\mathbf{X}' = R\frac{z}{\|\mathbf{T}\|}\mathbf{X} + \mathring{\mathbf{T}} \tag{3.2.4}$$

where

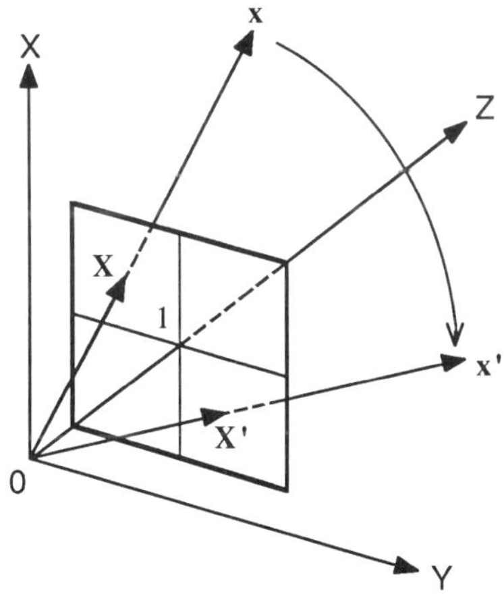

Fig. 3.1. Geometry and camera model of the setup.

$$\mathring{\mathbf{T}} = \frac{\mathbf{T}}{\|\mathbf{T}\|}.$$

Given n corresponding image vector pairs at two time instants, $\mathbf{X}_i$ and $\mathbf{X}'_i$, $i = 1, 2, \cdots, n$, the algorithm solves for the rotation matrix R. If the translation vector $\mathbf{T}$ does not vanish, the algorithm solves for the translational direction represented by a unit vector $\mathring{\mathbf{T}}$ and the relative depths $\frac{z_i}{\|\mathbf{T}\|}$ and $\frac{z'_i}{\|\mathbf{T}\|}$ for object points $\mathbf{x}_i$ and $\mathbf{x}'_i$, respectively. The magnitude of the translational vector, $\|\mathbf{T}\|$, and the absolute depths of the object points, z_i and z'_i, can not be determined by monocular vision. This can be seen from (3.2.4), which still holds when $\|\mathbf{T}\|$, z_i and z'_i are multiplied by any positive constant. In other words, multiplying the depths and $\|\mathbf{T}\|$ by the same scale factor does not change the images.

3.2.2 Algorithm

We shall first present the algorithm, and then justify each of the steps. Such an arrangement appears to make it easier to track the derivation and verify the algorithm. The first-time reader is advised to skim through the algorithm and check its correctness later when the algorithm is referred to during the derivation.

Step (i) Solving for E**.** Let $\mathbf{X}_i = (u_i, v_i, 1)^t$, $\mathbf{X}'_i = (u'_i, v'_i, 1)$, $i = 1, 2, \cdots, n$, be the corresponding image vectors of n $(n \geq 8)$ points, and

$$A = \begin{bmatrix} u_1u'_1 & u_1v'_1 & u_1 & v_1u'_1 & v_1v'_1 & v_1 & u'_1 & v'_1 & 1 \\ u_2u'_2 & u_2v'_2 & u_2 & v_2u'_2 & v_2v'_2 & v_2 & u'_2 & v'_2 & 1 \\ \vdots & \vdots & \vdots & \vdots & \vdots & \vdots & \vdots & \vdots & \vdots \\ u_nu'_n & u_nv'_n & u_n & v_nu'_n & v_nv'_n & v_n & u'_n & v'_n & 1 \end{bmatrix} \tag{3.2.5}$$

and

$$\mathbf{h} = (h_1, h_2, h_3, h_4, h_5, h_6, h_7, h_8, h_9)^t. \tag{3.2.6}$$

We solve for unit vector $\mathbf{h}$ in

$$\min_{\mathbf{h}} \|A\,\mathbf{h}\|, \quad \text{subject to: } \|\mathbf{h}\| = 1. \tag{3.2.7}$$

The solution of $\mathbf{h}$ is a unit eigenvector of A^tA associated with the smallest eigenvalue. (Alternatively, the above problem can be transformed to a linear least squares problem by setting a nonvanishing component of **h** to one and moving the corresponding column to the right hand side.) The matrix E is determined by

$$E = [\mathbf{E}_1 \ \ \mathbf{E}_2 \ \ \mathbf{E}_3] = \sqrt{2} \begin{bmatrix} h_1 & h_4 & h_7 \\ h_2 & h_5 & h_8 \\ h_3 & h_6 & h_9 \end{bmatrix}. \tag{3.2.8}$$

Step (ii) Determining a unit vector $\mathbf{T}_s$ with $\mathring{\mathbf{T}} = \pm \mathbf{T}_s$. Solve for unit vector $\mathbf{T}_s$ in

$$\min_{\mathbf{T}_s} \| E^t \mathbf{T}_s \|, \quad \text{subject to: } \|\mathbf{T}_s\| = 1. \tag{3.2.9}$$

The solution of $\mathbf{T}_s$ is a unit eigenvector of EE^t associated with the smallest eigenvalue. If

$$\sum_i (\mathbf{T}_s \times \mathbf{X}'_i) \cdot (E\,\mathbf{X}_i) < 0, \tag{3.2.10}$$

Then $\mathbf{T}_s \leftarrow -\mathbf{T}_s$. The summation in (3.2.10) is over several values of i's to suppress noise (usually three or four values of i will suffice).

Step (iii) Determining rotation matrix R. Without noise, it follows that

$$E = [\mathbf{T}_s]_\times R \tag{3.2.11}$$

or

$$R^t [-\mathbf{T}_s]_\times = E^t. \tag{3.2.12}$$

In the presence of noise, we find rotation matrix R in

$$\min_R \| R^t [-\mathbf{T}_s]_\times - E^t \|, \quad \text{subject to: } R \text{ is a rotation matrix.} \tag{3.2.13}$$

Alternatively, we can find R directly: Let

$$\begin{aligned} W &= [\mathbf{W}_1 \ \ \mathbf{W}_2 \ \ \mathbf{W}_3] \\ &= [\mathbf{E}_1 \times \mathbf{T}_s + \mathbf{E}_2 \times \mathbf{E}_3 \ \ \ \mathbf{E}_2 \times \mathbf{T}_s + \mathbf{E}_3 \times \mathbf{E}_1 \ \ \ \mathbf{E}_3 \times \mathbf{T}_s + \mathbf{E}_1 \times \mathbf{E}_2]. \end{aligned} \tag{3.2.14}$$

Without noise, $R = W$. In the presence of noise, we find rotation matrix R such that

$$\min_R \|R - W\|, \quad \text{subject to: } R \text{ is a rotation matrix.} \tag{3.2.15}$$

We can use either (3.2.13) or (3.2.15) to compute R. They both have the form

$$\min_R \|RC - D\|, \quad \text{subject to: } R \text{ is a rotation matrix} \tag{3.2.16}$$

where $C = [\mathbf{C}_1 \ \mathbf{C}_2 \ \mathbf{C}_3]$, $D = [\mathbf{D}_1 \ \mathbf{D}_2 \ \mathbf{D}_3]$. The solution of (3.2.16) is as follows:

Define a 4×4 matrix B by

$$B = \sum_{i=1}^{3} B_i^t B_i \tag{3.2.17}$$

where

$$B_i = \begin{bmatrix} 0 & (\mathbf{C}_i - \mathbf{D}_i)^t \\ \mathbf{D}_i - \mathbf{C}_i & [\mathbf{D}_i + \mathbf{C}_i]_\times \end{bmatrix}. \tag{3.2.18}$$

Let $\mathbf{q} = (q_0, q_1, q_2, q_3)^t$ be a unit eigenvector of B associated with the smallest eigenvalue. The solution of rotation matrix R in (3.2.16) is

$$R = \begin{bmatrix} q_0^2+q_1^2-q_2^2-q_3^2 & 2(q_1q_2 - q_0q_3) & 2(q_1q_3 + q_0q_2) \\ 2(q_2q_1 + q_0q_3) & q_0^2-q_1^2+q_2^2-q_3^2 & 2(q_2q_3 - q_0q_1) \\ 2(q_3q_1 - q_0q_2) & 2(q_3q_2 + q_0q_1) & q_0^2-q_1^2-q_2^2+q_3^2 \end{bmatrix}. \tag{3.2.19}$$

Step (iv) Checking $\mathbf{T} = 0$. If $\mathbf{T} \neq 0$, determine the sign of $\mathring{\mathbf{T}}$. Let α be a small threshold ($\alpha = 0$ without noise). If

$$\frac{\|\mathbf{X}'_i \times R\,\mathbf{X}_i\|}{\|\mathbf{X}'_i\| \, \|\mathbf{X}_i\|} \leq \alpha$$

for all $1 \leq i \leq n$, then report $\mathbf{T} \approx \mathbf{0}$. Otherwise determine the sign for $\mathring{\mathbf{T}}$ as follows. If

$$\sum_i (\mathbf{T}_s \times \mathbf{X}'_i) \cdot (\mathbf{X}'_i \times R\, \mathbf{X}_i) > 0, \tag{3.2.20}$$

then $\mathring{\mathbf{T}} = \mathbf{T}_s$. Otherwise $\mathring{\mathbf{T}} = -\mathbf{T}_s$. Similar to (3.2.10), summation (3.2.20) is over several values of i.

Step (v) If T ≠ 0, estimate relative depths. For i, $1 \le i \le n$, find relative depth

$$\mathbf{Z}_i = \left(\frac{z'_i}{\|\mathbf{T}\|}, \frac{z_i}{\|\mathbf{T}\|}\right)^t = (\tilde{z}'_i, \tilde{z}_i)^t \tag{3.2.21}$$

to minimize

$$\| [\mathbf{X}'_i \quad -R\,\mathbf{X}_i] \mathbf{Z}_i - \mathring{\mathbf{T}} \| \tag{3.2.22}$$

using standard least-squares method for linear equations.

A simple method to correct structure based on rigidity constraint is as follows (see Chapter 4 for more robust methods). The corrected relative 3-D position (scaled by $\|\mathbf{T}\|^{-1}$) of point i at time t_2 equals to $\tilde{\mathbf{x}}'_i = (R(\tilde{z}_i \mathbf{X}_i) + \mathring{\mathbf{T}} + \tilde{z}'_i \mathbf{X}'_i)/2$. Its relative 3-D position (scaled by $\|\mathbf{T}\|^{-1}$) at time t_1 equals to $\tilde{\mathbf{x}}_i = R^{-1}(\tilde{\mathbf{x}}'_i - \mathring{\mathbf{T}})$.

3.2.3 Justification of the Algorithm

We now justify each step of the algorithm.

For Step (i). Let $\mathbf{T}_s$ be a unit vector that is aligned with $\mathbf{T}$, *i.e.*,

$$\mathbf{T}_s \times \mathbf{T} = \mathbf{0}. \tag{3.2.23}$$

Pre-crossing both sides of (3.2.4) by $\mathbf{T}_s$ we get, using (3.2.1) and (3.2.2),

$$\frac{z'}{\|\mathbf{T}\|} \mathbf{T}_s \times \mathbf{X}' = \frac{z}{\|\mathbf{T}\|} [\mathbf{T}_s]_\times R\, \mathbf{X}. \tag{3.2.24}$$

Pre-multiplying both sides of (3.2.24) by $\mathbf{X}'^t$ (inner product between vectors), we get:

$$\mathbf{X}'^t [\mathbf{T}_s]_\times R\, \mathbf{X} = 0 \tag{3.2.25}$$

since $\mathbf{X}'^t (\mathbf{T}_s \times \mathbf{X}') = 0$ and $z > 0$. Geometrically, (3.2.25) means that three vectors $\mathbf{X}'$, $\mathbf{T}_s$ and $R\,\mathbf{X}$ are coplanar, which can be seen from (3.2.3). Define E to be

$$E = [\mathbf{T}_s]_\times R = [\mathbf{T}_s \times \mathbf{R}_1 \ \mathbf{T}_s \times \mathbf{R}_2 \ \mathbf{T}_s \times \mathbf{R}_3] = [\mathbf{E}_1 \ \mathbf{E}_2 \ \mathbf{E}_3] \tag{3.2.26}$$

where $R = [\mathbf{R}_1 \ \mathbf{R}_2 \ \mathbf{R}_3]$. From the definition of $\mathbf{T}_s$, the sign of E is arbitrary since the sign of $\mathbf{T}_s$ is arbitrary (as long as the sign of $\mathbf{T}_s$ and that of E match such that (3.2.26) holds). Using (3.2.26), the definition of E, we rewrite (3.2.25) as

$$\mathbf{X}'^t E \mathbf{X} = 0. \tag{3.2.27}$$

Our objective is to find E from the image vectors $\mathbf{X}$ and $\mathbf{X}'$. Each point correspondence gives one equation (3.2.27) which is linear and homogeneous in the elements of E. n point correspondences give n such equations. Let

$$E = [e_{ij}], \qquad \mathbf{E} = (e_{11}, e_{21}, \ \cdots, e_{33})^t.$$

Given n point correspondences, we rewrite (3.2.27) as linear equations in the elements of E and get

$$A\mathbf{E} = \mathbf{0}, \tag{3.2.28}$$

where the coefficient matrix A is given in (3.2.5). In the presence of noise, we use (3.2.7). The solution of $\mathbf{h}$ in (3.2.7) is then equal to $\mathbf{E}$ up to a scale factor provided rank$(A) = 8$. The rank of the $n \times 9$ matrix A cannot be larger than 8 since $\mathbf{E}$ is a nonzero solution of (3.2.28). Longuet-Higgins [3.11] gives a necessary and sufficient condition for the rank of A to fall below . Assuming the relative motion is due to motion of the camera, the condition is that the feature points do not lie on any quadratic surface that passes through the projection center of the camera at the two time instants. To satisfy this condition, at least 8 points are required. More points are needed to combat noise.

Since the sign of E is arbitrary, we need only to find the Euclidean norm of E to fully determine E (equivalently $\mathbf{E}$) from $\mathbf{h}$. Let $\mathbf{T}_s = (s_1, s_2, s_3)^t$. Noticing $\mathbf{T}_s$ is a unit vector and using (3.2.26), we get

$$\begin{aligned}\|E\|^2 &= \text{trace } \{EE^t\} \\ &= \text{trace } \{[\mathbf{T}_s]_\times R([\mathbf{T}_s]_\times R)^t\} \\ &= \text{trace } \{[\mathbf{T}_s]_\times([\mathbf{T}_s]_\times)^t\}\end{aligned}$$

$$= \|[\mathbf{T}_s]_\times\|^2$$

$$= 2(s_1^2 + s_2^2 + s_3^2) = 2.$$

So, $\mathbf{E} = \sqrt{2}\mathbf{h}$. This gives (3.2.8).

For Step (ii). We determine $\mathbf{T}_s$. From (3.2.26), $\mathbf{T}_s$ is orthogonal to all three columns of E. We get $E^t\mathbf{T}_s = \mathbf{0}$. With noise, we use (3.2.9).

It is easy to prove that the rank of E is always equal to 2. In fact, let $\mathbf{Q}_2$ and $\mathbf{Q}_3$ be such that $Q = [\mathbf{T}_s\ \mathbf{Q}_2\ \mathbf{Q}_3]$ is an orthonormal 3×3 matrix. $S = R^tQ$ is then also orthonormal. Post-multiplying the two sides of the first equation of (3.2.26) by S, we get

$$ES = [\mathbf{T}_s]_\times RS = [\mathbf{T}_s]_\times Q = [\ \mathbf{0}\quad \mathbf{T}_s\times\mathbf{Q}_2\ \mathbf{T}_s\times\mathbf{Q}_3].$$

We see the second and the third columns of ES are orthonormal, according to the definition of Q. Thus, $\text{rank}\{E\} = \text{rank}\{ES\} = 2$.

Since $\text{rank}\{E\} = 2$, the unit vector $\mathbf{T}_s$ is uniquely determined up to a sign by (3.2.9). To determine the sign of $\mathbf{T}_s$ such that (3.2.26) holds, we rewrite (3.2.24) using $E = [\mathbf{T}_s]_\times R$:

$$\frac{z'}{\|\mathbf{T}\|}\mathbf{T}_s\times\mathbf{X}' = \frac{z}{\|\mathbf{T}\|}E\mathbf{X}. \tag{3.2.29}$$

Since $z > 0$ and $z' > 0$ for all the visible points, from (3.2.29) we know the two vectors $\mathbf{T}_s\times\mathbf{X}'_i$ and $E\mathbf{X}_i$ have the same directions. If the sign of $\mathbf{T}_s$ is wrong, they have the opposite directions. Thus, if (3.2.10) holds, the sign of $\mathbf{T}_s$ should be changed.

For Step (iii). In steps (i) and (ii) we found E and $\mathbf{T}_s$ that satisfy (3.2.11). R can be determined directly by (3.2.14). We now prove W in (3.2.14) is equal to R without noise:

$$R = [\mathbf{R}_1\ \ \mathbf{R}_2\ \ \mathbf{R}_3]$$

$$= [\mathbf{E}_1\times\mathbf{T}_s + \mathbf{E}_2\times\mathbf{E}_3\ \ \mathbf{E}_2\times\mathbf{T}_s + \mathbf{E}_3\times\mathbf{E}_1\ \ \mathbf{E}_3\times\mathbf{T}_s + \mathbf{E}_1\times\mathbf{E}_2].$$

Using the identity equation $(\mathbf{a}\times\mathbf{b})\times\mathbf{c} = (\mathbf{a}\cdot\mathbf{c})\mathbf{b} - (\mathbf{b}\cdot\mathbf{c})\mathbf{a}$ and (3.2.26), we get

$$\mathbf{E}_1\times\mathbf{T}_s + \mathbf{E}_2\times\mathbf{E}_3$$

$$= (\mathbf{T}_s\times\mathbf{R}_1)\times\mathbf{T}_s + (\mathbf{T}_s\times\mathbf{R}_2)\times(\mathbf{T}_s\times\mathbf{R}_3)$$

$$= (\mathbf{T}_s \cdot \mathbf{T}_s)\mathbf{R}_1 - (\mathbf{R}_1 \cdot \mathbf{T}_s)\mathbf{T}_s + (\mathbf{T}_s \cdot (\mathbf{T}_s \times \mathbf{R}_3))\mathbf{R}_2 - (\mathbf{R}_2 \cdot (\mathbf{T}_s \times \mathbf{R}_3))\mathbf{T}_s$$

$$= \mathbf{R}_1 - (\mathbf{R}_1 \cdot \mathbf{T}_s)\mathbf{T}_s + (\mathbf{R}_2 \cdot (\mathbf{R}_3 \times \mathbf{T}_s))\mathbf{T}_s$$

$$= \mathbf{R}_1 - (\mathbf{R}_1 \cdot \mathbf{T}_s)\mathbf{T}_s + ((\mathbf{R}_2 \times \mathbf{R}_3) \cdot \mathbf{T}_s)\mathbf{T}_s$$

$$= \mathbf{R}_1 - (\mathbf{R}_1 \cdot \mathbf{T}_s)\mathbf{T}_s + (\mathbf{R}_1 \cdot \mathbf{T}_s)\mathbf{T}_s$$

$$= \mathbf{R}_1.$$

This proves that the first column of R is correct. Similarly we can prove that the remaining columns of R are correct.

In the presence of noise, however, the estimated E has errors, and so does the matrix determined by (3.2.14). In particular, W in (3.2.14) does not give a rotation matrix in general. For the same reason, generally, one cannot find a unit vector $\mathbf{T}_s$ and a rotation matrix R so that $[\mathbf{T}_s]_\times R = E$ if E has errors. This can be understood by considering degrees of freedom in a correct E (3 for rotation and 2 for a unit T_s), which is smaller than the degrees of freedom, 8, in a unit $\mathbf{h}$ in (3.2.7). In other words, in solving for $\mathbf{h}$ in (3.2.7), we neglect the constraint in $\mathbf{h}$. This is necessary to be able to derive a linear algorithm. The alternative steps (3.2.13) and (3.2.15) reconsider such a constraint through matrix fitting. We will return to this important issue in Chapter 4.

To solve the problem of (3.2.16), we represent the rotation matrix R in terms of a unit quaternion $\mathbf{q}$, which is briefly introduced in Appendix B. R and $\mathbf{q}$ are related by equation (3.2.19). We have (see Appendix 3.B)

$$\|RC - D\|^2 = \mathbf{q}^t B \mathbf{q} \tag{3.2.30}$$

where B is defined in (3.2.17) and (3.2.18). The problem of (3.2.16) is then reduced to the problem of minimization of a quadratic. The solution of the unit vector $\mathbf{q}$ in (3.2.30) is then a unit eigenvector of B associated with the smallest eigenvalue.

Note that R is uniquely determined in (3.2.12), since the rank of $[-\mathbf{T}_s]_\times$ is two and the positions of any two non-collinear vectors completely determine a rotation: If $R\mathbf{X}_1 = \mathbf{Y}_1$, $R\mathbf{X}_2 = \mathbf{Y}_2$, and $\mathbf{X}_1 \times \mathbf{X}_2 \neq \mathbf{0}$, then we have the third equation:

$$R(\mathbf{X}_1 \times \mathbf{X}_2) = \mathbf{Y}_1 \times \mathbf{Y}_2,$$

and $[X_1 \;\; X_2 \;\; X_1 \times X_2]$ has a full rank.

For Step (iv). Pre-crossing both sides of (3.2.3) by $\mathbf{X}'$, we get

$$\mathbf{0} = z\,\mathbf{X}' \times R\,\mathbf{X} + \mathbf{X}' \times \mathbf{T}. \tag{3.2.31}$$

If $\mathbf{T} = \mathbf{0}$, for any point $\mathbf{X}'$ we have (note $z > 0$)

$$\mathbf{X}' \times R\,\mathbf{X} = \mathbf{0}. \tag{3.2.32}$$

If $\mathbf{T} \neq \mathbf{0}$, $\mathbf{X}' \times \mathbf{T} \neq \mathbf{0}$ holds for all the points $\mathbf{X}'$ (except at most one). Therefore, (3.2.32) cannot hold for all points by virtue of (3.2.31). In the algorithm, we normalize the image vectors in (3.2.32) and give a tolerance threshold α in the presence of noise.

From (3.2.31), if $\mathring{\mathbf{T}} = \mathbf{T}_s$ then $\mathbf{T}_s \times \mathbf{X}'$ and $\mathbf{X}' \times R\,\mathbf{X}$ have the same directions. Otherwise they have opposite directions since $\mathring{\mathbf{T}} = -\mathbf{T}_s$. We use the sign of the inner product of the two vectors in (3.2.20) to determine the sign of $\mathring{\mathbf{T}}$.

For Step (v). The equations for the least-squares solution (3.2.21) follow directly from (3.2.4). The idea for correcting structure based on rigidity is as follows. Moving the recovered 3-D points at time t_1 using the estimated rotation and translation, their new positions should coincide with the recovered position at time t_2, if the data is noise free. However, in the presence of noise, the positions do not coincide. Here we adopt a simplistic way of removing this discrepancy: the midpoint between these two positions of a point at time t_2 is chosen as the corrected solution for the position of the point at time t_2. Moving the midpoint back gives the corrected 3-D position of the point at time t_1. A more detailed discussion of correcting structure can be found in Chapter 4, where noise distribution is taken into account to obtain a more robust estimate.

In summary, we have proved that if $\text{rank}(A) = 8$, the solution of R and $\mathring{T}$ is unique, and we have derived the close-form solution. Given 8 or more point correspondences, the algorithm first solves for the essential parameter matrix E. Then the motion parameters are obtained from E. Finally the spatial structure is derived from the motion parameters. All the steps of the algorithm make use of the redundancy in the data to combat noise. As the results of determining the signs in (3.2.10) and (3.2.20), the computations of three false solutions [3.10, 21, 28] are avoided. These steps for determining signs are stable in the presence of noise, since the decisions are made based on the signs of the inner product of the two vectors which are in the same or opposite direction without noise. Summations over several points in (3.2.10) and (3.2.20) suppress the effects of the cases where two noise-corrupted small vectors are used, whose inner products are close to zero and the signs are unreliable.

If $\mathbf{T} \neq \mathbf{0}$ and the spatial configuration is nondegenerate, the rank of A is 8. In this case, we can determine the unit vector $\mathbf{h}$ in (3.2.7) up to a sign, and determine R and $\overset{\circ}{\mathbf{T}}$ uniquely. If $\mathbf{T} = \mathbf{0}$, any unit vector $\mathbf{T}_s$ satisfies (3.2.24) and so matrix E, and correspondingly the unit vector $\mathbf{h}$, have two degrees of freedom (notice $\mathbf{T}_s$ and $\mathbf{h}$ are restricted to be unit vectors). Therefore, A in (3.2.5) has a rank less than or equal to 6. If $\mathbf{T} = \mathbf{0}$, relative depths of the points cannot be determined. However, the rotation parameters can be determined even if $\mathbf{T} = \mathbf{0}$.

The next section discusses how to estimate the reliability of the computed E and motion parameters.

3.3 Error Estimation

It is possible to determine the worst case bound on errors. Such a bound is useful only for applications with small errors such as computer round-off errors. Since computer word length is large enough for most applications, the worst case bound is generally tolerable. However, in problems where redundancy is utilized to combat noise and the errors in the data are not very small, the conventional worst case analysis usually renders an overly conservative bound. This can be visualized by considering the upper bound of a random variable with a Gaussian distribution. Since the bound is almost never reached, the utility of the bound is very limited. In many problems with redundant data, however, the error level of the solution is relatively stable for a fixed level of input noise. This stability is due to the redundancy in observations. For example, consider a Gaussian random variable with a small variance, which indicates error in a solution. The worst case error bound of this example is infinity. If the solution is derived from a large amount of data in a least squares sense, the variance of the error distribution in the solution is small. This makes it possible for us to estimate the expected amount of error in the solution. In this section we investigate how to estimate the expected error instead of deriving a worst case bound which is very large and almost never reached. The approach discussed in this section is applicable to problems where least-squares solution, minimum-norm solution or pseudoinverse is involved, since they essentially reduce to an eigenvalue and eigenvector problem.

3.3.1 Algorithm and Perturbation

The sources of errors in the image coordinates include spatial quantization, feature detector errors, point mismatching and camera miscalibration. In a system that is well calibrated so that systematic errors are negligible, errors in the image coordinates of a feature can be modeled by random variables. These errors cause the errors in the estimates of the motion parameters. Some spatial configurations of the points are relatively insensitive to the errors in the image coordinates of the points, but some are very sensitive. For example, if a spatial configuration of the points is degenerate mathematically but the errors in the measured image coordinates make them nondegenerate, any estimates under such a configuration is almost useless. If we move a single point slightly, so that the configuration stops being degenerate, such a configuration must be very sensitive to noise.

Formally, let the set of all the image points be denoted by P, and the errors in the image coordinates of these points be denoted by a random vector ε. The error e in the estimated motion parameters is a function of P and ε. Denoting this function by f, formally we can write

$$e = f(P, \varepsilon).$$

Our goal is to estimate the error e given the image point set P. However we don't know ε. If we can estimate the standard deviation of e (with ε as a random vector) given the noise-corrupted image point set P, we can use it to estimate the errors in the estimates. The image point set P corresponding to a degenerate or nearly degenerate spatial configuration should yield large estimates of e and those corresponding to a stable configuration should yield small estimates.

For the following discussion, we assume that the noise in the image coordinates has a zero mean and a known variance. For example, the spatial quantization noise can be well modeled by a uniform distribution with the range corresponding to the size of the pixels. The variance of feature detector error can also be estimated empirically. We also assume that noise in the original image plane measurement is uncorrelated between different points and different components. This assumption of uncorrelatedness is not exactly true in reality. However the correlation can be regarded negligible. We estimate the standard deviation of errors in the motion parameters on the basis of first order perturbation, i.e., we estimate the "linear terms" of the errors.

For conciseness, we introduce more notational convention: I_m denotes an $m \times m$ identity matrix. A matrix A without noise is denoted by A itself and its elements are denoted by the corresponding small letters a_{ij}, *i.e.*, $A = [a_{ij}]$. The noise-corrupted version of A is denoted by $A(\varepsilon)$. The noise matrix of A is denoted by Δ_A. That is

$$A(\varepsilon) = A + \Delta_A .$$

Similarly for vectors, we use $\boldsymbol{\delta}$ with corresponding subscript to denote the noise vector

$$\mathbf{X}(\varepsilon) = \mathbf{X} + \boldsymbol{\delta}_{\mathbf{X}}.$$

Γ with the corresponding subscript is used to denote the covariance matrix of the noise vector (considering only the first order error, the mean of the error is zero):

$$\Gamma_{\mathbf{x}} = \mathbb{E}\{\boldsymbol{\delta}_{\mathbf{x}} \boldsymbol{\delta}_{\mathbf{x}}^t\}$$

where $\mathbb{E}$ denotes expectation. A matrix $A = [\mathbf{A}_1 \ \mathbf{A}_2 \ \cdots \ \mathbf{A}_n]$ is associated with a corresponding vector $\mathbf{A}$ with

$$\mathbf{A} = \begin{bmatrix} \mathbf{A}_1 \\ \mathbf{A}_2 \\ \cdot \\ \cdot \\ \cdot \\ \mathbf{A}_n \end{bmatrix}.$$

In other words, the associated vector of a matrix consists of the column vectors that are lined up. Similarly, Γ_A denotes the corresponding covariance matrix of the vector $\mathbf{A}$ associated with matrix A. $\boldsymbol{\delta}_A$ denotes the perturbation vector associated with the perturbation matrix Δ_A. The symbol "$\triangleq$" is used in equations to define new variables when the variable to be defined is obvious.

Assuming two variables a and b with small errors:

$$a(\varepsilon) = a + \delta_a, \qquad b(\varepsilon) = b + \delta_b$$

we have

$$a(\varepsilon)b(\varepsilon) = ab + \delta_a b + a\delta_b + \delta_a\delta_b \triangleq ab + \delta_{ab}.$$

The error in $a(\varepsilon)b(\varepsilon)$ is

$$\delta_{ab} = \delta_a b + a\delta_b + \delta_a\delta_b \cong \delta_a b + a\delta_b. \tag{3.3.1}$$

In the last approximation we keep the linear terms of the error and ignore the higher order terms. Later in this book we use the sign "$\cong$" for the equations that are equal in the linear terms ("$\approx$" for the approximate equality in the usual sense). Considering a small perturbation in the original data, we analyze the linear terms of the corresponding (first order) perturbation of the final results to estimate the errors. In our problem, noise or error arises from image coordinates. The final results are the motion parameters calculated by the algorithm presented in the previous section.

3.3.2 Perturbation of Eigenvalues and Eigenvectors

The algorithm presented in Section 3.2 involves the calculation of the eigenvectors of a symmetrical matrix. With small perturbation in the matrix, we need to known the corresponding perturbation in its eigenvectors. We have the following theorem.

Theorem 3.1. Let $A = [a_{ij}]$ be an $n \times n$ symmetrical matrix and H be an orthonormal matrix such that

$$H^{-1}AH = \mathrm{diag}\{\lambda_1, \lambda_2, \cdots, \lambda_n\}$$

where $\mathrm{diag}\{\lambda_1, \lambda_2, \cdots, \lambda_n\}$ denotes the diagonal matrix with the corresponding diagonal elements. Let the eigenvalues be sorted in nondecreasing order. Without loss of generality, consider the eigenvalue λ_1. Assuming λ_1 is a simple eigenvalue, it follows that

$$\lambda_1 < \lambda_2 \le \lambda_3 \le \cdots \le \lambda_n.$$

Denote

$$H = [\mathbf{h}_1 \; \mathbf{h}_2 \cdots \mathbf{h}_n].$$

Let $\mathbf{X}$ be an eigenvector of A associated with λ_1. $\mathbf{X}$ is then a vector in $\mathrm{span}\{\mathbf{h}_1\}$, the linear space spanned by $\mathbf{h}_1$. Let $\mathbf{X}(\varepsilon)$ be the eigenvector of the perturbed matrix, $A(\varepsilon) = A + \Delta_A$, associated with the perturbed eigenvalue $\lambda_1(\varepsilon)$. $\mathbf{X}(\varepsilon)$ can be written as

$$\mathbf{X}(\varepsilon) = \mathbf{X} + \boldsymbol{\delta}_{\mathbf{x}}$$

with $\boldsymbol{\delta}_{\mathbf{x}} \in \operatorname{span}\{\mathbf{h}_2, \mathbf{h}_3, \cdots, \mathbf{h}_n\}$. Letting ε be the maximum absolute value of the elements in $\Delta_A = [\delta_{a_{ij}}]$, we have

$$\Delta_A = \varepsilon B$$

where $B = [b_{ij}]$, with $b_{ij} = \delta_{a_{ij}}/\varepsilon$. Therefore $|b_{ij}| \le 1$, $1 \le i \le n$, $1 \le j \le n$. Then for any sufficiently small ε, the perturbation of λ_1 can be expressed by a convergent series in ε:

$$\delta_{\lambda_1} \triangleq \lambda_1(\varepsilon) - \lambda_1 = p_1\varepsilon + p_2\varepsilon^2 + p_3\varepsilon^3 + \cdots \tag{3.3.2}$$

and the perturbation vector $\boldsymbol{\delta}_{\mathbf{x}}$ can be expressed by a convergent vector series in the space $\operatorname{span}\{\mathbf{h}_2, \mathbf{h}_3, \cdots, \mathbf{h}_n\}$. In other words, letting $H_2 = [\mathbf{h}_2, \mathbf{h}_3, \cdots, \mathbf{h}_n]$, for any sufficiently small positive ε, there exist (n-1)-dimensional vectors $\mathbf{g}_1, \mathbf{g}_2, \mathbf{g}_3, \cdots$ such that

$$\boldsymbol{\delta}_{\mathbf{x}} = \varepsilon H_2\mathbf{g}_1 + \varepsilon^2 H_2\mathbf{g}_2 + \varepsilon^3 H_2\mathbf{g}_3 + \cdots . \tag{3.3.3}$$

The liner term (in ε) in (3.3.2) is given by

$$p_1\varepsilon = \mathbf{h}_1^t \Delta_A \mathbf{h}_1.$$

The linear term (in ε) in (3.3.3) is given by

$$\varepsilon H_2\mathbf{g}_1 = H \Delta H^t \Delta_A \mathbf{X}$$

where

$$\Delta = \operatorname{diag}\{0, (\lambda_1 - \lambda_2)^{-1}, \cdots, (\lambda_1 - \lambda_n)^{-1}\}. \tag{3.3.4}$$

That is, suppressing the second and higher order terms (considering first order perturbation), for the eigenvalue

$$\delta_{\lambda_1} \cong \mathbf{h}_1^t \Delta_A \mathbf{h}_1,$$

and for the eigenvector

$$\delta_{\mathbf{x}} \cong H \Delta H^t \Delta_A \mathbf{X}.$$

Proof. See Appendix 3.A.

The above theorem gives the first order perturbation of the eigenvector associated with a simple eigenvalue λ_1. A similar result holds for other simple eigenvalues. For example, to give the first order perturbation of the eigenvector $\mathbf{X}_2$ associated with a simple eigenvalue λ_2, we just need to modify the matrix Δ in $\delta_{\mathbf{x}_2} \cong H \Delta H^t \Delta_A \mathbf{X}_2$:

$$\Delta = \text{diag}\{(\lambda_2 - \lambda_1)^{-1}, 0, (\lambda_2 - \lambda_3)^{-1}, \cdots, (\lambda_2 - \lambda_n)^{-1}\}.$$

3.3.3 Error Estimation for the Algorithm

According to the above theorem, if the perturbation matrix Δ_A can be estimated, the corresponding perturbation in the eigenvectors of A can be estimated. In the steps (i), (ii) and (iii) of the algorithm we need to find eigenvectors of the corresponding matrices. The problem now is to estimate the perturbation of the corresponding matrices from the perturbation in the image coordinates. Again, we use the first order approximation to estimate these perturbations in the matrices.

For Step (i). Suppose that the components of the image vectors $\mathbf{X}_i = (u_i, v_i, 1)^t$ and $\mathbf{X}'_i = (u'_i, v'_i, 1)^t$ have errors. (The third component 1 in the image vectors is accurate.) Let u_i, v_i, u'_i and v'_i have additive errors δ_{u_i}, δ_{v_i}, $\delta_{u'_i}$ and $\delta_{v'_i}$, respectively, for $1 \le i \le n$. From (3.2.5) we get:

$$\Delta_A^t = \begin{bmatrix} \delta_{u_1}u'_1 + \delta_{u'_1}u_1 & \delta_{u_2}u'_2 + \delta_{u'_2}u_2 & \cdots & \delta_{u_n}u'_n + \delta_{u'_n}u_n \\ \delta_{u_1}v'_1 + \delta_{v'_1}u_1 & \delta_{u_2}v'_2 + \delta_{v'_2}u_2 & \cdots & \delta_{u_n}v'_n + \delta_{v'_n}u_n \\ \delta_{u_1} & \delta_{u_2} & \cdots & \delta_{u_n} \\ \delta_{v_1}u'_1 + \delta_{u'_1}v_1 & \delta_{v_2}u'_2 + \delta_{u'_2}v_2 & \cdots & \delta_{v_n}u'_n + \delta_{u'_n}v_n \\ \delta_{v_1}v'_1 + \delta_{v'_1}v_1 & \delta_{v_2}v'_2 + \delta_{v'_2}v_2 & \cdots & \delta_{v_n}v'_n + \delta_{v'_n}v_n \\ \delta_{v_1} & \delta_{v_2} & \cdots & \delta_{v_n} \\ \delta_{u'_1} & \delta_{u'_2} & \cdots & \delta_{u'_n} \\ \delta_{v'_1} & \delta_{v'_2} & \cdots & \delta_{v'_n} \\ 0 & 0 & \cdots & 0 \end{bmatrix}.$$

Assume the errors between the different points and different components in the image coordinates are uncorrelated, and they have the same variance σ^2 (general cases with correlation can be formulated in a similar way). With this assumption we get

$$\Gamma_{A'} = \sigma^2 \ \mathrm{diag}\{P_1, P_2, \cdots, P_n\} \tag{3.3.5}$$

where P_i, $1 \le i \le n$, is a 9×9 sub-matrix:

$$P_i = \begin{bmatrix} \mathbf{X}'_i\mathbf{X}'^t_i & \mathbf{0} & \mathbf{0} \\ \mathbf{0} & \mathbf{X}'_i\mathbf{X}'^t_i & \mathbf{0} \\ \mathbf{0} & \mathbf{0} & \mathbf{X}'_i\mathbf{X}'^t_i \end{bmatrix} + \begin{bmatrix} u_i u_i J & u_i v_i J & u_i J \\ u_i v_i J & v_i v_i J & v_i J \\ u_i J & v_i J & J \end{bmatrix}$$

where

$$J \triangleq \begin{bmatrix} 1 & 0 & 0 \\ 0 & 1 & 0 \\ 0 & 0 & 0 \end{bmatrix}.$$

Consider the error in the solution **h** to the problem in (3.2.7). From Theorem 3.1 and (3.2.6), we get (note that **h** is an eigenvector of A^tA instead of A):

$$\begin{aligned} \delta_{\mathbf{h}} &\cong H\Delta H^t \Delta_{A^tA}\mathbf{h} \\ &= H\Delta H^t [h_1 I_9 \ h_2 I_9 \ \cdots \ h_9 I_9]\delta_{A^tA} \\ &\triangleq G_{\mathbf{h}}\delta_{A^tA}. \end{aligned} \tag{3.3.6}$$

In the above equations, we have rewritten the matrices Δ_{A^tA} as δ_{A^tA} and moved this perturbation factor to the right. In this way, the perturbation of the eigenvector is linear transformation (by matrix $G_{\mathbf{h}}$) of the perturbation vector δ_{A^tA}. Notice $\Gamma_{A'}$ $(= \Gamma_A^t)$ in (3.3.5). We need to relate δ_{A^tA} in (3.3.6) to $\delta_{A'}$. Similar to (3.3.1), using first order approximation, it follows that

$$\Delta_{A^tA} \cong A^t\Delta_A + \Delta_A^t A. \tag{3.3.7}$$

Letting

$$A^t = [a_{ij}]^t \triangleq [\mathbf{A}_1 \ \mathbf{A}_2 \ \cdots \ \mathbf{A}_n]$$

we write

$$\delta_{A^tA} \cong G_{A^tA}\,\delta_{A^t} \tag{3.3.8}$$

where G_{A^tA} can be easily determined from (3.3.7):

$$G_{A^tA} = [F_{ij}] + [G_{ij}]$$

where $[F_{ij}]$ and $[G_{ij}]$ are matrices with 9 by n sub-matrices F_{ij} and G_{ij}, respectively. $F_{ij} = a_{ji}I_9$. G_{ij} is a 9×9 matrix with the i-th column being the column vector $\mathbf{A}_j$ and all other columns being zeros. From (3.3.6) and (3.3.8) we get

$$\delta_{\mathbf{h}} \cong G_{\mathbf{h}}\delta_{A^tA} \cong G_{\mathbf{h}}G_{A^tA}\,\delta_{A^t} \triangleq D_{\mathbf{h}}\delta_{A^t}.$$

Then

$$\Gamma_{\mathbf{h}} \cong D_{\mathbf{h}}\Gamma_{A^t}D_{\mathbf{h}}^t. \tag{3.3.9}$$

From (3.2.8) we immediately obtain the covariance matrix for E:

$$\Gamma_E = 2\Gamma_{\mathbf{h}} \cong 2\,D_{\mathbf{h}}\Gamma_{A^t}D_{\mathbf{h}}^t. \tag{3.3.10}$$

To recapitulate, starting from the covariance matrix of the perturbation in A^t, we get the covariance matrix of the perturbation in the eigenvector of A^tA. This is done for E in equation (3.2.8). For $\mathbf{T}_s$ in equation (3.2.9) and for $\mathbf{q}$ in equation (3.2.16) and (3.2.30), the approaches are similar. For the perturbation vectors of the remaining parameters, we get the linear expression in terms of δ_E. For example, if we get $D_{\mathbf{T}_s}$ such that $\delta_{\mathbf{T}_s} \cong D_{\mathbf{T}_s}\delta_E$, we have $\Gamma_{\mathbf{T}_s} \cong D_{\mathbf{T}_s}\Gamma_E D_{\mathbf{T}_s}^t$.

The solution in step (i) needs the eigenvector of A^tA associated with the smallest eigenvalue. The smallest eigenvalue is a simple zero eigenvalue when rank$\{A\} = 8$ (with a nondegenerate configuration), When rank$\{A\} < 8$ (*i.e.*, when a degenerate configuration occurs), the solution $\mathbf{h}$ in step (i) is very sensitive to noise. As can be seen from (3.3.4), the second diagonal entry of Δ is infinite when $\lambda_1 = \lambda_2$. This makes the estimated errors infinite.

In most applications, we do not know the noise-free A. We only know the noise-corrupted A: $A(\varepsilon)$. We have to use $A(\varepsilon)$ to estimate A. In the presence of noise, generally, the rank of $A(\varepsilon)$ is full mathematically and the smallest eigenvalue of $A(\varepsilon)^tA(\varepsilon)$ is a small positive value. If noise is reasonably small, when rank$(A) < 8$ we have $\lambda_1 \approx \lambda_2$. Then large estimates of errors are still generated. From a slightly different point of view, we can regard A as a "noise-corrupted'' matrix by adding $-\Delta_A$ to the matrix $A(\varepsilon)$. Now the error is the deviation of the true solution from the noise-corrupted solution. This observation justify our use of the noise-corrupted A to estimate errors.

For Step (ii). $\mathbf{T}_s$ is the unit eigenvector of EE^t associated with the smallest eigenvalue. As we did earlier, we need $\boldsymbol{\delta}_{EE^t}$ to use the theorem. From

$$\Delta_{EE^t} \cong E\Delta_E^t + \Delta_E E^t$$

it is easy to find D_{EE^t} such that

$$\boldsymbol{\delta}_{EE^t} \cong D_{EE^t}\boldsymbol{\delta}_E .$$

In fact,

$$D_{EE^t} = [F_{ij}] + [G_{ij}]$$

where $[F_{ij}]$ and $[G_{ij}]$ are matrices with 3×3 sub-matrices F_{ij} and G_{ij}, respectively. F_{ij} is a 3×3 matrix with i-th column being $\mathbf{E}_j$ (see (3.2.8)) and all other columns being zeros, and $G_{ij} = e_{ij} I_3$.

We have, using Theorem 3.1 and the same notation for Δ and H,

$$\boldsymbol{\delta}_{\mathbf{T}_s} \cong H\Delta H^t \Delta_{EE^t}\mathbf{T}_s \cong H\Delta H^t [s_1 I_3 \;\; s_2 I_3 \;\; s_3 I_3]\boldsymbol{\delta}_{EE^t}$$

$$= H\Delta H^t [s_1 I_3 \;\; s_2 I_3 \;\; s_3 I_3] D_{EE^t}\boldsymbol{\delta}_E \triangleq D_{\mathbf{T}_s}\boldsymbol{\delta}_E$$

where $\mathbf{T}_s = (s_1, s_2, s_3)^t$. $\boldsymbol{\delta}_{\hat{\mathbf{T}}}$ is the same as $\boldsymbol{\delta}_{\mathbf{T}_s}$ except for a possible sign change depending on (3.2.10). Thus, it follows that

$$\Gamma_{\mathbf{T}_s} = \Gamma_{\hat{\mathbf{T}}} = D_{\mathbf{T}_s}\Gamma_E D_{\mathbf{T}_s}^t .$$

For conciseness, we define a new vector $\mathbf{K}$ that combines the vector $\mathbf{T}_s$ with the vector $\mathbf{E}$:

$$\mathbf{K} \triangleq \begin{bmatrix} \mathbf{T}_s \\ \mathbf{E} \end{bmatrix} = \begin{bmatrix} \mathbf{T}_s \\ \mathbf{E}_1 \\ \mathbf{E}_2 \\ \mathbf{E}_3 \end{bmatrix} .$$

So,

$$\delta_{\mathbf{K}} \cong \begin{bmatrix} D_{\mathbf{T}_s} \\ I_9 \end{bmatrix} \delta_E \triangleq D_{\mathbf{K}} \delta_E .$$

For Step (iii). From (3.2.17) and using first order approximation, it follows that

$$B(\varepsilon) = B + \Delta_B = \sum_{i=1}^{3} (B_i + \Delta_{B_i})^t (B_i + \Delta_{B_i})$$

$$\cong \sum_{i=1}^{3} B_i^t B_i + \sum_{i=1}^{3} (B_i^t \Delta_{B_i} + \Delta_{B_i}^t B_i)$$

where

$$\Delta_B \cong \sum_{i=1}^{3} (B_i^t \Delta_{B_i} + \Delta_{B_i}^t B_i). \tag{3.3.11}$$

If we use (3.2.15) to solve for R, from (3.2.14) we get

$$\delta_{\mathbf{W}_1} \cong \mathbf{E}_1 \times \delta_{\mathbf{T}_s} - \mathbf{T}_s \times \delta_{\mathbf{E}_1} - \mathbf{E}_3 \times \delta_{\mathbf{E}_2} + \mathbf{E}_2 \times \delta_{\mathbf{E}_3}$$

$$\delta_{\mathbf{W}_2} \cong \mathbf{E}_2 \times \delta_{\mathbf{T}_s} - \mathbf{T}_s \times \delta_{\mathbf{E}_2} - \mathbf{E}_1 \times \delta_{\mathbf{E}_3} + \mathbf{E}_3 \times \delta_{\mathbf{E}_1}$$

$$\delta_{\mathbf{W}_3} \cong \mathbf{E}_3 \times \delta_{\mathbf{T}_s} - \mathbf{T}_s \times \delta_{\mathbf{E}_3} - \mathbf{E}_2 \times \delta_{\mathbf{E}_1} + \mathbf{E}_1 \times \delta_{\mathbf{E}_2}$$

or

$$\delta_W \cong \begin{bmatrix} [\mathbf{E}_1]_\times & -[\mathbf{T}_s]_\times & -[\mathbf{E}_3]_\times & [\mathbf{E}_2]_\times \\ [\mathbf{E}_2]_\times & [\mathbf{E}_3]_\times & -[\mathbf{T}_s]_\times & -[\mathbf{E}_1]_\times \\ [\mathbf{E}_3]_\times & -[\mathbf{E}_2]_\times & [\mathbf{E}_1]_\times & -[\mathbf{T}_s]_\times \end{bmatrix} \delta_{\mathbf{K}}$$

$$\triangleq G_W \delta_{\mathbf{K}} \cong G_W D_{\mathbf{K}} \delta_E \triangleq D_W \delta_E .$$

Letting $W = [w_{ij}]$, from (3.3.11) we get

$$\delta_B \cong G_B \delta_W \cong G_B D_W \delta_E \triangleq D_B \delta_E \tag{3.3.12}$$

where

$$G_B = 2 \begin{bmatrix} F_t \\ F_b \end{bmatrix}$$

with

$$F_t = \begin{bmatrix} w_{11}-1 & w_{21} & w_{31} & w_{12} & w_{22}-1 & w_{32} & w_{13} & w_{23} & w_{33}-1 \\ 0 & 0 & 0 & 0 & 0 & -1 & 0 & 1 & 0 \\ 0 & 0 & 1 & 0 & 0 & 0 & -1 & 0 & 0 \\ 0 & -1 & 0 & 1 & 0 & 0 & 0 & 0 & 0 \\ 0 & 0 & 0 & 0 & 0 & -1 & 0 & 1 & 0 \\ w_{11}-1 & w_{21} & w_{31} & w_{12} & w_{22}+1 & w_{32} & w_{13} & w_{23} & w_{33}+1 \\ 0 & -1 & 0 & -1 & 0 & 0 & 0 & 0 & 0 \\ 0 & 0 & -1 & 0 & 0 & 0 & -1 & 0 & 0 \end{bmatrix},$$

$$F_b = \begin{bmatrix} 0 & 0 & 1 & 0 & 0 & 0 & -1 & 0 & 0 \\ 0 & -1 & 0 & -1 & 0 & 0 & 0 & 0 & 0 \\ w_{11}+1 & w_{21} & w_{31} & w_{12} & w_{22}-1 & w_{32} & w_{13} & w_{23} & w_{33}+1 \\ 0 & 0 & 0 & 0 & 0 & -1 & 0 & -1 & 0 \\ 0 & -1 & 0 & 1 & 0 & 0 & 0 & 0 & 0 \\ 0 & 0 & -1 & 0 & 0 & 0 & -1 & 0 & 0 \\ 0 & 0 & 0 & 0 & 0 & -1 & 0 & -1 & 0 \\ w_{11}+1 & w_{21} & w_{31} & w_{12} & w_{22}+1 & w_{32} & w_{13} & w_{23} & w_{33}-1 \end{bmatrix}.$$

For the case where (3.2.13) is used to solve for R, the relationship between $\boldsymbol{\delta}_B$ and $\boldsymbol{\delta}_E$ is relegated to Appendix 3.C.

Having obtained the expression of $\boldsymbol{\delta}_B$, we are ready to give the covariance matrix of $\mathbf{q}$. Since $\mathbf{q}$ is a unit eigenvector of B associated with the smallest eigenvalue, using Theorem 3.1, we have

$$\begin{aligned} \boldsymbol{\delta}_{\mathbf{q}} &\cong H\Delta H^t \Delta_B \mathbf{q} = H\Delta H^t [q_0 I_4 \;\; q_1 I_4 \;\; q_2 I_4 \;\; q_3 I_4] \boldsymbol{\delta}_B \\ &\cong H\Delta H^t [q_0 I_4 \;\; q_1 I_4 \;\; q_2 I_4 \;\; q_3 I_4] D_B \boldsymbol{\delta}_E \triangleq D_{\mathbf{q}} \boldsymbol{\delta}_E \end{aligned} \tag{3.3.13}$$

Using the relation between $\mathbf{q}$ and R, i.e., equation (3.2.19), we get the first order perturbation vector of R:

$$\delta_R \cong 2 \begin{bmatrix} q_0 & q_1 & -q_2 & -q_3 \\ q_3 & q_2 & q_1 & q_0 \\ -q_2 & q_3 & -q_0 & q_1 \\ -q_3 & q_2 & q_1 & -q_0 \\ q_0 & -q_1 & q_2 & -q_3 \\ q_1 & q_0 & q_3 & q_2 \\ q_2 & q_3 & q_0 & q_1 \\ -q_1 & -q_0 & q_3 & q_2 \\ q_0 & -q_1 & -q_2 & q_3 \end{bmatrix} \delta_{\mathbf{q}}$$

$$\triangleq G_R \, \delta\mathbf{q} \cong D_R D_{\mathbf{q}} \delta_E \triangleq D_R \, \delta_E . \tag{3.3.14}$$

As in step (i), in steps (ii) and (iii) we estimate the errors by using the perturbed E and B to substitute for the noise-free E and B.

In summary, the perturbation vectors of the parameters $\mathbf{q}$ and R are expressed in terms of linear transformations of perturbation of E. The covariance matrix of the perturbation of E is given in (3.3.10). The covariance matrices of $\mathbf{q}$ and R are then

$$\Gamma_{\mathbf{q}} = D_{\mathbf{q}} \Gamma_E D_{\mathbf{q}}^t ,$$

$$\Gamma_R = D_R \Gamma_E D_R^t .$$

From the covariance matrix of the perturbation, we can estimate the Euclidean norm of the perturbation vector and the perturbation matrix by the square root of the trace of the corresponding covariance matrix of the perturbations.

$$\|\delta_{\hat{\mathbf{T}}}\| \approx \sqrt{\mathrm{trace}\{\Gamma_{\hat{\mathbf{T}}}\}},$$

$$\|\Delta_R\| = \|\delta_R\| \approx \sqrt{\mathrm{trace}\{\Gamma_R\}}.$$

Similarly we get the estimate of perturbation in $\mathbf{q}$. Since the Euclidean norm of the orthonormal matrix R is equal to $\sqrt{3}$, the relative perturbation in R is defined by $\|\Delta_R\|/\sqrt{3}$.

The problem of estimating errors in the relative depths can be formulated in a similar manner. However, as indicated by our simulation, the variance of the errors in the relative depths is considerably larger than those of the motion parameters. This is what we have expected, since for each 3-D point we just get two observations. Therefore the estimated mean of the errors is not a reliable estimate of the error in depth.

3.4 Error Analysis

In reality, the perspective projections of feature points are corrupted by noise. The noise includes the feature detector errors, matching errors, quantization errors and system calibration errors. All those errors result in inexact solution for the motion parameters and 3-D structure of the scene. It is observed that computer roundoff errors are generally far less significant than those mentioned above, provided a double precision (about 64 bit representation) is used for real numbers. Therefore, we assume that the noise is introduced solely through the perturbations in the measured image coordinates of the projected of feature points.

However, with the same noise level, the resulting errors are not always the same for different scene structures, motion parameters, and system setups. The question is how they are related and to what degree they affect the reliability of the estimates. We will discuss three categories of factors that affect the reliability of the estimates:

(1) structure of the scene,

(2) motion,

(3) parameters of imaging systems.

Our analysis is mainly based on the algorithm presented in this chapter. We also provide algorithm-independent perspectives.

3.4.1 Structure of the Scene

The 9-dimensional unit vector $\mathbf{h}$ is determined up to a sign if and only if the rank of A in (3.2.7) is equal to 8. If a set of feature points is such that the rank of corresponding A is less than 8, we say that the structure is degenerate.

In the presence of noise, generally the rank of A is mathematically full even if the actual structure is degenerate. If the structure is nearly degenerate, the solution of (3.2.7) is conceivably not reliable. Therefore, in the presence of noise, we should consider the numerical condition of the matrix A. The previous section has presented a method of determining such a condition and gives an estimate of the errors.

Obviously, if the cluster of projections of feature points is confined to a small portion of images, image resolution is used to only a limited degree. This limits the reliability of the solutions. Therefore, the feature points should cover as much of the images as possible.

In the discussion of Subsection 3.4.2, we will see that long displacement vectors will result in more reliable solutions. Given a certain amount of motion, the scene should be close to the camera so that it yields large displacements in image plane. This condition is actually related to the numerical condition of matrix A.

Another factor is the number of feature points. It is very effective to reduce the error in the solutions by using more points than the minimally required 8. Since a severely noise-corrupted image vector can pull the solution away from the correct one by a large amount, it is desirable to use only reliable matches for motion parameter estimation.

It is clear that the relative depths can be reliably determined by minimizing (3.2.22) only if $\mathbf{X}'_i$ and $R\,\mathbf{X}_i$ are linearly independent. That is, $\mathbf{X}'_i \times R\,\mathbf{X}_i \neq \mathbf{0}$. When $\mathbf{T} \neq \mathbf{0}$, all points satisfy this except possibly one. In fact $\mathbf{X}'_i \times R\,\mathbf{X}_i = \mathbf{0}$ if and only if $\mathbf{T} \times R\,\mathbf{X}_i = \mathbf{0}$ from (3.2.3). Let $\mathbf{X}_p$ be such that $\mathbf{T} \times R\,\mathbf{X}_p = \mathbf{0}$. If $\mathbf{X}_p$ happens to be the image vector of a feature point, the depth of this point can not be determined. For those points whose projections are close to $\mathbf{X}_p$, the corresponding depths can not be reliably determined in the presence of noise. If the rotation angle is equal to zero, the point $\mathbf{X}_p$ corresponds to the focus of expansion or contraction. For this point, the projection is the same before and after motion.

3.4.2 Motion Parameters

As mentioned earlier, a motion can be represented by a rotation followed by a translation. In the discussion that follows, we divide motion parameters into three categories: (1) magnitude of translation, (2) direction of translation, (3) parameters of rotation.

Magnitude of translation. If the magnitude of a translation vector is equal to zero, the solution of the translation direction is arbitrary (since $\mathbf{T}_s$ in (3.2.23) is an arbitrary unit vector) and the depths of the feature points can not be determined. When $\|\mathbf{T}\|$ is close to zero, the direction of translation, $\mathring{\mathbf{T}}$, can not be reliably determined and therefore, neither can the depths of feature points whose solution needs $\mathring{\mathbf{T}}$ in (3.2.22).

When $\mathbf{T} = \mathbf{0}$, the rank of A in (3.2.5) is always no larger than 6 ($E = [\mathbf{T}_s]_\times R$ has two degrees of freedom since $\mathbf{T}_s$ is an arbitrary unit vector). R can still be determined by picking up any $\mathbf{h}$ satisfying (3.2.7), since in the definition of E in $E = [\mathbf{T}_s]_\times R$, $\mathbf{T}_s$ is just a unit vector satisfying (3.2.23).

Direction of translation. This is the most interesting factor associated with the reliability of the solutions. From (3.2.25) and the algorithm, it can be seen that the translation direction is determined from the fact that $\mathbf{T}\cdot(\mathbf{X}'\times R\,\mathbf{X}) = 0$, or in other words, $\mathbf{T}$ is orthogonal to the cross product $\mathbf{X}'\times R\,\mathbf{X}$. Fig. 3.2(a) illustrates the spatial relations between three vectors $\mathbf{x}'$, $\mathbf{x}$ and $\mathbf{T}$. Figs. 3.2(b) and (c) show the

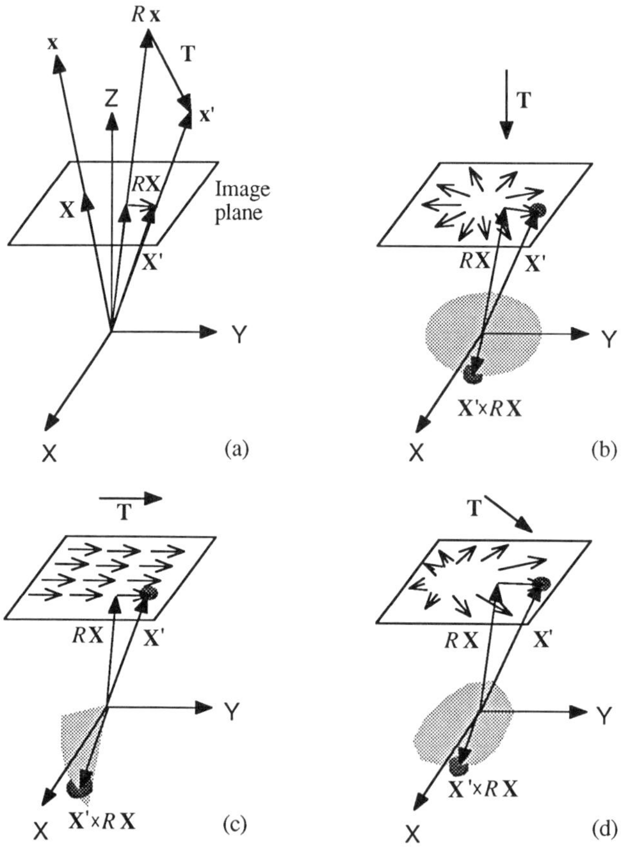

Fig. 3.2. Effects of perturbation versus translation direction.

cases where the translation vector is orthogonal and parallel, respectively, to the image plane. Fig. 3.2(d) shows a general case. Usually, the projections of feature points cover a considerably large area of the image around the optical z-axis. In the case of Fig. 3.2(b) it is clear that the vectors $\mathbf{X}'\times R\,\mathbf{X}$ occupy the area in xy-plane around the origin (shown shaded). However for the case of Fig. 3.2(c) the vectors $\mathbf{X}'\times R\,\mathbf{X}$ are confined to a small shaded area in xz-plane. For the general case, the area of $\mathbf{X}'\times R\,\mathbf{X}$ is shown shaded in Fig. 3.2(d). The algorithm determines the direction of $\mathbf{T}$ through $\mathbf{T}\cdot(\mathbf{X}'\times R\,\mathbf{X}) = 0$, one equation for each of n points. This means that $\mathbf{T}$ is orthogonal to n vectors in the shaded area. With perturbation in $\mathbf{X}$ and $\mathbf{X}'$, the product $\mathbf{X}'\times R\,\mathbf{X}$ will be slightly perturbed away from the original position and it may leave the plane of the shaded area. This causes the errors in the estimated $\mathring{\mathbf{T}}$. Since the shaded area in Fig. 3.2(b) extends in all directions around the origin but that in Fig. 3.2(c) is confined to a small area on one side of origin, statistically the former allows a more reliable estimate of $\mathring{\mathbf{T}}$ than the latter.

On the other hand, the perturbation of Fig. 3.2(b) generally will not leave the shaded area as much as that of Fig. 3.2(c). This can be seen in the following.

First, we study the perturbation in $\mathbf{X}'\times R\,\mathbf{X}$ due to image plane error. Suppose that the vector $\mathbf{X}'$ is perturbed in the image plane. The area indicating the result of perturbation is shown in Fig. 3.2 by a small dark disk around $\mathbf{X}'$ in the image plane. The corresponding perturbed vector $\mathbf{X}'\times R\,\mathbf{X}$ is roughly represented by a small dark disk around $\mathbf{X}'\times R\,\mathbf{X}$. Since this disk is orthogonal to $R\,\mathbf{X}$, it is nearly parallel to the image plane if $R\,\mathbf{X}$ is not far away from the optical axis. Similarly, the perturbation of $\mathbf{X}'\times R\,\mathbf{X}$ due to perturbation in $R\,\mathbf{X}$ is orthogonal to $\mathbf{X}'$, and thus it is nearly parallel to the image plane if $\mathbf{X}'$ is not far away from the optical axis. Thus, the perturbations of $\mathbf{X}'\times R\,\mathbf{X}$ due to individual perturbation of either $\mathbf{X}'$ or $R\,\mathbf{X}$ are always nearly *parallel* to the image plane provided $\mathbf{X}'$ and $R\,\mathbf{X}$ are not far from the optical axis.

Second, we know that perturbation of the product of two vectors is approximately the sum of the individual perturbations. For example, if we denote the perturbation of $\mathbf{a}$ by $\boldsymbol{\delta}_{\mathbf{a}}$ and that of $\mathbf{b}$ by $\boldsymbol{\delta}_{\mathbf{b}}$, it follows that

$$(\mathbf{a} + \boldsymbol{\delta}_{\mathbf{a}})\times(\mathbf{b} + \boldsymbol{\delta}_{\mathbf{b}}) - \mathbf{a}\times\mathbf{b} = \boldsymbol{\delta}_{\mathbf{a}}\times\mathbf{b} + \mathbf{a}\times\boldsymbol{\delta}_{\mathbf{b}} + \boldsymbol{\delta}_{\mathbf{a}}\times\boldsymbol{\delta}_{\mathbf{b}} \cong \boldsymbol{\delta}_{\mathbf{a}}\times\mathbf{b} + \mathbf{a}\times\boldsymbol{\delta}_{\mathbf{b}}. \tag{3.4.1}$$

The last two terms in (3.4.1) are perturbations of $\mathbf{a}\times\mathbf{b}$ due to individual perturbation of **a** and **b**, respectively. For example, the perturbation due to that of $\boldsymbol{\delta}_\mathbf{a}$ is $(\mathbf{a} + \boldsymbol{\delta}_\mathbf{a})\times\mathbf{b} - \mathbf{a}\times\mathbf{b} = \boldsymbol{\delta}_\mathbf{a}\times\mathbf{b}$. Therefore, the above analysis on individual perturbation of $\mathbf{X}'$ and $R\mathbf{X}$ accounts for the major perturbation of $\mathbf{X}'\times R\mathbf{X}$.

With the above two points we discussed, we are ready to see the perturbation of $\mathbf{X}'\times R\mathbf{X}$. In Fig. 3.2(b), the shaded area is *parallel* to the image plane, while in Fig. 3.2(c) it is *orthogonal* to the image plane. Thus, the perturbation of $\mathbf{X}'\times R\mathbf{X}$, represented by a small disk, lies almost inside the shaded area in Fig. 3.2(b) but sticks out from the shaded area in Fig. 3.2(c). Since $\mathring{\mathbf{T}}$ is determined by being orthogonal to all those $\mathbf{X}'\times R\mathbf{X}$'s in the least squares sense, statistically the perturbations of $\mathbf{X}'$ and $R\mathbf{X}$ in the case of Fig. 3.2(c) will cause larger errors in the estimated $\mathring{\mathbf{T}}$ than those in the case of Fig. 3.2(b). It can be easily seen from Fig. 3.2(c) that the major perturbation of the estimated translation direction is in z-component.

Both the shape of the shaded areas and the orientation of the perturbation disks imply that a translation orthogonal to the image plane allows more stable estimation of translation direction $\mathring{\mathbf{T}}$ than a translation parallel to the image plane. This relationship can also be explained in more intuitive way. A translation in depth will cause less changes in the images than a translation parallel to the image plane but with the same translational magnitude. In other words, z-component of translation is very sensitive to errors in the image data. Therefore, z-component of the translation cannot be reliably determined. With a relatively large perturbation in z-component, the direction of a translation orthogonal to image plane is not as significantly affected as the direction of a translation parallel to the image plane.

Parameters of rotation. First, it is necessary to discuss briefly the error measurement of rotation parameters. There exist a variety of ways of representing rotation [3.2]. For example, (1) an axis of rotation and an angle about the axis, (2) three rotation angles about three fixed (or moving) coordinate axes, respectively, (3) rotation matrix, (4) quaternion. We need a measurement for the errors of rotation that does not very much depend on the actual rotation parameters. Consider the relative error of rotation axis and rotation angle in (1). The relative error of rotation angle θ is $|\tilde{\theta} - \theta|/\theta$, where $\tilde{\theta}$ is the estimated θ. It is infinity when $\theta = 0$. If rotation angle is zero or nearly zero, the error in rotation axis is not important at all. Therefore, the error in terms of parameters of (1) is not desirable.

Similarly, relative error of the parameters in (2) is not what we need. The relative error of rotation matrix R, $\|\tilde{R} - R\| / \|R\|$, where $\tilde{R}$ is the estimated R, and relative error of rotation quaternion do not suffer from the problems mentioned above. Here, we use the relative error of rotation matrix to indicate the errors in rotation unless we must use rotation axis and rotation angle for other reasons. Since $\|\hat{R}I - RI\| = \|\hat{R} - R\|$, the geometrical meaning of the relative error in a rotation matrix is that it measures the root mean-square error in the 3 unit vectors of a rotated orthonormal frame.

The correlation between rotation and translation is very complicated. The following gives some properties of their relationship.

First, let us consider how rotation can be separated from translation. A rotation about the optical axis is easily distinguished from translation by the algorithm since no translation gives a similar displacement field in the image. What about rotation about an axis parallel to the image plane, say x-axis? Let us consider two cases. In the first case, the viewer rotates about a vertical axis through his or her body. In the second case, the viewer translates to the side in the direction of the above rotation. If one is looking at a forefrontal wall (small depth variation), the displacement field on his retina is very similar in the above two cases. This implies that it is difficult to distinguish the translation from rotation. In fact, there exists a slight difference between rotation and translation in terms of projections as shown in Fig. 3.3. The linear algorithm uses this kind of difference since the direction of image vectors determine the essential parameter matrix E in (3.2.27). However, the difference is not very large, especially for short displacement vectors or at the center of images. Thus, the algorithm may easily confuse the translation with the rotation in the presence of noise. As a result, the solution is more sensitive to noise in case of translation parallel to the image plane than in case of translation orthogonal to the image plane. Similarly, rotation about an axis parallel to the image plane is more sensitive to noise than other

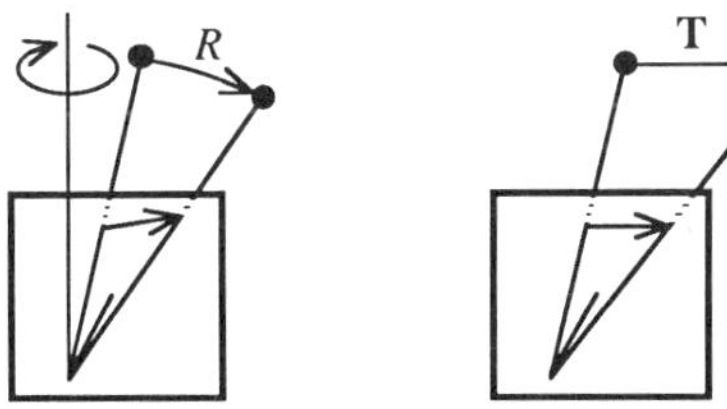

Fig. 3.3. Rotation and translation generate different displacement fields. Such differences are large near the peripheral areas of images.

rotations. However, since in most cases the displacement is mainly caused by translation and so the effects caused by translation are dominant. If the translation direction cannot be reliably determined, generally rotation cannot either, since R is determined using translation. Therefore, an unstable case for the estimation of translation is also unstable for the estimation of rotation.

After translation is determined, do different rotations imply different reliability of the estimated rotational parameters? If we consider the relative errors in terms of rotation axis and rotation angle, different types of rotations do imply different reliabilities of these two parameters. As shown in Fig. 3.4, different perturbations in the image vectors affect, by different amounts, the rotation axis **n** and rotation angle θ. Figs. 3.4(a) and (b) correspond to the case where rotation axis is orthogonal to the image plane. Figs. 3.4(c) and (d) correspond to the case where rotation axis is parallel to the image plane. The perturbation, represented in Fig. 3.4(a) by a two-way arrow in the

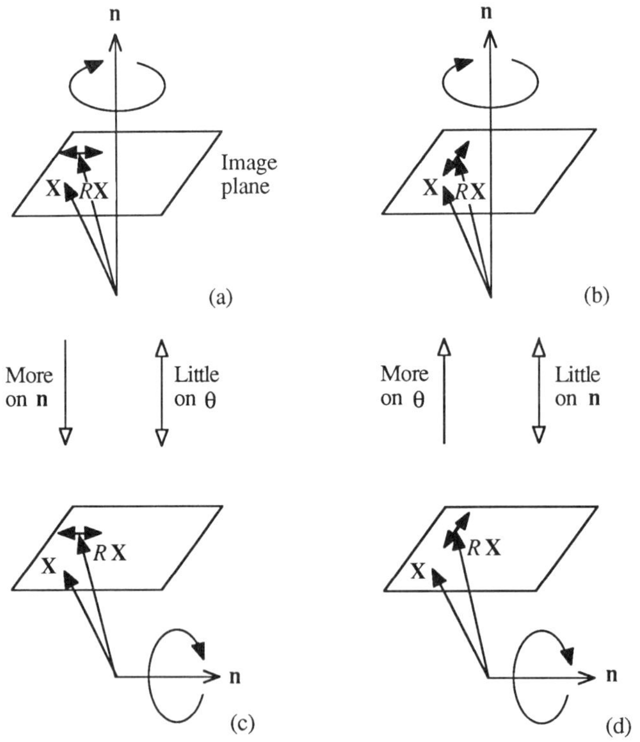

Fig. 3.4. Effects of perturbation versus rotation axis.

image plane, has smaller effect on rotation axis than that in Fig. 3.4(c), while both cases (a) and (c) have little effect on rotation angle. The perturbation in Fig. 3.4(b) has larger effect on rotation angle than that in Fig. 3.4(d), while both cases (b) and (d) have little effect on rotation axis. Summerizing from Fig. 3.4: comparing the two cases where rotation axis is orthogonal to the image plane and where rotation axis is parallel to the image plane, the rotation axis can be more reliably estimated for the former and the rotation angle can be more reliably estimated for the latter.

The above opposite effects on rotation axis and rotation angle make the error in R less sensitive to the type of motion. On the other hand, rotation is determined after translation. The errors in the estimated translation parameters also cause errors in rotation parameters. When the errors in translation are the main source of error in rotation parameters, the effects caused by different rotations are not significant. Simulations presented in Section 3.4 confirm that the errors in motion parameters are not sensitive to rotation parameters.

From Figs. 3.2, 3.3 and 3.4, it is clear that in the presence of noise long displacement vectors will generally result in more reliable solutions than short ones. To yield long displacement vectors, the motion should be large and the scene should be close to the image sensor.

3.4.3 System Parameters

Let the resolution and focal length be fixed. The remaining geometrical parameter of the imaging system of our model is the image size, or equivalently, the field of view.

Suppose that the image size is reduced, say, by a factor of 2. For the image to cover roughly the same scene, the scene has to be moved away from camera in the $\mathbf{Z}$ direction such that it is about twice as far as before. This reduces the variation in depth, which leads to increased instability in the estimates. If the scene is not moved farther away, the camera will cover a smaller area of the scene, which will usually reduce the variation of depth. Furthermore, the scene covering a smaller portion of the image is equivalent to reducing image resolution. In a word, a small image size, or a narrow field of view, yields unstable estimates. Another important drawback of narrowing the field of view is shown in Fig. 3.3. A large field of view is crucial for distinguishing rotation from translation, since the difference is more significant in the peripheral areas of images. Therefore, a reduction in image size will particularly worsen the estimates for the case where translation is parallel to the image plane.

With conventional imaging sensors, the image size is fixed. The focal length is the parameter that changes the field of view. Reducing image size is equivalent to increasing focal length and vise versa.

The following section quantitatively demonstrates the relationships discussed in this section.

3.5 Performance

In this section, we examine the performance of the methods through results of the experiments. The first series of simulations is designed to demonstrate the performance of the motion estimation algorithm introduced in Section 3.2, compared with other existing linear methods. The second series is intended to test the error estimation discussed in Section 3.3. The third series is meant to show the dependency of the errors on motion and system parameters. Finally, some results are presented for images of real world scenes.

In the simulations, the feature points of the scene are generated randomly according to a uniform distribution in a cube of $s \times s \times s$ (s is called *object size*) the center of which is called the center of the object before motion. The distance between the projection center of the camera and center of the object is called *object distance*. The image is a square whose side length is called *image size*. The field of view is determined by the size of the image and the focal length. Feature points undergo a rotation about an axis through the origin (projection center) and then a translation such that after the motion the object points are still in the field of view. The image coordinates of the points are quantized according to the resolution of the camera. If the resolution is $m \times m$, each of the horizontal and vertical coordinates has uniformly spaced m levels. The positions of these levels correspond to the locations of the pixels. The image coordinates are rounded-off to the nearest levels before they are used by the motion estimation algorithm. These round off errors result in the errors in the motion parameters and the relative depths computed by the algorithm. Simulations show that reducing the resolution by a factor of two roughly doubles the errors. Other additional random errors can be simulated by a reduced image resolution with the similar variance of quantization noise. Except for rotation angle, all the errors shown in this section are relative errors. Relative error of a matrix, or a vector, is defined by the norm of the error matrix, or vector, divided by the Euclidean norm of the correct matrix, or vector, respectively. Since no ambiguity can arise, in the remainder of this book, relative errors are often simply referred to as errors. Unless stated otherwise, in this section the object size is 10, object distance is 11 units, image size is 2, image resolution is 256 by 256 and the focal length is one unit.

3.5.1 Performance

First, to demonstrate the importance of the steps designed for noisy data, we show the performance difference between an algorithm designed for noise-free data and the algorithm here. The algorithm in Longuet-Higgins [3.10] and that in Tsai & Huang [3.21] are two typical algorithms primarily for noise-free data. We extract some steps from these two algorithms to constitute a sample algorithm for noise-free data, which we call the L-T algorithm. Since the way to compute the ratios of the components of $\mathbf{T}$ (from $\tilde{\mathbf{Q}}\mathbf{Q}$ of equation (17) in [3.10]) is not specifically given, for the L-T algorithm the unit vector $\overset{\circ}{\mathbf{T}}$ is determined by the algorithm in Tsai & Huang [3.21]. Rotation matrix R is determined by the method in [3.10], since it is computationally simpler. The remaining steps of the L-T algorithm are the same as the algorithm presented here. Such an algorithm represents the typical algorithms that are designed primarily for noise free data. We compare the performance of the algorithm presented here with the L-T algorithm on arbitrarily chosen motion parameters.

In Fig. 3.6, the rotation is about an axis (–0.2, 1, 0.2) by an angle of 8°. The projection of translation on to xz-plane changes from $\mathbf{X}$ to $-\mathbf{Z}$ direction (with a magnitude 3.0, evenly spaced 21 translation directions from the $\mathbf{X}$ to the $-\mathbf{Z}$ in xz-plane, with horizontal index from 0 to 20). The y-component is always equal to 1. 12 point correspondences are used. Fig. 3.5 shows two examples of displacement fields, corresponding to horizontal index 0 and 20, respectively. As can be seen, the projections of points do not spread over the entire image plane and, therefore, the field of view is not completely used. In other words, the effective resolution and field of view are actually reduced. This arrangement puts the test under more general situation since it is not always true that points can spread over the entire image. Fig. 3.6(a) shows the errors of R and Fig. 3.6(b) shows that of $\mathbf{T}$ averaging over 100 trials (random trial here always means randomly generated points). Significant improvement by our algorithm over the

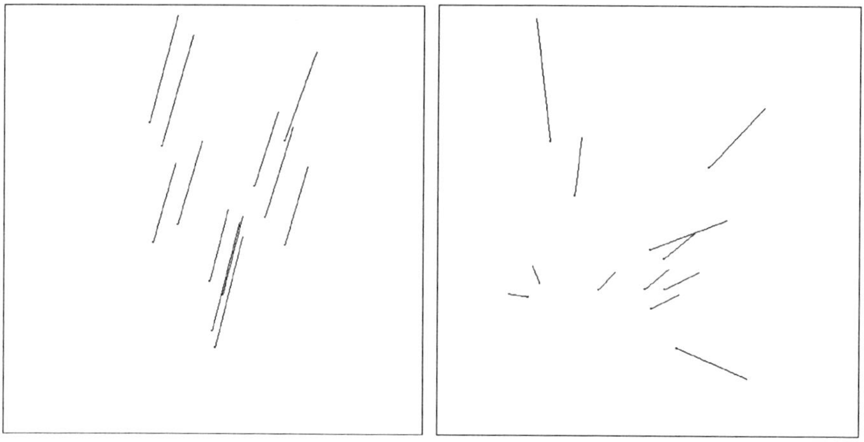

(a) (b)

Fig. 3.5. Two examples of displacement fields for the data shown in Fig. 3.6. **(a)** Displacement field corresponding to the horizontal index 0 in Fig. 3.6. **(b)** Displacement field corresponding to the horizontal index 20 in Fig. 3.6.

L-T algorithm is evident. Fig. 3.6 also shows the errors of the L-T algorithm for the same motion parameters but no round-off is performed for image coordinates (noise-free). Idealy the error should be almost equal to zero. However it is not so for horizontal index 0 and 20. The reason for this can be easily seen from equations (22)-(24) in [3.21]. That is, some special cases are not considered in [3.21]. This partially accounts for the different amount of improvement for different translations.

The performance of the algorithm in other cases is demonstrated in the remainder of this section.

3.5.2 Error Estimation

For error estimation, we assume that the round-off error is uniformly distributed between plus half and minus half of the pixel size. Therefore, the variance of the errors in the image coordinates is $\sigma^2 = p^2/12$, where p is the spacing between the quantization levels in the images (pixel size). This variance is used for error estimation.

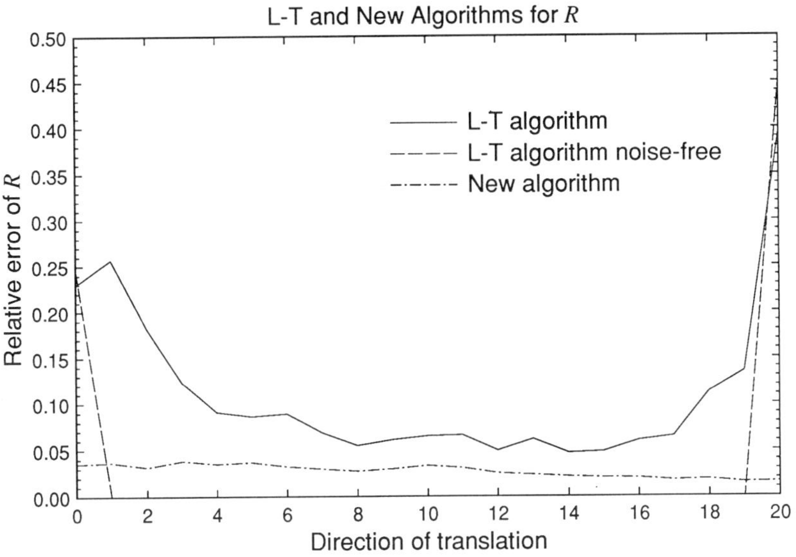

Fig. 3.6(a)

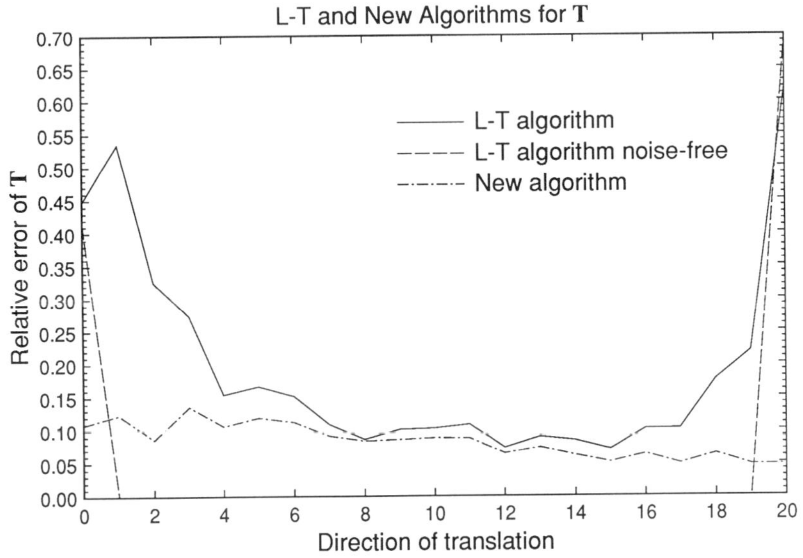

Fig. 3.6(b)

Fig. 3.6. Relative errors of the L-T algorithm and our algorithm. **(a)** Relative errors of R. **(b)** Relative errors of **T**. Rotation axis: (–0.2, 1, 0.2). Rotation angle: 8°. For horizontal axis from 0 to 20, the projection of translation on to xz-plane changes from **X** to **–Z** direction (with magnitude 3.0, evenly spaced 21 translation directions). The y-component is always equal to 1.

Different motion parameters with different image resolutions are simulated. If the image resolution is doubled, the errors in the solution is reduced roughly by a factor of two. Fig. 3.7 shows the results of a typical sequence of trials with 9 point correspondences. 20 random trials are shown in the order of their generation. Fig. 3.7(a) shows error estimation for relative errors of E. As can be seen from the figure, the estimated errors are strongly correlated with the actual errors. The estimated errors are especially important to detect a nearly degenerate configuration, e.g., trial No. 16 in Fig. 3.7(a) where relatively unreliable results are generated by the algorithm. Figs. 3.7(b) and (c) give the relative errors in the translation $\mathbf{T}$ and the rotation matrix R, respectively. The very similar curves of errors in E, $\mathbf{T}$ and R indicate that the steps after estimating E are stable. Main errors are attributed to the estimate of E. In other words, the accuracy of E dominates the accuracy of the final motion parameters.

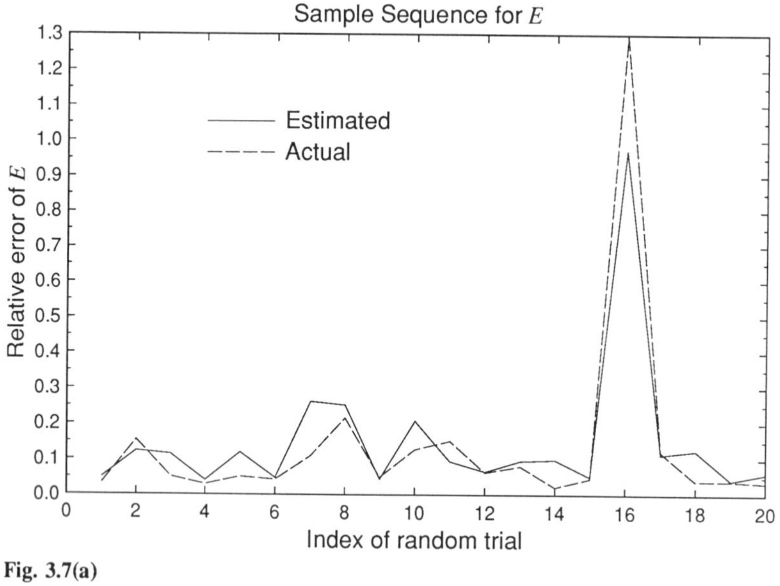

Fig. 3.7(a)

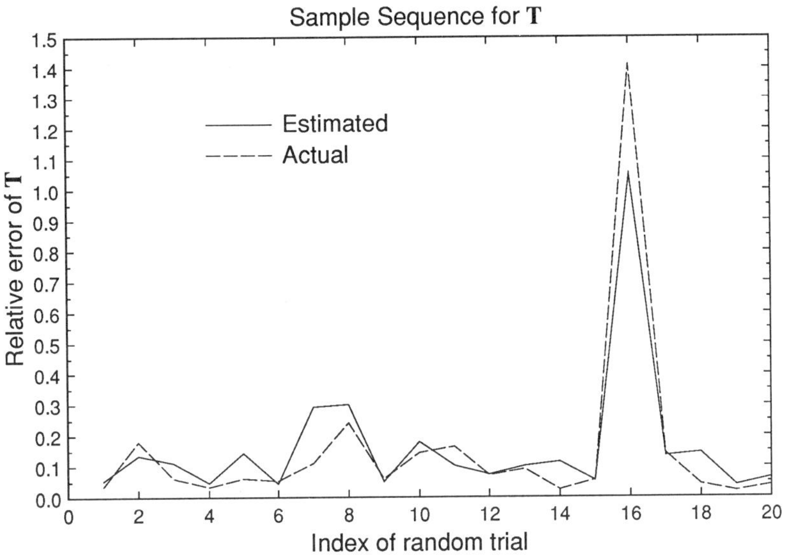

Fig. 3.7(b)

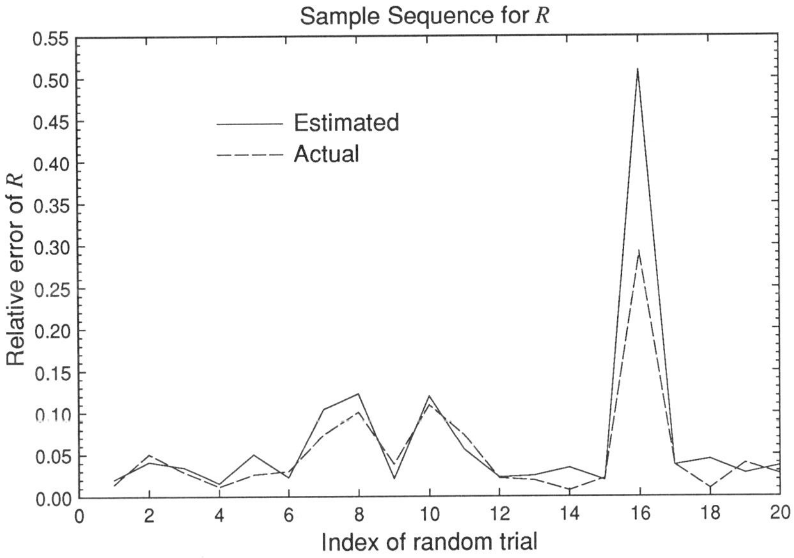

Fig. 3.7(c)

Fig. 3.7. Actual relative errors and estimated relative errors of **(a)** E, **(b)** $\mathbf{T}$, and **(c)** R. Rotation axis: (1, 0.9, 0.8). Rotation angle: 5°. Translation: (0.5, –0.5, –3.0).

The average performance of the error estimation as well as our motion estimation algorithm is presented in Fig. 3.8. Average relative errors (solid curves) are recorded over 20 random trials, with the same motion as that in Fig. 3.7, for different numbers of point correspondences used. (The sequence with 9 point correspondences is presented in Fig. 3.7). In Fig. 3.8 the dashed curves indicate the mean absolute difference between the estimated error and the actual error (called deviation of error estimation here), and the dot-dashed curves indicate the bias (difference between the mean of the estimated errors and the mean of the actual errors) over these 20 trials. As seen from Fig. 3.8, the errors decrease very quickly when the number of points increases beyond the required minimum of 8. This indicates that it is very effective to reduce the error by using a few more points in addition to the minimally required 8. It can also be seen that the mean deviation between the estimated error and the actual error is about a half of the actual error with the exception of the cases where the number of points is equal to 8. When the number of point correspondences is 8, there is a reasonably high probability for the randomly generated points to form a nearly degenerate configuration. When the point configuration is degenerate or nearly degenerate, the difference between the estimated error and the actual error is then large. This is one of the reasons for the large deviations and bias in the 8-point case. Some individual simulations still show a good agreement between the estimated errors and the actual errors in the 8-point case. Fig. 3.9 shows the mean relative errors in the relative depths, averaged over 100 trials, under the same motion.

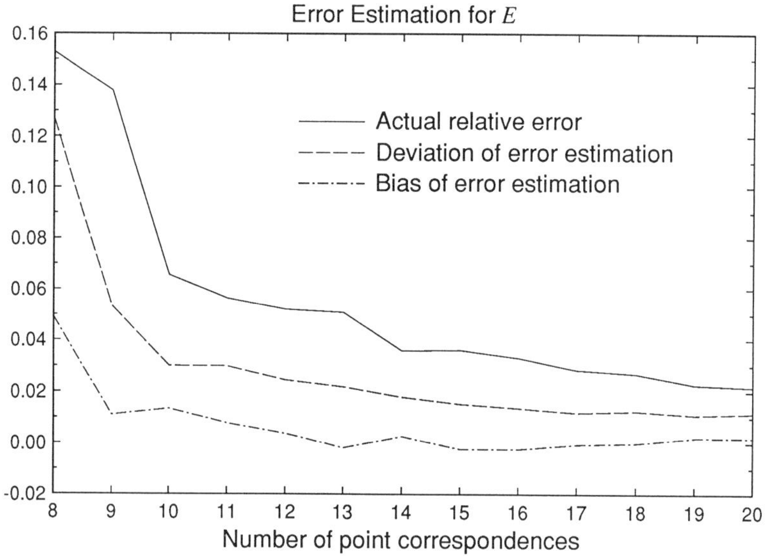

Fig. 3.8(a)

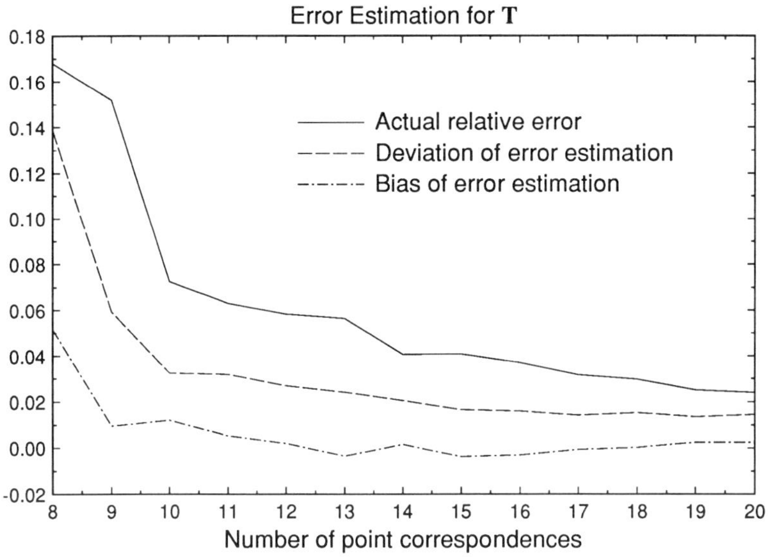

Fig. 3.8(b)

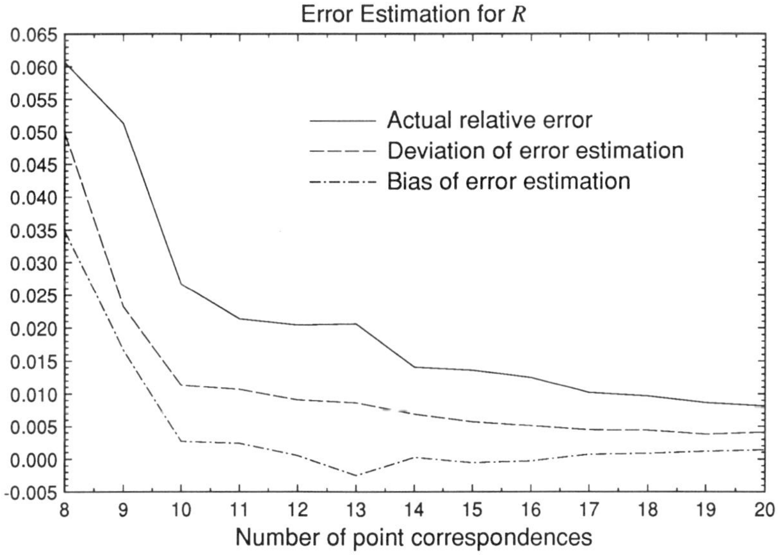

Fig. 3.8(c)

Fig. 3.8. Statistical record of error estimation. Actual relative error, deviation of error estimation and bias of error estimation for **(a)** E, **(b)** $\mathbf{T}$, and **(c)** R vs. Number of Point Correspondences. Rotation axis: (1, 0.9, 0.8). Rotation angle: 5°. Translation: (0.5, –0.5, –3.0).

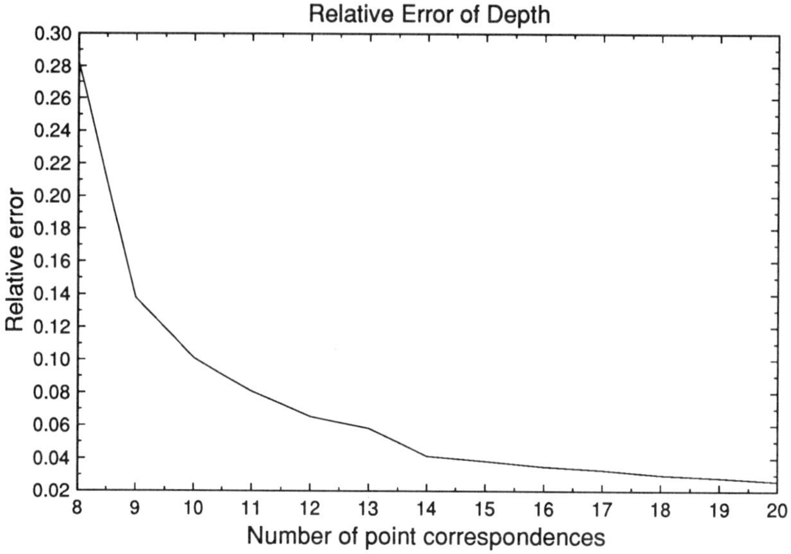

Fig. 3.9. Relative error of relative depths vs. number of point correspondences. Same motion as in Fig. 3.8.

3.5.3 Error versus Motion and System Parameters

The experiments will demonstrate remarkable dependency of the errors in solution on the values of the motion and system parameters. For the simulations presented in this section, 12 point correspondences are used. 100 trials are recorded for the computation of the average relative errors.

Magnitude of translation. With rotation axis (1, 0, 0) and rotation angle 5°, the translation direction being equal to $(k, 0, k)$ where k is such that the length of translation vector changes from 0.5 up to 4.5 with 20 evenly spaced values (corresponding to horizontal indexing from 0 to 20), the average relative errors of the estimates are shown in Fig. 3.10. It is very clear that the errors in rotation matrix R is almost unaffected by merely changing the magnitude of translation. However the errors in translation direction and relative depths drastically decrease as the magnitude of translation increases. This is consistent with the discussion in Section 3.4.

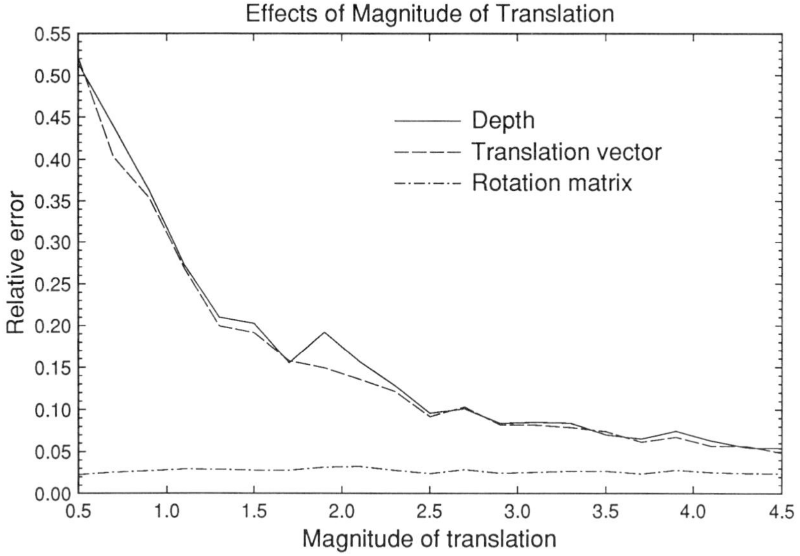

Fig. 3.10. Relative errors versus magnitude of translation. Rotation axis: (1, 0, 0). Rotation angle: 5°. Translation direction direction: $(k, 0, k)$. k is such that the length of translation vector changes from 0.5 up to 4.5 with 20 evenly spaced values along the horizontal axis.

Direction of translation. In Fig. 3.11, with the magnitude of translation fixed to be 3, the direction of the translation changes from (1, 0, 0) to (0, 0, 1) with evenly spaced 20 directions. The rotation angle is 5°. Three rotation axes (1, 0, 0), (0, 1, 0) and (0, 0, 1) are used in Figs. 3.11(a), (b) and (c), respectively. Despite of different rotation axes (it has been discussed in Section 3.4 and will be shown below that rotation parameters has no significant effects), the errors for rotation matrix, translation direction and relative depths all significantly decrease as the translation direction changes from being parallel to being orthogonal to image plane. The reasons for this relationship have been discussed earlier.

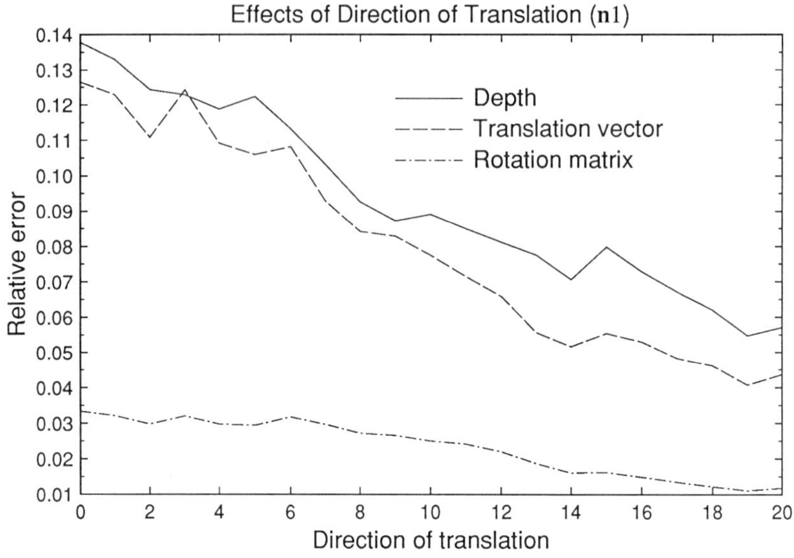

Fig. 3.11(a)

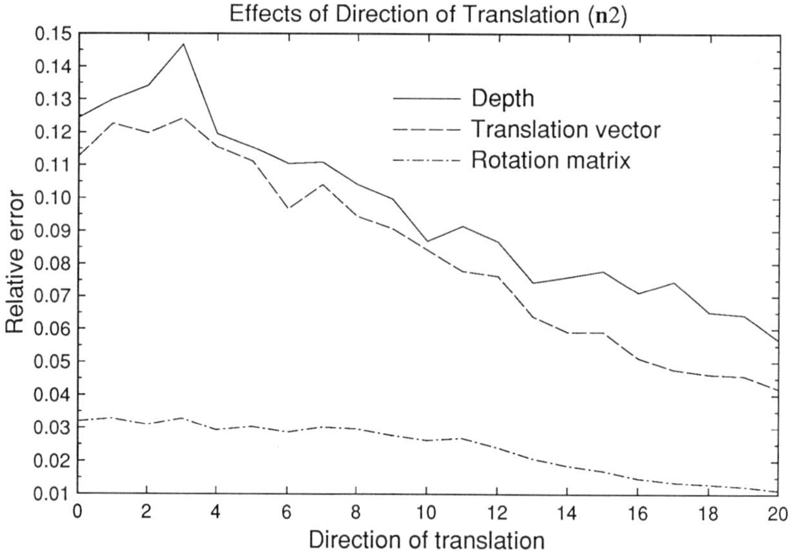

Fig. 3.11(b)

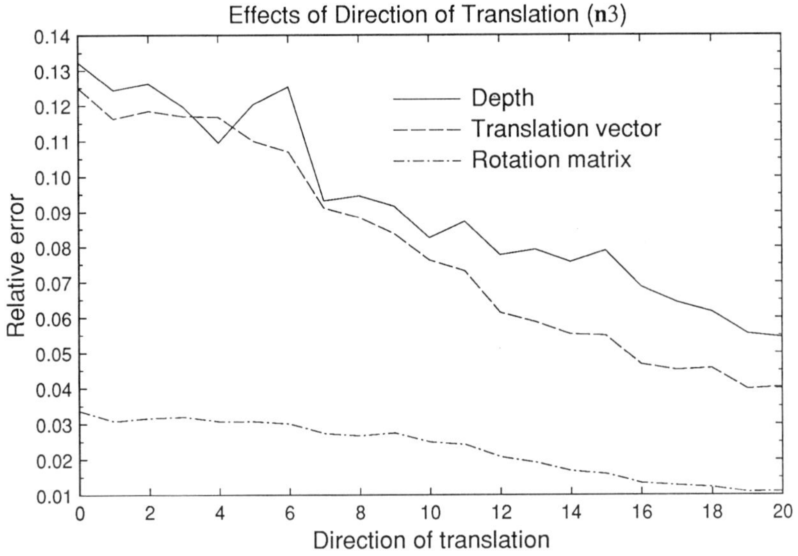

Fig. 3.11(c)

Fig. 3.11. Relative errors versus direction of translation. Rotation axis— **(a)**: (1, 0, 0); **(b)**: (0, 1, 0); **(c)**: (0, 0, 1). Rotation angle: 5°. With horizontal index from 0 to 20, the direction of the translation changes from (1, 0, 0) to (0, 0, 1) with evenly spaced 20 directions. The magnitude of translation is fixed to be 3.

Rotation parameters. We mentioned earlier that the rotation parameters generally do not significantly affect the errors in solutions. In Fig. 3.12, rotation angle changes from 0° to 30° and with different translation vectors and rotation axes. It is indicated by the figures that errors in R, $\hat{\mathbf{T}}$ and relative depths are not significantly affected by the change of rotation angle. Similarly Fig. 3.13 shows that rotation axis does not significantly influence those errors. However, as discussed earlier, when the error of rotation axis and rotation angle are considered separately (instead of error in R as a whole), the rotation axis has different effects. This can be seen from Fig. 3.14, which shows the errors of rotation axis and rotation angle with the same motion parameters as those in Fig. 3.13. These results do indicate that when the rotation axis changes from being parallel to the image plane to being orthogonal to image plane, the errors in rotation axis slightly decrease while those in rotation angle slightly increase.

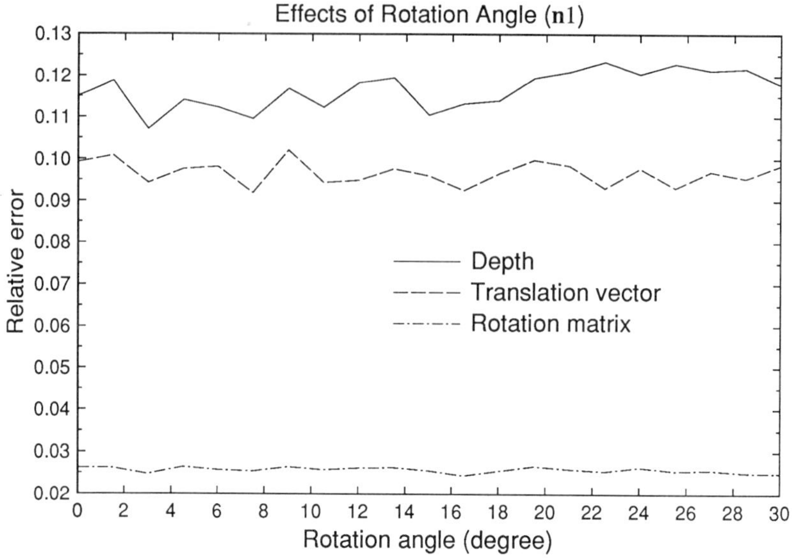

Fig. 3.12(a)

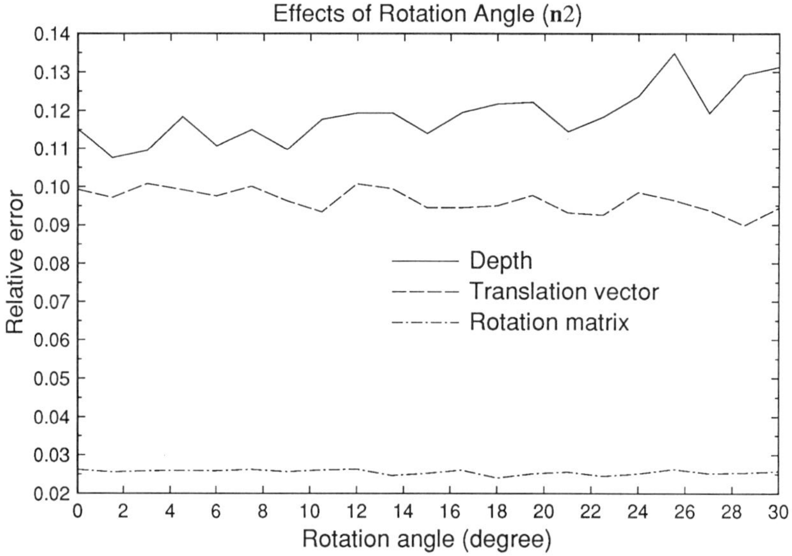

Fig. 3.12(b)

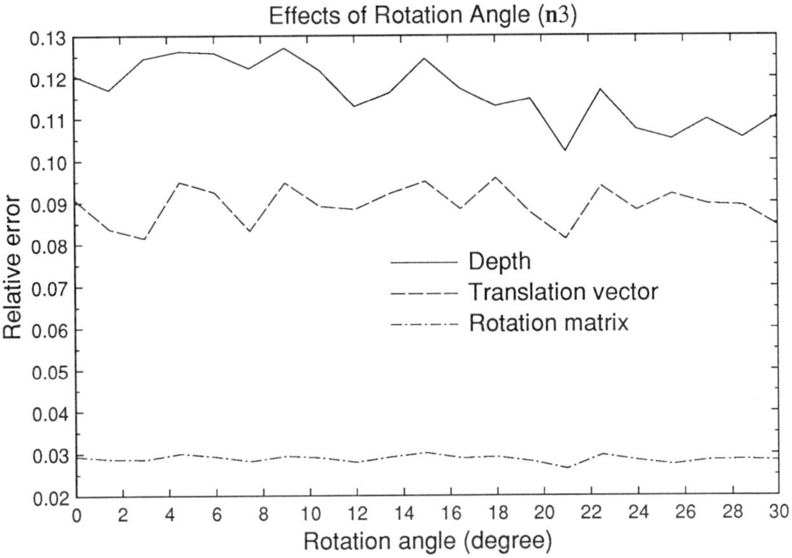

Fig. 3.12(c)

Fig. 3.12. Relative errors versus rotation angle. **(a)** Rotation axis: (1, 0, 0). Translation: (1.732, 0, −1.732). **(b)** Rotation axis: (0, 0, 1). Translation: (1.732, 0, −1.732). **(c)** Rotation axis: (1, 1, 1). Translation: (1.732, 1.732, −1.732).

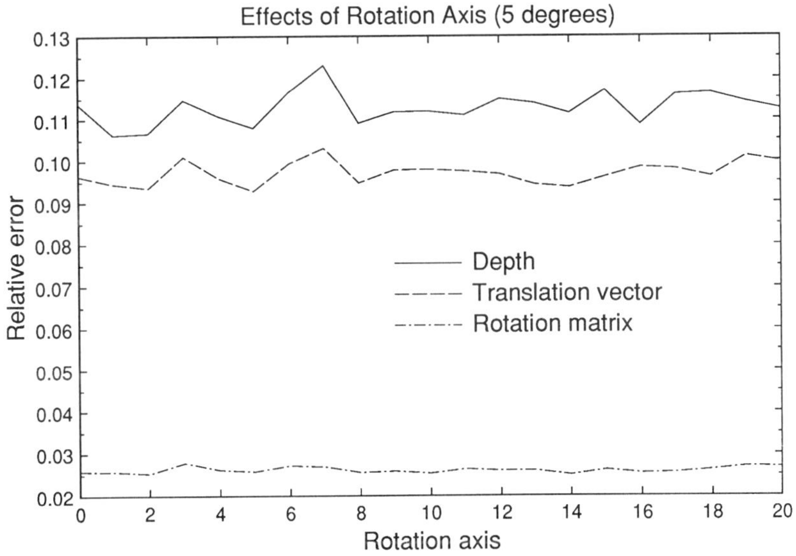

Fig. 3.13(a)

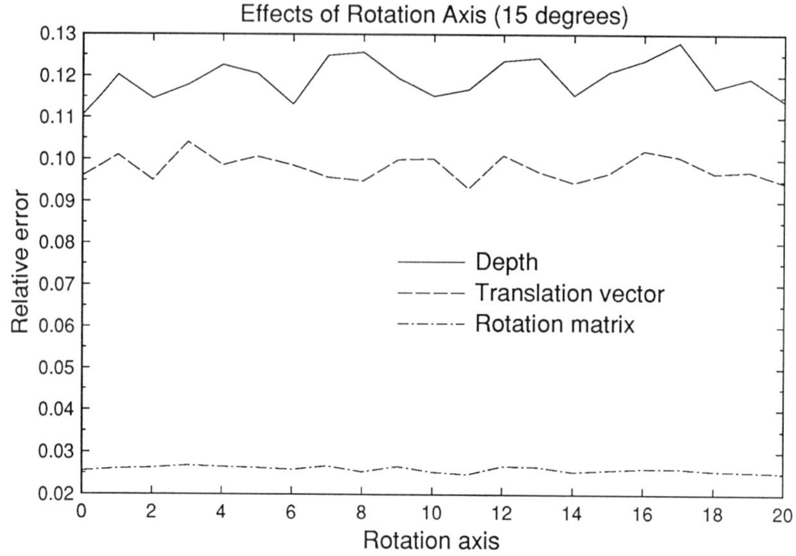

Fig. 3.13(b)

Fig. 3.13. Relative errors versus rotation axis. Translation: (1.732, 0, −1.732). With horizontal index from 0 to 20, rotation axis changes from (1, 0, 0) to (0, 0, 1) in xz-plane at 20 evenly spaced directions. Rotation angle: **(a)** 5°; **(b)** 15°.

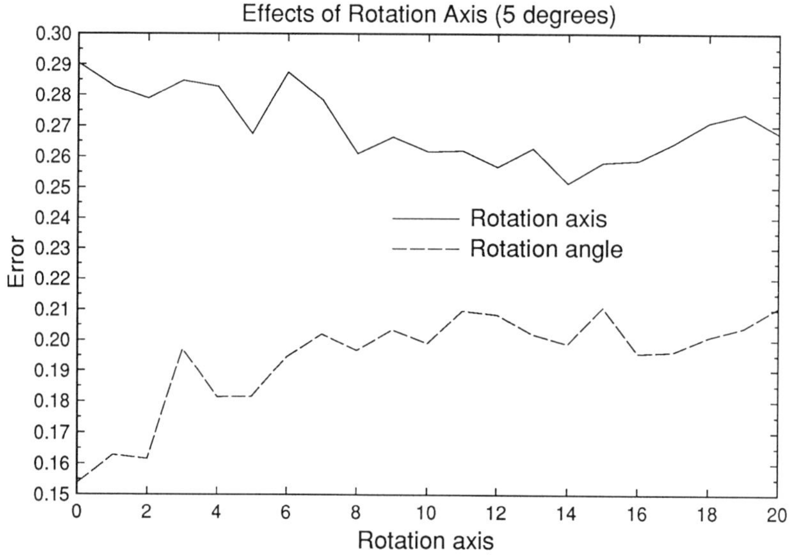

Fig. 3.14(a)

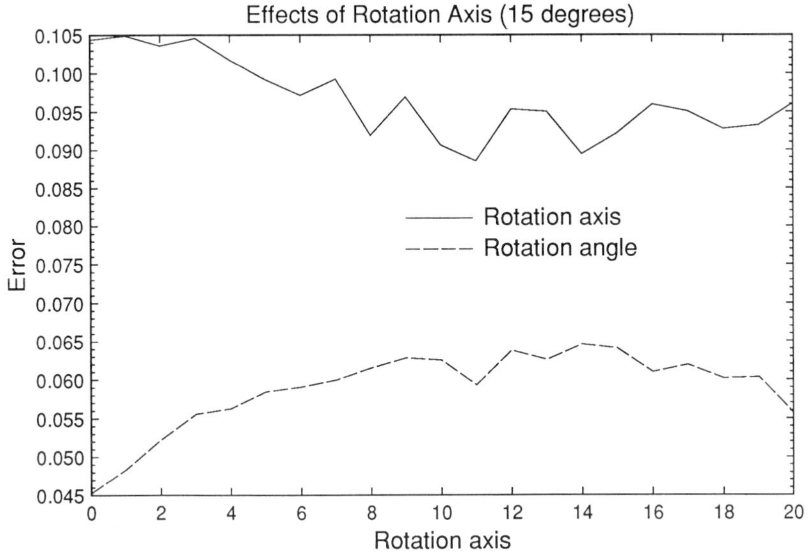

Fig. 3.14(b)

Fig. 3.14. Relative errors of rotation axis and absolute errors rotation angle (in radian) versus rotation axis. Translation: (1.732, 0, −1.732). With horizontal index from 0 to 20, rotation axis changes from (1, 0, 0) to (0, 0, 1) in xz-plane at 20 evenly spaced directions. Rotation angle: **(a)** 5°; **(b)** 15°.

Field of view. To demonstrate the effects of decreasing image size (field of view), the image size is reduced by a factor of 2. To make sure that the same scene is visible and cover roughly the same area of the images as earlier, the object is moved away from the camera such that object distance is doubled. Other parameters are kept unchanged. Then with the same motion parameters as in Fig. 3.11, the errors become those shown in Fig. 3.15. Comparing with Fig. 3.11, the corresponding curves in Fig. 3.15 increases more drastically as horizontal index decreases and the corresponding errors are much larger. This means that the reduction of image size particularly worsens the cases where the translation direction is nearly parallel to image plane. All of those are again consistent with the discussion in Subsection 3.4.3.

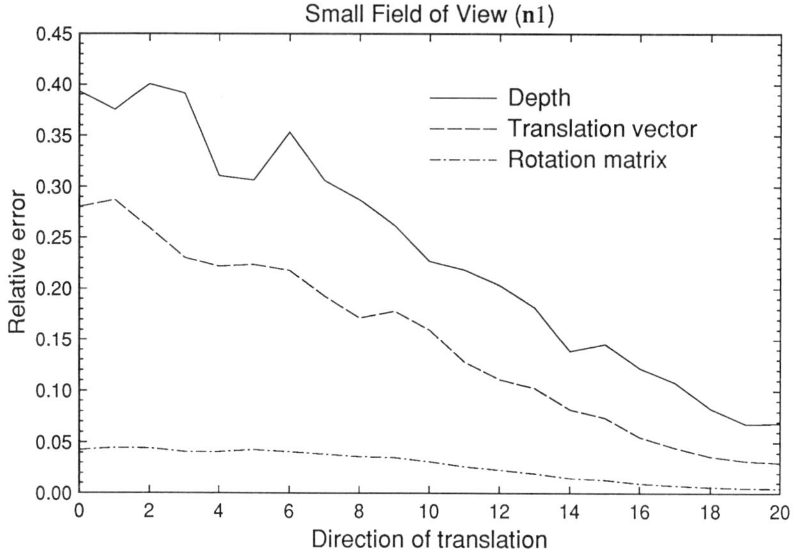

Fig. 3.15(a)

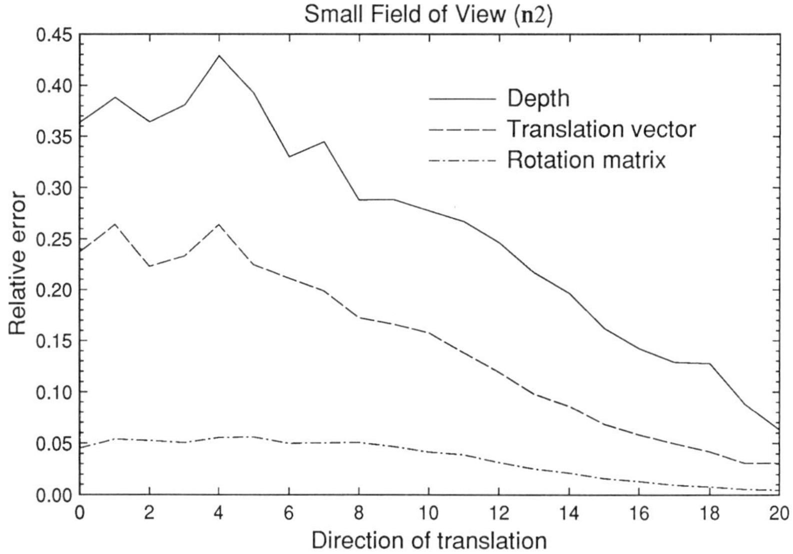

Fig. 3.15(b)

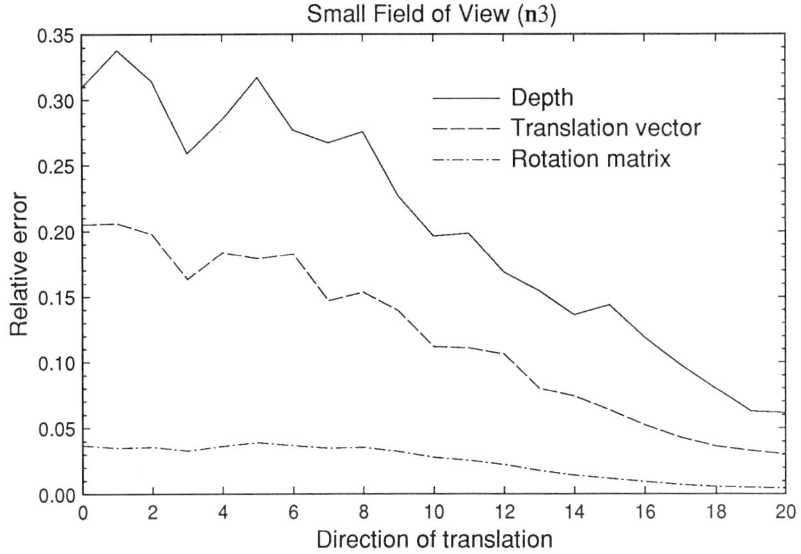

Fig. 3.15(c)

Fig. 3.15. Relative errors versus direction of translation with smaller image size (same resolution) compared with that in Fig. 3.11. Rotation axis— (**a**): (1, 0, 0); (**b**): (0, 1, 0); (**c**): (0, 0, 1). Rotation angle: 5°. With horizontal index from 0 to 20, the direction of the translation changes from (1, 0, 0) to (0, 0, 1) with evenly spaced 20 directions. The magnitude of translation is fixed to be 3.

Resolution. Fig. 3.16 shows the relations between the errors in the estimates and the image resolution. We see that reducing the resolution by a factor of two roughly doubles the errors, which is expected according to our discussion on error estimation.

The simulations shown in this subsection support the qualitative observations discussed in Section 3.4. We have also seen, from these numerical examples, to what degree these parameters affect the accuracy of the solution.

3.5.4 Real Images

Here we present the result with the Path scene shown in Fig. 2.33 in the previous chapter. The 169 displacement vectors shown in Fig. 2.34 are used as point correspondences. The computed motion parameters are shown in the following table. The accuracy of the algorithm can be indicated by the image error defined in (2.6.1).

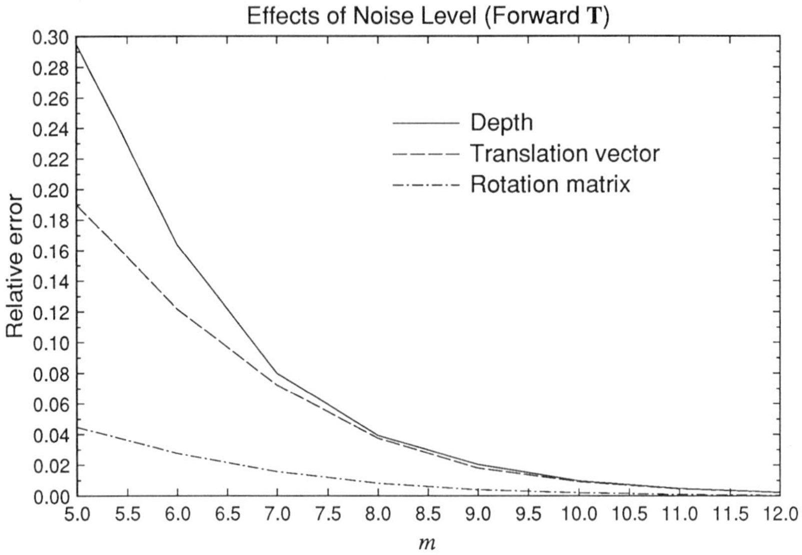

Fig. 3.16(a)

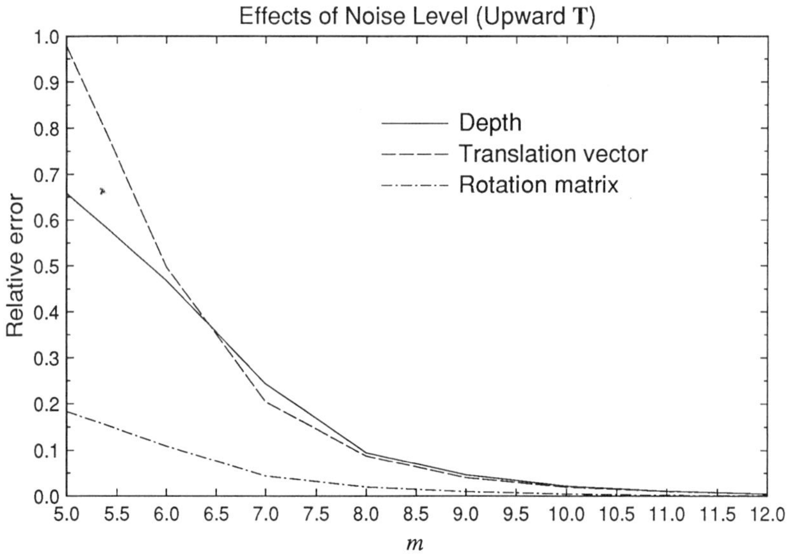

Fig. 3.16(b)

Fig. 3.16. Relative errors versus image resolution. Translation — **(a)**: (0, 0, −3), orthogonal to image plane; **(b)**: (3, 0, 0), parallel to image plane. Rotation axis: (1, 1, 1). Rotation angle: 5°. Image resolution: $2^m \times 2^m$.

Since a large number of point correspondences are used, the image error shown in the table is within a pixel width, which is very satisfactory. The image error shown in Table 2.3 is even smaller than that in Table 3.1 here, since the result shown in Table 2.3 has used optimization discussed in Chapter 4. Notice that the motion of the Path scene has a dominant longitudinal translation. As we discussed in Section 3.4, a lateral translation results in a significantly less accurate solution, especially with a narrow field of view. Under this latter unstable situation, the optimization to be discussed in Chapter 4 is crucial.

Table 3.1. Data and results for the Path scene using the linear algorithm

Parameters	x	y	z
Translation	0.053	-0.061	0.9968
Rotation axis	0.24	0.64	-0.72
Rotation angle		0.36°	
Image error		0.00078	
Pixel width		0.00094	

3.6 Summary

We have investigated in this chapter the problem of determining motion and structure of a rigid object from the point correspondences in two views. A linear algorithm is presented which computes the closed-form solution. It exploits the redundancy available in the data to combat noise. The approach to error estimation is based on first order perturbation. The results of simulations show a strong correlation between the estimated and the actual errors. Degeneracy or near degeneracy is identified and the reliability of the solution in the presence of noise is indicated by the estimated error.

Based on the qualitative analysis and the quantitative simulations, the following observations can be made for reliable motion estimation.

(1) The focal length of the image sensor should be short, or equivalently the field of view should be large, to obtain more reliable estimates.

(2) The magnitude of translation should be large in order to obtain reliable translational direction and accurate structure of the scene.

(3) A translation orthogonal to the image plane allows more stable estimation.

(4) The scene should be close to the imaging sensor and the motion should be large to yield long displacement vectors in the image plane.

(5) Rotation parameters are not significant to the reliability of the solutions (assuming rotation error is represented by the error of rotation matrix). However, a rotation about an axis orthogonal to the image plane generally results in a slightly more reliable estimate for the rotation axis and a less reliable estimate for rotation angle compared with the rotation about an axis parallel to the image plane.

3.7 Bibliographical Notes

The possibility of recovering motion and structure of scene from monocular views is known to photogrammetrists for quite long. It attracted investigation from computer vision area around early 80's. A few iterative algorithms were proposed, including those by Jain & Nagel [3.9], Roach & Aggarwal [3.18], Bruss & Horn [3.3], Adiv [3.1] and Mitiche & Aggarwal [3.14].

Two linear algorithms were developed independently by Longuet-Higgins [3.10], and Tsai & Huang [3.21]. Longuet-Higgins derived the algorithm using vector equations while Tsai & Huang used singular value decomposition, but they all used the same linear equation to solve for the essential parameters. In these linear algorithms, four candidate solutions are computed and the correct one is identified as the one that yields all positive depths. Yen & Huang [3.26] reported a vector-geometric approach to this problem. Since these algorithms were designed primarily for noise-free cases, high sensitivity to noise has been reported [3.4, 21]. On the other hand, closed-form solutions from optical flow have been presented by Zhuang et al [3.27, 29] and Waxm et al [3.22].

Since then, improvements have been made in reducing the sensitivity to noise while still keeping the algorithm linear. The post-consideration of the constraint in E through a matrix fitting was independently reported by Faugeras, Lustman & Toscani, [3.6] and Weng, Huang & Ahuja [3.23]. The latter algorithm is almost the same as the one presented here, which eliminates the need to compute three false solutions. Other measures taken to improve the stability include the summation while determining the signs in (3.2.13) and (3.2.23) as well as the least-squares method in solving the depths (3.2.25) [3.23, 24]. Another line of improvement involves iterations and, therefore, is discussed in Chapter 4.

Recently, an interesting result was provided by Spetsakis & Aliomonos [3.20] and Philip [3.17]. That is, **T** given by (3.2.12) (which is the same as the first column of the orthogonal matrix of the singular value decomposition of E) plus R determined by the Tsai & Huang's method in [3.21] gives the solution to a one-step minimization: determining the translation vector **T** and a rotation matrix R to minimize $\|[\mathbf{T}]_{\times}R - E\|$ from an estimated E. Since the objective function in (3.2.15) has the same objective function as this one-step minimization, this result also implies that the solution of our two-step minimization in (3.2.12) and (3.2.15) coincides with that of the one-step minimization.

Longuet-Higgins [3.11] has derived a necessary and sufficient condition on the spatial configurations that cause the failure of the existing linear algorithms: rank(A) < 8. Zhuang, Huang & Haralick [3.28] presented another necessary and sufficient condition, called surface assumption, for such a degeneracy. It should be mentioned that rank(A) < 8 does not mean that the corresponding case is not solvable. Some recent studies [3.7, 16] have been concentrating on the possible solutions under rank(A) < 8. Surface conditions that inherently allow multiple solutions can be found in [3.12, 13, 15].

Appendices

Appendix 3.A Perturbation of Eigenspace

Theorem 3.1. Let $A = [a_{ij}]$ be an n by n symmetrical matrix and H be an orthonormal matrix such that

$$H^{-1}AH = \text{diag}\{\lambda_1, \lambda_2, \cdots, \lambda_n\}. \tag{3.A.1}$$

Let the eigenvalues be ordered according to increasing magnitudes. Without loss of generality, consider the eigenvalue λ_1. Assume λ_1 is a simple eigenvalue.

$$\lambda_1 < \lambda_2 \le \lambda_3 \le \cdots \le \lambda_n.$$

Let

$$H = [\mathbf{h}_1 \ \mathbf{h}_2 \ \cdots \ \mathbf{h}_n]$$

and $\mathbf{X}$ be an eigenvector of A associated with λ_1. Let $\mathbf{X}(\varepsilon)$ be the eigenvector of the perturbed matrix $A(\varepsilon) = A + \Delta_A$ associated with the perturbed eigenvalue $\lambda_1(\varepsilon)$. $\mathbf{X}(\varepsilon)$ can be written as

$$\mathbf{X}(\varepsilon) = \mathbf{X} + \delta_{\mathbf{x}}$$

with $\delta_{\mathbf{x}} \in \text{span}\{\mathbf{h}_2, \mathbf{h}_3, \ \cdots, \mathbf{h}_n\}$. Letting ε be the maximum absolute value of the elements in $\Delta_A = [\delta_{a_{ij}}]$ gives

$$\Delta_A = \varepsilon B$$

where $B = [b_{ij}]$, with $b_{ij} = \delta_{a_{ij}}/\varepsilon$. Therefore $|b_{ij}| \le 1$, $1 \le i \le n$, $1 \le j \le n$. Then for sufficiently small ε, the perturbation of λ_1 can be expressed by a convergent series in ε:

$$\delta_{\lambda_1} \triangleq \lambda_1(\varepsilon) - \lambda_1 = p_1\varepsilon + p_2\varepsilon^2 + p_3\varepsilon^3 + \ \cdots \tag{3.A.2}$$

and the perturbation vector $\delta_{\mathbf{x}}$ can be expressed by a convergent vector series in the space $\text{span}\{\mathbf{h}_2, \mathbf{h}_3, \ \cdots, \mathbf{h}_n\}$. In other words, letting $H_2 = [\mathbf{h}_2, \mathbf{h}_3, \ \cdots, \mathbf{h}_n]$, then for sufficiently small ε, there exist (n-1)-dimensional vectors $\mathbf{g}_1, \mathbf{g}_2, \mathbf{g}_3, \ \cdots$ such that

$$\delta_{\mathbf{x}} = \varepsilon H_2\mathbf{g}_1 + \varepsilon^2 H_2\mathbf{g}_2 + \varepsilon^3 H_2\mathbf{g}_3 + \ \cdots . \tag{3.A.3}$$

The linear term (in ε) in (3.A.2) is given by

$$p_1\varepsilon = \mathbf{h}_1^t \Delta_A \mathbf{h}_1. \tag{3.A.4}$$

The linear term (in ε) in (3.A.3) is given by

$$\varepsilon H_2\mathbf{g}_1 = H \Delta H^t \Delta_A \mathbf{X}$$

where

$$\Delta = \text{diag}\{0, (\lambda_1 - \lambda_2)^{-1}, \ \cdots, (\lambda_1 - \lambda_n)^{-1}\}.$$

That is, suppressing the second and higher order terms (*i.e.*, considering first order perturbation), for the eigenvalue we have

$$\delta_{\lambda_1} \cong \mathbf{h}_1^t \Delta_A \mathbf{h}_1$$

and for the eigenvector:

$$\delta_{\mathbf{x}} \cong H \Delta H^t \Delta_A \mathbf{X}.$$

Proof. Under the assumption of simple eigenvalue and the definition of ε, there exists a positive δ such that if $\varepsilon < \delta$, λ_1 is represented by a scalar convergent power series in ε with constant term being λ_1:

$$\lambda_1(\varepsilon) = \lambda_1 + \varepsilon p_1 + \varepsilon p_2 + \cdots \tag{3.A.5}$$

and the eigenvector of $A(\varepsilon)$, $\mathbf{X}(\varepsilon)$, associated with $\lambda_1(\varepsilon)$ is represented by a convergent (vector) power series in ε with the constant vector term being $\mathbf{X}$:

$$\mathbf{X}(\varepsilon) = \mathbf{X} + \varepsilon \mathbf{z}_1 + \varepsilon^2 \mathbf{z}_2 + \cdots \triangleq \mathbf{X} + \delta_{\mathbf{x}} \tag{3.A.6}$$

(see, for example, Wilkinson [3.25]). $\delta_{\mathbf{x}}$ can be represented in the basis $\mathbf{h}_1, \mathbf{h}_2, \cdots, \mathbf{h}_n$. The component of $\delta_{\mathbf{x}}$ on $\mathbf{h}_1$ can always be included into $\mathbf{X}$ since $\mathbf{X}$ is any vector collinear with $\mathbf{h}_1$. So, $\delta_{\mathbf{x}} \in$ span $\{\mathbf{h}_2, \mathbf{h}_3, \cdots, \mathbf{h}_n\}$. That is, there exist (n-1)-dimensional vectors $\mathbf{g}_1, \mathbf{g}_2, \mathbf{g}_3, \cdots$ such that (3.A.3) holds, or

$$\mathbf{X}(\varepsilon) = \mathbf{X} + \varepsilon H_2 \mathbf{g}_1 + \varepsilon^2 H_2 \mathbf{g}_2 + \varepsilon^3 H_2 \mathbf{g}_3 + \cdots.$$

We have

$$(A + \varepsilon B)\mathbf{X}(\varepsilon) = \lambda_1(\varepsilon)\mathbf{X}(\varepsilon).$$

Substituting $\lambda_1(\varepsilon)$ using (3.A.5) and $\mathbf{X}(\varepsilon)$ using (3.A.6) gives

$$(A + \varepsilon B)(\mathbf{X} + \varepsilon H_2 \mathbf{g}_1 + \varepsilon^2 H_2 \mathbf{g}_2 + \varepsilon^3 H_2 \mathbf{g}_3 + \cdots)$$

$$= (\lambda_1 + \varepsilon p_1 + \varepsilon p_2 + \cdots)\mathbf{X}(\varepsilon). \tag{3.A.7}$$

(3.A.7) holds for all $\varepsilon < \delta$. Therefore the coefficient vectors of ε on the both sides should be equal:

$$AH_2 \mathbf{g}_1 + B\mathbf{X} = \lambda_1 H_2 \mathbf{g}_1 + p_1 \mathbf{X}$$

or, using (3.A.1):

$$H_2 \operatorname{diag}\{\lambda_2, \lambda_3, \cdots, \lambda_n\} \mathbf{g}_1 - \lambda_1 H_2 \mathbf{g}_1 + B\mathbf{X} = p_1 \mathbf{X}. \tag{3.A.8}$$

Premultiplying both sides by $\mathbf{X}^t$ and noticing $\mathbf{X}^t H_2 = \mathbf{0}$, we get

$$\mathbf{X}^t B \mathbf{X} = p_1 \|\mathbf{X}\|^2$$

or

$$p_1 = \mathbf{h}_1^t B \mathbf{h}_1.$$

Then (3.A.4) follows immediately. Premultiplying both sides of (3.A.8) by H_2^t and noticing H is an orthonormal matrix, we get

$$\mathrm{diag}\{\lambda_1 - \lambda_2, \lambda_1 - \lambda_3, \cdots, \lambda_1 - \lambda_n\}\mathbf{g}_1 = H_2^t B \mathbf{X}.$$

We have

$$\mathbf{g}_1 = \mathrm{diag}\{(\lambda_1 - \lambda_2)^{-1}, (\lambda_1 - \lambda_3)^{-1}, \cdots, (\lambda_1 - \lambda_n)^{-1}\}H_2^t B \mathbf{X}.$$

The linear term in (3.A.3) is then

$$\varepsilon H_2 \mathbf{g}_1 = \varepsilon H_2\, \mathrm{diag}\{(\lambda_1 - \lambda_2)^{-1}, (\lambda_1 - \lambda_3)^{-1}, \cdots, (\lambda_1 - \lambda_n)^{-1}\}H_2^t B \mathbf{X}$$

$$= H \Delta H^t \varepsilon B \mathbf{X} = H \Delta H^t \Delta_A \mathbf{X}.$$

□

Appendix 3.B Quaternions

We need to prove (3.2.30)

$$\|RC - D\|^2 = \mathbf{q}^t B \mathbf{q} \tag{3.B.1}$$

where $C = [\mathbf{C}_1\ \mathbf{C}_2\ \cdots\ \mathbf{C}_n]$, $D = [\mathbf{D}_1\ \mathbf{D}_2\ \cdots\ \mathbf{D}_n]$, R is a rotation matrix and B is defined by

$$B = \sum_{i=1}^{n} B_i^t B_i$$

where

$$B_i = \begin{bmatrix} 0 & (\mathbf{C}_i - \mathbf{D}_i)^t \\ \mathbf{D}_i - \mathbf{C}_i & [\mathbf{D}_i + \mathbf{C}_i]_\times \end{bmatrix}. \tag{3.B.2}$$

In (3.B.2) the rotation matrix R and unit $\mathbf{q}$ are related by (3.2.19).

We first introduce quaternions. A quaternion $\mathbf{q}$ consists of a scalar component q_0 and a component of a 3-dimensional vector $\mathbf{Q} = (q_1, q_2, q_3)^t$:

$$\mathbf{q} = q_0 + \mathbf{Q}.$$

Two quaternions are equal if and only if the corresponding components are equal. A vector quaternion is a quaternion with zero scalar component. A scalar quaternion is a quaternion with zero vector component. For convenience, we regard a scalar as a scalar quaternion and a 3-dimensional vector as vector quaternion. The conjugate of a quaternion $\mathbf{q}$, denoted by $\tilde{\mathbf{q}}$, is defined by $\tilde{\mathbf{q}} = q_0 - \mathbf{Q}$. The addition of two quaternions is defined by

$$(q_0 + \mathbf{Q}) + (p_0 + \mathbf{P}) = (q_0 + p_0) + (\mathbf{Q} + \mathbf{P}).$$

The multiplication of two quaternions, denoted by "$*$", is defined by

$$(q_0 + \mathbf{Q}) * (p_0 + \mathbf{P}) = (q_0 p_0 - \mathbf{Q} \cdot \mathbf{P}) + (q_0 \mathbf{P} + p_0 \mathbf{Q} + \mathbf{Q} \times \mathbf{P})$$

where "$\cdot$" and "$\times$" are vector dot product and cross product operations, respectively. It is easy to prove that quaternion multiplication is not commutative (unless a quaternion is scalar), but it is associative, and it is distributive over additions. The conjugate of $\mathbf{p} * \mathbf{q}$ is equal to $\tilde{\mathbf{q}} * \tilde{\mathbf{p}}$. Define the norm of a quaternion, $\|\cdot\|$, by a nonnegative value such that $\|\mathbf{q}\|^2 = \mathbf{q} * \tilde{\mathbf{q}}$. It is easy to prove $\mathbf{q} * \tilde{\mathbf{q}} = \sum_{i=0}^{4} q_i^2$. Therefore, the norm of a quaternion is equal to the Euclidean norm of the corresponding 4-dimensional space. A quaternion with a unit norm is called unit quaternion. Therefore, for unit quaternion $\mathbf{q}$ we have $\mathbf{q} * \tilde{\mathbf{q}} = 1$.

For a rotation about unit axis $\mathbf{n} = (n_x, n_y, n_z)$ by an angle θ, letting $\mathbf{q}$ be a unit quaternion such that

$$\mathbf{q} = \cos(\theta/2) + \sin(\theta/2)\mathbf{n} \tag{3.B.3}$$

and R be the corresponding rotation matrix:

$$\begin{bmatrix} (n_x^2-1)(1-\cos\theta)+1 & n_x n_y(1-\cos\theta)-n_z\sin\theta & n_x n_z(1-\cos\theta)+n_y\sin\theta \\ n_y n_x(1-\cos\theta)+n_z\sin\theta & (n_y^2-1)(1-\cos\theta)+1 & n_y n_z(1-\cos\theta)-n_x\sin\theta \\ n_z n_x(1-\cos\theta)-n_y\sin\theta & n_z n_y(1-\cos\theta)+n_x\sin\theta & (n_z^2-1)(1-\cos\theta)+1 \end{bmatrix}$$

then [3.2, 8]

$$R\mathbf{X} = \mathbf{q} * \mathbf{X} * \tilde{\mathbf{q}} \tag{3.B.4}$$

for any 3-dimensional vector $\mathbf{X}$. If θ is incremented by 360°, the sign of $\mathbf{q}$ in (3.B.3) is changed but the rotation is not changed. Obviously $-\mathbf{q}$ also satisfies (3.B.4). Therefore $\mathbf{q}$ and $-\mathbf{q}$ represent the same rotation. In fact, letting Q be the group of unit quaternions under quaternion product, and letting $S = \{1, -1\}$, the factor (or quotient) group Q/S and group $SO(3)$ of rotations in Euclidean space R^3 are isomorphic. (3.B.3) with S defines an isomorphism.

It is convenient to convert quaternion multiplications to matrix multiplications and regard a quaternion as a 4-dimensional column vector when it is operated with matrices.

$$\mathbf{p} * \mathbf{q} = \begin{bmatrix} p_0 \\ \mathbf{P} \end{bmatrix} * \begin{bmatrix} q_0 \\ \mathbf{Q} \end{bmatrix} = \begin{bmatrix} p_0 & -\mathbf{P}^t \\ \mathbf{P} & p_0 I + [\mathbf{P}]_\times \end{bmatrix} \begin{bmatrix} q_0 \\ \mathbf{Q} \end{bmatrix} \triangleq [\mathbf{p}]_l \mathbf{q} \tag{3.B.5}$$

where $[\cdot]_l$ is a mapping from a quaternion to a 4×4 matrix (l stands for $l$eft multiplication). Similarly, we define the mapping $[\cdot]_r$:

$$\mathbf{q} * \mathbf{p} = \begin{bmatrix} q_0 \\ \mathbf{Q} \end{bmatrix} * \begin{bmatrix} p_0 \\ \mathbf{P} \end{bmatrix} = \begin{bmatrix} p_0 & -\mathbf{P}^t \\ \mathbf{P} & p_0 I - [\mathbf{P}]_\times \end{bmatrix} \begin{bmatrix} q_0 \\ \mathbf{Q} \end{bmatrix} \triangleq [\mathbf{p}]_r \mathbf{q}. \tag{3.B.6}$$

Now we are ready to prove (3.B.1). In the following, vectors are augmented to vector quaternions when they are operated with quaternions. From (3.B.4) it follows that

$$\|D - RC\|^2 = \sum_{i=1}^{n} \|\mathbf{D}_i - R\mathbf{C}_i\|^2 = \sum_{i=1}^{n} \|\mathbf{D}_i - \mathbf{q} * \mathbf{C}_i * \tilde{\mathbf{q}}\|^2$$

$$= \sum_{i=1}^{n} \|(\mathbf{D}_i * \mathbf{q} - \mathbf{q} * \mathbf{C}_i) * \tilde{\mathbf{q}}\|^2 = \sum_{i=1}^{n} \|(\mathbf{D}_i * \mathbf{q} - \mathbf{q} * \mathbf{C}_i)\|^2 \tag{3.B.7}$$

since $\tilde{\mathbf{q}}$ is a unit quaternion and $\|\mathbf{p} * \tilde{\mathbf{q}}\|^2 = \mathbf{p} * \tilde{\mathbf{q}} * \mathbf{q} * \tilde{\mathbf{p}} = \mathbf{p} * \tilde{\mathbf{p}} = \|\mathbf{p}\|^2$. Using our matrix notation, we have

$$\mathbf{D}_i * \mathbf{q} - \mathbf{q} * \mathbf{C}_i = [\mathbf{D}_i]_l \mathbf{q} - [\mathbf{C}_i]_r \mathbf{q}$$

$$= ([\mathbf{D}_i]_l - [\mathbf{C}_i]_r])\mathbf{q} = B_i \mathbf{q} \tag{3.B.8}$$

where B_i is readily determined by our mapping (3.B.5) and (3.B.6) and it is presented in (3.B.2). Finally, using (3.B.7) and (3.B.8) gives

$$\|D - RC\|^2 = \sum_{i=1}^{n} \|\mathbf{D}_i - R\,\mathbf{C}_i\|^2 = \sum_{i=1}^{n} \|B_i \mathbf{q}\|^2$$

$$= \sum_{i=1}^{n} \mathbf{q}^t B_i^t B_i \mathbf{q} = \mathbf{q}^t \left(\sum_{i=1}^{n} B_i^t B_i \right) \mathbf{q} \triangleq \mathbf{q}^t B\, \mathbf{q}.$$

In the algorithm, we have $n = 3$. The readers are referred to [3.2, 8, 19, 5] for more discussions.

Appendix 3.C Alternative Fitting

If we use (3.2.13) to solve R, using (3.2.18) and (3.3.11) gives

$$\delta_B \cong G_B\, \delta_{\mathbf{K}} \cong G_B D_{\mathbf{K}} \delta_E \triangleq D_B\, \delta_E$$

where

$$G_B = 2 \begin{bmatrix} F_{tl} & F_{tr} \\ F_{bl} & F_{br} \end{bmatrix}$$

where (note $\mathbf{T}_s = (s_1, s_2, s_3)^t$)

$$F_{tl} = \begin{bmatrix} 2s_1+e_{23}-e_{32} & 2s_2+e_{31}-e_{13} & 2s_3+e_{12}-e_{21} & e_{11} & e_{21}-s_3 \\ -e_{22}-e_{33} & e_{12} & e_{13} & 0 & 0 \\ e_{21} & e_{11}-e_{33} & e_{23} & -s_2 & t_1 \\ e_{31} & e_{32} & -e_{11}-e_{22} & -s_3 & 0 \\ -e_{22}-e_{33} & e_{12} & e_{13} & 0 & 0 \\ 2s_1+e_{32}-e_{23} & 2s_2+e_{13}-3e_{31} & 2s_3+3e_{21}-e_{12} & e_{11} & e_{21}+3s_3 \\ e_{31} & -e_{32} & e_{22}-e_{11} & -s_3 & 0 \\ -e_{21} & e_{11}-e_{33} & e_{23} & s_2 & -s_1 \end{bmatrix}$$

$$F_{tr} = \begin{bmatrix} e_{31}+s_2 & e_{12}+s_3 & e_{22} & e_{32}-s_1 & e_{13}-s_2 & e_{23}+s_1 & e_{33} \\ 0 & s_2 & -s_1 & 0 & s_3 & 0 & -s_1 \\ 0 & 0 & 0 & 0 & 0 & s_3 & -s_2 \\ s_1 & 0 & -s_3 & s_2 & 0 & 0 & 0 \\ 0 & s_2 & -s_1 & 0 & s_3 & 0 & -s_1 \\ e_{31}-3s_2 & e_{12}-s_3 & e_{22} & e_{32}+s_1 & e_{13}+s_2 & e_{23}-s_1 & e_{33} \\ s_1 & 0 & s_3 & -s_2 & 0 & 0 & 0 \\ 0 & 0 & 0 & 0 & 0 & s_3 & -s_2 \end{bmatrix}$$

$$F_{bl} = \begin{bmatrix} e_{21} & -e_{11}-e_{33} & e_{23} & -s_2 & s_1 \\ e_{31} & -e_{32} & e_{22}-e_{11} & -s_3 & 0 \\ 2s_1+3e_{32}-e_{23} & 2s_2+e_{13}-e_{31} & 2s_3+e_{21}-3e_{12} & e_{11} & e_{21}+s_3 \\ e_{33}-e_{22} & e_{12} & -e_{13} & 0 & 0 \\ e_{31} & e_{32} & -e_{11}-e_{22} & -s_3 & 0 \\ -e_{21} & e_{11}-e_{33} & e_{23} & s_2 & -s_1 \\ e_{33}-e_{22} & e_{12} & -e_{13} & 0 & 0 \\ 2s_1+e_{32}-3e_{23} & 2s_2+3e_{13}-e_{31} & 2s_3+e_{21}-e_{12} & e_{11} & e_{21}+s_3 \end{bmatrix}$$

$$F_{br} = \begin{bmatrix} 0 & 0 & 0 & 0 & 0 & s_3 & -s_2 \\ s_1 & 0 & s_3 & -s_2 & 0 & 0 & 0 \\ e_{31}-s_2 & e_{12}-3s_3 & e_{22} & e_{32}+3s_1 & e_{13}+s_2 & e_{23}-s_1 & e_{33} \\ 0 & s_2 & -s_1 & 0 & -s_3 & 0 & s_1 \\ s_1 & 0 & -s_3 & s_2 & 0 & 0 & 0 \\ 0 & 0 & 0 & 0 & 0 & s_3 & -s_2 \\ 0 & s_2 & -s_1 & 0 & -s_3 & 0 & s_1 \\ e_{31}-s_2 & e_{12}-s_3 & e_{22} & e_{32}+s_1 & e_{13}+3s_2 & e_{23}-3s_1 & e_{33} \end{bmatrix}$$

4. Optimization

Although the linear algorithm discussed in Chapter 3 is computationally fast, the solution is not optimal in the presence of noise. The methods discussed in this chapter aim at global optimality to significantly improve the accuracy of the solution. First, some situations are identified in which the solution of the linear algorithm is very unstable and thus the optimization is especially crucial. Then, methods for optimal estimation are investigated with two types of noise model: 3-D noise and 2-D image plane noise. A two-step computational approach is introduced for the nonlinear optimization. The first step is computing preliminary solution using a linear algorithm. The second step is iteratively improving this preliminary solution to reach an optimal solution. Then, other related issues are investigated, which include error bound, error estimation, as well as sequential and batch processing techniques.

4.1 Motivations

The optimization is motivated by the following observations on linear algorithms.

(a) With certain types of motion, even pixel level perturbations (such as digitization noise of conventional CCD video cameras) may override the information characterized by the epipolar constraint, which is a key constraint used for determining motion and structure by linear algorithms. The epipolar constraint restricts only one of the two components in image point displacement. The other component is related to the depth of the point and the motion. If this component is also used for motion estimation, the accuracy of the estimated motion parameters can be considerably improved.

(b) Existing linear algorithms give closed-form solution to motion parameters. However, the constraints in the intermediate parameter matrix (essential matrix E) are not fully used. It is useful to examine the constraints in the intermediate parameter matrix and derive necessary and sufficient conditions for the intermediate parameter matrix to satisfy the constraints. The use of these constraints can improve the accuracy of the solution in the presence of noise.

The above considerations are unified under a general framework of optimal estimation: Given the noise-contaminated point correspondences, we need the best estimator for motion and structure parameters. The following aspects of motion and structure estimation are investigated:

(1) The approaches to optimal estimation with known or unknown noise distributions.

(2) Accessing the accuracy of the optimal solutions. Without an assessment of the accuracy, the estimates are often unusable. In the previous chapter, we have discussed this issue for the solution of the linear algorithm. Since the optimal solutions are computed iteratively, the accuracy of the solutions has to be estimated by a different method.

(3) Theoretical error bound. Given an algorithm that computes a solution from noise-contaminated data, a fundamental question to ask is: Can one design an algorithm that gives solutions with higher accuracy? The questions of this type address the inherent stability issue of motion estimation. Since noise in data are random and the accuracy of the solutions is also random, the question should be investigated in terms of statistics. We formulate the theoretical error bound for this problem and compare the bound with the actual errors. This study also enables us to quantitatively assess the inherent stability problem of estimating motion from small image disparities such as motion from optical flow.

(4) Data processing techniques. Except for some simple problems where optimal solutions can be directly solved for, nonlinear optimization problems are generally solved through iterative procedures. The type of algorithms used for nonlinear optimization is crucial to determine whether the optimal solution can be obtained. The general framework of Kalman filtering and its simple computational procedures have attracted many applications in optimization. Kalman filter is a sequential processing algorithm in the sense that the estimates are updated sequentially based on last estimates and current single observation. It is different from a batch processing algorithm

which computes the solution based on all observations as a whole. We analyze sequential and batch processing algorithms and their performance differences for nonlinear problems. We will see that one is penalized by using sequential processing for nonlinear problems. In general, the performance of the iterated extended Kalman filtering algorithm is inferior to that of a batch processing algorithm. The performance differences between the two algorithms are quite large in our nonlinear problem of motion and structure estimation. For the cases where estimates are required sequentially while data are obtained, a recursive-batch approach is proposed to trade-off between performance and efficiency.

4.2 Stability of Linear Algorithms

First, we briefly review the problem of motion and structure analysis and discuss two coordinate systems: camera-centered and object-centered.

In the previous chapter, we let the coordinate system be fixed on the camera, as shown in Fig. 4.1. This coordinate system is called camera-centered. In this coordinate system, corresponding to a feature point located at $\mathbf{x} = (x, y, z)^t$, the image plane vector in the image is defined as

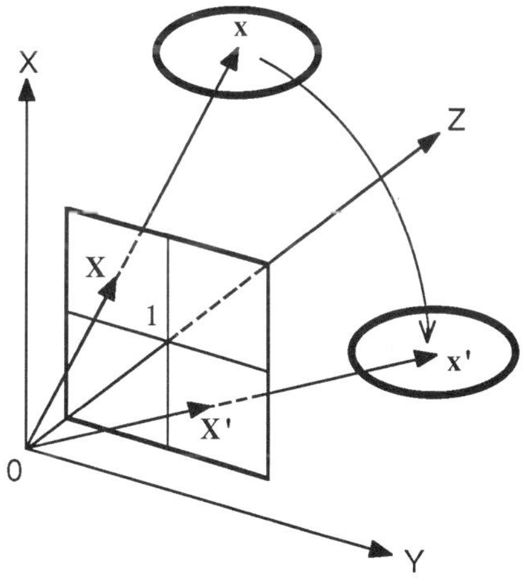

Fig. 4.1. A view from the camera-centered coordinate system: a moving scene observed by a stationary camera.

$$\mathbf{u} = (u, v)^t = (fx/z, fy/z)^t$$

which is the perspective projection of $\mathbf{x}$ onto the image plane. Since (u, v) can be measured in terms of f, without loss of generality, we assume $f = 1$ in the *normalized camera model* as shown in Fig. 4.1. Then,

$$\mathbf{u} = (u, v)^t = (x/z, y/z)^t.$$

Image vector of the point $\mathbf{x}$ is defined by

$$\mathbf{X} = (u, v, 1)^t = (x/z, y/z, 1)^t$$

in the 3-D coordinate system. It is easy to see that the image plane vector $\mathbf{u}$ consists of the first two components of the corresponding image vector $\mathbf{X}$. In the camera-centered coordinate system, the object undergoes a motion from time t_1 to t_2, relative to the camera. The relative motion of the scene is represented by a rotation followed by a translation. Letting R be the rotation matrix and $\mathbf{T}$ be the translation vector, and letting $\mathbf{x}$ move to $\mathbf{x}'$ under the motion, we have

$$\mathbf{x}' = R\mathbf{x} + \mathbf{T}. \tag{4.2.1}$$

Similarly, define the image plane vector $\mathbf{u}'$ of the image vector $\mathbf{X}'$.

$$\mathbf{X}' = (u', v', 1)^t = (x'/z', y'/z', 1)^t.$$

We define another coordinate system, called object-centered coordinate system, is fixed on the object or the scene. In this coordinate system, the camera undergoes a motion and the object is fixed. Let the object-centered coordinate system be fixed with the scene and coincide with the camera coordinate system before motion. To result in the same images as before, the motion of the camera can be represented by a "reverse motion": a translation $-\mathbf{T}$ followed by a rotation R^t in the object-centered coordinate system. Namely, any point $\mathbf{p}$ on the camera is moved to $\mathbf{p}'$, and they are related by

$$\mathbf{p}' = R^t(\mathbf{p} - \mathbf{T})$$

in the object-centered coordinate system (see Fig. 4.2). We will mostly consider the camera-centered coordinate system, in which the object is moving and the camera is stationary. When it is necessary, we will mention the object-centered system, in which the camera is moving.

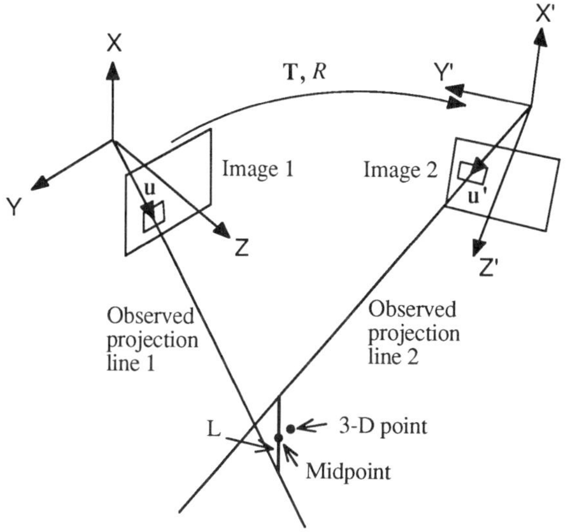

Fig. 4.2. A view from the object-centered coordinate system: a moving camera takes two images of the scene.

4.2.1 The Linear Algorithms

Different versions of the linear algorithms have been published in the literature [4.17, 38, 49, 40, 10]. Although those algorithms use different ways to determine unknowns, they share the same key structure: determining intermediate parameters, called essential parameters, based on the epipolar constraint. To be specific, we re-examine the algorithm presented in the previous chapter as an example.

Suppose $\mathbf{T} \neq \mathbf{0}$ (otherwise, the structure cannot be determined but rotation can still be determined). Let $\alpha = \|\mathbf{T}\|^{-1}$, $\tilde{\mathbf{x}} = \alpha \mathbf{x}$ and $\tilde{\mathbf{x}}' = \alpha \mathbf{x}'$, where α is a positive global scale factor that cannot be determined from monocular images unless additional information is available. Given correspondences for image vectors at the two time instants, $(\mathbf{X}, \mathbf{X}')$, the linear algorithm solves for the rotation matrix R and the unit translation vector $\mathring{\mathbf{T}} = \alpha \mathbf{T}$. For each point, the algorithm solves for $\tilde{\mathbf{x}}$ and $\tilde{\mathbf{x}}'$.

Notice that $\mathbf{x} = z\,\mathbf{X}$ and $\mathbf{x}' = z'\,\mathbf{X}'$. Letting $\tilde{z} = \alpha z$, $\tilde{z}' = \alpha z'$ denote relative depths, (4.2.1) yields

$$\tilde{z}'\,\mathbf{X}' = R\tilde{z}\;\mathbf{X} + \mathring{\mathbf{T}}. \tag{4.2.2}$$

After solving for R and $\mathring{\mathbf{T}}$, for each image point the linear algorithm solves for least-squares solutions of $\tilde{z}$ and $\tilde{z}'$ based on (4.2.2). We get $\tilde{\mathbf{x}} = \tilde{z}\,\mathbf{X}$ and $\tilde{\mathbf{x}}' = \tilde{z}'\,\mathbf{X}'$. However, two observed projection lines generally do not intersect due to noise (see Fig. 4.2). From the view point of the camera-centered coordinate system, this is equivalent to the fact that the observed object is not exactly rigid due to noise. In the linear algorithm presented in Chapter 3, we use a simple solution to this problem. The geometrical interpretation of that correction procedure is very intuitive from the view point of the object-centered coordinate system: Among the line segments that connect two observed projection lines in Fig. 4.2, let L be the shortest one. The point in the estimated structure is the midpoint of L as shown in Fig. 4.2. The recovered structure is therefore exactly rigid.

4.2.2 The Epipolar Constraint

The key constraint that the linear algorithms employ to solve for motion parameters is the epipolar constraint: $\mathbf{X}'$, $R^t\mathbf{X}$ and $\mathring{\mathbf{T}}$ must be linearly dependent (or coplanar) according to (4.2.2), or, equivalently, the vector triple product vanishes:

$$(\mathbf{X}')^t(\mathring{\mathbf{T}}\times(R\,\mathbf{X})) = 0. \tag{4.2.3}$$

Its geometrical illustration is shown in Fig. 4.3. The plane on which $\mathbf{X}'$, $\mathbf{T}$ and $R\mathbf{X}$ lie is called the epipolar plane of the point. Its intersection with the image plane is called the epipolar line of the point. In Chapter 3, E is defined to be

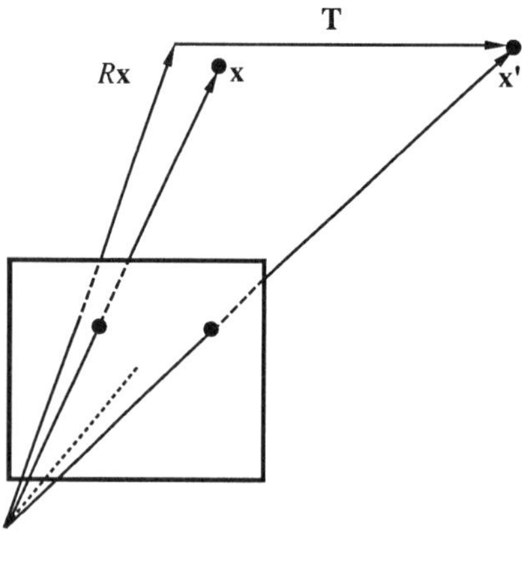

Fig. 4.3. Epipolar constraint: $R\mathbf{X}$, $\mathbf{T}$ and $\mathbf{X}'$ are coplanar.

$$E = [\mathbf{T}_s]_\times R \tag{4.2.4}$$

where $\mathbf{T}_s$ is a unit vector such that $\mathbf{T} \times \mathbf{T}_s = \mathbf{0}$. Equation (4.2.3) can then be rewritten as

$$(\mathbf{X}')^t E \mathbf{X} = 0. \tag{4.2.5}$$

which is linear in the elements of matrix E. Using 8 or more point correspondences, the algorithms first solve for E based on (4.2.5) and then solve for motion parameters from E. It is important to note that E is solved as an unconstrained matrix.

We now show that only one component of the image position of a point is used by the epipolar constraint. In fact, consider two unit vectors in the image plane at a point $\mathbf{X}'$, one is aligned with the epipolar line (denoted by δ_a), and the other is perpendicular to the epipolar line (denoted by δ_p). Any perturbed image position of $\mathbf{X}'$ can be represented by $\mathbf{X}'(\varepsilon) = \mathbf{X}' + a\,\delta_a + b\,\delta_p$ for some real numbers a and b. Since $\delta_a^t(\mathring{\mathbf{T}} \times R\,\mathbf{X}) = 0$, from (4.2.3) it follows that

$$\mathbf{X}'(\varepsilon)^t(\mathring{\mathbf{T}} \times R\,\mathbf{X}) = \delta_p^t(\mathring{\mathbf{T}} \times R\,\mathbf{X}).$$

In other words, the perturbation of $\mathbf{X}'$ along the epipolar line direction does not affect the value of $\mathbf{X}'(\varepsilon)^t(\mathring{\mathbf{T}} \times R\,\mathbf{X})$. The location of the points on the epipolar line is irrelevant to the epipolar constraint. This location is related to the depth of the point as well as motion parameters.

We have proved in Chapter 3 that based on the epipolar constraint, the rotation and translation parameters can be solved uniquely from image vectors of a nondegenerate configuration of 3-D points. The questions to ask are:

(1) The essential matrix E has only 5 degrees of freedom (2 for unit vector $\mathring{\mathbf{T}}$ and 3 for rotation matrix R). How can the constraint in E be used to improve accuracy in the presence of noise?

(2) How reliably the motion parameters can be estimated using just the epipolar constraint?

(3) Can another component (along the epipolar line) of the image points be used, in addition to the epipolar constraint, to improve the reliability of the estimated motion and structure parameters in the presence of noise?

These issues are investigated in the following subsections.

4.2.3 Using the Constraint in Matrix E

According to the definition in Equation (4.2.4), E has only 5 degrees of freedom. E should be the product of a skew symmetric matrix ($S^t = -S$) and a rotation matrix R (orthonormal with determinant 1).

Theorem 4.1. Given a 3×3 matrix E, the necessary and sufficient condition for existing a rotation matrix R and a skew symmetric matrix S, such that $E = SR$, is that one of the singular values of E is equal to 0 and the remaining two are equal.

Proof. See Appendix 4.A.

For singular values and singular value decomposition, see, e.g., [4.28, 5, 15, 26].

Corollary. Given a 3×3 matrix E, the necessary and sufficient condition for existing a rotation matrix R and a unit vector $\mathring{\mathbf{T}}$, such that $E = [\mathring{\mathbf{T}}]_\times R$, is that the singular values of E are 0, 1, 1, respectively.

Proof. See Appendix 4.A.

The constraint on the singular values of $E^t E$, stated in Theorem 4.1, can be written as polynomial equations in terms of elements of E. These yield more equations, in addition to (4.2.5). Those additional equations can be used to investigate the solutions when fewer than 8 points are available. However, polynomial equations introduce spurious solutions. In the case where 8 or more points are available, E can be determined by linear equations (4.2.5) without considering the constraint in E. R and $\mathring{\mathbf{T}}$ then are determined from the estimated E. This is the method used in the linear algorithms. In the presence of noise, E solved from the linear equations generally does not satisfy the conditions in Theorem 4.1. This causes errors in the solutions of R and $\mathring{\mathbf{T}}$. However, those solutions are generally good enough to serve as an initial guess for iterative improvement. The constraint in E can be employed by iteratively improving the computed R and $\mathring{\mathbf{T}}$ to minimize the weighted sum of $[(\mathbf{X}')^t(\mathring{\mathbf{T}}\times R\,\mathbf{X})]^2$. The weight is the reciprocal of the error variance of $(\mathbf{X}')^t(\mathring{\mathbf{T}}\times R\,\mathbf{X})$. Assume the components of $\mathbf{u}$ and $\mathbf{u}'$ have additive uncorrelated zero mean noise with variance σ^2, it is shown in Appendix 4.B that the variance of the first order error of $(\mathbf{X}')^t(\mathring{\mathbf{T}}\times R\,\mathbf{X})$ is given by

$$\sigma^2(\|R^t(\mathring{\mathbf{T}}\times\mathbf{X}')\|^2_{z=0} + \|\mathring{\mathbf{T}}\times R\,\mathbf{X}\|^2_{z=0}) \tag{4.2.6}$$

where $\|(a, b, c)\|^2_{z=0} \triangleq a^2 + b^2$. Thus, the objective function of the improvement based on epipolar constraint is

$$\sum_{i=1}^{n} \frac{[(\mathbf{X}_i')^t(\mathring{\mathbf{T}} \times R\,\mathbf{X}_i)]^2}{\sigma^2(\|R^t(\mathring{\mathbf{T}} \times \mathbf{X}_i')\|_{z=0}^2 + \|\mathring{\mathbf{T}} \times R\,\mathbf{X}_i\|_{z=0}^2)}. \tag{4.2.7}$$

Minimizing the residuals based on the epipolar constraint, such as the above objective function, is called epipolar improvement.

Some comments are in order here. (1) Since the weights include the unknowns, R and $\mathring{\mathbf{T}}$, one cannot use those weights in solving the weighted least-squares solution of (4.2.5) without using iterations. (2) If we want to ensure that the equations to be solved are all linear, the constraints in E stated in the above theorem cannot be used either, since those constraints are nonlinear. (3) If those constraints are used together with the linear equation (4.2.5), generally fewer than 8 points are need to solve for motion parameters. But this again requires solving nonlinear equations. (4) More constraints beside the epipolar constraint can be used in the presence of noise. The epipolar constraint alone cannot always ensure reliable solutions, which we will now discuss in the next section.

4.2.4 A Type of Motion

Reliability of the estimated motion and structure depends on many factors, e.g., structure of the scene, motion parameters, and imaging system parameters. The effects of those factors on the reliability of solution have been investigated qualitatively and quantitatively in Chapter 3. Here we will identify a type of motion where even small errors will override the information used by the epipolar constraint.

Using linear algorithms, the motion parameters are determined based on the epipolar constraint (4.2.3). That is, the rotation matrix R is such that there exists a vector $\mathbf{T}$ which is coplanar with $R\,\mathbf{X}$ and $\mathbf{X}'$ for all the point pairs.

Let us consider a type of pure translation and the corresponding type of pure rotation. For the type of translation, the translation vector and the optical axis is orthogonal. For the type of pure rotation, the rotation axis is orthogonal to both the optical axis and the translation vector of the pure translation. Without loss of generality, let the translation direction be aligned with y-axis, the rotation axis be aligned with x-axis. Fig. 4.4 shows examples of the displacement fields of the pure translation and the pure rotation, respectively. It is clear that the translation produces horizontal displacement vectors and the rotation produces almost horizontal ones. We analyze this

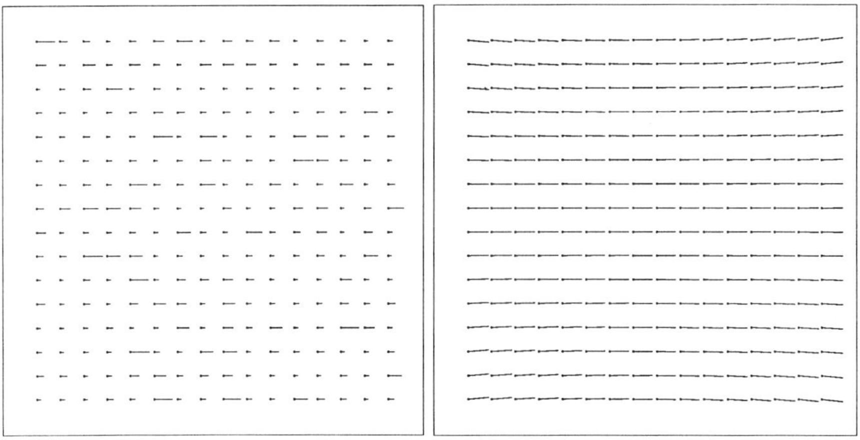

(a) (b)

Fig. 4.4. Two image plane displacement fields: from a horizontal translation **(a)** and a rotation about a vertical axis **(b)**. The field from the translation is exactly horizontal. The field from the rotation is almost horizontal since the maximum vertical displacement is just 2.9 pixels. Image size: 0.7. Resolution: 512×512.

property quantitatively. For horizontal pure translation $\mathbf{T} = (0, t_2, 0)^t$, $t_2 \neq 0$, from $\mathbf{x}' = \mathbf{x} + \mathbf{T}$ it follows that

$$u' = x'/z' = x'/z = u,$$

$$v' = y'/z' = (y + t_2)/z = v + t_2/z \neq v. \tag{4.2.8}$$

The displacement vector on image plane is equal to

$$\mathbf{u}' - \mathbf{u} = (u' - u, v' - v)^t = (0, v' - v)^t.$$

The left hand side of (4.2.3) for this case is

$$(\mathbf{X}')^t (\mathring{\mathbf{T}} \times R\,\mathbf{X}) = u' - u$$

which should be equal to 0. Therefore the epipolar constraint for this pure horizontal translation is that the image plane displacement vector $\mathbf{u}' - \mathbf{u}$ is horizontal.

For the pure rotation, the rotation matrix is given by

$$R = \begin{bmatrix} 1 & 0 & 0 \\ 0 & \cos\theta & -\sin\theta \\ 0 & \sin\theta & \cos\theta \end{bmatrix}.$$

From $\mathbf{x}' = R\,\mathbf{x}$ it follows that

$$u' = \frac{x'}{z'} = \frac{x}{y\sin\theta + z\cos\theta} = \frac{u}{v\sin\theta + \cos\theta},$$

$$v' = \frac{y'}{z'} = \frac{y\cos\theta - z\sin\theta}{y\sin\theta + z\cos\theta} = \frac{v\cos\theta - \sin\theta}{v\sin\theta + \cos\theta}.$$

Then, it follows that

$$u' - u = \frac{u}{v\sin\theta + \cos\theta} - u = \frac{u(1 - \cos\theta - v\sin\theta)}{v\sin\theta + \cos\theta}, \tag{4.2.9}$$

$$v' - v = \frac{v\cos\theta - \sin\theta}{v\sin\theta + \cos\theta} - v = -\frac{(1 + v^2)\sin\theta}{v\sin\theta + \cos\theta}.$$

Since generally $u' - u \neq 0$, the pure rotation does not exactly satisfy the epipolar constraint of the horizontal translation. However, the value of $u' - u$ is very close to zero: Assume the image size is s (image is a $s \times s$ square) with $m \times m$ resolution (pixels). The pixel size is then s/m. The vertical displacement in (4.2.9) in terms of the number of pixels is then

$$(u' - u)\frac{m}{s} = \frac{mu(1 - \cos\theta - v\sin\theta)}{s(v\sin\theta + \cos\theta)}. \tag{4.2.10}$$

Based on (4.2.10), Fig. 4.5 shows the value of the vertical displacement, in terms of number of pixels, at the center of a quadrant of an image with 512×512 resolution, for different horizontal displacement, $v' - v$, and different image sizes, s. For example, for image size 1 (equivalent to a 35mm wide angle lens of a 35mm camera), the vertical displacement is just about 2 pixels for a large 40-pixel horizontal displacement. For small motions with 4-pixel image displacement (similar or even smaller motion is required for optical flow approaches), the vertical displacement is less than a half pixel even for a very wide field of view. Since the vertical displacement is so small for the rotation, the corresponding displacement field almost satisfies the epipolar constraint for the pure translation discussed earlier. This implies that in the presence of small errors in the image coordinates (e.g., in a magnitude of one or two pixels), pure

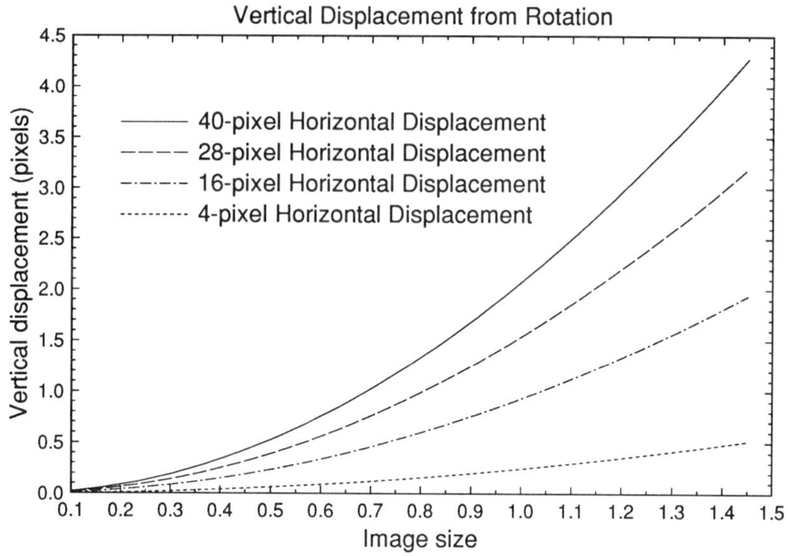

Fig. 4.5. Vertical displacement of a rotation about x-axis versus image size for different horizontal displacement. Position of the image point: the center of a upper right quadrant of the image plane. Image resolution: 512×512.

translation can be interpreted by a rotation as far as the epipolar constraint is concerned, and vice versa. In other words, the epipolar constraint cannot disambiguate the translation from the corresponding rotation in the presence of even small image digitization noise.

Now, let us consider the factor of decreasing the field of view, or equivalently, the image size with a unit focal length. Assume a constant resolution (or equivalently the same number of pixels). If the number of vertical displacement, in terms of number of pixels, decreases as the image size decreases, we can see that it is even harder to disambiguate the above two types. As the image size decreases, the pixel size and the allowed amount of motion decrease. They are two competing factors. From Fig. 4.5 it can be seen that the vertical displacement, in terms of number of pixels, approaches zero as the image size decreases with constant image resolution and constant image horizontal displacement. In fact, the vertical displacement, in terms of pixels, decreases quadratically as the image size decreases. This fact can be derived from (4.2.10). For a small rotation with angle θ, the horizontal displacement is in the same order as θ (remember we have unit focal length). To fix the amount of horizontal displacement in the image plane with respect to image size s, let $\theta = s/k$, where k is a constant. When s goes to zero, so does θ. For a point fixed relatively in the image frame, $|u| = s/h_u$, $|v| = s/h_v$, where h_u and h_v are constants. Letting s go to zero, the absolute value of (4.2.10) is in the order of

$$\frac{muv\theta}{s} = \frac{ms^2}{h_u h_v k}$$

which is quadratic in s. We can see that the number of pixels of vertical displacement goes to zero quadratically as the image size s approaches zero. Therefore, the vertical displacement decreases much faster than the pixel size. This result implies that motion estimation based on the epipolar constraint is inherently very unreliable with a small field of view.

In summary, we have shown that for a type of translations there exists the corresponding type of rotations such that the displacement field of translation can be interpreted by the rotation without significantly violating the epipolar constraint. In the presence of even small pixel level noise, the displacement of the translation can be interpreted by appropriate rotation and vise versa. If only epipolar constraint is used, small pixel level errors will cause large errors in estimated R. The vector $\mathbf{T}$ determined from $\mathbf{X}$ and $R\mathbf{X}$ will also be contaminated by the large errors in $R\mathbf{X}$. Up to here, we have answered the second question at the end of Subsection 4.2.2: There are cases where the motion parameters can not be estimated reliably using just the epipolar constraint.

There are two ways to improve the estimated motion parameters. The first one is to use a large number of points (or a dense displacement field). If the noise has zero mean and is highly random, the errors tend to be overcome in the solution of an overdetermined system. This is true for the cases of our computer simulations where the noise is generated by zero mean pseudo-random numbers. However, from a displacement field (or point correspondences) computed automatically by an algorithm, the errors are often biased, and the amount of bias is usually unknown. This fact makes the overdetermination less effective.

The second way also gives an answer to the third question raised at the end of Subsection 4.2.2, and will be discussed in the following subsection.

4.2.5 Beyond the Epipolar Constraint

From (4.2.3) it can be seen that the depths of the object points are excluded from the epipolar constraint. This is desirable to the linear algorithms since the depths of the points are unknown. The epipolar constraint uses only one component of the coordinates of the image points: $\mathbf{X}'$ can be any image vector in the epipolar plane where $\mathring{\mathbf{T}}$ and $R\mathbf{X}$ lie. The position of $\mathbf{X}'$ on the epipolar plane is related to the depth of the point. For example, for the pure horizontal translation we have $v' - v = t_2/z$ from (4.2.8). However, the other component left by the epipolar constraint is important for determining motion parameters due to the following two properties, which constitute what we call parallactic effect as shown in Fig. 4.6.

(1) Under a rotation, all the points on a projection line (passing through the origin and an image point) project on to a common image point after the rotation.

(2) Under a translation, those points on a projection line have different projections after the translation: the closer the point is to the image plane, the larger the displacement of the point is in the image plane.

These two properties can be proved analytically. For pure rotation with rotation matrix R, we have $\mathbf{x}' = R\mathbf{x}$, or for image vectors

$$z'\mathbf{X}' = zR\mathbf{X}.$$

With any positive constant α, we consider a new vector $\alpha\mathbf{x}$, with the same rotation, $\alpha\mathbf{x}' = R\,\alpha\mathbf{x}$, or for image vectors,

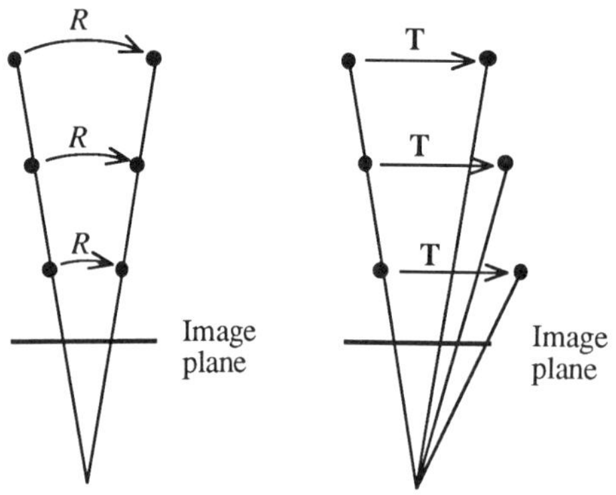

Fig. 4.6. The parallactic effect: rotation and translation yield different displacement fields (a two-dimensional illustration). For a rotation, all the points on a projection line have the same projections after the motion. For a translation, those points on a projection line have different projections after the motion.

$$\alpha z'\mathbf{X}' = \alpha z R\,\mathbf{X}.$$

Therefore $(\mathbf{x}, \mathbf{x}')$ and $(\alpha\mathbf{x}, \alpha\mathbf{x}')$ have the same image vectors, $(\mathbf{X}, \mathbf{X}')$.

For the second property, let $\mathbf{T} = (t_x, t_y, t_z)^t$. From $\mathbf{x}' = \mathbf{x} + \mathbf{T}$, some algebraic manipulations give

$$\|\mathbf{u}' - \mathbf{u}\|^2 = [(t_x - t_z u)^2 + (t_y - t_z v)^2]/(z + t_z)^2. \tag{4.2.11}$$

Notice that $z + t_z = z' > 0$. With $\mathbf{u}$ and $\mathbf{T}$ fixed, the larger the z, the smaller the magnitude of the image displacement is.

From (4.2.11) we know that the magnitude of the image displacement is inversely proportional to the depth z'. As long as $1/z' = 1/(z + t_z)$ has a large variation among points, the displacement field is quite different between a translation and a rotation. These differences are useful for distinguishing a rotation from a translation. For example, in Fig. 4.4 the image displacement vectors are almost horizontal for both rotation and translation. However, the lengths of the displacement vectors are quite different for translation, and are similar for rotation. It is impossible to interpret a translation by a rotation if both components of the points are used.

Aiming at a closed-form solution, the linear algorithms have to exclude information that is related to point depths. Only the component of the point that is independent with the depth is used (based on the epipolar constraint). Mathematically, this is enough to determine motion parameters in the absence of noise. However, in the presence of noise, excluding information that is related to the structure of the scene may result in less reliable estimates of motion parameters, since this information is also related to motion (Figs. 4.4 and 4.6). This observation partially accounts for the unreliability of the solutions of linear algorithms for the cases where translation is parallel to image plane as we see in Chapter 3. The optimization discussed later in this chapter makes use of both components of the image points in an integrated way. As shown by simulations, this significantly improves accuracy. We first investigate optimization based on maximum likelihood in the next section.

4.3 Maximum Likelihood Estimation

In reality, the image coordinates of an object point as well as the corresponding displacement vector in image plane are the results from a feature detector and the corresponding matcher, whose accuracy is influenced by a variety of factors including lighting, structure of the scene, accuracy of camera calibration, image resolution and the performance of the feature matching algorithms. Thus, the observed 2-D image plane vectors $\mathbf{u}_i$ of image 1 and $\mathbf{u}_i'$ of image 2 are noise-contaminated versions of the true ones. Let $(\mathbf{u}_i, \mathbf{u}_i')$ be the observed value of a pair of random vectors $(\mathbf{U}_i, \mathbf{U}_i')$. (With n point correspondences over two time instants, we add subscripts i to denote the i-th point and a subscript-free letter denotes a general example of the vectors.) What we obtain is a sequence of the observed image vector pairs

$$\mathbf{u} \triangleq (\mathbf{u}_1^t, (\mathbf{u}_1')^t, \mathbf{u}_2^t, (\mathbf{u}_2')^t, \cdots, \mathbf{u}_n^t, (\mathbf{u}_n')^t)^t$$

of a sequence of random vector pairs

$$\mathbf{U} \triangleq (\mathbf{U}_1^t, (\mathbf{U}_1')^t, \mathbf{U}_2^t, (\mathbf{U}_2')^t, \cdots, \mathbf{U}_n^t, (\mathbf{U}_n')^t)^t.$$

We need to estimate the motion parameter vector $\mathbf{M}$ and the 3-D positions of the feature points (scene structure)

$$\mathbf{X} \triangleq (\mathbf{x}_1^t, (\mathbf{x}_1')^t, \mathbf{x}_2^t, (\mathbf{x}_2')^t, \cdots, \mathbf{x}_n^t, (\mathbf{x}_n')^t)^t.$$

Let the conditional density of $\mathbf{U}$ conditioned on $\mathbf{M} = \mathbf{m}$ and $\mathbf{X} = \mathbf{x}$ be $p_{\mathbf{U}|\mathbf{M},\mathbf{X}}(\mathbf{u}|\mathbf{m},\mathbf{x})$. To eliminate possible bias in motion estimates, we do not assume a priori distribution of the true motion parameters. In other words, true motion parameters are deterministic values. Under this framework, the maximum likelihood estimator can be used. The maximum likelihood estimates of motion parameters, $\mathbf{m}^*$, and scene structure $\mathbf{x}^*$ are such that the conditional density $p_{\mathbf{U}|\mathbf{M},\mathbf{X}}(\mathbf{u}|\mathbf{m},\mathbf{x})$ reaches the maximum. Namely,

$$p_{\mathbf{U}|\mathbf{M},\mathbf{X}}(\mathbf{u}|\mathbf{m}^*,\mathbf{x}^*) \geq p_{\mathbf{U}|\mathbf{M},\mathbf{X}}(\mathbf{u}|\mathbf{m},\mathbf{x}) \qquad (4.3.1)$$

holds for all possible motion parameters $\mathbf{m}$ and scene structure $\mathbf{x}$. The motivation for using maximum likelihood criterion is that, among others, under fairly general conditions the estimator is consistent (it converges in probability to the correct value when the number of observations approaches infinity), asymptotically unbiased, asymptotically Gaussian, and asymptotically efficient (as the number of observations approaches infinity, it is unbiased, has finite covariance, and there is no other unbiased estimate whose covariance is smaller) [4.9, 37, 46, 21].

To find a maximum likelihood estimate, we need the conditional density $p_{\mathbf{U}|\mathbf{M},\mathbf{X}}(\mathbf{u}\,|\,\mathbf{m},\mathbf{x})$. Given **M** and **X**, what is the distribution of the observed projection of 3-D points? To answer this question, we need to investigate error distribution of the observed image points.

4.3.1 Gaussian Distribution

Gaussian distribution is commonly used for modeling noise. Intuitively, if the errors arise from many sources and are influenced by the sum of many random error terms, the distribution is roughly Gaussian by the central limit theorem. Now assume each image coordinate of the observed projection has an additive zero mean Gaussian noise. Although the actual Gaussian distribution can be very general with a non-diagonal covariance matrix, such a covariance matrix is usually very difficult to obtain in practice. Therefore, we assume the conditional distributions, given motion parameters and scene structure, are independent between different points. When the distance between two points is not very small, compared to image resolution, such an assumption of independence is reasonable. Then, the conditional density of the observed projection, given the motion parameters **M** and the structure of the scene **X**, may be written as

$$p_{\mathbf{U}|\mathbf{M},\mathbf{X}}(\mathbf{u}\,|\,\mathbf{m},\mathbf{x}) = \prod_{i=1}^{n} p_i(\mathbf{u}_i\,|\,\mathbf{m},\mathbf{x})\, p_i'(\mathbf{u}_i'\,|\,\mathbf{m},\mathbf{x}) \tag{4.3.2}$$

where $\mathbf{u}_i$ is the observed projection of the 3-D feature point i in the first image, $\mathbf{u}_i'$ is that of i in the second image, p_i is the Gaussian density with mean equal to the exact projected location of the given feature point, $\mathbf{h}_i(\mathbf{m},\mathbf{x})$, in the first image. p_i' is the Gaussian density for the second image. That is

$$p_i(\mathbf{u}_i\,|\,\mathbf{m},\mathbf{x}) = \frac{1}{2\pi\sigma^2}\exp\left\{-\frac{\|\mathbf{u}_i - \mathbf{h}_i(\mathbf{m},\mathbf{x})\|^2}{2\sigma^2}\right\}, \tag{4.3.3}$$

$$p_i'(\mathbf{u}_i'\,|\,\mathbf{m},\mathbf{x}) = \frac{1}{2\pi\sigma^2}\exp\left\{-\frac{\|\mathbf{u}_i' - \mathbf{h}_i'(\mathbf{m},\mathbf{x})\|^2}{2\sigma^2}\right\}. \tag{4.3.4}$$

It is easy to extend (4.3.3) and (4.3.4) such that the variance is different for different points, which may be useful when such a variance is available. For each image, the line passing through the observed image point and the center of projection is called the observed projection line (Fig. 4.2). Fig. 4.7 shows the value of $p_i(\mathbf{u}_i \mid \mathbf{m}, \mathbf{x}) p'_i(\mathbf{u}'_i \mid \mathbf{m}, \mathbf{x})$ in a horizontal plane as an intensity image. It shows the distribution of the 3-D points given its images. Such a distribution in 3-D depends on the relative positions of the 3-D points relative to the camera. It can be imagined from Fig. 4.7 that the shape of iso-probability surface is roughly like a distorted ellipsoid with a swallowed end on the farther side to the cameras.

The maximum likelihood estimates of the motion and structure parameters are such that the conditional density reaches the maximum. Maximizing the left hand side of (4.3.2) is equivalent to maximizing its logarithm:

$$\ln p_{\mathbf{U}\mid\mathbf{M},\mathbf{X}}(\mathbf{u}\mid\mathbf{m},\mathbf{x})$$
$$= -\sum_{i=1}^{n}\{\|\mathbf{u}_i - \mathbf{h}_i(\mathbf{m},\mathbf{x})\|^2 + \|\mathbf{u}'_i - \mathbf{h}'_i(\mathbf{m},\mathbf{x})\|^2\}/(2\sigma^2) - 2n\ln(2\pi\sigma^2). \tag{4.3.5}$$

Maximizing the right-hand side of (4.3.5) is equivalent to minimizing

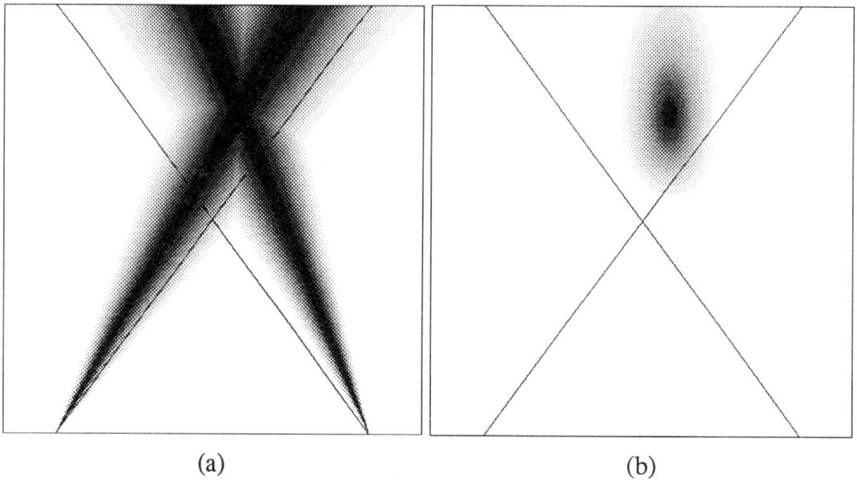

Fig. 4.7. A two-dimensional illustration of point distribution in space, based on Gaussian error distribution in image planes. Darker areas have higher probability (the half tone images here have only a few printable gray tones). Two diagonal lines across the figure are optical axes of two cameras. **(a)** Two observed projection lines, with error distribution, intersect. **(b)** Two observed projection lines, with error distribution, determine the distribution of the point in 3-D.

$$\sum_{i=1}^{n} \{ \|\mathbf{u}_i - \mathbf{h}_i(\mathbf{m},\mathbf{x})\|^2 + \|\mathbf{u}_i' - \mathbf{h}_i'(\mathbf{m},\mathbf{x})\|^2 \}, \tag{4.3.6}$$

which is just the sum of discrepancies between the observed projection and the inferred projection. In fact, given motion parameters $\mathbf{m}$, the structure $\mathbf{x}$ that minimizes the value of (4.3.6) can be estimated analytically. That is, we can compute

$$g_i(\mathbf{m}) \triangleq \min_{\mathbf{x}} \{ \|\mathbf{u}_i - \mathbf{h}_i(\mathbf{m},\mathbf{x})\|^2 + \|\mathbf{u}_i' - \mathbf{h}_i'(\mathbf{m},\mathbf{x})\|^2 \}. \tag{4.3.7}$$

We have

$$\min_{\mathbf{m},\,\mathbf{x}} \{ \sum_{i=1}^{n} (\|\mathbf{u}_i - \mathbf{h}_i(\mathbf{m},\mathbf{x})\|^2 + \|\mathbf{u}_i' - \mathbf{h}_i'(\mathbf{m},\mathbf{x})\|^2) \}$$

$$= \min_{\mathbf{m}} \{ \sum_{i=1}^{n} \min_{\mathbf{x}} \{ \|\mathbf{u}_i - \mathbf{h}_i(\mathbf{m},\mathbf{x})\|^2 + \|\mathbf{u}_i' - \mathbf{h}_i'(\mathbf{m},\mathbf{x})\|^2 \} \}$$

$$= \min_{\mathbf{m}} \{ \sum_{i=1}^{n} g_i(\mathbf{m}) \}.$$

So computationally, structure $\mathbf{x}$ will not be included in the parameter space of iteration. This space decomposition between structure and motion parameters drastically reduces the amount of computation. Otherwise it is computationally extremely expensive (iteration on n points needs 3n-dimensional parameter space!). This technique will be discussed in more detail in Subsection 4.3.3. Since the optimal structure $\mathbf{x}$ can be determined directly from motion parameters, we can exclude $\mathbf{x}$ from the notation for parameters to be estimated. That is, symbolically, the parameters to be determined by nonlinear optimization are just $\mathbf{m}$. Computationally, this space decomposition drastically speeds up the algorithm by reducing the nonlinear optimization space, as shown in Fig. 4.8.

The value of (4.3.6) measures the differences in the observed images and the inferred images. Even without the assumption of Gaussian noise distribution, the objective function of (4.3.6) is desirable. In Chapter 2, we have defined the *image plane error*, or *image error* for short, as the square root of the value of (4.3.6) divided by $2n$:

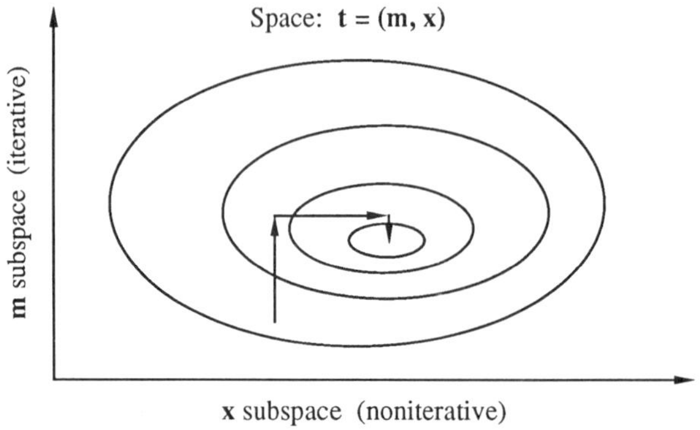

Fig. 4.8. Space decomposition of **t** into **m** and **x**. Iterative algorithm is used only in a small subspace corresponding to **m**. Given each **m**, the best **x** is computed directly. Since the dimension of x is very large, this decomposition drastically reduces computational cost.

$$\left\{ \frac{1}{2n} \sum_{i=1}^{n} (\| \mathbf{u}_i - \mathbf{h}_i(\mathbf{m},\mathbf{x}) \|^2 + \| \mathbf{u}_i' - \mathbf{h}_i'(\mathbf{m},\mathbf{x}) \|^2) \right\}^{1/2} . \tag{4.3.8}$$

As we mentioned, the image error can be easily extended to a weighted version when we have knowledge about the reliability of each point.

4.3.2 Uncertainty Polyhedron Distribution

Let us consider another case where the magnitude of noise is confined to a small range. Digitization noise is an example of such a case. Let the true projection of a point be confined to a rectangle centered at the observed image position of the point. Such a rectangle is called uncertainty rectangle. The true 3-D position of the point is confined to an infinite pyramid defined by the center of projection as apex and the uncertainty rectangle as a cross-section as shown in Fig. 4.9. The second view defines another pyramid for the point. The 3-D point must lie inside the volume of intersection of these two pyramids. Such an intersection of the pyramids is called *uncertainty polyhedron.*

The observed projection lines of two images generally do not intersect due to noise (see Fig. 4.2). However, the two pyramids must intersect since each contains the feature point. In other words, the uncertainty polyhedron has a positive volume.

When the observed image positions, determined by motion parameter vector **m** and structure **x**, of a point in two images are exactly correct the corresponding uncertainty polyhedron has a volume $w(\mathbf{m}, \mathbf{x})$. If those image positions are perturbed by noise, the volume of the uncertainty polyhedron will generally decrease to a value

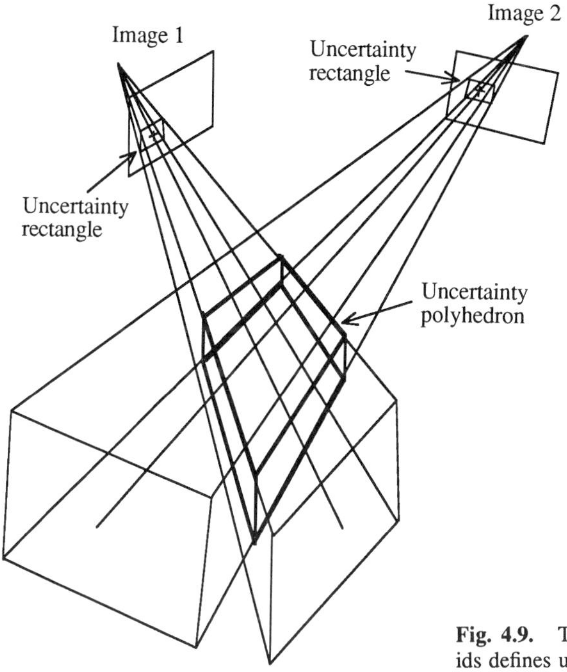

Fig. 4.9. The intersection of two pyramids defines uncertainty polyhedron.

$v(\mathbf{u}, \mathbf{m}, \mathbf{x})$, where $\mathbf{u}$ is a noisy observation vector. If the perturbation is so large that the two pyramids do not intersect, the volume of the uncertainty polyhedron v is equal to zero. Therefore, we assume that the probability for a point to be confined to the uncertainty polyhedron with a volume v is equal to v/w. Obviously, the actual probability density is not exactly constant within each uncertainty polyhedron. However, the variation of probability density within each polyhedron is negligible, since uncertainty polyhedrons are very small compared to the camera-centered coordinates of the points and so is this type of density variation. Therefore, the probability can be adequately characterized by the volume defined above.

For simplicity of computation, we assume that the conditional probability for a point to lie in the intersection is independent from point to point. For the observed point i, the volumes v and w discussed above are denoted by v_i and w_i, respectively. The event observed is that the feature points are all confined to the corresponding uncertainty polyhedrons. Then the conditional density in (4.3.1) should be discrete conditional probability distribution. Thus the conditional probability that all points lie in the corresponding intersection can be written as

$$p_{\mathbf{U}|\mathbf{M},\mathbf{X}}(\mathbf{u} \mid \mathbf{m}, \mathbf{x}) = \prod_{i=1}^{n} \frac{v_i(\mathbf{u}, \mathbf{m}, \mathbf{x})}{w_i(\mathbf{m}, \mathbf{x})}. \tag{4.3.9}$$

Now, let us assume that the cameras are perturbed by a small amount away from the position where the image was taken. Then, the volume, v_i, of the intersection of the two pyramids changes. The value w_i remains almost unchanged. Any perturbation from the correct position generally decreases the volume of intersection for most points, although it might increase the volume of the intersection for a few points. If the perturbation is large enough, the intersection volume for a point becomes zero and so the conditional probability becomes zero. Therefore, a perturbation of the camera position away from the correct position generally decreases the conditional density in (4.3.9). If the perturbation is large enough, the conditional density becomes zero. The maximum likelihood estimates of the motion parameters and structure are such that the conditional probability reaches the maximum.

From (4.3.9) it follows that

$$\ln p = \sum_{i=1}^{n} \ln(v_i / w_i).$$

Maximizing $p_{\mathbf{U}|\mathbf{M},\mathbf{X}}(\mathbf{u} \mid \mathbf{m}, \mathbf{x})$ is equivalent to minimizing

$$-\ln p = \sum_{i=1}^{n} \ln(w_i / v_i)$$

and so, a least-squares procedure can be used to reach an optimal solution.

The parameter space includes motion parameters $\mathbf{m}$ and the structure parameters $\mathbf{x}$. For computational efficiency, we need to eliminate $\mathbf{x}$ from search space. It is easy to see that the value $w_i(\mathbf{m}, \mathbf{x})$ is not very sensitive to a small perturbation of the true 3-D point. That is, $w_i(\mathbf{m}, \mathbf{x})$ can be estimated without given $\mathbf{x}$: The volume w_i can be estimated as the corresponding volume of uncertainty polyhedron of the midpoint of the shortest line segment L that connects two observed projection lines (see Fig. 4.2). $v_i(\mathbf{u}, \mathbf{m}, \mathbf{x})$ can actually be determined from $\mathbf{u}$ and $\mathbf{m}$ only (see next section). In other words, the conditional probability can be estimated without determining the 3-D position of the points $\mathbf{x}$. Therefore, like the case of Gaussian distribution, 3-D structure **x** is eliminated from the parameter space of iteration. After the motion parameters are estimated by the maximum likelihood estimates, the 3-D position of each feature point is naturally estimated by the centroid of the corresponding uncertainty polyhedron. The desirable properties of mean estimate have been discussed extensively [4.27, 37, 13]. Since the uncertainty polyhedron for the maximum likehood estimate is always nonempty, we can always calculate the centroid.

If the estimated parameters are such that the corresponding conditional probability in (4.3.9) is equal to zero, the results must not be a maximum likelihood estimate since the conditional probability in (4.3.9) for the true parameters is not zero (for implementation, we make sure that the volume for a mere touch is positive). Therefore the maximum likelihood must correspond to a conditional probability that is nonzero.

If the conditional density corresponding to the final estimates is positive (which is the case for all the simulations performed), the results define a solution consistent with the given images. In other words, we obtain a structure of the scene and the corresponding motion parameters such that the projection of each inferred point is within the range of noise from the observed projection.

If the given motion parameter has large errors so that a point does not lead to any intersection for the corresponding pyramids, the conditional probability in (4.3.9) is then equal to zero in a small neighborhood. In such a case, the conventional iterative procedure for the maximum will lose direction. Thus, the iterative algorithm needs a good estimate of motion parameters to start with. The initial guess provided by the linear algorithm does not always guarantee a nonzero intersection for every point. Therefore we need a model of noise distribution for which the distribution is not restricted to a finite area. The corresponding objective function of Gaussian distribution can be used as intermediate objective function before using the uncertainty polyhedron model.

4.4 A Two-Step Approach and Computation

The optimization discussed in the last section is nonlinear. Computationally we need schemes to ensure that the global optimal solution can be reached reliably. A two-step approach is proposed here. First, a linear algorithm is applied. Then the solution of the linear algorithm is used as an initial guess solution for the second step in which an iterative algorithm improves the initial guess to minimize an objective function. This two-step approach has the following advantages:

(1) A solution is generally guaranteed. The linear algorithm always gives a solution provided that degeneracy does not occur. Unless the noise level is very high, this solution is close to the true one. As long as the initial guess is within the convergent region to a globally optimal point, iteration leads to the optimal solution.

(2) The approach allows flexible design of the objective functions. If a good initial guess is available, the design of the nonlinear algorithm can emphasize the stability of the derived estimates, and the objective function can be chosen with more flexibility.

(3) The approach yields reliable solutions. The linear algorithms use only the epipolar constraint and so, the solution is sensitive to noise and the reliability of solution varies with motion types. The optimization in the second step employs more global constraints and achieves significant improvement over the first step.

(4) The computation is fast. Generally, a linear algorithm is fast, and a nonlinear algorithm is slow. When a linear algorithm is followed by a nonlinear algorithm, the amount of computation is not simply equal to the sum of those needed by each algorithm individually. Since the linear algorithm provides good initial guess, the time taken by the nonlinear algorithm to reach a solution is greatly reduced.

In the following we discuss some computational aspects of the algorithms.

First, for Gaussian distribution, equation (4.3.6) involves both motion parameters and 3D position of every feature point. The maximum is over all the possible motion parameters and scene structures. The parameter space for iteration is huge and computation is very expensive. However, as mentioned, we do not have to iterate on the structure of the scene. That is, we can derive the closed-form expression for $\mathbf{x}$ that gives $g_i(\mathbf{m})$ in (4.3.7). From motion parameters and the observed projections of point i, two observed projection lines are determined. These two observed projection lines do not intersect in general (see Fig. 4.2). If the true 3-D point is on the observed projection line of the first image, the discrepancy $\|\mathbf{u}_i - \mathbf{h}_i(\mathbf{m},\mathbf{x})\|^2$ is equal to zero, but $\|\mathbf{u}'_i - \mathbf{h}'_i(\mathbf{m},\mathbf{x})\|^2$ is generally not. If the true 3-D point is on the other observed projection line, $\|\mathbf{u}'_i - \mathbf{h}'_i(\mathbf{m},\mathbf{x})\|^2$ is equal to zero while $\|\mathbf{u}_i - \mathbf{h}_i(\mathbf{m},\mathbf{x})\|^2$ is not. Given the motion parameters, we need to find a 3-D point for each feature point such that the corresponding term $\|\mathbf{u}_i - \mathbf{h}_i(\mathbf{m},\mathbf{x})\|^2 + \|\mathbf{u}'_i - \mathbf{h}'_i(\mathbf{m},\mathbf{x})\|^2$ is minimized. If the desired point lies in the shortest line segment L that connects the two observed projection lines (see Fig. 4.2), an exact solution of the optimal point requires solving a fourth order polynomial equation. Appendix 4.C shows that, by a reasonable approximation, we can get a closed-form solution. The optimal point is generally not far from the midpoint of the line segment L, unless the object distance and the viewing angle differ a lot for two images. In simulations, only an extremely small performance difference is observed if the midpoint is used as the optimal point instead of the analytical solution presented in Appendix 4.C. For computational efficiency, we may just use the midpoint of the line L as an approximated optimal point.

With regard to the model of uncertainty polyhedron, we need to calculate the volumes v and w for each point. The shape of the intersection of two arbitrary pyramids at arbitrary orientations is very irregular. Exact computation through exhaustive procedure will be very inefficient. We need a simple and efficient way to calculate the volume of the intersection.

The region of intersection is circumscribed by a regular diamond-like polyhedron (called regular diamond) shown in Fig. 4.10. This regular diamond has three symmetrical axes. Two image vectors give

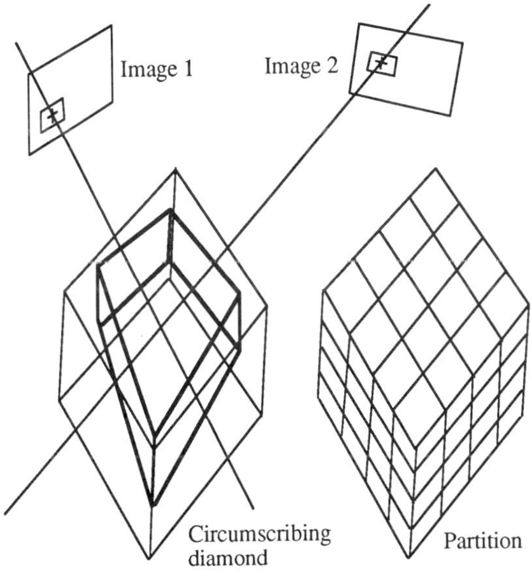

Fig. 4.10. Circumscribing diamond and partitions.

two axes of the circumscribing diamond. The third axis is perpendicular to the other two axes as shown in Fig. 4.10. The center of the circumscribing diamond is located at the 3-D position of the point. The size of the circumscribing diamond is such that it just circumscribes the largest possible intersection. The circumscribing diamond is divided into eight octants along its three axes. The division is similar to octree subdivision except that the axes here are not orthogonal. The 8 octants are recursively subdivided into 8 suboctants, until the desired resolution is reached. Then intersection can be checked using octree techniques [4.42] as follows.

First consider the volume of the intersection v. Starting from the top level, if the eight corners of the current octants are all visible to the uncertainty rectangle in both images (visible means the projection fall in the uncertainty rectangle), the current octant is in the intersection and the volume of the intersection is incremented by the volume of the current octant. If at least one of the eight corners is not visible, we check the suboctants of the current octant. Recursively repeat the same procedure. To compute the volume of w, we move the uncertainty rectangle in two images such that the center of the rectangle is the projection of the center of the circumscribing diamond. Then we calculate the intersection with regard to the moved uncertainty rectangles. For a diamond with $m \times m \times m$ unit cells (smallest partition units) the order of the computation is m^2 [4.42]. If m is small, a straight forward checking for each unit cells suffices. In fact, the computation of the volume of the intersections does not have to be very accurate. The performance would not improve further if m is large enough. Simulation indicates that the performance does not improve much beyond $m = 8$.

For both Gaussian model and uncertainty polyhedron model, we need to compute nonlinear least squares solution. For uncertainty polyhedron model, the maximization of a product of n factors corresponds to the maximization of the sum of their logarithms, which in turn corresponds to the minimization of a sum of squares as we did for the Gaussian distribution. Therefore the maximum likelihood problem is reduced to a nonlinear least squares problem for both models. It can be solved through iterations starting with a good initial guess.

The maximum likelihood estimation of the two noise models discussed in the last section leads to minimizing an objective function which can be expressed by the sum of squares. That is, based on observation vector $\mathbf{u}$ we want to determine $\mathbf{m}$ that minimizes the norm of a residual vector $\mathbf{f}(\mathbf{u}, \mathbf{m})$:

$$\min_{\mathbf{m}} \{ \|\mathbf{f}(\mathbf{u}, \mathbf{m})\| \}.$$

For example, in the case of Gaussian noise, let

$$\mathbf{h}(\mathbf{m}, \mathbf{x}) \triangleq (\mathbf{h}_1^t, (\mathbf{h}_1')^t, \mathbf{h}_2^t, (\mathbf{h}_2')^t, \cdots, \mathbf{h}_n^t, (\mathbf{h}_n')^t)^t. \tag{4.4.1}$$

With the best $\mathbf{x}$ computed from $\mathbf{u}$ and $\mathbf{m}$ as $\mathbf{y}(\mathbf{u}, \mathbf{m})$, the residual vector to be minimized is

$$\mathbf{f}(\mathbf{u}, \mathbf{m}) = \mathbf{h}(\mathbf{m}, \mathbf{y}(\mathbf{u}, \mathbf{m})) - \mathbf{u} \tag{4.4.2}$$

where $\mathbf{f}$ is a $4n$ dimensional vector for n point correspondences. Without noise, $\mathbf{u}$ is exact: $\mathbf{u} = \mathbf{h}(\mathbf{m}, \mathbf{x})$.

Given an estimated $\mathbf{m}$, $\mathbf{m}_i$, expanding $\mathbf{f}(\mathbf{m})$ at $\mathbf{m}_i$ yields

$$\mathbf{f}(\mathbf{m}) \approx \mathbf{f}(\mathbf{m}_i) + \frac{\partial \mathbf{f}(\mathbf{m}_i)}{\partial \mathbf{m}} (\mathbf{m} - \mathbf{m}_i). \tag{4.4.3}$$

Denoting

$$J_i \triangleq \frac{\partial \mathbf{f}(\mathbf{m}_i)}{\partial \mathbf{m}}, \tag{4.4.4}$$

Equation (4.4.3) gives

$$J_i(\mathbf{m} - \mathbf{m}_i) \approx -\mathbf{f}(\mathbf{m}_i) + \mathbf{f}(\mathbf{m}). \tag{4.4.5}$$

Vector $\mathbf{m}$ that minimizes $\|J_i(\mathbf{m} - \mathbf{m}_i) + \mathbf{f}(\mathbf{m}_i)\| \approx \|\mathbf{f}(\mathbf{m})\|$ is

$$\mathbf{m} = \mathbf{m}_i - (J_i^t J_i)^{-1} J_i^t \, \mathbf{f}(\mathbf{m}_i). \tag{4.4.6}$$

Since $J_i^t J_i$ may be singular, $J_i^t J_i$ is replaced by $D_i + J_i^t J_i$, where D_i is a diagonal matrix with non-negative diagonal elements. Then we get a sequence of approximation to a minimum solution

$$\mathbf{m}_{i+1} = \mathbf{m}_i + (D_i + J_i^t J_i)^{-1} J_i^t \, \mathbf{f}(\mathbf{m}_i).$$

This is called the Levenberg-Marquardt method (or L-M method for short) [4.16, 20, 25, 19]. The special case in which $D_i = 0$ for all i is called the Gauss', or the Gauss-Newton method [4.25, 19].

To compute matrix J_i, one needs to calculate the partial derivatives of the image residual at each point with respect to every independent motion parameter. An example of independent motion parameters and the related derivatives are discussed in Appendix 4.D. In fact, some procedures use finite difference method and so, do not need analytical expressions of the equations. The finite difference analogues of the Levenberg-Marquardt and Gauss algorithms by Brown [4.7] is one of the examples.

4.5 Minimum Variance Estimation

In Section 4.3, the maximum likelihood estimation with Gaussian noise distribution leads to minimizing the image error. The computation of the maximum likelihood estimate requires the knowledge of actual noise distribution and the solution of the likelihood equation. This causes difficulties in practice. First, the distribution of noise is usually unknown. Second, even if the noise distribution is known, the solution of maximum likelihood estimator is often very difficult to obtain analytically. In this section, we will discuss optimal estimator for general error distributions.

Let $\mathbf{m}$ be the parameter vector to be estimated and $\hat{\mathbf{m}}(\mathbf{u})$ be the estimator based on the observation vector $\mathbf{u}$. The error vector of $\hat{\mathbf{m}}$ is $\boldsymbol{\delta}_{\mathbf{m}} \triangleq \hat{\mathbf{m}} - \mathbf{m}$. Given the parameter vector $\mathbf{m}$, the estimated parameter $\hat{\mathbf{m}}$ is random due to random noise in the observation vector $\mathbf{u}$. Two widely used criteria are that the estimator is unbiased

$$\mathbb{E}\hat{\mathbf{m}} = \mathbf{m}$$

($\mathbb{E}$ denotes expectation) and it minimizes the expected squared norm of the error vector

$$\mathbb{E}\|\boldsymbol{\delta}_{\mathbf{m}}\|^2 = \mathbb{E}\|\hat{\mathbf{m}} - \mathbf{m}\|^2 = \int \|\hat{\mathbf{m}} - \mathbf{m}\|^2 \, p_{\mathbf{U}|\mathbf{M}}(\mathbf{u}\,|\,\mathbf{m}) d\mathbf{u} \tag{4.5.1}$$

where integration $\int d\mathbf{u}$ is over the space of $\mathbf{u}$ where $p_{\mathbf{U}|\mathbf{M}}(\mathbf{u}\,|\,\mathbf{m})$ is defined. The value of (4.5.1) provides a measure of the spread of the error. The estimator that minimizes (4.5.1) is called minimum variance estimator (or least-squares estimator, minimum mean square estimator). Let us first consider a linear problem.

Gauss-Markov Theorem. Suppose

$$\mathbf{y} = A\,\mathbf{m} + \boldsymbol{\delta}_{\mathbf{y}} \tag{4.5.2}$$

where $\boldsymbol{\delta}_{\mathbf{y}}$ is a random vector with zero mean, $\mathbb{E}\boldsymbol{\delta}_{\mathbf{y}} = \mathbf{0}$, and covariance matrix

$$\Gamma_{\mathbf{y}} = \mathbb{E}\boldsymbol{\delta}_{\mathbf{y}}\boldsymbol{\delta}_{\mathbf{y}}^t.$$

The unbiased, linear minimum variance estimator of $\mathbf{m}$ that minimizes $\mathbb{E}\|\hat{\mathbf{m}} - \mathbf{m}\|^2$ is

$$\hat{\mathbf{m}} = (A^t \Gamma_{\mathbf{y}}^{-1} A)^{-1} A^t \Gamma_{\mathbf{y}}^{-1} \mathbf{y} \tag{4.5.3}$$

with error covariance matrix

$$\Gamma_{\hat{\mathbf{m}}} \triangleq \mathbb{E}(\hat{\mathbf{m}} - \mathbf{m})(\hat{\mathbf{m}} - \mathbf{m})^t = (A^t \Gamma_{\mathbf{y}}^{-1} A)^{-1}.$$

For proofs, see, e.g., [4.18, 33, 13].

In (4.5.2), $\mathbf{y}$ is usually an observation vector, and $\boldsymbol{\delta}_{\mathbf{y}}$ is the error vector of $\mathbf{y}$. This estimator corresponds to the least-squares estimator with weight matrix $\Gamma_{\mathbf{y}}^{-1}$, i.e., the estimator minimizes

$$(\mathbf{y} - A\,\mathbf{m})^t \Gamma_{\mathbf{y}}^{-1} (\mathbf{y} - A\,\mathbf{m}).$$

If the errors in the components of $\mathbf{y}$ are uncorrelated and have the same variance σ^2, i.e., $\Gamma_{\mathbf{y}} = \sigma^2 I$. Then the minimum variance estimator of $\mathbf{m}$ is given by

$$\hat{\mathbf{m}} = (A^t A)^{-1} A^t \mathbf{y}$$

with error covariance matrix

$$\Gamma_{\hat{\mathbf{m}}} = \mathbb{E}(\hat{\mathbf{m}} - \mathbf{m})(\hat{\mathbf{m}} - \mathbf{m})^t = \sigma^2 (A^t A)^{-1}. \tag{4.5.4}$$

Obviously, $\hat{\mathbf{m}}$ is just the least-squares solution that minimizes $\|A\,\mathbf{m} - \mathbf{y}\|^2$.

Consider again the maximum likelihood estimation for motion and structure parameters with Gaussian noise distribution. From (4.4.3) and (4.4.5), including the higher order terms $o(\|\mathbf{m} - \mathbf{m}_i\|)$, it follows that

$$-\mathbf{f}(\mathbf{m}_i) + J_i \mathbf{m}_i = J_i \mathbf{m} - \mathbf{f}(\mathbf{m}) + o(\|\mathbf{m} - \mathbf{m}_i\|).$$

Denoting the left-hand side of the above equation as $\mathbf{y}_i$, and J_i as A_i, we get

$$\mathbf{y}_i = A_i \mathbf{m} - \mathbf{f}(\mathbf{m}) + o(\|\mathbf{m} - \mathbf{m}_i\|). \tag{4.5.5}$$

Neglecting the higher order term $o(\|\mathbf{m} - \mathbf{m}_i\|)$, (4.5.5) is of the form of (4.5.2). Noise term, $-\mathbf{f}(\mathbf{m})$, is the noise in the coordinates of the image points as defined in (4.4.2). Instead of assuming independent Gaussian noise, we just assume the noises are uncorrelated and have the zero mean and equal variance. By the Gauss-Markov Theorem, the Levenberg-Marquardt method gives unbiased linear minimum variance estimator based on the locally linearized equation of (4.5.5), neglecting the higher order terms. After the iteration correctly converges, the converged point $\mathbf{m}_i$ is not far from the true solution if noise is not very large. Thus, $o(\|\mathbf{m} - \mathbf{m}_i\|)$ is small, and Equation (4.5.5) is a good linear approximation of the system. The nonzero $o(\|\mathbf{m} - \mathbf{m}_i\|)$ accounts for the nonlinear nature of the problem. If the convergence occurs far from the true solution, e.g., the iteration is trapped at a local minimum, $o(\|\mathbf{m} - \mathbf{m}_i\|)$ is generally large and the linearized model does not well characterize the system.

In summary, assuming zero mean and uncorrelated noise, if just the locally linearized model is considered (neglecting $o(\|\mathbf{m} - \mathbf{m}_i\|)$ in (4.5.5), which is a good linear approximation of the system for correctly converged solution), the estimator is unbiased linear minimum variance estimator. However, due to nonlinear nature of the problem, such properties just approximately hold in general.

The objective of minimizing the image error is introduced when maximum likelihood estimation of Gaussian noise distribution is investigated in Section 4.3. Here, we have established the optimality of this objective for general noise distribution. Therefore it can be expected that minimizing the image errors leads to good estimates for other noise distributions, e.g., the uncertainty polyhedron model introduced in Subsection 4.3.2. In simulations, very similar performance has been observed when algorithms of two noise models are applied to data with uniform noise. Such results are predicted by the discussion in this section. Since minimizing the image errors is easier to implement than the optimal solution for the uncertainty polyhedron model, and computationally less expensive, it is recommended for general applications where the exact noise distribution is unknown.

4.6 Error Estimation and Error Bounds

From the noise-contaminated point correspondences, the optimal estimators discussed above give solutions to the motion parameters that are best according to the criteria chosen. We need to investigate the following two issues. (1) How can we access the accuracy of the solutions? Without such an estimated accuracy, it is often risky to use those solutions in practice. (2) What is the theoretical lower bound of the error? How close can the actual error of the algorithm approach the lower bound? These two questions are investigated in this section. Since the errors in the measurements are random, the accuracy or performance of the estimator (or algorithm) should be investigated in statistical sense.

4.6.1 Error Estimation

The error estimation problem has been discussed for linear algorithms in Chapter 3. There, the aim is basically to estimate errors in the least-squares solution of a linear system $A\,\mathbf{m} = \mathbf{y}$, where both matrix A and $\mathbf{y}$ are contaminated by noise. The problem here is different since the function is not linear and only $\mathbf{y}$ is contaminated by noise. The reliability of the solution depends not only on noise level, but also on the structure of the scene, motion parameters and the parameters of the sensor system. Different components of the motion parameters may often have quite different accuracies. Therefore, a method for automatic error estimation is especially useful for the problem here.

The minimum variance estimation discussed above leads to a method for estimating errors in the estimates. By the Gauss-Markov Theorem, the covariance matrix of the error vector $\hat{\mathbf{m}} - \mathbf{m}$ is give by

$$\Gamma_{\hat{\mathbf{m}}} \triangleq \mathbb{E}(\hat{\mathbf{m}} - \mathbf{m})(\hat{\mathbf{m}} - \mathbf{m})^t = (A^t \Gamma_{\mathbf{y}}^{-1} A)^{-1}.$$

For the nonlinear problem investigated here, the matrix A corresponds to

$$J = \frac{\partial \mathbf{f}(\mathbf{u}, \mathbf{m})}{\partial \mathbf{m}}$$

evaluated at the finally estimated parameter $\hat{\mathbf{m}}$ and actual observation $\mathbf{u}$. For uncorrelated noise with uniform noise, $\Gamma_{\mathbf{y}} = \sigma^2 I$, the covariance matrix is simply

$$\Gamma_{\hat{\mathbf{m}}} = \mathbb{E}(\hat{\mathbf{m}} - \mathbf{m})(\hat{\mathbf{m}} - \mathbf{m})^t = \sigma^2 (J^t J)^{-1}. \tag{4.6.1}$$

The trace of the covariance matrix gives the expected squared norm of error vector

$$\text{trace}\{\Gamma_{\hat{\mathbf{m}}}\} = \mathbb{E}(\hat{\mathbf{m}} - \mathbf{m})^t (\hat{\mathbf{m}} - \mathbf{m}) = \mathbb{E}\|\hat{\mathbf{m}} - \mathbf{m}\|^2.$$

Since the optimization discussed in Section 4.5 does not require the knowledge of exact noise distribution, the error estimation discussed here does not either.

The elements in J are partial derivatives. Appendix 4.D presents an example of motion parameters and some of the related derivatives. The partial derivative can also be estimated by finite difference, which is easier to program. We have implemented both analytical and finite difference versions. The estimated errors showed only negligible differences between those two versions.

As we know, motion parameters can be represented in many ways. Sometimes one needs to know the errors in terms of the required representation. Generally, for a representation $\mathbf{m}' = \mathbf{g}(\mathbf{m})$ we have

$$\hat{\mathbf{m}}' - \mathbf{m}' \cong \frac{\partial \mathbf{g}(\hat{\mathbf{m}})}{\partial \mathbf{m}} (\mathbf{m} - \hat{\mathbf{m}}).$$

Therefore with $\Gamma_{\hat{\mathbf{m}}}$, the covariance matrix of the error vector of $\hat{\mathbf{m}}'$ can be estimated by

$$\Gamma_{\hat{\mathbf{m}}'} = \mathbb{E}(\hat{\mathbf{m}}' - \mathbf{m}')(\hat{\mathbf{m}}' - \mathbf{m}')^t = \left[\frac{\partial \mathbf{g}(\hat{\mathbf{m}})}{\partial \mathbf{m}}\right] \Gamma_{\hat{\mathbf{m}}} \left[\frac{\partial \mathbf{g}(\hat{\mathbf{m}})}{\partial \mathbf{m}}\right]^t.$$

4.6.2 Error Bounds

Bias and covariance of error in the estimator provide two measures of the quality of an estimator. Suppose $\hat{\mathbf{m}}$ is an estimator of $\mathbf{m}$. Its expected mean is $\mathbf{b}(\mathbf{m})$,

$$\mathbb{E}\hat{\mathbf{m}} = \mathbf{b}(\mathbf{m})$$

and its covariance matrix is $C_{\hat{\mathbf{m}}}$,

$$C_{\hat{\mathbf{m}}} = \mathbb{E}(\hat{\mathbf{m}} - \mathbf{b}(\mathbf{m})(\hat{\mathbf{m}} - \mathbf{b}(\mathbf{m}))^t.$$

Error vector of $\hat{\mathbf{m}}$, $\hat{\mathbf{m}} - \mathbf{m}$ can be decomposed into two components:

$$\hat{\mathbf{m}} - \mathbf{m} = (\hat{\mathbf{m}} - \mathbf{b}(\mathbf{m})) + (\mathbf{b}(\mathbf{m}) - \mathbf{m}) \triangleq \mathbf{a}_v + \mathbf{a}_b$$

where $\mathbf{a}_v$ can be called variation vector and $\mathbf{a}_b$ can be called bias vector. Since $\mathbf{a}_v$ has a zero mean, and is uncorrelated with $\mathbf{a}_b$ ($\mathbf{a}_b$ is a fixed vector), we can express the correlation matrix of error vector $\hat{\mathbf{m}} - \mathbf{m}$ by two terms,

$$\Gamma_{\hat{\mathbf{m}}} = \mathbb{E}(\mathbf{a}_v + \mathbf{a}_v)(\mathbf{a}_v + \mathbf{a}_b)^t = \mathbb{E}\mathbf{a}_v\mathbf{a}_b^t + \mathbb{E}\mathbf{a}_b\mathbf{a}_b^t.$$

The first matrix is $C_{\hat{\mathbf{m}}}$ and the second is the correlation matrix of bias vector $\mathbf{a}_b = \mathbf{b}(\mathbf{m}) - \mathbf{m}$, denoted by $B_{\hat{\mathbf{m}}}$. We have

$$\Gamma_{\hat{\mathbf{m}}} = C_{\hat{\mathbf{m}}} + B_{\hat{\mathbf{m}}} \tag{4.6.2}$$

which gives

$$\mathrm{trace}\{\Gamma_{\hat{\mathbf{m}}}\} = \mathrm{trace}\{C_{\hat{\mathbf{m}}}\} + \mathrm{trace}\{B_{\hat{\mathbf{m}}}\}. \tag{4.6.3}$$

Therefore, if the estimator is unbiased, the trace of the covariance matrix of the estimator directly gives the expected error. Otherwise it just indicates a smaller part of the expected error since both terms on the right-hand side of (4.6.3) are nonnegative.

There exist theoretical lower bounds on the covariance matrix of any estimator. The Cramér-Rao bound is one of them.

Cramér-Rao bound [4.9, 30]. Suppose $\mathbf{m}$ is a parameter vector of probability density $p(\mathbf{u}, \mathbf{m})$. $\hat{\mathbf{m}}$ is an estimator of $\mathbf{m}$ based on measurement $\mathbf{u}$ with $\mathbb{E}\hat{\mathbf{m}} = \mathbf{b}(\mathbf{m})$. Let $\mathbf{y}^t = \dfrac{\partial \ln p(\mathbf{u}, \mathbf{m})}{\partial \mathbf{m}}$ and

$$F = \mathbb{E}\mathbf{y}\mathbf{y}^t.$$

The matrix F is called the Fisher information matrix. Denote

$$B = \frac{\partial \mathbf{b}(\mathbf{m})}{\partial \mathbf{m}}.$$

Then

$$\mathbb{E}(\hat{\mathbf{m}} - \mathbf{b}(\mathbf{m}))(\hat{\mathbf{m}} - \mathbf{b}(\mathbf{m}))^t \geq BF^{\dagger}B^t \tag{4.6.4}$$

where the inequality means that the larger side minus the smaller side is a nonnegative definite matrix. $F^{\dagger}$ is the pseudo-inverse of F. The equality (4.6.4) holds if and only if

$$\hat{\mathbf{m}} - \mathbf{b}(\mathbf{m}) = BF^{\dagger}\left[\frac{\partial \ln p(\mathbf{u}, \mathbf{m})}{\partial \mathbf{m}}\right]^t$$

almost everywhere.

The proof of the Cramér-Rao bound can be found in [4.30, 48], in [4.33] for the case $B = I$, and in [4.9] for the scaler case. Appendix 4.E presents our proof, which seems simpler.

If F is invertible, the pseudo-inverse is the same as inverse (see [4.24, 28, 29, 18] for discussions on pseudo-inverse).

The Cramér-Rao bound provides a lower bound on the variance of the estimator, biased or unbiased. However, we are often interested in the expected errors instead of variance. From (4.6.2) it follows that

$$\Gamma_{\hat{\mathbf{m}}} \geq C_{\hat{\mathbf{m}}}.$$

Therefore, the Cramér-Rao bound also provides a lower bound on the expected errors. The bias vector $\mathbf{b}(\mathbf{m})$ is often unknown in many real world problems. The same is true for the nonlinear problem invested here. Letting $B = I$, the Cramér-Rao bound provides a lower bound on the expected errors of any unbiased estimator. Therefore, we can compare the expected errors with that of a ''best possible'' unbiased estimator using the Cramér-Rao bound.

The evaluation of the Cramér-Rao bound requires the knowledge of noise distribution. For Gaussian noise, the Cramér-Rao bound is easy to compute. With parameters $\mathbf{m}$ (the best $\mathbf{x}$ can be computed from $\mathbf{m}$) and observation $\mathbf{u}$, equations (4.3.5), (4.4.1) and (4.4.2) yield

$$\ln p_{\mathbf{U}|\mathbf{M}}(\mathbf{u} \mid \mathbf{m}) = \frac{\mathbf{f}(\mathbf{m})^t\,\mathbf{f}(\mathbf{m})}{2\sigma^2} - 2n\ln(2\pi\sigma^2),$$

$$\frac{\partial \ln p_{\mathbf{U}|\mathbf{M}}(\mathbf{u} \mid \mathbf{m})}{\partial \mathbf{m}} = \sigma^{-2}\mathbf{f}(\mathbf{m})^t\,\frac{\partial \mathbf{f}(\mathbf{m})}{\partial \mathbf{m}}.$$

Using the similar donation of (4.4.4), let

$$J = \frac{\partial \mathbf{f}(\mathbf{m})}{\partial \mathbf{m}}.$$

We get the expression of the Fisher information matrix:

$$F = \mathbb{E}\left[\frac{\partial \ln p_{\mathbf{U}|\mathbf{M}}(\mathbf{u}\,|\,\mathbf{m})}{\partial \mathbf{m}}\right]^t \left[\frac{\partial \ln p_{\mathbf{U}|\mathbf{M}}(\mathbf{u}\,|\,\mathbf{m})}{\partial \mathbf{m}}\right]$$

$$= \sigma^{-4} J^t \mathbb{E}\{\mathbf{f}(\mathbf{m})\mathbf{f}(\mathbf{m})^t\} J$$

$$= \sigma^{-4} J^t (\sigma^2 I) J = \sigma^{-2} J^t J.$$

Then, for an unbiased estimator $\hat{\mathbf{m}}$ with identically, independently distributed zero mean Gaussian noise added to true image projections, the Cramér-Rao bound gives

$$\Gamma_{\hat{\mathbf{m}}} \geq F^{-1} = \sigma^2 (J^t J)^{-1} \tag{4.6.5}$$

assuming $J^t J$ has a full rank. More generally, if the noise has a covariance matrix C, then the Fisher information matrix is given by $F = J^t C^{-1} J$.

When the minimum attainable variance is larger than the Cramér-Rao bound, other tighter bounds can be derived. For example, Bhattacharyya bound gives another lower bound of covariance [4.48, 33, 37]. In fact, the Cramér-Rao bound is actually a special case of the Bhattacharyya bound. Since the Bhattacharyya bound involves higher order derivatives of probability density, the computation is more involved. If the actual errors are close to the Cramér-Rao bound (this is true in the experiments presented in Subsection 4.6.2), the more general Bhattacharyya bound is obviously very close to the Cramér-Rao bound.

In Section 4.8, simulations show that for the optimal solution, the actual bias is small and the actual error is very close to the Cramér-Rao bound for unbiased estimators. In other words, the error is very close to those that would result from the "best possible" unbiased estimator.

It is interesting to compare the expressions of error estimation and the Cramér-Rao bound. The estimated error in (4.6.1) has the same form as the bound in (4.6.5). However, they are different. Matrix J in (4.6.1) is evaluated with the estimated $\mathbf{m}$ and the noise-contaminated observation $\mathbf{u}$ while J in (4.6.5) is evaluated with the true $\mathbf{m}$ and noise-free $\mathbf{u}$. Therefore, the estimated error and the corresponding Cramér-Rao bound generally render different values. In fact, the estimated error indicates the expected amount of perturbation of solution, away from the *current* solution, which is caused by the amount of perturbation away from the *actual* observations. If the performance of an algorithm is so poor that the estimated parameters have large errors (e.g., a false local minimum), the matrix J does not well characterize the actual system because it is evaluated with bad parameters. In this case, the estimated error may significantly underestimate the actual error. Therefore, for a nonlinear problem, a correct convergence is important to the reliability of error estimation. On the other hand, the Cramér-Rao bound is independent of actual algorithms and actual values of noise.

4.7 Batch and Sequential Methods

If all the data acquired are processed simultaneously, the processing method is called batch processing. If a new solution is computed after every piece of new data is acquired, and the new solution is computed based on the previous solution and the new data, the method is called sequential (or recursive) processing. Due to popularity of a sequential processing technique called Kalman filtering, sequential processing method has been used for many applications, including for cases where sequential processing is not necessary. Therefore, it is very important to analyze and compare the performances of batch and sequential techniques. This section shows that although both types of technique are mathematically equivalent for linear problems, the performance of the batch techniques is generally better than that of the sequential techniques for nonlinear problems. For the cases where estimates are needed while new data are collected, a *recursive-batch technique* is proposed to improve the performance for nonlinear problems.

As discussed in Section 4.5, for a linear system $\mathbf{y} = A\,\mathbf{m} + \boldsymbol{\delta}_{\mathbf{y}}$, the unbiased linear minimum variance estimator is given by (4.5.3). The dimension of vector $\mathbf{y}$ corresponds to the number of observations. Given n observations, each corresponds to a scalar equation in (4.5.2), equation (4.5.3) gives the estimate based on these n observations. This solution is commonly referred to as "batch solution" because the data are processed simultaneously. A batch approach is not very suitable if observations are collected sequentially and estimates are needed before all observations are obtained, because the old data have to be reprocessed completely and old estimates are discarded. If new estimates are computed from new data and the old estimates, the computation is more efficient. This type of solution is called "sequential solution". However, we will see in the following that for a nonlinear problem, $\mathbf{y} = \mathbf{h}(\mathbf{m}) + \boldsymbol{\delta}_\mathbf{y}$, where $\mathbf{h}(\mathbf{m})$ is a nonlinear function, the batch solution and the sequential solution are quite different.

4.7.1 Linear Systems

In the interest of generality, we first extend the problem to allow the parameters to change, or evolve, when observations are acquired sequentially. Since the parameters are time-varying, the system is called a time-varying or dynamic system and the parameters are sometimes called the state of the system. The formulation for the evolution of the parameters may also have errors. The system dynamic model (or plant model) is formulated by

$$\mathbf{m}_{k+1} = \Phi_k \mathbf{m}_k + \boldsymbol{\eta}_k \tag{4.7.1}$$

and measurement $\mathbf{y}_k$ is determined by the measurement model

$$\mathbf{y}_k = A_k \mathbf{m}_k + \boldsymbol{\varepsilon}_k \tag{4.7.2}$$

$k = 0, 1, 2, \cdots$, where $\mathbf{m}_k$ is the parameter vector (or state) at time k, Φ_k and A_k are matrices, $\boldsymbol{\eta}_k$ and $\boldsymbol{\varepsilon}_k$ are random errors with zero means, $\mathbb{E}\boldsymbol{\eta}_i\boldsymbol{\eta}_j^t = \delta_{ij} Q_i$, $\mathbb{E}\boldsymbol{\varepsilon}_i\boldsymbol{\varepsilon}_j = \delta_{ij} C_i$, and $\boldsymbol{\eta}_i$ and $\boldsymbol{\varepsilon}_j$ are all uncorrelated for $i \geq 0$ and $j \geq 0$. ($\delta_{ij} = 0$ for $i \neq j$ and $\delta_{ii} = 1$ for all i.)

We want to determine the unbiased minimum variance estimate of $\mathbf{m}_k$ after k observations and the covariance matrix of the estimate. We will first show that it can be solved by a batch approach provided Φ_k is invertible. If Φ_k is not invertible, the independent parameters in $\mathbf{m}_{k+1}$ are fewer than that in $\mathbf{m}_k$ for the noise-free system model. That is, the previous parameters cannot be determined from the current parameters based on (4.7.1) even if $\boldsymbol{\eta}_k = \mathbf{0}$. This is usually not the case. (In fact, Φ_k^{-1} can be extended to be any $\mathbf{m}_k$ such that $\mathbf{m}_{k+1} = \Phi_k(\mathbf{m}_k)$. To $\mathbf{m}_{k+1}$, all such $\mathbf{m}_k$ is the same. Since that leads to the same $\mathbf{m}_{k+1}$.) Equation (4.7.1) yields

$$\mathbf{m}_k = \Phi_k^{-1}\mathbf{m}_{k+1} - \Phi_k^{-1}\eta_k \triangleq \Omega_k\mathbf{m}_{k+1} + \mu_k. \tag{4.7.3}$$

Then

$$\mathbb{E}\mu_k\mu_k^t = \Phi_k^{-1}C_k(\Phi_k^{-1})^t.$$

By induction on k it can be proved that any past parameters can be related to the current parameters by

$$\mathbf{m}_{n-k} = (\prod_{i=n-k}^{n-1}\Omega_i)\mathbf{m}_n + \sum_{i=1}^{k}(\prod_{j=n-k}^{n-i-1}\Omega_j)\mu_{n-i}, \qquad k < n, \tag{4.7.4}$$

where $\prod_{i=l}^{n}\Omega_i = I$, if $n < l$. The system measurement model (4.7.2) can be expressed for the current parameter vector $\mathbf{m}_n$ by using (4.7.4):

$$\begin{aligned}\mathbf{y}_{n-k} &= A_{n-k}\mathbf{m}_{n-k} + \varepsilon_{n-k}\\ &= A_{n-k}(\prod_{i=n-k}^{n-1}\Omega_i)\mathbf{m}_n + A_{n-k}\sum_{i=1}^{k}(\prod_{j=n-k}^{n-i-1}\Omega_j)\mu_{n-i} + \varepsilon_{n-k}\\ &\triangleq \Theta_{n-k}\mathbf{m}_n + \mathbf{v}_{n-k}\end{aligned} \tag{4.7.5}$$

where $\{\mathbf{v}_i\}$ are random errors with zero mean and covariance

$$\begin{aligned}&\mathbb{E}\mathbf{v}_{n-k}\mathbf{v}_{n-l}^t\\ &= \mathbb{E}[A_{n-k}\sum_{i=1}^{k}(\prod_{j=n-k}^{n-i-1}\Omega_j)\mu_{n-i} + \varepsilon_{n-k}][A_{n-l}\sum_{i=1}^{l}(\prod_{j=n-l}^{n-i-1}\Omega_j)\mu_{n-i} + \varepsilon_{n-l}]^t\\ &= A_{n-k}[\sum_{i=1}^{\min\{k,\,l\}}(\prod_{j=n-k}^{n-i-1}\Omega_j)\Phi_{n-i}^{-1}C_{n-i-1}(\Phi_{n-i}^{-1})^t(\prod_{j=n-k}^{n-i-1}\Omega_j)^t]A_{n-l}^t\\ &\quad + \delta_{kl}Q_{n-k}\end{aligned}$$

for $0 < k \le n$ and $0 < l \le n$. Equation (4.7.5) expresses each previous observation in terms of the current parameters $\mathbf{m}_n$ with additive noise term $\mathbf{v}_{n-k}$. Therefore, the problem has been converted to the form in (4.5.2). The unbiased linear minimum variance estimator, or equivalently, weighted least squares solution, of the current parameter $\mathbf{m}_n$ and the corresponding covariance matrix can be immediately determined by the Gauss-Markov theorem. This is a batch solution.

The linear time-varying system can be solved more naturally and efficiently by a sequential approach, due to the dynamic nature (or time-varying parameters) of the system (4.7.1). A sequential least-square approach called the Kalman filtering [4.14, 12, 21, 4, 33, 22, 13] is widely used. The following is a version of the Kalman filtering algorithm.

Measurement update:

Kalman gain: $$G_k = P_{k,k-1}A_k^t(A_kP_{k,k-1}A_k^t + C_k)^{-1}, \quad (4.7.6)$$

current estimate: $$\hat{\mathbf{m}}_k = \hat{\mathbf{m}}_{k,k-1} + G_k(\mathbf{y}_k - A_k\hat{\mathbf{m}}_{k,k-1}),$$

current covariance: $$P_k = P_{k,k-1} - G_kA_kP_{k,k-1}. \quad (4.7.7)$$

Time update:

predicted estimate: $$\hat{\mathbf{m}}_{k+1,k} = \Phi_k\hat{\mathbf{m}}_k,$$

predicted covariance: $$P_{k+1,k} = \Phi_kP_k\Phi_k^t + Q_k. \quad (4.7.8)$$

The above computations are done for $k = 0, 1, 2, \cdots n$, initialized by $\hat{\mathbf{m}}_{0,-1} = \mathbf{a}$ and $P_{0,-1} = \mathbb{E}(\hat{\mathbf{m}}_{0,-1} - \mathbf{a})(\hat{\mathbf{m}}_{0,-1} - \mathbf{a})^t$. If a priori information for $\mathbf{m}_0$ is available, $\mathbf{m}_0$ is viewed as a random vector, assuming $\mathbf{m}_0$ is uncorrelated with $\{\eta_k\}$ and $\{\varepsilon_k\}$. Then $\mathbf{a}$ should be the expected $\mathbf{m}_0$: $\mathbf{a} = \mathbb{E}\mathbf{m}_0$. When there is no a priori information about $\mathbf{m}_0$, one can let $\mathbf{a} = \mathbf{0}$ and $P_{0,-1} = \infty I$. If $\mathbf{y}_k$ is an l-dimensional vector, the update is called rank-l update. If rank-one update is used, the matrix inverse in (4.7.6) is just a scalar division.

The Kalman filtering algorithm can be derived by either probabilistic or deterministic methods. Both types methods are unified under Hilbert space optimization [4.18, 13]. By the Kalman algorithm, $\hat{\mathbf{m}}_k$ is the linear minimum variance estimate of $\mathbf{m}_k$ that minimizes $\|\mathbf{m}_k - \hat{\mathbf{m}}_k\|$ based on observations $\mathbf{y}_i$, $0 < i < k$, and a priori information of $\mathbf{m}_0$ (initial conditions). According to the property of minimum-variance linear estimator, the minimum variance estimate $\hat{\mathbf{m}}_k$ also minimizes $\mathbb{E}(\mathbf{m}_k - \hat{\mathbf{m}}_k)P(\mathbf{m}_k - \hat{\mathbf{m}}_k)^t$ for any nonnegative definite weighting matrix P. With $P_{0,-1} = \infty I$, the estimate is the same as that of the batch solution discussed above. The Kalman filtering algorithm can also be derived by solving for $\hat{\mathbf{m}}_i$ and $\hat{\eta}_i$ that minimize

$$(\hat{\mathbf{m}}_0 - \mathbf{a})^tP_{0,-1}^{-1}(\hat{\mathbf{m}}_0 - \mathbf{a})$$

$$+ \sum_{i=0}^{n}(\mathbf{y}_i - A_i\hat{\mathbf{m}}_i)^t C_i^{-1}(\mathbf{y}_i - A_i\hat{\mathbf{m}}_i) + \sum_{i=0}^{n-1}\hat{\boldsymbol{\eta}}_i^t Q_i^{-1}\hat{\boldsymbol{\eta}}_i \tag{4.7.9}$$

subject to the constraint $\mathbf{m}_{k+1} = \Phi_k \mathbf{m}_k + \boldsymbol{\eta}_k$.

Let us see a numerical example computed by the program used in Section 4.8. Let the parameters be time-invariant, $\mathbf{m} = (1, 2, -3, 4, -5)^t$. Then $\Phi_k = I$, $\mathbf{m}_k = \mathbf{m}$, $C_k = \mathbf{0}$ for all k. The elements of A is randomly generated by a pseudo-random variable with uniform distribution in [−5, 5]. Additive noise $\boldsymbol{\varepsilon}_k$ is uniformly distributed in [−0.35, 0.35]. Batch solution is computed after each observation. The sequential solution uses rank-one update, starting with $\hat{\mathbf{m}} = 0$ and $P_{0,-1} = 10^6 I$. Table 4.1 shows the batch estimates, sequential estimates and the square root of the diagonal elements of P_k denoted by batch, Kalman and σ, respectively. Batch solution needs at least 5 observations to determine the unknown $\mathbf{m}$. It can be seen that batch and sequential solution are virtually identical after 5 observations are obtained, which is expected. Since zero initial guess are provided for the Kalman filter, the corresponding estimates are roughly the minimum-norm least-squares solutions before 5 observations are obtained. The estimated error (σ) is close to the actual errors.

In summary, the batch solution and the sequential solution are the same for linear systems without a priori knowledge of the parameters. The batch solution requires the inversion of a symmetrical matrix. Kalman filter equations provide a very simple procedure for computer implementation, and require very few computations for each observation. Therefore, for a linear problem, the sequential Kalman filtering approach has computational advantages, especially for time-varying parameters.

4.7.2 Nonlinear Systems

However, like most real world problems, the problem of estimating motion parameters and structure of the scene is nonlinear. A general nonlinear system with time-invariant parameters can be expressed as

$$\mathbf{f}(\mathbf{h}, \mathbf{m}) = \mathbf{0}$$

where $\mathbf{h}$ is a noise-free observation vector and $\mathbf{m}$ is the parameter to be estimated. With actual observation $\mathbf{h}^*$ and estimate $\mathbf{m}^*$, expending $\mathbf{f}(\mathbf{h}, \mathbf{m})$ at $(\mathbf{h}^*, \mathbf{m}^*)$ yields

Table 4.1. Batch and sequential (Kalman) solutions for a linear system

k	method	m_1	m_2	m_3	m_4	m_5
	batch	0.000000	0.000000	0.000000	0.000000	0.000000
0	Kalman	0.602833	-0.465899	0.483534	0.621420	-0.101110
	σ	836.556635	905.929243	898.263943	825.246345	995.768827
	batch	0.000000	0.000000	0.000000	0.000000	0.000000
1	Kalman	-0.192245	-1.178379	-0.705888	1.850552	0.307604
	σ	741.932844	837.118217	687.466976	569.273132	975.748945
	batch	0.000000	0.000000	0.000000	0.000000	0.000000
2	Kalman	1.275420	1.397069	-3.359516	3.536210	-5.139505
	σ	711.270108	750.703640	571.789527	515.066029	581.637859
	batch	0.000000	0.000000	0.000000	0.000000	0.000000
3	Kalman	1.437180	1.782845	-3.386216	3.738357	-4.837958
	σ	648.698646	282.086846	569.758585	363.871322	206.373193
	batch	0.944562	1.824841	-2.701296	4.039422	-4.891733
3	Kalman	0.944562	1.824841	-2.701296	4.039422	-4.891733
	σ	0.036618	0.169134	0.288711	0.042517	0.083798
	batch	0.968310	1.966517	-2.945285	4.005723	-4.958069
4	Kalman	0.968310	1.966517	-2.945285	4.005723	-4.958069
	σ	0.023878	0.034287	0.044706	0.015993	0.031754
	batch	0.971435	1.968931	-2.947189	4.000380	-4.955265
5	Kalman	0.971435	1.968931	-2.947189	4.000380	-4.955265
	σ	0.023441	0.034106	0.044620	0.013975	0.031491
	batch	0.971990	1.971881	-2.951717	4.000719	-4.957658
6	Kalman	0.971990	1.971881	-2.951717	4.000719	-4.957658
	σ	0.023114	0.027070	0.031251	0.013770	0.026615
	batch	0.984279	2.009192	-3.008340	3.990893	-4.977205
7	Kalman	0.984279	2.009192	-3.008340	3.990893	-4.977205
	σ	0.011978	0.012339	0.029373	0.015618	0.015011

$$\mathbf{f}(\mathbf{h}^*, \mathbf{m}^*) + \frac{\partial \mathbf{f}(\mathbf{h}^*, \mathbf{m}^*)}{\partial \mathbf{h}} (\mathbf{h} - \mathbf{h}^*) + \frac{\partial \mathbf{f}(\mathbf{h}^*, \mathbf{m}^*)}{\partial \mathbf{m}} (\mathbf{m} - \mathbf{m}^*) \approx \mathbf{0}. \tag{4.7.10}$$

Letting

$$J = \frac{\partial \mathbf{f}(\mathbf{h}^*, \mathbf{m}^*)}{\partial \mathbf{m}}, \tag{4.7.11}$$

$$G = \frac{\partial \mathbf{f}(\mathbf{h}^*, \mathbf{m}^*)}{\partial \mathbf{h}},$$

$\mathbf{y} = -\mathbf{f}(\mathbf{h}^*, \mathbf{m}^*)$ and $\boldsymbol{\varepsilon} = -G(\mathbf{h} - \mathbf{h}^*)$, neglecting higher order terms, (4.7.10) becomes

$$\mathbf{y} = J(\mathbf{m} - \mathbf{m}^*) + \boldsymbol{\varepsilon}. \tag{4.7.12}$$

$\mathbf{h} - \mathbf{h}^*$ gives the difference between the noise-free observation $\mathbf{h}$ and actual observation $\mathbf{h}^*$. Therefore $\mathbf{h} - \mathbf{h}^*$ corresponds to observation noise. Letting $\mathbb{E}(\mathbf{h} - \mathbf{h}^*)(\mathbf{h} - \mathbf{h}^*)^t = Q_{\mathbf{h}}$ gives

$$\mathbb{E}\varepsilon\varepsilon^t = GQ_{\mathbf{h}}G^t.$$

By the Gauss-Markov Theorem, obtaining the linear minimum variance estimate of $(\mathbf{m} - \mathbf{m}^*)$ in (4.7.12) yields

$$\hat{\mathbf{m}} = \mathbf{m}^* + (J^t(GQ_{\mathbf{h}}G^t)^{-1}J)^{-1}(GQ_{\mathbf{h}}G^t)^{-1}J^t\mathbf{y}. \tag{4.7.13}$$

This is a generalized version of (4.4.6) for the case where noise is not white. Since the problem is nonlinear, J is evaluated at the estimated parameter $\mathbf{m}^*$. A sequence of approximations of $\mathbf{m}$ is obtained by iterations using (4.7.13), after each iteration replacing $\mathbf{m}^*$ by $\hat{\mathbf{m}}$. After the iterations converge, the estimate $\hat{\mathbf{m}}$ is an approximated minimum variance estimate of $\mathbf{m}$, or equivalently weighted nonlinear least-squares approximate solution that minimizes

$$\mathbf{f}(\hat{\mathbf{m}})^t(GQ_{\mathbf{h}}G^t)^{-1}\mathbf{f}(\hat{\mathbf{m}}).$$

This is a batch solution for time-invariant nonlinear systems.

For time-varying parameters, similar techniques in Subsection 4.7.1 can be used to express previous observations in terms of the current parameter (see (4.7.5)). The system model (4.7.1) should be replaced by

$$\mathbf{m}_{k+1} = \Phi_k(\mathbf{m}_k) + \boldsymbol{\eta}_k \tag{4.7.14}$$

where $\Phi_k(\mathbf{m}_k)$ is a nonlinear function. If Φ_k is invertible, it follows that

$$\mathbf{m}_k = \Phi_k^{-1}(\mathbf{m}_{k+1} - \boldsymbol{\eta}_k) \approx \Phi_k^{-1}(\mathbf{m}_{k+1}) - \nabla\Phi_k^{-1}(\mathbf{m}_{k+1})\,\boldsymbol{\eta}_k.$$

This equation plays the role of (4.7.3). Iterations can be performed to obtain approximated minimum variance estimate of the current parameters $\mathbf{m}_k$. Therefore, the batch approach can also be applied to nonlinear system with time-varying parameters.

For nonlinear Kalman filtering, conventionally a nonlinear measurement model is assumed:

$$\mathbf{y}_k = \mathbf{h}_k(\mathbf{m}_k) + \boldsymbol{\varepsilon}_k \tag{4.7.15}$$

where $\mathbf{h}_k(\mathbf{m})$ is a nonlinear function. Equation (4.7.15) is a special case of (4.7.9) in which the observation can be explicitly expressed as a function of parameters and measurement errors. From (4.7.15) it follows that

$$\mathbf{y}_k \approx \mathbf{h}_k(\hat{\mathbf{m}}_{k,k-1}) + \frac{\partial \mathbf{h}_k(\hat{\mathbf{m}}_{k,k-1})}{\partial \mathbf{m}}(\mathbf{m}_k - \hat{\mathbf{m}}_{k,k-1}) + \boldsymbol{\varepsilon}_k .$$

Neglecting higher order terms gives

$$\mathbf{y}_k - \mathbf{h}_k(\hat{\mathbf{m}}_{k,k-1}) + \frac{\partial \mathbf{h}_k(\hat{\mathbf{m}}_{k,k-1})}{\partial \mathbf{m}}\hat{\mathbf{m}}_{k,k-1} = \frac{\partial \mathbf{h}_k(\hat{\mathbf{m}}_{k,k-1})}{\partial \mathbf{m}}\mathbf{m}_k + \boldsymbol{\varepsilon}_k . \tag{4.7.16}$$

Each scalar equation (or a subset of scalar equations) of (4.7.16) gives an equation similar to (4.7.2). That is, if the left-hand side is consider as ''observation'', the Kalman algorithm can be used to compute the estimate of $\mathbf{m}_k$ sequentially. Such an trivial extension of Kalman algorithm is called extended Kalman filter (EKF) [4.4, 33, 22].

Unlike the case of linear system, for nonlinear system the batch solution and sequential solution are not the same. J in (4.7.12) for a nonlinear problem corresponds to A in (4.5.2) for a linear problem. While A is constant, J is not. J defined in (4.7.11) is a function of observation $\mathbf{h}^*$ and current estimate $\mathbf{m}^*$.

Now, we investigate the sequential solutions. Due to the sequential nature of the Kalman solution, the estimate of $\mathbf{m}_k$ is based on the estimate of $\mathbf{m}_{k-1}$, its covariance matrix and a new observation. Once the previous parameter $\mathbf{m}_{k-1}$ is estimated, it is not improved after new observations $\{\mathbf{y}_i\}$, $i > k$, are obtained. Let us consider the nonlinear system of (4.7.15). The objective function (4.7.9) for Kalman filter to minimize is replaced by the corresponding nonlinear version:

$$(\hat{\mathbf{m}}_0 - \mathbf{a})^t P_{0,-1}^{-1} (\hat{\mathbf{m}}_0 - \mathbf{a}) +$$

$$\sum_{i=0}^{n}(\mathbf{y}_i - J_i \hat{\mathbf{m}}_i)^t C_i^{-1}(\mathbf{y}_i - J_i \hat{\mathbf{m}}_i) + \sum_{i=0}^{n-1} \hat{\boldsymbol{\eta}}_i^t Q_i^{-1} \hat{\boldsymbol{\eta}}_i \tag{4.7.17}$$

subject to $\mathbf{m}_{k+1} = \Phi_k(\hat{\mathbf{m}}_k) + \boldsymbol{\eta}_k$, and

$$J_i = \frac{\partial \mathbf{h}_i(\hat{\mathbf{m}}_i^{(i)})}{\partial \mathbf{m}}$$

where $\hat{\mathbf{m}}_i^{(i)}$ is the estimated $\mathbf{m}_i$ based on the observations $\{\mathbf{y}_k\}$, $0 \le k \le i$. To see what (4.7.17) implies more clearly, assume the parameter is time-invariant ($\Phi_k(\hat{\mathbf{m}}_k) = I$, $\eta_k = \mathbf{0}$), $C_i = I$ and $P_{0,-1} = \mathbf{0}$. Equation (4.7.17) becomes

$$\sum_{i=0}^{n} \|\mathbf{y}_i - J_i \hat{\mathbf{m}}\|^2 = \sum_{i=0}^{n} \|\mathbf{y}_i - \frac{\partial \mathbf{h}_i(\hat{\mathbf{m}}^{(i)})}{\partial \mathbf{m}} \hat{\mathbf{m}}\|^2. \tag{4.7.18}$$

To improve the performance for nonlinear systems, the Kalman filtering algorithm needs iterations (called iterated extended Kalman filtering (IEKF) algorithm). For each observation $\mathbf{y}_i$: After $\mathbf{m}$ is estimated from $\mathbf{y}_i$, it is used to evaluate J_i. $\mathbf{y}_i$ and the improved J_i in turn give a new estimate of $\mathbf{m}$. Such iteration is carried on until no improvement for $\mathbf{m}$ occurs. Then $\hat{\mathbf{m}}^{(i)}$ is determined. For small i, $\hat{\mathbf{m}}^{(i)}$ has a large error since just $i+1$ observations are available. Therefore, J_i evaluated at $\hat{\mathbf{m}}^{(i)}$ gives a system matrix that is evaluated far from the true parameter. This results in inaccurate system equations. Once those inaccurate system equations are generated, they will not be updated later when new observations are collected. They are included in the objective function (4.7.18), further preventing the estimated parameter $\mathbf{m}$ from approaching the correct parameters while new data are obtained. The more nonlinear $\mathbf{h}(\mathbf{u}, \mathbf{m})$ is in terms of $\mathbf{m}$, the worse J_i is. Therefore, sequential methods generally are not suitable to be used to solve a nonlinear equation from an arbitrary initial guess for $\mathbf{m}$.

The performance of the IEKF algorithm depends on the initial guess solution $\mathbf{a}$ and the initial covariance matrix $P_{0,-1}$. Suppose the initial guess of $\mathbf{m}$ is quite close to the true solution, J_i is evaluated at a good estimate of $\mathbf{m}$ for small i. Then, J_i fairly well estimates the system matrix. However, if the initial covariance matrix $P_{0,-1}$ is large, a few noise-contaminated observations may incorrectly change the initial guess by a large amount. As a consequence, the following J_i's are evaluated at the bad parameters. For the following reason, it is very difficult for the IEKF to gradually pull those bad parameters back using new observations. The error covariance matrix P_k and $P_{k+1,k}$ are correct only for linear models. In fact, they are independent of the actual observations for a linear models. For nonlinear models, since J_k's are evaluated with estimated parameters, P_k and $P_{k+1,k}$ are very inaccurate for early observations. After certain number of observations, P_k and $P_{k+1,k}$ significantly underestimate the corresponding errors due to nonlinearity of the system. This makes Kalman gain very small. The filter "closes eyes" and almost does not improve the parameters with new observations. Therefore the Kalman filter requires a huge number of observations for a very little improvement for the parameters. The errors in the estimated parameters remain large even after a large number of observations are obtained. Table 4.2 lists the effects of initial condition $\hat{\mathbf{m}}_{0,-1}$ and the associated covariance matrix $P_{0,-1}$ on the estimated parameter $\hat{\mathbf{m}}_n$ for a nonlinear time-varying system in general. In all the cases, the covariance matrix P_n is always small. That is, P_n may significantly under-estimate the errors of the estimated parameters.

The batch method treats J_i differently. For the model discussed above, the objective of the batch solution is to minimize

$$\sum_{i=0}^{n} \|\mathbf{y}_i - J_i \hat{\mathbf{m}}\|^2 = \sum_{i=0}^{n} \|\mathbf{y}_i - \frac{\partial \mathbf{h}_i(\hat{\mathbf{m}})}{\partial \mathbf{m}} \hat{\mathbf{m}}\|^2 \tag{4.7.19}$$

through iterative improvement. In each iteration, the estimated parameters are modified based on all observations (unlike the sequential algorithm which modifies the parameters with each observation). Therefore, such a modification is more reliable. Further, J_i is always updated through iterations. As a local minimum is reached, all J_i's are evaluated at this final solution.

For Kalman filtering, divergence is said to occur when the error covariance matrix computed by the filter becomes unjustifiably small compared with the actual error in the estimate. Divergence of Kalman filter has been observed for nonlinear systems and it has been attributed to system nonlinearity [4.32, 11, 4, 22, 34, 6]. Although it is

Table 4.2. Characteristics of IEKF for nonlinear systems

$\hat{\mathbf{m}}_{0,-1}$	$P_{0,-1}$	$\hat{\mathbf{m}}_n$	P_k
bad	large	divergent	small
bad	small	not improved	small
good	large	divergent	small
good	small	improved	small

known that states and the associated covariance are no longer sufficient statistics in a nonlinear system with Gaussian noise, the underlying problems of Kalman filtering for nonlinear systems, with either Gaussian or non-Gaussian noise, have not been fully analyzed. As a consequence, extended Kalman filtering has been widely used as a general tool to solve nonlinear equations and results in poor performance. Our analysis is not based on Gaussian noise and is applicable to any noise distribution. We have analyze the reasons for the poorer performance, and, in particular, Equations (4.7.18) and (4.7.19) display a fundamental difference between the sequential and batch solutions. They have revealed alarming degradation from a batch method to a sequential method for a nonlinear system, no matter whether the noise is Gaussian or not.

In summary, the divergence and poorer performance of Kalman filtering, compared with batch approaches, are consequences of (a) fluctuations of parameters when updating using early observations cause system matrix J_i to be evaluated at bad parameters, even if the initial parameters are good; (b) system matrix J_i is not updated using newer observations $\mathbf{y}_k$, $k > i$. The performance difference between a sequential and a batch approach is especially large for a highly nonlinear systems with a small or moderate number of observations.

4.7.3 Recursive-Batch Solution

The above analysis leads to the conclusion that, for nonlinear systems, a batch approach will out-perform a sequential approach in general. Therefore, whenever possible, batch techniques should be used to achieve best performance. However, if estimates are required while data are collected, or the number of observations is extremely large, batch methods are computationally expensive since repeated iterations cover all the data. Because the parameters may be well estimated by a batch method using a relatively large number of observations, very old observations contribute very little to the current estimates. Therefore, we introduce a technique called recursive-batch method. By this technique, current estimate $\mathbf{m}_k$ is computed by a batch method based on most recent l observations, the previous estimate $\mathbf{m}_{k-l}$ and its uncertainty P_{k-l}. The method is recursive since the new batch estimates are computed from previous batch estimates. The number l should be sufficiently large such that performance differences between the sequential and batch methods are negligible for those observations older than $\mathbf{y}_{k-l+1}$. With an infinitely long or very long observation sequence and an appropriately chosen l, the recursive-batch solution will perform significantly better than a sequential solution and the computational efficiency will be superior to a straight batch approach.

To formulate the recursive-batch solution in more detail, let us consider the system dynamic equations (4.7.14) and (4.7.15). First consider estimating $\mathbf{m}_k$, for $k < l$. Since no more than l observations are available, batch techniques are used based on all observations $\{\mathbf{y}_i\}$, $0 \le i < k$. For each batch solution, start with an initial estimate for $\mathbf{m}_k$ which is inferred from the estimate of $\hat{\mathbf{m}}_{k-1}$ based on (4.7.14) by

$$\hat{\mathbf{m}}_k = \Phi_{k-1}(\hat{\mathbf{m}}_{k-1}).$$

The batch solution is then computed using iterative improvement from the initial guess. In fact, the Gauss-Markov and the Kalman methods both can be used to compute batch solution. To use the former, relate observations with $\mathbf{m}_k$ using (4.7.5). Then the Gauss-Markov theorem is used to determine the iteration of $\mathbf{m}_k$. For time-invariant parameters, the observations are already related to the constant parameters. Therefore, the Gauss-Markov method is simple. For time-varying parameters, the computation of (4.7.13) is relatively complicated. Since the Kalman algorithm gives the same solution for a linear problem, Kalman equations can be used to compute a batch solution provided that J_i is evaluated using the fixed parameters within each batch iteration, (The conventional equations of Kalman filtering can lead to erroneous results due to finite wordlength in computer implementation. The numerical stability can be improved by using square root Kalman filtering algorithms [4.4, 21, 13].) To use Kalman equations for the batch solutions, we need to start at $\mathbf{m}_{0,-1}$. Initialize $\mathbf{m}_{0,-1}$ from (4.7.14) by

$$\mathbf{m}_{0,-1} = \Phi_0^{-1}(\Phi_1^{-1}(\cdots \Phi_{k-2}^{-1}(\hat{\mathbf{m}}_{k-1}) \cdots)).$$

The initial covariance matrix is assigned with cI where c is a large positive constant. Use the extended Kalman filtering equations to get an estimate of $\mathbf{m}_k$. $\mathbf{m}_k$ is used to provide another initial guess for $\mathbf{m}_{0,-1}$ with a large initial covariance cI. Such iterations are performed until no improvement on $\mathbf{m}_k$ occur. To prevent the effect of (b) mentioned above J_k is not evaluated at the current estimated $\mathbf{m}_k$. Instead, it is evaluated at $\tilde{\mathbf{m}}_k$ witch is computed from $\mathbf{m}_{0,-1}$ by

$$\tilde{\mathbf{m}}_k \triangleq \Phi_{k-1}(\cdots \Phi_0(\mathbf{m}_{0,-1}) \cdots).$$

For estimating $\mathbf{m}_k$, $k \geq l$, the computational steps are very similar, except that only observations $\mathbf{y}_i$, $k-l+1 \leq i \leq k$ are used for batch solutions. Initial guess for $\mathbf{m}_{k-l+1}$ is determined from the estimate of $\mathbf{m}_{k-l}$ based on (4.7.14). The initial covariance matrix associated with the initial guess for $\mathbf{m}_{k-l+1}$ is determined from the covariance matrix of $\mathbf{m}_{k-l}$ by (4.7.7) and (4.7.8). As before, both Gauss-Markov or Kalman filtering equations can be used to compute the batch solution. In Chapter 8, we will return to recursive-batch method and show an example of application.

4.8 Numerical Examples

To further verify the analyses presented above and demonstrate the performances of the algorithms, simulations have been conducted for estimating motion and structure parameters from point correspondences through two monocular images. The algorithms have also been tested on images of real world scenes.

In the simulations, the focal length is one unit. The image frame forms a $s \times s$ square. Unless stated otherwise, $s = 0.70$, which is roughly equivalent to a 50mm normal lens of a 35mm camera, and 12 point correspondences are used. The object points are generated randomly and uniformly between depths 6 and 16. Only those points that are visible before and after motion are used. The image coordinates of the points are digitized according to the resolution of the camera. This digitization errors result in the errors in the motion parameters and the relative depths calculated by the algorithm. The linear algorithm presented in Chapter 3 was used to generate initial guess solutions.

4.8.1 Whether a Good Initial Guess Is Necessary

First, to investigate whether the initial guess provided by the linear algorithm indeed helps the second optimization step, a fixed guess is used as the initial guess. The subroutine ZXSSQ in IMSL library is used to find motion parameters by minimize the image errors starting with the fixed guess. The fixed initial guess has a zero rotation angle and a translation vector of (1, 1, 1). The image resolution is 256 by

256 and the sign reversal for the translation and rotation angle is performed when the iteration does not converge. Different motion parameters are selected randomly. Among 36 examples with 12 point correspondences, 16 of them do not converge, or converge to a wrong answer. A wrong answer means that the error of translation is larger than 100% or the error of rotation matrix is larger than 50%. Remaining cases give correct solutions with the translation error less than 10% and the rotation error less than 5%. This shows that a good initial guess is necessary to ensure a correct solution.

If the noise is moderate (resolution is 128×128 or higher), the initial guess provided by the linear algorithm in the first step is generally good enough to ensure correct convergence in the second step. When the noise is large, for example the resolution is 64×64 or lower, the initial guess given by the linear algorithm is severely contaminated by noise and the iteration may not converge. If this happens, the initial guess should be modified so that the iteration skips to another parameter region. One simple way of modification is just changing the sign of translation direction or rotation angle. A more reliable way is searching for an initial guess on a coarse to fine grid in the parameter space. For example, the rotation angle can be set to zero (since the rotation is not very large) and for the direction of translation, the parameters ϕ and ψ in (4.D.1) are searched starting with a very coarse grid (say, 2×2). When the corresponding motion parameters render all positive depths and small image errors, the search for an initial guess on the grid terminates and the motion parameters are improved through iteration. Otherwise, the search grid for an initial guess is refined.

4.8.2 Epipolar Improvement versus Optimization

As we discussed in Subsection 4.2.3, the linear algorithm does not use the constraint in E when solving for E based on a set of linear equations (4.2.5). After the solution is obtained by the linear algorithm, the solution can be improved by using the constraint on E to minimize the weighted sum in (4.2.7). This is called the epipolar improvement. The solution is obtained by the Levenberg-Marquardt method for nonlinear least-squares problems. The errors after the epipolar improvement are compared with those of minimizing the image errors (using also the Levenberg-Marquardt method). For longitudinal translation (parallel to optical axis), both methods give very similar errors, since the solutions of the linear algorithm are already fairly good.

The performance shows difference for unstable lateral translations (parallel to image plane). Fig. 4.11 shows typical average errors over 100 random trials. For the simulations, the magnitude of translation varies according to the image size such that the scene will be visible in both images. Fig. 4.11 shows that the epipolar improvement over the linear algorithm is significant. As shown in Fig. 4.11, the errors after minimizing the image errors are significantly smaller than those after the epipolar improvement, especially for small field of view. As we discussed in Subsection 4.2.4 that the epipolar constraint is weak especially for small field of view. Therefore, the use of two components in image point coordinates does significantly improve the accuracy.

4.8.3 Sequential versus Batch Solutions

The problem of minimizing the image errors in (4.3.8) is a nonlinear problem. We compute batch solutions using the Levenberg-Marquardt method. The sequential solution is obtained by iterated extended Kalman filter (IEKF). Iterations are performed for each point correspondence (4 observations, i.e., rank-four update) to improve the performance of the regular IEKF (which just iterates on one observation): After 4 observations, u, v, u', v', are used to update the motion

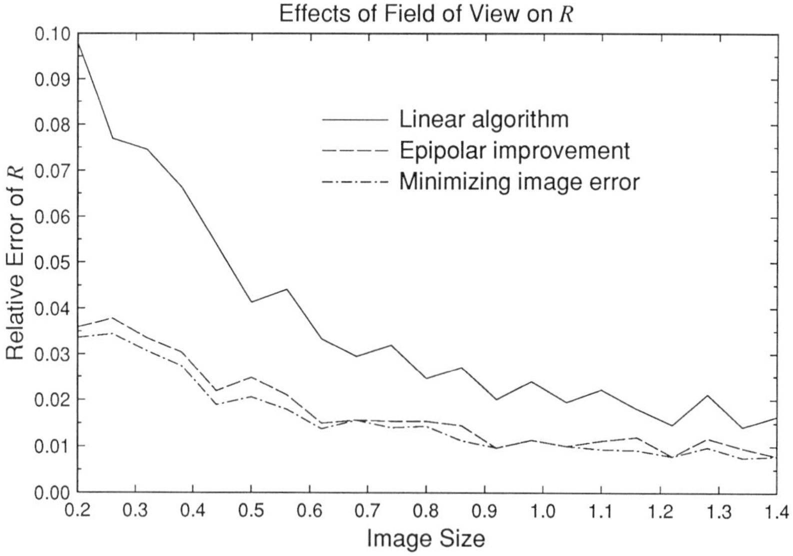

Fig. 4.11(a)

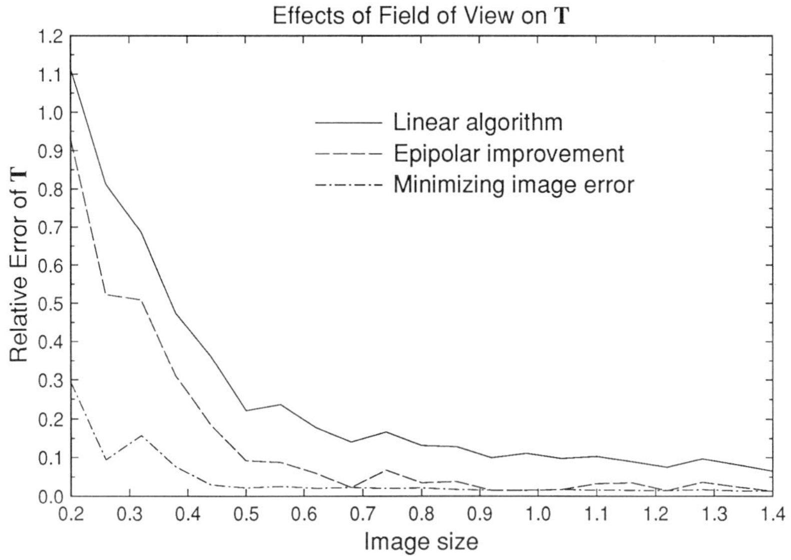

Fig. 4.11(b)

Fig. 4.11. Epipolar improvement versus minimizing the image errors. Rotation axis: (1, 1, 1). Rotation angle: 1°. Translation: $(t_x, 0, 0)$, where t_x varies linearly from 0.1 (for image size 0.2) to 0.7 (for image size 1.4). 100 random trials. **(a)** Relative errors of rotation matrix R. **(b)** Relative errors of translation vector $\mathbf{T}$.

parameters, J_i is then re-evaluated at the updated motion parameters. The 4 observations are again used to update the motion parameters based on the updated J_i. Such iterations are performed for each point correspondence until no improvement for the parameters occurs.

In the simulations, the IEKF always diverges if the algorithm starts with a zero rotation angle, an arbitrary translation direction and a large covariance matrix. The solutions given by the IEKF have more than 100% relative errors. This is expected based on our analysis. With a good initial guess provided by the linear algorithm, the solutions of the IEKF vary significantly with the initial covariance matrix $P_{0,-1}$. If the diagonal matrix $P_{0,-1}$ are assigned with large diagonal values, the IEKF often diverges. Only with small diagonal values, the IEKF can relatively reliably improve the initial guesses. Fig. 4.12 shows typical sequences of IEKF sequential updating. With 12 point correspondences, totally 12×4 = 48 observations are obtained. In Fig. 4.12(a), with a very good initial guess and a relatively small diagonal covariance matrix $P_{0,-1}$, the IEKF improves the initial guess. The actual relative errors slowly approach the expected errors indicated by P_k. In Fig. 4.12(b) a large diagonal covariance matrix $P_{0,-1}$

is used. The final errors are significantly larger than those in Fig. 4.12(a) because the Kalman filter updates the estimates without much a priori information due to a large initial covariance. A few early noise-contaminated observations cause premature updating of the estimated parameters, which causes J_k to be evaluated at deteriorated parameters. In Fig. 4.12(c), the IEKF diverges from a zero initial guess and a large diagonal initial covariance $P_{0,-1}$ (if a small $P_{0,-1}$ is used, the errors will remain almost the same as that of the initial guess).

It is interesting to observe the expected errors determined from P_k. With a good initial guess, the system is linearized at the good estimate of the solution. Therefore, the expected errors are consistent with the nature of the problem: Since at least 5 point correspondences are required to determine the motion parameters, the expected error should not be small until at least the 5th point is observed (starting at 17th observation). Fig. 4.12(a) shows that the expected errors reduce significantly at the 17th observation (from a good initial guess)!

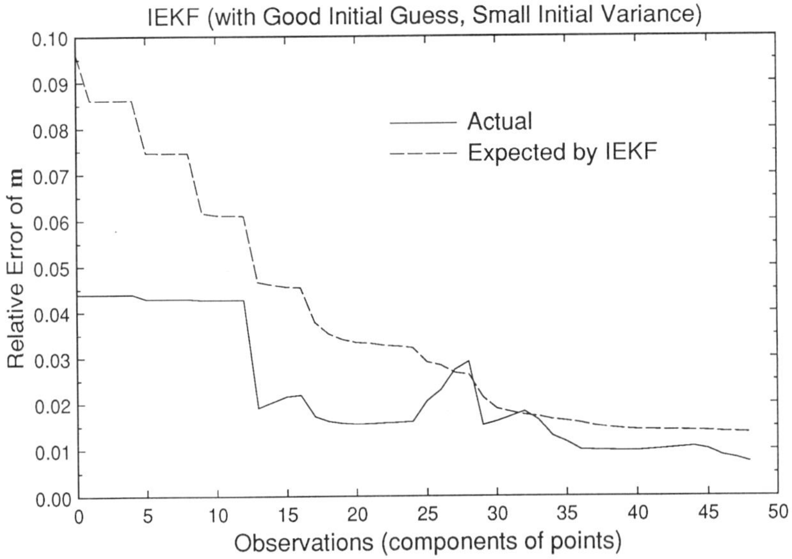

Fig. 4.12(a)

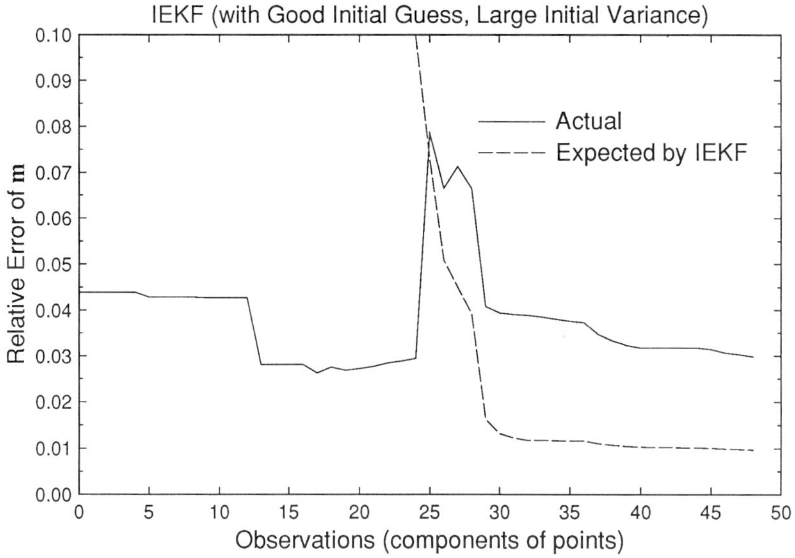

Fig. 4.12(b)

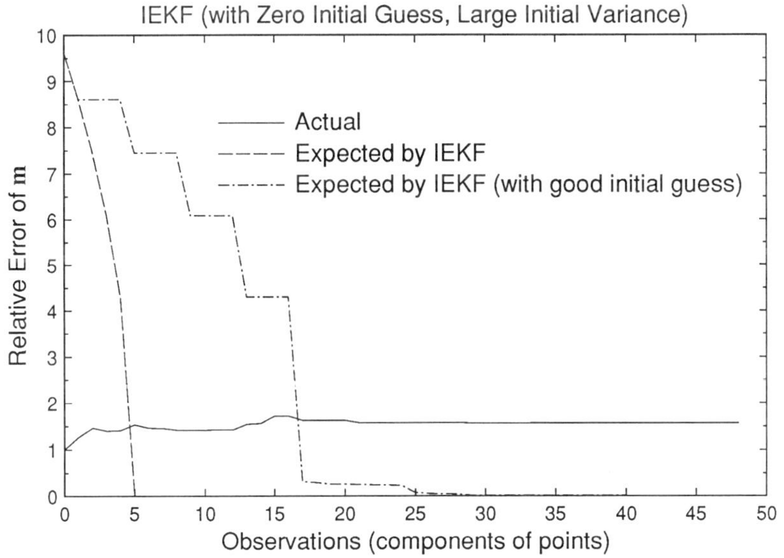

Fig. 4.12(c)

Fig. 4.12. IEKF for the nonlinear problem: a sample sequence. "Expected by IEKF": error predicted by P_k. **(a)** With good initial guesses and a small diagonal $P_{0,-1}$. **(b)** With the same good initial guesses as in (a) and a large $P_{0,-1}$. (The covariance of IEKF with less than 25 observations is shown in (c) under "Expected by IEKF with good initial guess"). **(c)** With zero initial guesses and a large $P_{0,-1}$. Rotation axis (1, 1, 1), angle 3°. Translation: (1.73, 1.73, −1.73).

However, from a zero initial guess, the system matrix is very inaccurate, and, so, it does not well reflect the actual system behavior. As indicated in Fig. 4.12(c), the expected error becomes very small after the 5th observation. In our problem, there are 5 independent motion parameters. Each point correspondence gives 4 observations (u, v, u', v'). 5 observations can be provided by only 2 point correspondences. However, 2 point correspondences cannot determine the motion at all. The expected covariance matrix gives incorrect information after the 5th observation, as if this is a linear system! The filter captures the behavior of the system through local linearization. The local linearization is effective only if the location of the linearization is good. If the location is very bad (e.g., near zero initial guess) the linearized system cannot characterize the behavior of the nonlinear system. That is why in in Fig. 4.12(c), with a zero initial guess and large diagonal initial $P_{0,-1}$, the Kalman filter acts like a linear filter for a linear system and completely fails. This example confirms that Kalman filter does not perform well for a nonlinear problem with an arbitrary initial guess, and the corresponding error variance given by the Kalman algorithm is misleading. In general, Kalman filter is not suitable to be used as a tool to find a solution for a nonlinear problem unless a good initial guess is available and a proper initial covariance is used.

In the following, we compare the performance of the Kalman filter with that of a batch solution. The performances of the linear algorithm, a sequential algorithm (IEKF) and a batch algorithm (L-M method) are compared in Fig. 4.13. Both IEKF and L-M method use the solutions of the linear algorithm as initial guesses. The sequential algorithm improves over linear algorithm for most of cases. However, for some cases the results are worse than the initial guesses (see some average errors in Fig. 4.13). The initial covariance matrix $P_{0,-1}$ of IEKF is carefully selected (same as that in Fig. 4.12(a)). A larger $P_{0,-1}$ will result in more divergent cases and a smaller one will impede parameter improvement. If the initial covariance is even smaller for the IEKF, the degradation from initial guesses will be smaller, but the improvement will also be smaller. The image errors in Fig. 4.13(d) show that the batch solution find the minimum very consistently, but the IEKF does not. We can see that the batch optimization significantly improves the solution of the linear algorithm and performs significantly better than the IEKF sequential algorithm.

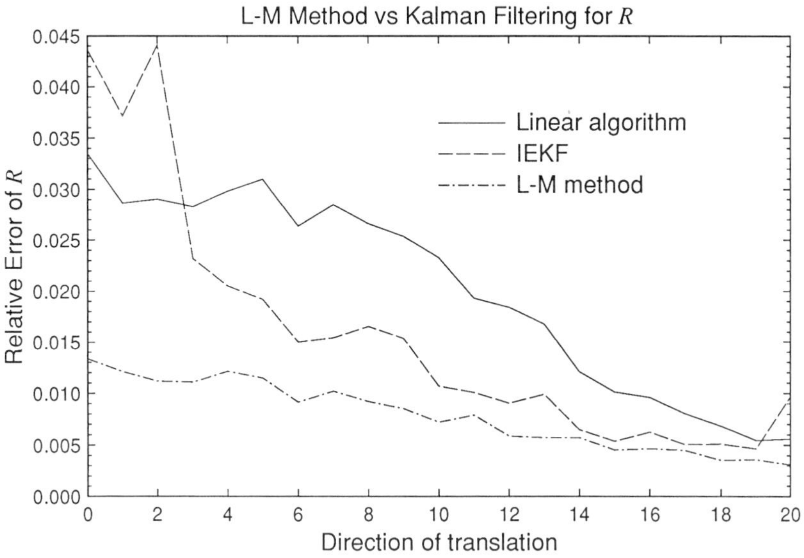

Fig. 4.13(a)

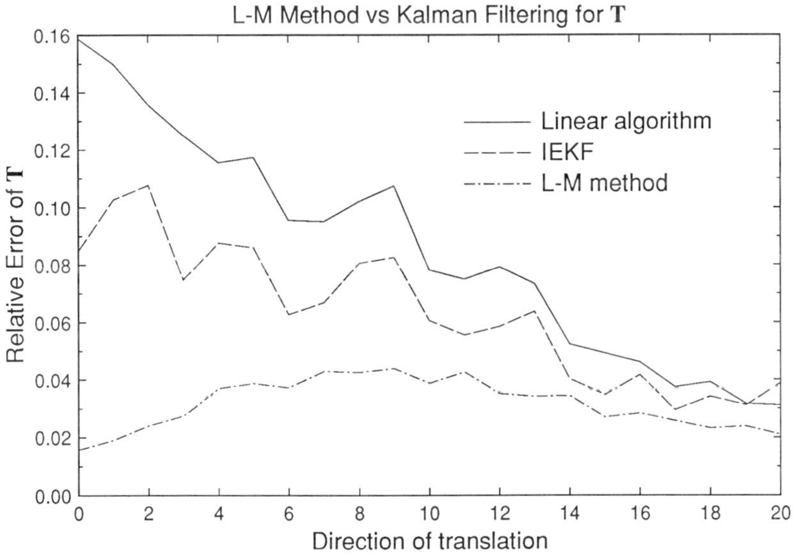

Fig. 4.13(b)

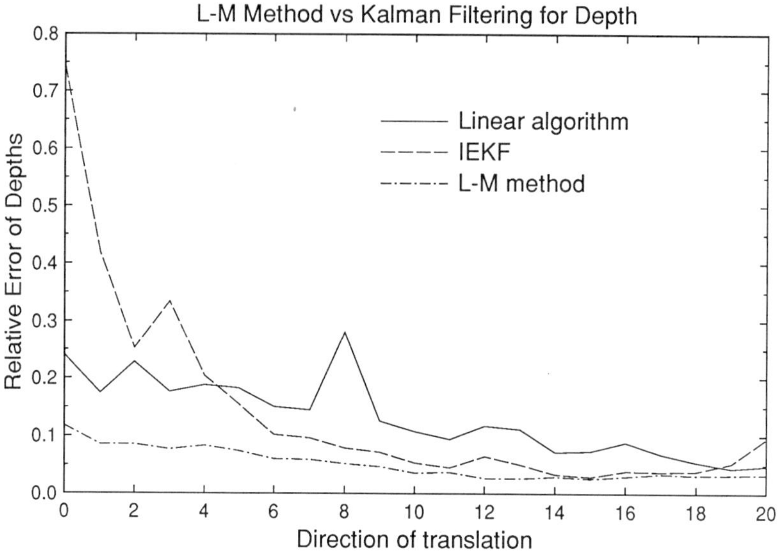

Fig. 4.13(c)

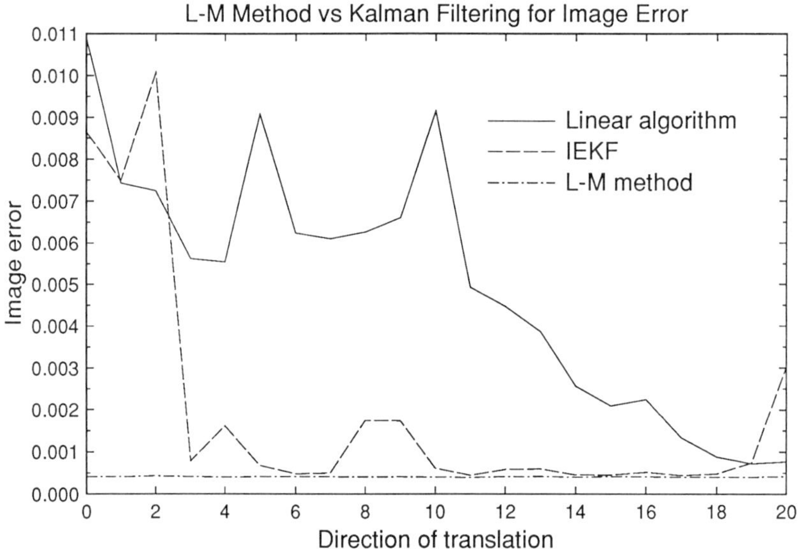

Fig. 4.13(d)

Fig. 4.13. Relative Errors of the linear algorithm, batch solution (the Levenberg-Marquardt method) and sequential solution (IEKF). Rotation axis: (1, 1, 1). Rotation angle: 3°. For horizontal index from 0 to 20, the direction of translation changes from (1, 0, 0) to (0, 0, 1) in xz-plane at evenly spaced 21 steps. The length of the translation vector is equal to 2.1 units. 100 random trials. **(a)** R; **(b)** **T**; **(c)** depths; **(d)** image errors.

To show the effects of iterations on the performance of the IEKF, experiments were conducted with different number of iterations for the IEKF. Fig. 4.14 shows an example of average relative error using different numbers of iterations. It can be seen that the iterations do significantly improve the performance of extended Kalman filter ("not iterated" in Fig. 4.14 means that no iterations are performed).

The improvement of the batch optimization over the linear algorithm is shown in Fig. 4.15 for different image resolutions. It can be seen that the average image errors are always about equal to the average errors of in the image coordinates. This is true even for extremely low resolution, 32×32. This implies that global minimum solutions are reliably obtained.

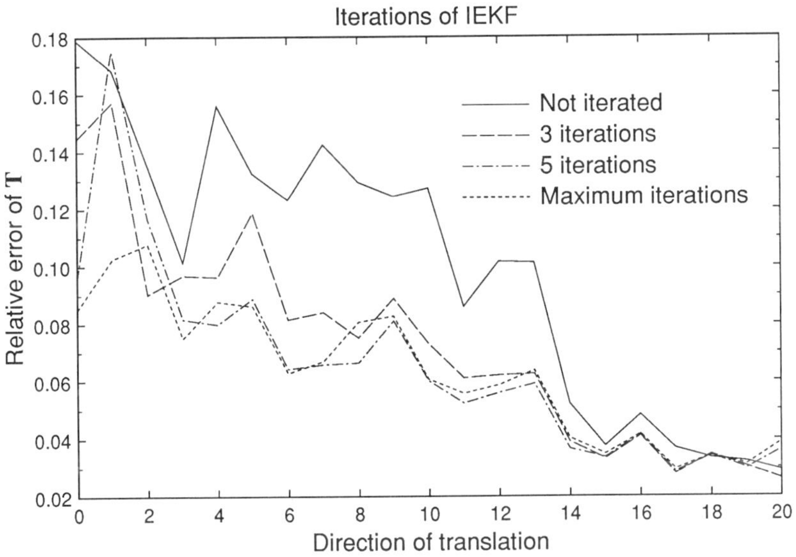

Fig. 4.14. Effects of iterations in Kalman filtering. Same motion as in Fig. 4.13. For horizontal index from 0 to 20, the direction of translation changes from (1, 0, 0) to (0, 0, 1) in xz-plane at evenly spaced 21 steps. The length of the translation vector is equal to 2.1 units. 100 random trials.

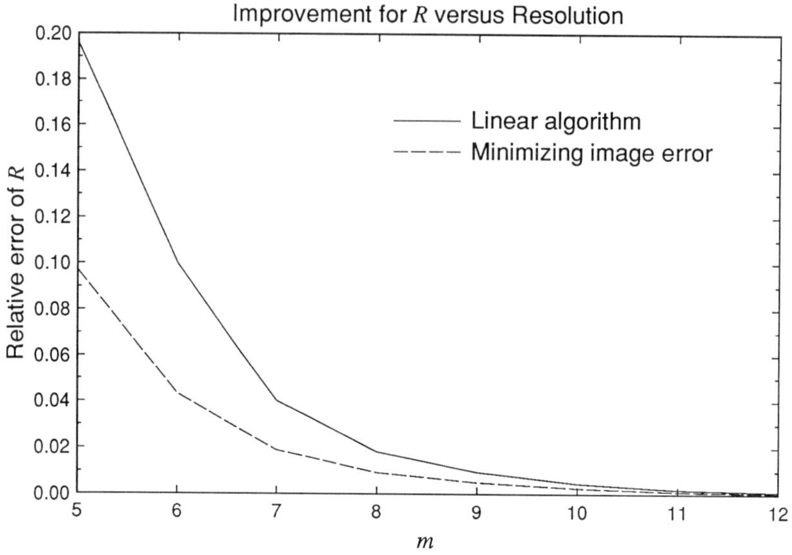

Fig. 4.15(a)

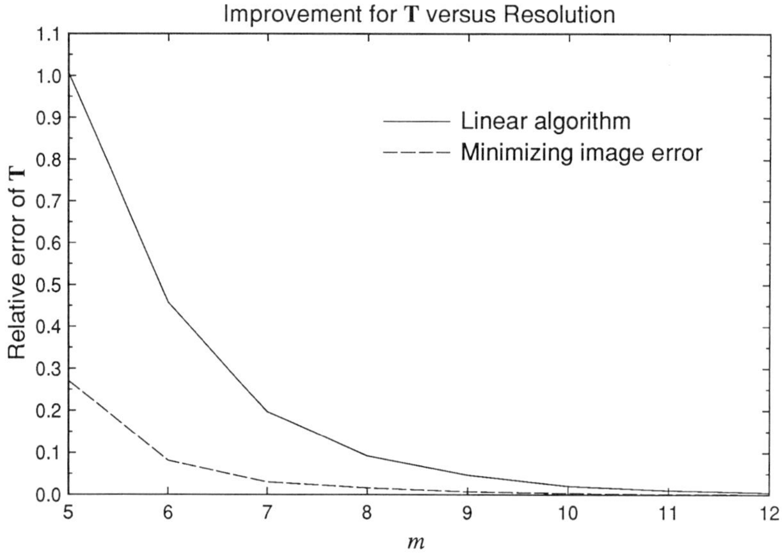

Fig. 4.15(b)

Fig. 4.15(c)

Fig. 4.15. Improvement of the batch optimization over the linear algorithm versus image resolutions ($2^m \times 2^m$). Rotation axis: (1, 1, 1). Rotation angle: 5°. Translation: (3, 0, 0). Image size: 1; 100 random trials. **(a)** R; **(b) T**; **(c)** image errors.

4.8.4 Uncertainty Polyhedron Model

Fig. 4.16 presents the comparison between the model 1: the Gaussian distribution (minimizing the image errors) and model 2: the uncertainty polyhedron distribution. The actual noise added is digitization noise. Very similar performances are observed for these two models except for the image errors, although model 2 seems more appropriate for the actual noise. This implies that the errors are not very sensitive to the moderate changes in the assumed noise distribution. This is also expected, since the minimization of the image errors is optimal (minimum variance) for locally linearized system with general distribution (see Section 4.5). Since the approach of minimizing the image errors has its optimality for general noise distribution and is less computationally expensive than the uncertainty polyhedron model, it can be generally used for applications where the distribution of noise is unknown.

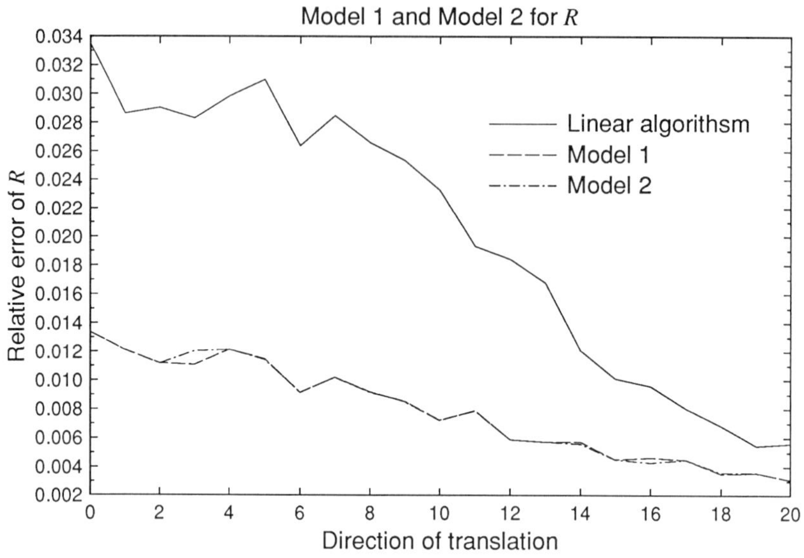

Fig. 4.16(a)

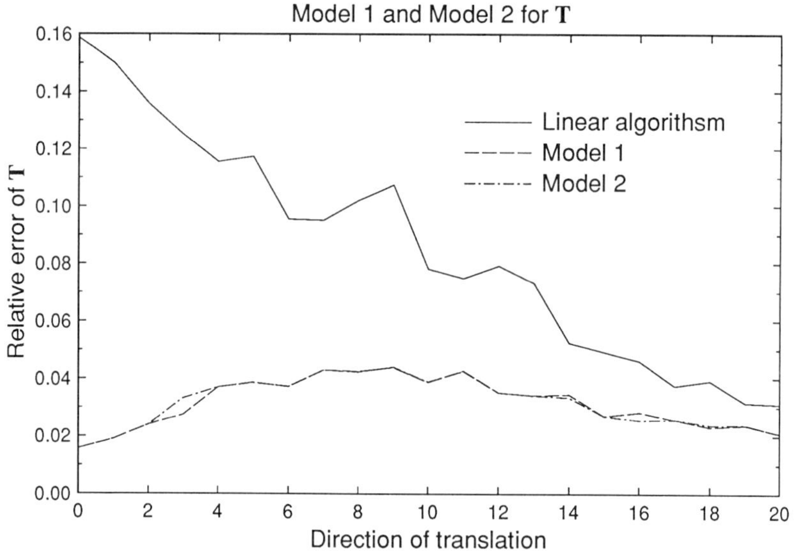

Fig. 4.16(b)

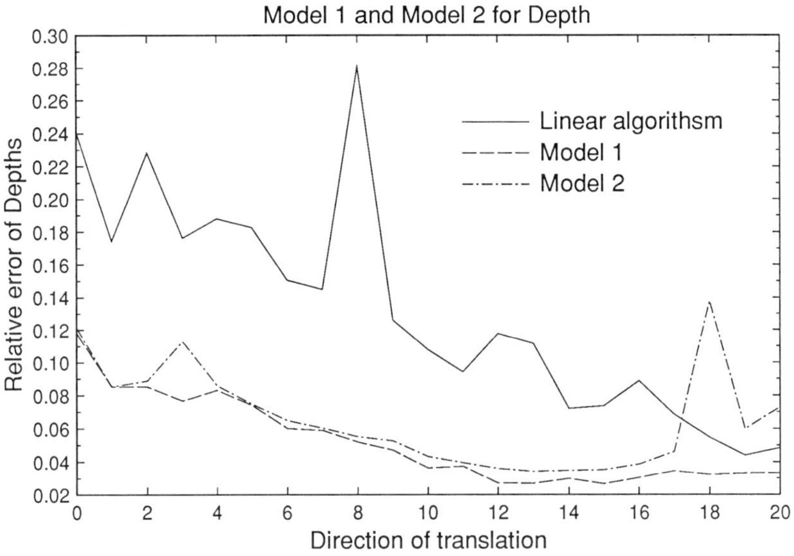

Fig. 4.16(c)

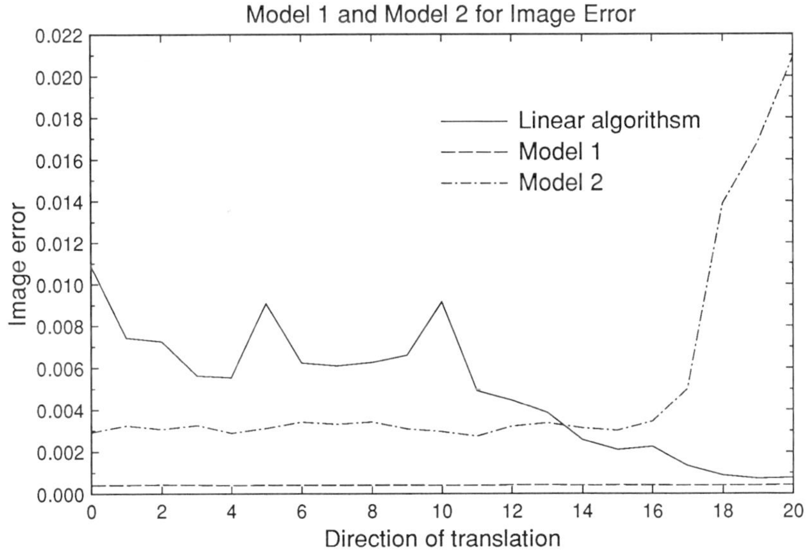

Fig. 4.16(d)

Fig. 4.16. Relative Errors of the linear algorithm, model 1 (minimizing the image errors) and model 2 (uncertainty polyhedron). Rotation axis: (1, 1, 1). Rotation angle: 3°. For horizontal index from 0 to 20, the direction of translation changes from (1, 0, 0) to (0, 0, 1) in xz-plane at evenly spaced 21 steps. The length of the translation vector is equal to 2.1 units. 100 random trials. **(a)** R; **(b)** **T**; **(c)** depths; **(d)** image errors.

4.8.5 Error Estimation and Error Bound

As discussed in Section 4.6, the covariance matrix of the errors in the parameters can be estimated by (4.6.1), where J is evaluated at the estimated parameters obtained by batch optimization. Since the errors are random, it is never possible to determine the error completely. The estimated error covariance matrix provides a rough estimate of the spread of the errors. Since the difference between the estimated error and the actual error is also random, the performance of the error estimation should be evaluated based on statistical behavior. Fig. 4.17 shows average relative errors, average deviation of the error estimation (the absolute difference between the estimated relative errors and the actual relative errors) and bias of error estimation (the average difference between the estimated relative errors and the actual relative errors) over 40 random trials. As can be seen from the figure, the average deviation is generally less than a half of the magnitude of the actual relative errors. The bias is also small. Figs 4.17(c) and (d) show sample sequences of estimated errors and actual errors for the case with 9 point correspondences.

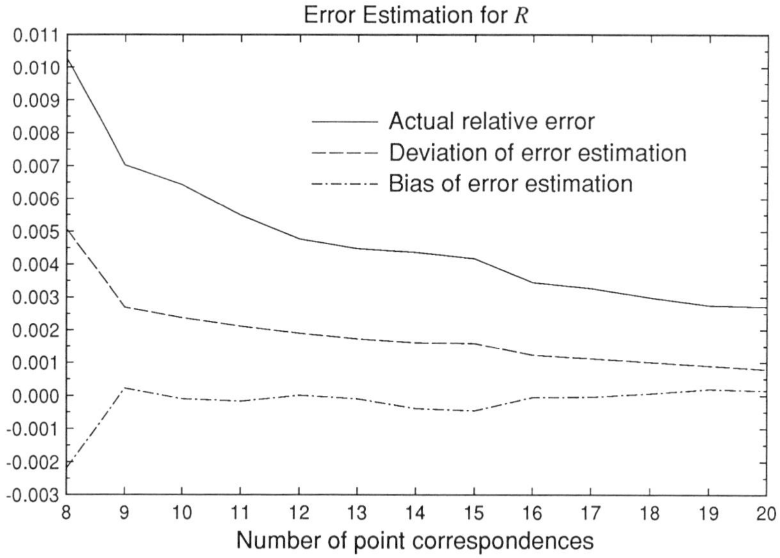

Fig. 4.17(a)

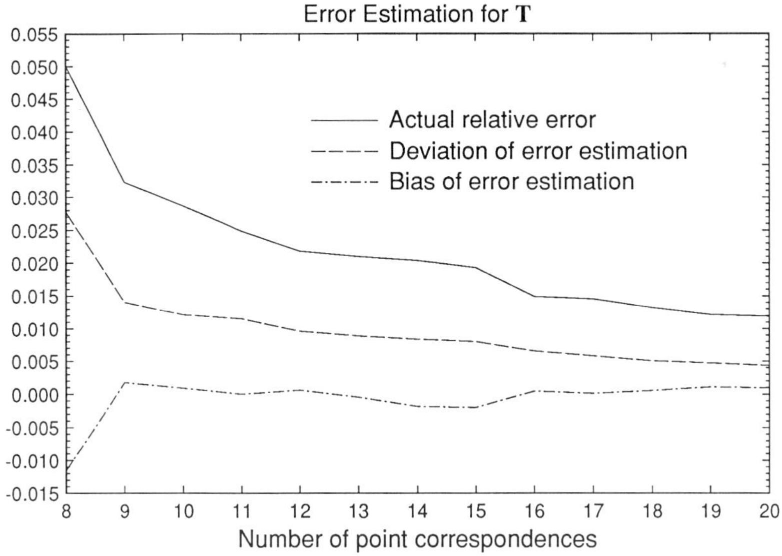

Fig. 4.17(b)

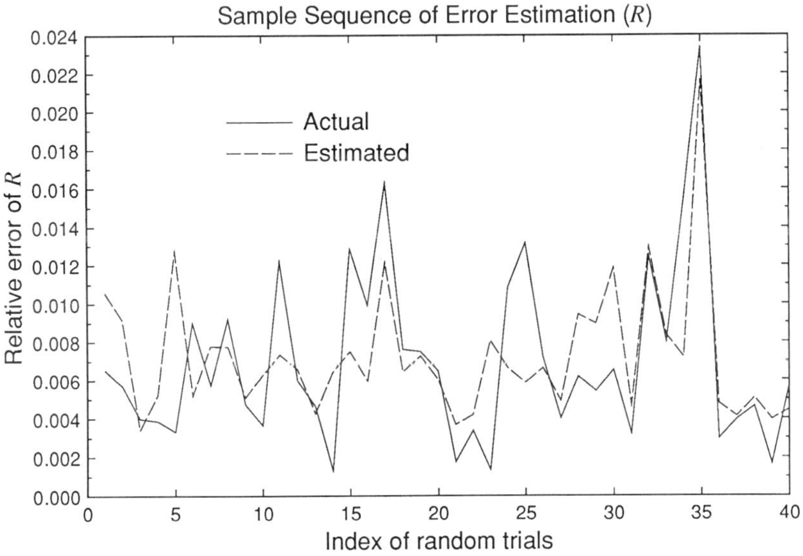

Fig. 4.17(c)

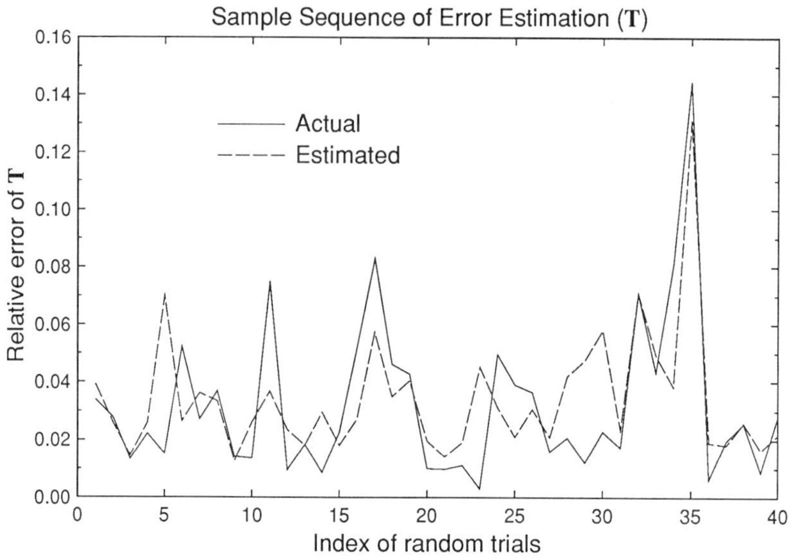

Fig. 4.17(d)

Fig. 4.17. Statistical record of error estimation. Actual relative error, deviation of error estimation and bias of error estimation for **(a)** R, **(b)** **T** vs. Number of Point Correspondences. Sample sequences of estimated errors and actual errors with 9 point correspondences: **(c)** R, **(d)** **T**. Rotation axis: (1, 0.9, 0.8). Rotation angle: 5°. Translation: (0.5, –0.5, –3.0). 40 random trials.

Fig. 4.18 shows the comparison between the actual relative errors and the corresponding Cramér-Rao bound with Gaussian noise. Two types of noise are simulated, Gaussian and uniform (with same variance). As shown in Fig. 4.18, the actual relative errors in the estimated solutions are very close to the Cramér-Rao bound with Gaussian noise. In other words, the errors of the algorithm are very close to that of a best possible unbiased estimator with Gaussian noise. Figs 4.18(b) and (c) show that the errors are similar for the two types of noise. This again implies that the distribution of noise does not significantly influence the actual errors as long as the variance is kept the same. Fig. 4.18 also shows the relative absolute bias of the estimates (the norm of the bias matrix, or vector, divided by the norm of the true matrix, or vector). The bias is small relative to the actual errors. In Fig. 4.18, for the case of 8 point correspondences the actual (average) errors are exceptionally larger than the Cramér-Rao bound. This is due to the fact that the preliminary estimates of the linear algorithm have large errors from a minimal 8 point correspondences. These large errors may cause the nonlinear optimization procedure to converge to a false solution. Thus again shows that a good preliminary initial guess is very important to obtain a correct optimal solution.

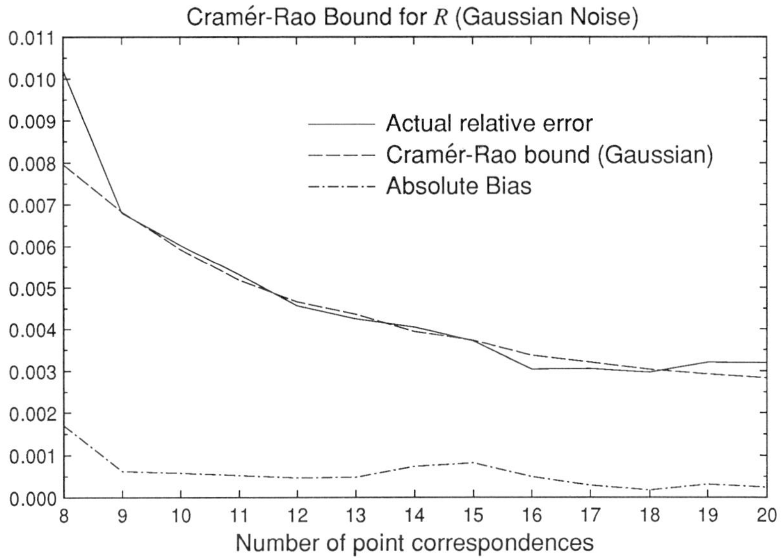

Fig. 4.18(a)

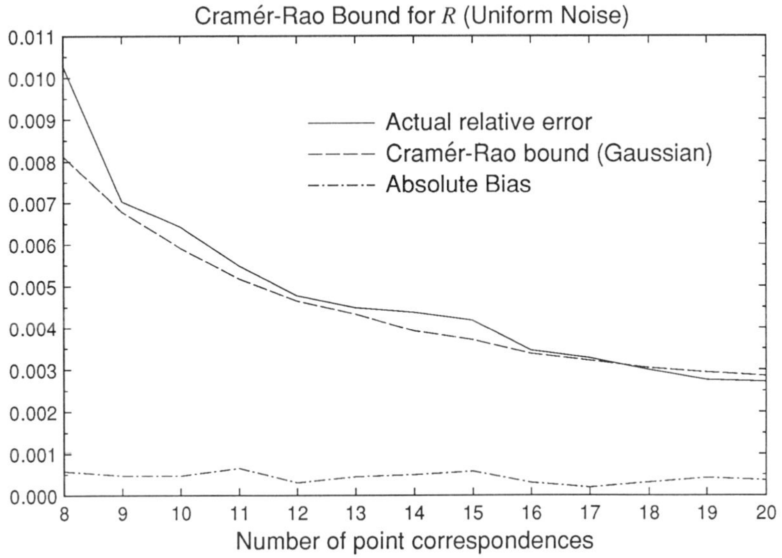

Fig. 4.18(b)

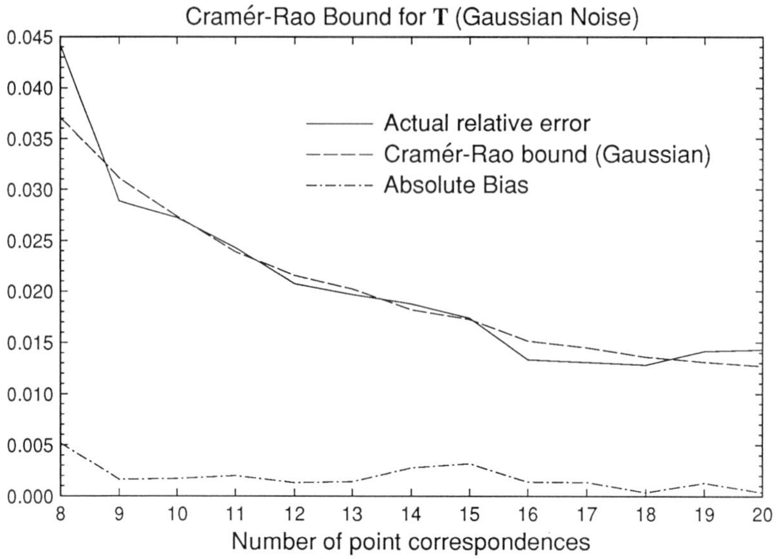

Fig. 4.18(c)

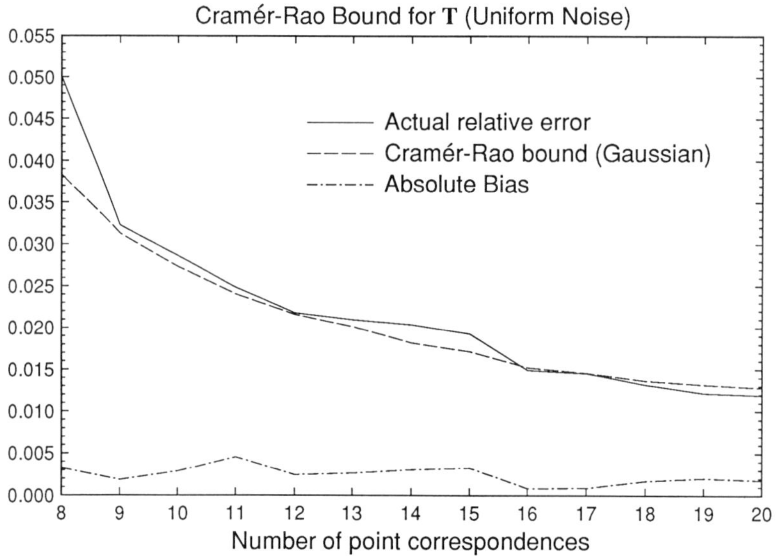

Fig. 4.18(d)

Fig. 4.18. Actual errors, the Cramér-Rao bound with Gaussian noise, and the absolute bias of the estimator vs. number of point correspondences. Comparison for R: **(a)** Gaussian noise added; **(b)** uniform noise added. Comparison for **T**: **(c)** Gaussian noise added; **(d)** uniform noise added. Rotation axis: (1, 0.9, 0.8). Rotation angle: 5°. Translation: (0.5, –0.5, –3.0). 40 random trials.

The performance of the algorithm virtually reach the theoretical lower bounds with Gaussian noise, which provides quantitative insight into the impact of errors which may seem negligible, such as digitization errors. From Fig. 4.18 we know that to give an estimated translation direction with under 2.0% expected error, the variance of the errors in the image points cannot be larger than those of digitization errors of 256×256 images.

4.8.6 Inherent Limitation of Optical Flow

As discussed in Section 2.1, the discrete approaches are applicable to both small or large interframe motions, while the continuous approaches are applicable to only small motions. The restriction of small motion for continuous approaches arise primarily from two facts: (1) Optical flow, by definition, is the projection of 3-D velocity onto image plane. The formulation of computing optical flow is in terms of velocity. (2) The mathematical formulation of computing motion parameters from optical flow is also in terms of motion velocity [4.8, 50, 1, 39]. However, what actually observed is the displacements between images. Only in the case of small motion, can displacement be approximated by velocity.

Although the restriction of small interframe motion simplifies both computing image matching (optical flow as a result) and computing motion parameters from optical flow (motion velocity as a result), the reliability of the computed motion parameters in the presence of noise is inherently limited. A small amount of motion is easily overridden by the errors in the estimated optical flow, even if the optical flow can be estimated in subpixel accuracy. In other words, the signal to noise ratio in the estimated optical flow is low. Many researchers have been trying to computing motion parameters from optical flow [4.8, 1, 39, 50] but difficulties have been reported in stability. There may be a lot one can do to improve those existing algorithms that compute motion parameters and structure from optical flow. However, the bottom line is that theoretical lower error bound is large.

As shown in Fig. 4.18, the algorithm (discrete approach) has essentially reached the theoretical error bound. With a relatively large interframe motion, the error bound is small (e.g., about 2% in the direction of translation in Fig. 4.18(b)). What about small motions typically used for optical flow? We consider a setup: The image is a unit square with 512×512 pixels. We assume that the image positions of the points are contaminated by additive white Gaussian noise with a variance equal to that of the uniform distribution in the range of ±1 pixel. The magnitudes of translation are such that the maximum disparities caused by translation in the image plane are 2, 4, 8 and 16 pixels, respectively. The Cramér-Rao bounds of the relative error in the estimated translation, averaging over 10 random point sets, are shown in Fig. 4.19. It can be seen that, under a small motion with 2-pixel maximum disparity (average disparity is roughly 1 pixel), the errors in translation are bounded below by 60% even using a large number of points (70). For a small motion with 4-pixel maximum disparity the error bound is still large (about 38% with 70 points). Recall that the bound here is meant for exact algorithms (discrete approaches) and does not include any approximations that a continuous approach may use. In other words, these bounds apply to any algorithms. Therefore, it is inherently very unreliable trying to recover motion parameters from a small motion with a disparity of a few pixels. The data shown here quantitatively predict the inherent stability problem for motion and structure from optical flow (small motions).

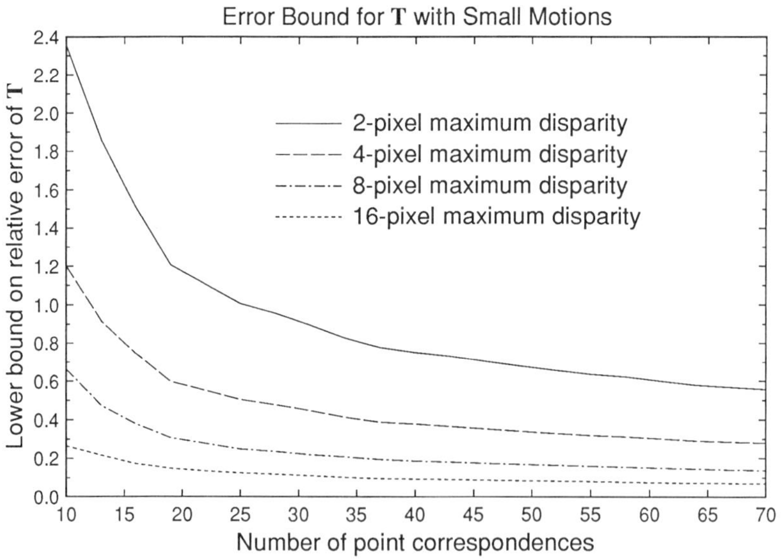

Fig. 4.19. The Cramér-Rao bound for relative errors in translation under small motions. 10 random trials. Translation: $(k, k, 0)$. The value of k is such that the maximum disparity caused by translation is d-pixels, $d = 2, 4, 8, 16$. Rotation axis: (1, 0.9, 0.8). Rotation angle: 5°.

4.8.7 Real World Images

We apply the algorithms of optimization and error estimation to the images shown in Chapter 2.

For the Mac scene shown in Fig. 2.16, the 182 displacement vectors shown in Fig. 2.24(a) are used as point correspondences to compute the optimal motion parameters, using the solution of the linear algorithm as an initial guess. The results are shown in Table 2.1. The image error is equal to about half a pixel width as shown in Table 2.1, which seems to be satisfactory. The computed optimal motion parameters for the Desk scene in Fig. 2.30 and the Path scene in Fig. 2.33 are presented in Tables 2.2 and 2.3, respectively.

Assume the errors in the coordinates of the matching points are uncorrelated. The estimated variance of the errors is given by the squared image errors. The estimated errors of the computed motion parameters for the Mac scene are shown in Table 4.3 and those for the Path scene are shown in Table 4.4.

Table 4.3. Estimated errors in the solution for the Mac scene

Parameters	x	y	z
Error of $\hat{\mathbf{T}}$	0.0026	0.0011	0.012
Error of Rotation axis	0.0091	0.021	0.023
Error of rotation angle		0.14°	
Relative error of $\hat{\mathbf{T}}$		0.012	
Relative error of rotation axis		0.032	

Table 4.4. Estimated errors in the solution for the Path scene

Parameters	x	y	z
Error of $\hat{\mathbf{T}}$	0.0023	0.0031	0.00031
Error of Rotation axis	0.068	0.044	0.036
Error of rotation angle		0.012°	
Relative error of $\hat{\mathbf{T}}$		0.0039	
Relative error of rotation axis		0.088	

4.9 Summary

In this chapter we first discusses a type of motion for which the algorithms based on the epipolar constraint are very sensitive to noise.

The analysis leads to the conclusion that it is important to use both components of image positions of points to determine motion parameters in the presence of noise. The simulations showed that the use of both components (i.e., minimizing the image errors) significantly improves the accuracy of the estimates compared with the use of only one component (i.e., using the epipolar constraint only).

Both components of image position of every point as well as constraints in parameters are taken into account in a unified way by the approach to optimal estimation presented. The maximum likelihood estimator with independent Gaussian noise leads to minimizing the image errors. With uncorrelated general noise distribution, minimizing the image errors corresponds to minimum variance estimation for the locally linearized system. Therefore, minimizing the image errors is a good objective when the noise distribution is unknown.

The optimal estimation leads to remarkable improvement from the preliminary estimates computed by the linear algorithm. The stability of the motion estimation problem has been investigated in terms of theoretical limits. The Cramér-Rao lower error bound has been determined for the problem, and experiments showed that the actual errors are quite close to the Cramér-Rao bound for unbiased estimators with Gaussian noise. This close-to-limit achievement for our nonlinear optimization problem is mainly due to: (1) the closed-form solution that provides a good initial guess, (2) the optimal estimation that makes good use of all the information available, and (3) the effective batch processing. An approach to estimating errors in the optimal estimates is introduced and implemented. This provides a general framework of error estimation for iterative optimization algorithms. The analysis and experiments on batch processing (the Levenberg-Marquardt or the Gauss-Newton method) and sequential processing (IEKF) lead to the flowing conclusions: For nonlinear problems, the performance of the IEKF algorithm is inferior to that of the Levenberg-Marquardt algorithm, and the covariance matrices given by the IEKF may significantly underestimate the actual errors. To improve the computational efficiency for those nonlinear problems with very long observation sequences, a recursive-batch processing approach is proposed.

4.10 Bibliographical Notes

Most of the early iterative algorithms for motion analysis used objective functions that measure the residual errors of some particular equations. Roach & Aggarwal [4.31] used the sum of squared residuals of a set of nonlinear equations as an objective function. Using optical flow, Bruss & Horn [4.8] as well as Adiv [4.1] minimize some measure of the discrepancy between the measured flow and that predicted from the computed motion parameters. Mitiche & Aggarwal [4.23] employed a direct rigidity criterion with depths of all the points as parameters. Yasumoto & Medioni [4.47] include, as the term of regularization, the magnitude of the rotational component into their objective function.

Optimal motion estimation has been explicitly investigated recently. Weng, Huang & Ahuja proposed a two-step approach to optimal motion and structure estimation [4.41], and at the same time, Toscani & Faugeras [4.36] proposed a reconstruction-and-reprojection technique. Interestingly, the two objective functions [4.36, 41] stem from the same criterion. Maximum likelihood formulation has been used in [4.43], and the necessity of using both components of an image point was reported in [4.44] together with a method of minimum variance estimation for the problem. Aisbett [4.3] and Spetsakis & Aloimonos [4.35] minimize equation errors (epipolar improvement) to approach optimality.

Adiv [4.2] has investigated the stability of motion using optical flow from an algorithm-independent point of view, and points out the ambiguity of a type of motion similar to what we discuss here in Subsection 4.2.4. Weng, Ahuja & Huang [4.44] demonstrated the effect of depths in disambiguating such type of unstable motions. The unstability of motion from small interframe motions was demonstrated quantitatively by evaluating the Cramér-Rao bound [4.45].

Appendices

Appendix 4.A Decomposability

Theorem 4.1. Given a 3×3 matrix E, the necessary and sufficient condition for existing a rotation matrix R and a skew symmetric matrix S, such that $E = SR$, is that one of the singular values of E is equal to 0 and the remaining two are equal.

Proof. Necessity: Let $E = SR$, where $S = [s_{ij}]$ is a skew symmetric matrix and R is a rotation matrix. Letting $\mathbf{T} = (-s_{23}, s_{13}, -s_{12})^t$, we have $[\mathbf{T}]_\times = S$. If $\mathbf{T} = 0$, then $E = \mathbf{0}$ and all three singular values of E are equal to 0, and the necessity is true. Now assume $\mathbf{T} \neq \mathbf{0}$. Define an orthonormal matrix

$$H = R^{-1}[\mathring{\mathbf{T}}, \mathring{\mathbf{T}}_2, \mathring{\mathbf{T}}_3]$$

where $\mathring{\mathbf{T}} = \mathbf{T}/\|\mathbf{T}\|$, and $\mathring{\mathbf{T}}_2$, $\mathring{\mathbf{T}}_3$ are such that $[\mathring{\mathbf{T}}, \mathring{\mathbf{T}}_2, \mathring{\mathbf{T}}_3]$ is orthonormal. We have

$$EH = [\mathbf{T}]_\times[\mathring{\mathbf{T}}, \mathring{\mathbf{T}}_2, \mathring{\mathbf{T}}_3] = [0, \mathbf{V}_2, \mathbf{V}_3]$$

where $\mathbf{V}_i = \mathbf{T}\times\mathring{\mathbf{T}}_i$, $i = 2, 3$. Obviously $\mathbf{V}_2$ and $\mathbf{V}_3$ are orthogonal and $\|\mathbf{V}_i\| = \|\mathbf{T}\|$, $i = 2, 3$. Therefore,

$$H^t E^t EH = \text{diag}\{0, \|\mathbf{T}\|^2, \|\mathbf{T}\|^2\}.$$

Therefore, the singular values of E are equal to 0, $\|\mathbf{T}\|$ and $\|\mathbf{T}\|$, respectively.

Sufficiency: Suppose E^tE has three singular values $0, \lambda, \lambda$. Then there exist two orthogonal matrices H and Q such that

$$E = Q\,\text{diag}\{0, \lambda, \lambda\}H^t. \tag{4.A.1}$$

If $\det(H) = -1$, one can change the sign of the first column of H so that $\det(H) = 1$ but (4.A.1) still holds true. The same is true for Q. Thus, we can let H and Q in (4.A.1) be rotation matrices. On the other hand,

$$\text{diag}\{0, \lambda, \lambda\} = \begin{bmatrix} 0 & 0 & 0 \\ 0 & 0 & -\lambda \\ 0 & \lambda & 0 \end{bmatrix} \begin{bmatrix} 1 & 0 & 0 \\ 0 & 0 & 1 \\ 0 & -1 & 0 \end{bmatrix} \triangleq SR. \tag{4.A.2}$$

(The use of this seemingly odd equation was guided by our original proof which is more natural but less compact) From (4.A.1) and (4.A.2), we get

$$E = Q\,\text{diag}\{0, \lambda, \lambda\}H^t = QSRH^t = \{QSQ^t\}\{QRH^t\} \tag{4.A.3}$$

where in the last expression, the first part is a skew symmetric matrix and the second part is a rotation matrix. This completes the proof.

Corollary. Given a 3×3 matrix E, the necessary and sufficient condition for existing a rotation matrix R and a unit vector $\mathring{\mathbf{T}}$, such that $E = [\mathring{\mathbf{T}}]_\times R$, is that the eigenvalues of $E^t E$ are 0, 1, 1, respectively.

Proof. Procedure of the proof is very similar to the above. For necessity let $\|\mathbf{T}\| = 1$. For sufficiency we have $\lambda = 1$. Let $[\mathring{\mathbf{T}}]_\times = QSQ^t$ in (4.A.3).

$$2\|\mathring{\mathbf{T}}\|^2 = \text{trace}\{(QSQ^t)(QSQ^t)^t\}$$

$$= \text{trace}\{QSS^tQ^t\} = \text{trace}\{SS^t\} = 2.$$

Therefore $\mathring{\mathbf{T}}$ is a unit vector. □

Appendix 4.B Weights

We need to derive (4.2.6). Equation (4.2.3) is equivalent to

$$\mathring{\mathbf{T}}^t(\mathbf{X}' \times R\,\mathbf{X}) = 0.$$

With additive noise

$$\mathbf{X}'(\varepsilon) = \mathbf{X}' + \delta_{\mathbf{X}'}$$

and

$$\mathbf{X}(\varepsilon) = \mathbf{X} + \delta_{\mathbf{X}}$$

we have

$$\eta \triangleq \mathring{\mathbf{T}}^t(\mathbf{X}'(\varepsilon) \times R\,\mathbf{X}(\varepsilon)) = \mathring{\mathbf{T}}^t((\mathbf{X}' + \delta_{\mathbf{X}'}) \times R\,(\mathbf{X} + \delta_{\mathbf{X}}))$$

$$\cong \mathring{\mathbf{T}}^t(\mathbf{X}' \times R\,\mathbf{X} + \mathbf{X}' \times R\,\delta_{\mathbf{X}} + \delta_{\mathbf{X}'} \times R\,\mathbf{X})$$

where the second order term $\delta_{\mathbf{X}'} \times (R\,\delta_{\mathbf{X}})$ is neglected. Since $\mathring{\mathbf{T}}^t(\mathbf{X}' \times R\,\mathbf{X}) = \mathbf{0}$, it follows that

$$\eta \cong \mathring{\mathbf{T}}^t(\mathbf{X}' \times R\,\delta_{\mathbf{X}}) + \mathring{\mathbf{T}}^t(\delta_{\mathbf{X}'} \times R\,\mathbf{X})$$

$$= (\mathring{\mathbf{T}} \times \mathbf{X}')^t(R\,\delta_{\mathbf{X}}) - (\mathring{\mathbf{T}} \times R\,\mathbf{X})^t\delta_{\mathbf{X}'}$$

$$= (R^t(\mathring{\mathbf{T}} \times \mathbf{X}'))^t\delta_{\mathbf{X}} - (\mathring{\mathbf{T}} \times R\,\mathbf{X})^t\delta_{\mathbf{X}'}.$$

The first two components of $\mathbf{X}$ are image plane coordinates and the last component is exactly 1. Assume the first two components of its error vector $\delta_{\mathbf{X}}$ are uncorrelated random noise with zero mean and variance σ^2. That is,

$$\mathbb{E}\delta_X\delta_X^t = \mathrm{diag}\{\sigma^2, \sigma^2, 0\}.$$

Similarly assume $\mathbf{X}'$ has additive zero mean uncorrelated nose with variance σ^2. Then

$$\begin{aligned}\mathbb{E}\eta\eta^t &\cong (R^t(\mathring{\mathbf{T}}\times\mathbf{X}'))^t\,\mathrm{diag}\{\sigma^2, \sigma^2, 0\}R^t(\mathring{\mathbf{T}}\times\mathbf{X}')\\ &+ (\mathring{\mathbf{T}}\times R\,\mathbf{X})^t\,\mathrm{diag}\{\sigma^2, \sigma^2, 0\}(\mathring{\mathbf{T}}\times R\,\mathbf{X})\\ &= \sigma^2(\|R^t(\mathring{\mathbf{T}}\times\mathbf{X}')\|_{z=0}^2 + \|\mathring{\mathbf{T}}\times R\,\mathbf{X}\|_{z=0}^2)\end{aligned}$$

where we define $\|(a, b, c)\|_{z=0}^2 \triangleq a^2 + b^2$.

Appendix 4.C Triangulation with Noise

As shown in Fig. 4.20, we need to find the point $\mathbf{x}_\lambda = \mathbf{x}_0 + \lambda(\mathbf{x}_1 - \mathbf{x}_0)$, whose image is $\mathbf{u}_\lambda = \mathbf{h}(\mathbf{m}, \mathbf{x})$ in the first image and $\mathbf{u}'_\lambda = \mathbf{h}'(\mathbf{m}, \mathbf{x})$ in the second, that minimizes

$$\|\mathbf{u}_\lambda - \mathbf{u}\|^2 + \|\mathbf{u}'_\lambda - \mathbf{u}'\|^2. \tag{4.C.1}$$

Let line segment L be a line that is the shortest line segment connecting two observed projection lines. L intersects the observed projection line 1 at $\mathbf{x}_0$ and intersect the observed projection line 2 at $\mathbf{x}_1$. We determine the parameter λ such that (4.C.1) is minimized. In the

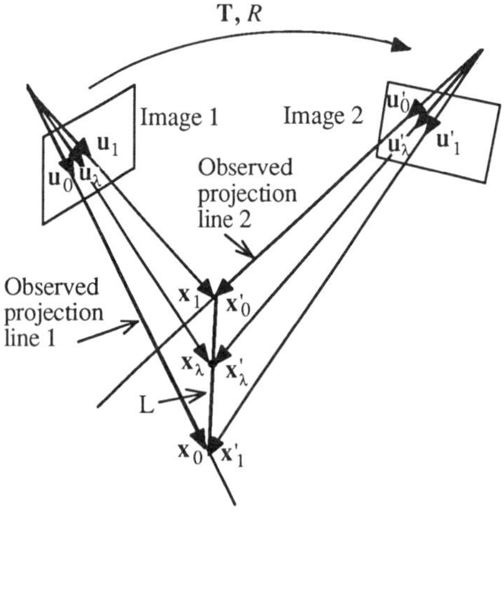

Fig. 4.20. Find a point $\mathbf{x}_\lambda$ that minimizes (4.C.1) for the point.

coordinate system of the image 1, we have

$$\mathbf{x}_0 = (x_0, y_0, z_0), \quad \mathbf{x}_1 = (x_1, y_1, z_1),$$

$$\mathbf{x}_\lambda = \mathbf{x}_0 + \lambda(\mathbf{x}_1 - \mathbf{x}_0) = (x_\lambda, y_\lambda, z_\lambda). \tag{4.C.2}$$

From (4.C.2) we get the corresponding image vectors $\mathbf{u}_0 = \mathbf{x}_0/z_0$, $\mathbf{u}_1 = \mathbf{x}_1/z_1$ and $\mathbf{u}_\lambda = \mathbf{x}_\lambda/z_\lambda$. After simple algebraic manipulations, we get

$$\|\mathbf{u}_\lambda - \mathbf{u}\|^2 = \frac{\lambda^2 a^2}{z_0^2 z_\lambda^2} \tag{4.C.3}$$

where $a^2 = (x_1 z_0 - x_0 z_1)^2 + (y_1 z_0 - y_0 z_1)^2$. In the coordinate system of the image 2 denoting the corresponding variables by primes, we get the similar results:

$$\|\mathbf{u}'_\lambda - \mathbf{u}'\|^2 = \frac{(1-\lambda)^2 (a')^2}{(z'_0)^2 (z'_\lambda)^2} \tag{4.C.4}$$

where $(a')^2 = (x'_1 z'_0 - x'_0 z'_1)^2 + (y'_1 z'_0 - y'_0 z'_1)^2$. From (4.C.3) and (4.C.4) we write the image error as a function of λ:

$$F(\lambda) = \|\mathbf{u}_\lambda - \mathbf{u}\|^2 + \|\mathbf{u}'_\lambda - \mathbf{u}'\|^2 = \frac{\lambda^2 a^2}{z_0^2 z_\lambda^2} + \frac{(1-\lambda)^2 (a')^2}{(z'_0)^2 (z'_\lambda)^2}. \tag{4.C.5}$$

Taking the derivative of $F(\lambda)$ and setting it equal to zero, we get

$$0 = \frac{1}{2}\frac{dF(\lambda)}{d\lambda} = \frac{b\lambda}{z_\lambda^3} - \frac{b'(1-\lambda)}{(z'_\lambda)^3} \tag{4.C.6}$$

where $b = a^2/z_0$ and $b' = (a')^2/z'_0$. Or,

$$\frac{b\lambda}{z_\lambda^3} - \frac{b'(1-\lambda)}{(z'_\lambda)^3} = 0. \tag{4.C.7}$$

Since $z_\lambda = z'_0 + \lambda(z_1 - z_0)$ and $z'_\lambda = z_0 + (1-\lambda)(z'_1 - z'_0)$, solving (4.C.7) for λ needs solving a fourth order polynomial equation. If we approximate z_λ and z'_λ in (4.C.7) by

$$z_\lambda \approx (z_0 + z_1)/2, \qquad z'_\lambda \approx (z'_0 + z'_1)/2$$

we get

$$\lambda \approx \frac{b' z_\lambda^3}{b(z'_\lambda)^3 + b' z_\lambda^3}.$$

This value of λ determines the point $\mathbf{x}_\lambda$ which is generally not the midpoint of L.

Appendix 4.D Matrix Derivatives

We can parameterize rotation by (α, β, γ), where α, β and γ are rotation angles around x-, y- and z-axes, respectively. The corresponding rotation matrix R is given by

$$\begin{bmatrix} \cos\gamma & -\sin\gamma & 0 \\ \sin\gamma & \cos\gamma & 0 \\ 0 & 0 & 1 \end{bmatrix} \begin{bmatrix} \cos\beta & 0 & \sin\beta \\ 0 & 1 & 0 \\ -\sin\beta & 0 & \cos\beta \end{bmatrix} \begin{bmatrix} 1 & 0 & 0 \\ 0 & \cos\alpha & -\sin\alpha \\ 0 & \sin\alpha & \cos\alpha \end{bmatrix}$$

$$= \begin{bmatrix} \cos\beta\cos\gamma & -\cos\alpha\sin\gamma + \sin\alpha\sin\beta\cos\gamma & \sin\alpha\sin\gamma + \cos\alpha\sin\beta\cos\gamma \\ \cos\beta\sin\gamma & \cos\alpha\cos\gamma + \sin\alpha\sin\beta\sin\gamma & -\sin\alpha\cos\gamma + \cos\alpha\sin\beta\sin\gamma \\ -\sin\beta & \sin\alpha\cos\beta & \cos\alpha\cos\beta \end{bmatrix}.$$

Letting $R = [r_{ij}]$, it is easy to get

$$\frac{\partial R}{\partial \alpha} = \begin{bmatrix} 0 & r_{13} & -r_{12} \\ 0 & r_{23} & -r_{22} \\ 0 & r_{33} & -r_{32} \end{bmatrix},$$

$$\frac{\partial R}{\partial \beta} = \begin{bmatrix} -\sin\beta\cos\gamma & \sin\alpha\cos\beta\cos\gamma & \cos\alpha\cos\beta\cos\gamma \\ -\sin\beta\sin\gamma & \sin\alpha\cos\beta\sin\gamma & \cos\alpha\cos\beta\sin\gamma \\ -\cos\beta & -\sin\alpha\sin\beta & -\cos\alpha\sin\beta \end{bmatrix},$$

and

$$\frac{\partial R}{\partial \gamma} = \begin{bmatrix} -r_{21} & -r_{22} & -r_{23} \\ r_{11} & r_{12} & r_{13} \\ 0 & 0 & 0 \end{bmatrix}.$$

$\mathring{\mathbf{T}}$ can be parameterized as

$$\mathring{\mathbf{T}} = \begin{bmatrix} \cos\psi\sin\phi \\ \sin\psi\sin\phi \\ \cos\phi \end{bmatrix}. \tag{4.D.1}$$

From image plane vectors $\mathbf{u}$ and $\mathbf{u}'$, the actual computed image point

$\mathbf{h}$ can be expressed as the function of R and $\mathring{\mathbf{T}}$. Then the partial derivatives of $\mathbf{h}$ with respective to motion parameters $\mathbf{m} = (\alpha, \beta, \gamma, \psi, \phi)^t$ can be computed.

Appendix 4.E Cramér-Rao Bound

Cramér-Rao bound [4.9, 30]. Suppose $\mathbf{m}$ is a parameter vector for probability density $p(\mathbf{u}, \mathbf{m})$. $\hat{\mathbf{m}}$ is an estimator of $\mathbf{m}$ based on measurement $\mathbf{u}$ with $\mathbb{E}\hat{\mathbf{m}} = \mathbf{b}(\mathbf{m})$. Let $\mathbf{y}^t = \dfrac{\partial \ln p(\mathbf{u}, \mathbf{m})}{\partial \mathbf{m}}$. Define

$$F = \mathbb{E}\mathbf{y}\mathbf{y}^t.$$

Matrix F is called the Fisher information matrix. Denote

$$B = \frac{\partial \mathbf{b}(\mathbf{m})}{\partial \mathbf{m}}.$$

Then

$$\mathbb{E}(\hat{\mathbf{m}} - \mathbf{b}(\mathbf{m}))(\hat{\mathbf{m}} - \mathbf{b}(\mathbf{m}))^t \geq BF^{\dagger}B^t.$$

The equality holds if and only if

$$\hat{\mathbf{m}} - \mathbf{b}(\mathbf{m}) = B\left\{\frac{\partial \ln p(\mathbf{u}, \mathbf{m})}{\partial \mathbf{m}}\right\}^t$$

almost everywhere.

Proof. We first prove an inequality. Denoting $C_{\mathbf{XY}} \triangleq \mathbb{E}\mathbf{X}\mathbf{Y}^t$ then

$$C_{\mathbf{XX}} \geq C_{\mathbf{XY}} C_{\mathbf{YY}}^{\dagger} C_{\mathbf{XY}}^t. \tag{4.E.1}$$

In fact, for any matrix M, we get

$$\begin{aligned}
\mathbf{0} &\leq \mathbb{E}(\mathbf{X} + M\mathbf{Y})(\mathbf{X} + M\mathbf{Y})^t \\
&= \mathbb{E}(\mathbf{X}\mathbf{X}^t + \mathbf{X}\mathbf{Y}^t M^t + M\mathbf{Y}\mathbf{X}^t + M\mathbf{Y}\mathbf{Y}^t M^t) \\
&= C_{\mathbf{XX}} + C_{\mathbf{XY}} M^t + M C_{\mathbf{XY}}^t + M C_{\mathbf{YY}} M^t.
\end{aligned} \tag{4.E.2}$$

Letting $M = -C_{\mathbf{XY}}^{\dagger}$ and using $C_{\mathbf{YY}}^{\dagger} C_{\mathbf{YY}} C_{\mathbf{YY}}^{\dagger} = C_{\mathbf{YY}}^{\dagger}$, Equation (4.E.2) gives (4.E.1) immediately. The equality of (4.E.1) holds if and only of $\mathbf{X} + M\mathbf{Y} = \mathbf{0}$ almost everywhere.

Let $\mathbf{X} = \hat{\mathbf{m}} - \mathbf{b}(\mathbf{m})$ and $\mathbf{Y}^t = \mathbf{y}^t = \dfrac{\partial \ln p(\mathbf{u}, \mathbf{m})}{\partial \mathbf{m}}$. Using (4.E.1), we just need to establish

$$\mathbb{E}\mathbf{X}\mathbf{Y}^t \triangleq C_{\mathbf{XY}} = B$$

to finish the proof. In fact,

$$\mathbb{E}\mathbf{y}^t = \mathbb{E}\frac{\partial \ln p}{\partial \mathbf{m}} = \int \frac{\partial \ln p}{\partial \mathbf{m}} p \; d\mathbf{u} = \int \frac{\partial p}{\partial \mathbf{m}} d\mathbf{u} = \frac{\partial}{\partial \mathbf{m}} \int p \; d\mathbf{u} = \frac{\partial}{\partial \mathbf{m}} 1 = \mathbf{0} \tag{4.E.3}$$

where assuming the integration and the differentiation can exchange the order (similar for the following). Also

$$\begin{aligned} B &= \frac{\partial b(\mathbf{m})}{\partial \mathbf{m}} = \frac{\partial \mathbb{E}\hat{\mathbf{m}}}{\partial \mathbf{m}} = \int \hat{\mathbf{m}} \frac{\partial p}{\partial \mathbf{m}} d\mathbf{u} \\ &= \int \hat{\mathbf{m}} \frac{\partial \ln p}{\partial \mathbf{m}} p \; d\mathbf{u} = \mathbb{E}\hat{\mathbf{m}} \frac{\partial \ln p}{\partial \mathbf{m}} = \mathbb{E}\hat{\mathbf{m}}\mathbf{y}^t . \end{aligned} \tag{4.E.4}$$

From (4.E.3) and (4.E.4) it follows that

$$\mathbb{E}\mathbf{X}\mathbf{Y}^t = \mathbb{E}(\hat{\mathbf{m}} - \mathbf{b}(\mathbf{m}))\mathbf{y}^t = \mathbb{E}\hat{\mathbf{m}}\mathbf{y}^t - \mathbb{E}\mathbf{b}(\mathbf{m})\mathbf{y}^t = B - \mathbf{b}(\mathbf{m})\mathbf{0}^t = B.$$

□

5. Planar Scenes

If the feature points used for motion analysis arise from a planar scene, the algorithm presented in Chapter 3 fails to give a solution because coplanar points correspond to a degenerate configuration. However, if it is known that the points are coplanar, the motion and structure can still be determined using this additional information. The subjects related to planar scenes are treated in this chapter. A linear algorithm is derived which gives a closed-form solution from point correspondences. We also study the related issues such as uniqueness, optimality and stability.

5.1 Planar Scene as a Degenerate Case

Coplanar point configurations can arise from a planar scene or object. Aerial images of plain and sea, images of wall, ceiling, floor and open road are some examples of planar scenes. Sometimes, the extracted and matched points can be coplanar although the actual surface where the points are located is not a plane.

As we discussed in Chapter 3, the configurations for which the algorithms fail to give a unique solution are called degenerate configurations. Viewing from object-centered coordinate system, a point configuration is degenerate to the linear algorithms if and only if there exists a quadratic surface that passes through all the object points and the projection centers of the camera at two time instants [5.4]. Letting $f(x, y, x) = 0$ be a linear equation that determined the object plane and $g(x, y, z) = 0$ represent a plane that passes through the two projection centers, then

$$f(x, y, x) g(x, y, z) = 0$$

is such a quadratic surface. Therefore, any configuration that consists of coplanar points is degenerate to the linear algorithms discussed in Chapter 3. In fact, as we will see in this chapter, even if we know that the point configuration is planar, there exist multiple interpretations, no matter how many points are available. Therefore, planar scenes are degenerate not only to the linear algorithms discussed in Chapter 3 but also to any algorithm for general surfaces, either linear or nonlinear. In other words, planar scenes are inherently degenerate.

Using the knowledge that the points are coplanar, we are able to determine the motion and orientation of the plane to some extent. As we will see, there are generally two valid interpretations, one veridical and the other illusive, from two views. And generally there is a unique interpretation from three views. We derive a linear algorithm which can solve the problem for all the cases that are solvable, using point correspondences.

In fact, the use of such a plane-based algorithm is not restricted to only planar scenes. If the point configuration is degenerate or nearly degenerate to the algorithms for general surfaces, the problem may still be solved by first segmenting the scene into planar surfaces and then applying the plane-based algorithm to each planar surface.

5.2 Motion from a Plane

We still use the coordinate system and the camera model illustrated in Fig. 3.1. Notice that the camera model is normalized so that focal length is unit. Visible points are located in the half-space $z > 0$.

5.2.1 Problem Statement

Let R and $\mathbf{T}$ be a rotation matrix and a translation vector, respectively, that represent a rigid motion from time t_1 to time t_2, i.e., the spatial vectors at the two time instants are related by

$$\mathbf{x}' = R\mathbf{x} + \mathbf{T}. \tag{5.2.1}$$

In terms of image vectors,

$$z'\mathbf{X}' = zR\mathbf{X} + \mathbf{T}. \tag{5.2.2}$$

If $\|\mathbf{T}\| \neq 0$, from (5.2.2) we get

$$\frac{z'}{\|\mathbf{T}\|}\mathbf{X}' = \frac{z}{\|\mathbf{T}\|}R\mathbf{X} + \mathring{\mathbf{T}} \tag{5.2.3}$$

where, using our notation,

$$\mathring{\mathbf{T}} = \frac{\mathbf{T}}{\|\mathbf{T}\|}. \tag{5.2.4}$$

Let the plane where the points are located at time t_1 be represented by

$$\mathbf{N}^t\mathbf{x} = 1 \tag{5.2.5}$$

where $\mathbf{N}$ is the normal vector of the plane. The distance d between the origin and the plane is $d = \|\mathbf{N}\|^{-1}$. By such a representation of object plane, we exclude the cases in which the plane goes through the origin, because in those cases the plane is invisible to the camera (the image of the plane is a straight line and the motion cannot be determined). Similarly, at time t_2, that plane is represented by

$$(\mathbf{N}')^t\mathbf{x}' = 1. \tag{5.2.6}$$

From (5.2.5) and (5.2.6), it follows that

$$(\|\mathbf{T}\|\mathbf{N}^t)\frac{z}{\|\mathbf{T}\|}\mathbf{X} = 1, \qquad (\|\mathbf{T}\|(\mathbf{N}')^t)\frac{z'}{\|\mathbf{T}\|}\mathbf{X}' = 1. \tag{5.2.7}$$

We want to determine the normalized depths $\frac{z}{\|\mathbf{T}\|}$ and $\frac{z'}{\|\mathbf{T}\|}$. Equivalently we determine the normalized normals

$$\tilde{\mathbf{N}} = \|\mathbf{T}\|\mathbf{N}, \qquad \tilde{\mathbf{N}}' = \|\mathbf{T}\|\mathbf{N}'. \tag{5.2.8}$$

From the normalized normal we can determine the normalized depth from (5.2.7).

Solving for $\mathbf{x}$ in (5.2.1) and substituting $\mathbf{x}$ in (5.2.5), we get the plane at time t_2

$$(R\mathbf{N})^t\mathbf{x}' = R\mathbf{N}\cdot\mathbf{T} + 1. \tag{5.2.9}$$

From (5.2.6) and (5.2.9), the normal of the plane at time t_2 is expressed by

$$\mathbf{N}' = \frac{R\mathbf{N}}{R\mathbf{N}\cdot\mathbf{T} + 1}. \tag{5.2.10}$$

From (5.2.8) and (5.2.10) we get the equation for the normalized normal at time t_2:

$$\tilde{\mathbf{N}}' = \frac{R\tilde{\mathbf{N}}}{R\tilde{\mathbf{N}}\cdot\mathring{\mathbf{T}} + 1}. \tag{5.2.11}$$

Given n corresponding image vector pairs at two time instants, $(\mathbf{X}_i, \mathbf{X}_i')$, $i = 1, 2, \cdots, n$, which represent the projections of 3-D points on the image plane, the algorithm solves for the rotation matrix R. If the translation vector $\mathbf{T}$ does not vanish, the algorithm solves for the translation direction (represented by a unit vector $\mathring{\mathbf{T}}$) and the normalized normal $\tilde{\mathbf{N}}$. The normalized depth $z_i \|\mathbf{T}\|^{-1}$ and $z_i' \|\mathbf{T}\|^{-1}$ for object points $\mathbf{x}_i$ and $\mathbf{x}_i'$, can then be determined from (5.2.7) and (5.2.11). By solution or interpretation, we always mean this normalized solution. The magnitude of the translation vector ($\|\mathbf{T}\|$), and the absolute depths of the object points (z_i and z_i') can not be determined by monocular vision. This is easy to be seen from (5.2.3), which still holds when $\|\mathbf{T}\|$, z and z' are multiplied by any positive constant. i.e., multiplying the point depths and $\|\mathbf{T}\|$ by the same scale factor does not change the images. Equivalently the magnitude of the normal vectors $\|\mathbf{N}\|$ and $\|\mathbf{N}'\|$ cannot be determined since we have (5.2.7). Once any one of those original unknowns is available according to some additional knowledge, the scale factor can be determined directly using the normalized version and then, all the rest of the unknowns can be determined completely. In summary, $\|\mathbf{T}\|$ inherently cannot be determined from monocular vision. We defined the normalized version of the solution above that correspond to a unit translation. In the remainder of this chapter, we call normalized solutions simply solutions.

5.2.2 Algorithm Overview

We first present an overview of the algorithm. We will define a 3×3 intermediate parameter matrix F, which is a function of the motion parameters and the normal of the object plane. The equations that relate the image coordinates of the points and the matrix F are linear in the elements of F. This allows us to solve for F based on these linear equations. Then the motion parameters R and $\mathring{\mathbf{T}}$, and the normalized normal of the object plane $\tilde{\mathbf{N}}$ are solved for from the intermediate matrix F. It turns out that the two images allow two different interpretations in general. Therefore, two solutions are computed from F. There are some special cases in which the number of solutions is not exactly two. In the absence of noise, our algorithm gives exact solutions. For the reliability in the presence of noise, the algorithm is also designed to utilize redundant information in the data. The complete algorithm is presented in Appendix 5.E, which the readers can refer to during the following derivation.

5.2.3 Intermediate Parameter Matrix F

We first introduce the intermediate parameter matrix. From (5.2.1) and (5.2.5), it follows that

$$\mathbf{x}' = (R + \mathbf{T}\mathbf{N}^t)\mathbf{x}. \tag{5.2.12}$$

We define the intermediate parameter matrix:

$$F = R + \mathbf{T}\mathbf{N}^t. \tag{5.2.13}$$

From (5.2.12) and (5.2.13) we get

$$z'\mathbf{X}' = Fz\mathbf{X}. \tag{5.2.14}$$

Since $z > 0$ and $z' > 0$, we get the relation between the intermediate parameter matrix F and the image vectors:

$$\mathbf{X}' \times F\mathbf{X} = \mathbf{0}. \tag{5.2.15}$$

Let $F = [e_{ij}]$ and $\mathbf{h} \mathbin{//} (e_{11}, e_{12}, e_{13}, e_{21}, \cdots, e_{33})^t$, where "//" means parallel, $\mathbf{a} \mathbin{//} \mathbf{b}$ if and only if $\mathbf{a}\times\mathbf{b} = \mathbf{0}$. Equation (5.2.15) can be rewritten as

$$\begin{bmatrix} \mathbf{X}^t & \mathbf{0} & -u'\mathbf{X}^t \\ \mathbf{0} & \mathbf{X}^t & -v'\mathbf{X}^t \\ v'\mathbf{X}^t & -u'\mathbf{X}^t & \mathbf{0} \end{bmatrix} \mathbf{h} = \mathbf{0}. \tag{5.2.16}$$

It easy to see that the third row of the above matrix is a linear combination of the other two. Therefore, at most two of the above three scalar equations are independent. Since the last can be derived from the first two, we keep the first two equations. Thus, each point correspondence gives two scalar equations and n point correspondences give $2n$ such scalar equations. We get

$$A\mathbf{h} = \mathbf{0} \tag{5.2.17}$$

where A is a $2n$ by 9 matrix with each point correspondence contributing two rows as in (5.2.16), and it is explicitly expressed in (5.E.1) of Appendix 5.E. If rank(A) is equal to 8, $\mathbf{h}$ can be solved up to a scale factor using (5.2.17). n must be at least 4 to make rank(A) = 8 possible. The following theorem gives a necessary and sufficient condition for the rank of A to be equal to 8.

Theorem 5.1. rank(A) = 8 if and only if there exists a set of four object points such that no image projections of any three points in this set are collinear in any of the two images.

Proof. See Appendix 5.A.

The above theorem gives a necessary and sufficient condition in terms of the images of the points. The following corollary gives a necessary and sufficient condition in terms of 3-D object points.

Corollary 5.1. rank(A) = 8 if and only if

(1) there exists a set of four points in the object plane such that no three points in this set are collinear in the object plane and

(2) if the object plane is extended, it does not go through the center of projection before and after motion.

Proof. See Appendix 5.A.

Assuming rank(A) = 8, we can determine $\mathbf{h}$ (and consequently F) up to a scale factor. Since rank(A) = 8, we have $F_s = sF$ for some scale factor s. Let $H = [\mathbf{h}_1\ \mathbf{h}_2\ \mathbf{h}_3]$ be a 3×3 orthogonal matrix such that

$$H^t F^t F H = \mathrm{diag}(\lambda_1, \lambda_2, \lambda_3), \tag{5.2.18}$$

where $\lambda_1 \le \lambda_2 \le \lambda_3$. The magnitude of the scale factor s can be determined by the second smallest eigenvalue, λ_2 of $F^t F$, which should be equal to one as we show bellow in (5.2.29). Let the second smallest eigenvalue of $F_s^t F_s$ be $\gamma_2 = s^2$. Thus, $F = \pm F_s / |s| = \pm F_s / \sqrt{\gamma_2}$, which determines F up to a sign. The sign can be determined from (5.2.14): Since $z > 0$, $z' > 0$, we have

$$\mathbf{X}' \cdot F\mathbf{X} > 0. \tag{5.2.19}$$

Otherwise, the sign of F should be changed.

5.2.4 General Case

We first give a guideline. From the definition of F in (5.2.13), we can see that $F\mathbf{V}_i = R\mathbf{V}_i$ if $\mathbf{N}^t\mathbf{V}_i = \mathbf{0}$. Geometrically, this implies that F is a rotation matrix for the vectors in the plane Π that passes through the origin and is parallel to the object plane. We can determine this plane Π by using the property that the length of vector is conserved under a rotation. What we first need to do is finding two different vectors $\mathbf{V}_1$ and $\mathbf{V}_2$ that belong to Π, which is done using the eigenvalue decomposition of F^tF. The rotation matrix R is then determined by the constraint that R rotates V_i to $F\mathbf{V}_i$, for i=1, 2. $\mathring{\mathbf{N}}$ is a unit vector that is orthogonal to both $\mathbf{V}_1$ and $\mathbf{V}_2$, and finally $\mathring{\mathbf{T}}$ is determined from R, $\mathring{\mathbf{N}}$ and F.

Let $V = [\mathbf{V}_1\ \mathbf{V}_2\ \mathbf{V}_3]$, where $\mathbf{V}_3 = \mathring{\mathbf{N}} \triangleq \dfrac{\mathbf{N}}{\|\mathbf{N}\|}$, $\mathbf{V}_1$ and $\mathbf{V}_2$ are such that V is an orthogonal matrix. Since $\mathbf{V}_1$ is orthogonal to $\mathbf{N}$, it follows that

$$F\mathbf{V}_1 = (R + \mathbf{T}\mathbf{N}^t)\mathbf{V}_1 = R\mathbf{V}_1 + \mathbf{T}\mathbf{N}^t\mathbf{V}_1 = R\mathbf{V}_1. \tag{5.2.20}$$

Similarly we get

$$F\mathbf{V}_2 = R\mathbf{V}_2. \tag{5.2.21}$$

Therefore $F\mathbf{V}_1$ and $F\mathbf{V}_2$ are orthonormal. We get

$$F\mathbf{V}_3 = R\mathbf{V}_3 + \mathbf{T}\mathbf{N}^t\mathbf{V}_3 = R\mathbf{V}_3 + \|\mathbf{N}\|\mathbf{T}. \tag{5.2.22}$$

Since $F\mathbf{V}_1$ and $F\mathbf{V}_2$ are orthonormal, we obtain

$$V^tF^tFV = \begin{bmatrix} 1 & 0 & a \\ 0 & 1 & b \\ a & b & c \end{bmatrix} \triangleq U \tag{5.2.23}$$

where (using (5.2.20)-(5.2.22))

$$a = F\mathbf{V}_1 \cdot F\mathbf{V}_3 = R\mathbf{V}_1 \cdot (R\mathbf{V}_3 + \|\mathbf{N}\|\mathbf{T}) = (R\mathbf{V}_1 \cdot \mathbf{T})\|\mathbf{N}\|, \tag{5.2.24.a}$$

$$b = F\mathbf{V}_2 \cdot F\mathbf{V}_3 = R\mathbf{V}_2 \cdot (R\mathbf{V}_3 + \|\mathbf{N}\|\mathbf{T}) = (R\mathbf{V}_2 \cdot \mathbf{T})\|\mathbf{N}\|, \tag{5.2.24.b}$$

$$c = F\mathbf{V}_3 \cdot F\mathbf{V}_3 = \|R\mathbf{V}_3 + \|\mathbf{N}\|\mathbf{T}\|^2$$

$$= 1 + 2(R\mathbf{N}\cdot\mathbf{T}) + \|\mathbf{T}\|^2\|\mathbf{N}\|^2. \tag{5.2.24.c}$$

From (5.2.23), the characteristic polynomial of U is

$$\det\{\lambda I_3 - U\} = \det\left\{\begin{bmatrix} \lambda - 1 & 0 & -a \\ 0 & \lambda - 1 & -b \\ -a & -b & \lambda - c \end{bmatrix}\right\} \tag{5.2.25}$$

$$= (\lambda - 1)(\lambda^2 - (1 + c)\lambda - (a^2 + b^2 - c)). \tag{5.2.26}$$

The eigenvalues of F^tF are obtained by solving for the roots of the characteristic polynomial in (5.2.26):

$$\lambda_2 = 1, \tag{5.2.27.a}$$

$$\lambda_1 = \frac{1 + c - \sqrt{(1 - c)^2 + 4(a^2 + b^2)}}{2}, \tag{5.2.27.b}$$

$$\lambda_3 = \frac{1 + c + \sqrt{(1 - c)^2 + 4(a^2 + b^2)}}{2}. \tag{5.2.27.c}$$

First consider $a^2 + b^2 = 0$. From (5.2.27) it follows that

$$\lambda_1 = \min(1, c), \quad \lambda_3 = \max(1, c), \quad \text{if } a^2 + b^2 = 0. \tag{5.2.28.a}$$

If $a^2 + b^2$ increases from zero, from (5.2.27) we see that λ_1 decreases and λ_3 increases. Therefore, it follows that

$$\lambda_1 < \min(1, c), \quad \lambda_3 > \max(1, c), \quad \text{if } a^2 + b^2 \neq 0. \tag{5.2.28.b}$$

Noticing $\lambda_2 = 1$, we always have

$$\lambda_1 \leq \lambda_2 = 1 \leq \lambda_3. \tag{5.2.29}$$

The multiple eigenvalues result in special solutions, which we will discuss in the next subsection. We first assume that the eigenvalues are distinct.

Let Π be a plane spanned by all the vector orthogonal to $\mathbf{N}$, i.e., $\Pi = \{\mathbf{V} \mid \mathbf{V}\cdot\mathbf{N} = 0\}$. In addition, we define another plane Π_H that is transformed from Π by orthogonal matrix H^t, where H is defined in (5.2.18), i.e., $\Pi_H = \{H^t\mathbf{V} \mid \mathbf{V}\cdot\mathbf{N} = 0\}$. For any vector $\mathbf{U} \in \Pi_H$, $\mathbf{U} = (u_x, u_y, u_z)$, there exists a vector $\mathbf{V}$ in Π such that

$$\mathbf{U} = H^t \mathbf{V}. \tag{5.2.30}$$

From (5.2.18) and (5.2.30), it follows that

$$\|F\mathbf{V}\|^2 = \mathbf{V}^t F^t F \mathbf{V} = \mathbf{U}^t H^t F^t F H \mathbf{U}$$

$$= \mathbf{U}^t \operatorname{diag}(\lambda_1, 1, \lambda_3)\mathbf{U} = \lambda_1 u_x^2 + u_y^2 + \lambda_3 u_z^2. \tag{5.2.31}$$

On the other hand,

$$\|R\mathbf{V}\|^2 = \|\mathbf{V}\|^2 = \|H\mathbf{U}\|^2 = \|\mathbf{U}\|^2 = u_x^2 + u_y^2 + u_z^2. \tag{5.2.32}$$

Since $\mathbf{V}$ is in Π, we get

$$F\mathbf{V} = (R + \mathbf{T}\mathbf{N}^t)\mathbf{V} = R\mathbf{V}. \tag{5.2.33}$$

Therefore the last terms in (5.2.31) and (5.2.32) are equal:

$$\lambda_1 u_x^2 + u_y^2 + \lambda_3 u_z^2 = u_x^2 + u_y^2 + u_z^2 \tag{5.2.34}$$

or equivalently

$$\sqrt{1-\lambda_1}\, u_x = \pm \sqrt{\lambda_3 - 1}\, u_z. \tag{5.2.35}$$

Each sign in (5.2.35) determines a plane. If $\lambda_1 \neq 1$ and $\lambda_3 \neq 1$, we get two planes:

$$\Pi_a = \{(u_x, u_y, u_z) \mid \sqrt{1-\lambda_1}u_x = \sqrt{\lambda_3 - 1}u_z\}, \tag{5.2.36.a}$$

$$\Pi_b = \{(u_x, u_y, u_z) \mid \sqrt{1-\lambda_1}u_x = -\sqrt{\lambda_3 - 1}u_z\}. \tag{5.2.36.b}$$

Since all the vectors in the plane Π_H satisfy (5.2.34), we known $\Pi_H = \Pi_a$ or $\Pi_H = \Pi_b$.

Geometrically, (5.2.31) implies that F transforms a unit sphere to an ellipsoid. Equation (5.2.34) indicates that we intersect a unit sphere with this ellipsoid to find a plane Π from which F is a length-preserving transformation. Since F is a rotation for vectors in Π, such an intersection does exist. However, since there are generally two planes whose intersection with the ellipsoid gives a circle, two planes Π_a and Π_b are determined. Π_H is one of them.

Now we are ready to determine the solutions. First assume $\Pi_H = \Pi_a$. Choosing any two unit vectors $\mathbf{U}_1$ and $\mathbf{U}_2$ from Π_a, from (5.2.30) we get two unit vectors in Π: $\mathbf{V}_i = H\mathbf{U}_i$, $i = 1, 2$. We can choose two orthonormal vectors $\mathbf{U}_1 = (\alpha, 0, \beta)^t$ and $\mathbf{U}_2 = (0, 1, 0)^t$, where

$$\alpha = \left[\frac{\lambda_3 - 1}{\lambda_3 - \lambda_1}\right]^{1/2}, \qquad \beta = \left[\frac{1 - \lambda_1}{\lambda_3 - \lambda_1}\right]^{1/2}, \tag{5.2.37}$$

from the plane Π_a. Using (5.2.30) we get two orthonormal vectors in plane Π:

$$\mathbf{V}_1 = H\,\mathbf{U}_1, \qquad \mathbf{V}_2 = H\,\mathbf{U}_2. \tag{5.2.38}$$

From (5.2.33) it follows that

$$R\,\mathbf{V}_i = F\,\mathbf{V}_i,\ i = 1, 2. \tag{5.2.39}$$

The rotation matrix R is determined uniquely by the positions of the two nonparallel vectors. In fact,

$$R\,[\mathbf{V}_1\ \mathbf{V}_2\ \mathbf{V}_1{\times}\mathbf{V}_2] = [F\,\mathbf{V}_1\ \ F\,\mathbf{V}_2\ \ F\,\mathbf{V}_1{\times}F\,\mathbf{V}_2].$$

Now consider the unit normal of the plane $\mathring{\mathbf{N}}$. Since $\mathbf{V}_1$ and $\mathbf{V}_2$ are two orthonormal vectors in plane Π which is orthogonal to $\mathbf{N}$, we determine $\mathring{\mathbf{N}}$ by

$$\mathring{\mathbf{N}} = \pm\mathbf{V}_1{\times}\mathbf{V}_2. \tag{5.2.40}$$

For the time being, we assume the positive sign is correct. Define

$$\tilde{\mathbf{T}} = \|\mathbf{N}\|\mathbf{T}. \tag{5.2.41}$$

We get $F\mathring{\mathbf{N}} = R\mathring{\mathbf{N}} + \tilde{\mathbf{T}}\mathring{\mathbf{N}}^t\mathring{\mathbf{N}} = R\mathring{\mathbf{N}} + \tilde{\mathbf{T}}$. Thus, we determine $\tilde{\mathbf{T}}$ by

$$\tilde{\mathbf{T}} = E\mathring{\mathbf{N}} - R\mathring{\mathbf{N}}. \tag{5.2.42}$$

From (5.2.2) we know that the solutions p and p' in

$$p'\mathbf{X}' = pR\,\mathbf{X} + \tilde{\mathbf{T}}$$

are positive. The above equation gives

$$\mathbf{0} = p\,\mathbf{X}'{\times}R\,\mathbf{X} + \mathbf{X}'{\times}\tilde{\mathbf{T}}.$$

That is, $\tilde{\mathbf{T}}{\times}\mathbf{X}'$ and $\mathbf{X}'{\times}R\,\mathbf{X}$ have the same directions. If and only if the sign of $\tilde{\mathbf{T}}$ is wrong, they have the opposite directions and

$$(\mathbf{X}' \times R\,\mathbf{X}) \cdot (\tilde{\mathbf{T}} \times \mathbf{X}') < 0 \tag{5.2.43}$$

holds. On the other hand, from (5.2.42) we know that the sign of $\mathring{\mathbf{N}}$ is wrong if and only if that of $\tilde{\mathbf{T}}$ is wrong. Thus, we change the sign of $\mathring{\mathbf{N}}$ (and that of $\tilde{\mathbf{T}}$ accordingly) if (5.2.43) is true.

From (5.2.4), (5.2.8) and (5.2.41), we get $\mathring{\mathbf{T}} = \tilde{\mathbf{T}} / \|\tilde{\mathbf{T}}\|$, and $\tilde{\mathbf{N}} = \|\tilde{\mathbf{T}}\| \mathring{\mathbf{N}}$. Note that $\tilde{\mathbf{T}}$ is not a zero vector, since otherwise F is orthogonal from (5.2.39) and Case (4) occurs.

If $\Pi_H = \Pi_b$, $\mathbf{U}_1 = (\alpha, 0, -\beta)$ and $\mathbf{U}_2 = (0, 1, 0)$ are the two orthonormal vectors chosen from Π_b. The remaining steps are the same as those in the case $\Pi_H = \Pi_a$.

Since we have either $\Pi_H = \Pi_a$ or $\Pi_H = \Pi_b$, there are two possible candidate solutions. The following theorem says that they are both valid solutions to $F = R + \mathring{\mathbf{T}} \tilde{\mathbf{N}}^t$.

Theorem 5.2. If the eigenvalues of $F^t F$ are distinct, there are exactly two solutions for R, $\mathring{\mathbf{T}}$ and $\mathring{\mathbf{N}}$ in the equation

$$R + \mathring{\mathbf{T}} \tilde{\mathbf{N}}^t = F \tag{5.2.44}$$

with the constraints that R is a rotation matrix, $\mathring{\mathbf{T}}$ a unit vector and the signs of $\mathring{\mathbf{T}}$ and $\tilde{\mathbf{N}}$ determined by (5.2.43).

Proof. By the above derivation we know that there are at most two possible solutions corresponding to $\Pi_H = \Pi_a$ and $\Pi_H = \Pi_b$, respectively. We now prove that these two possible solutions are both solutions of (5.2.44). According to the above steps, we get the first solution $(R, \mathring{\mathbf{T}}, \tilde{\mathbf{N}})$. Since $\tilde{\mathbf{N}}$ determined in (5.2.40) is orthogonal to $\mathbf{V}_i$, for $i = 1, 2$, we have

$$(R + \mathring{\mathbf{T}} \tilde{\mathbf{N}}^t) \mathbf{V}_i = R\,\mathbf{V}_i = F\,\mathbf{V}_i. \tag{5.2.45}$$

The last equation holds because R is determined from this, and such a R can always find since $\mathbf{V}_1$ and $\mathbf{V}_2$ are orthonormal and so are $F\,\mathbf{V}_1$ and $F\,\mathbf{V}_2$ (see (5.2.33)). From (5.2.42), it follows that

$$\begin{aligned}(R + \mathring{\mathbf{T}} \tilde{\mathbf{N}}^t) \mathring{\mathbf{N}} &= R\,\mathring{\mathbf{N}} + \mathring{\mathbf{T}} \|\tilde{\mathbf{T}}\| = R\,\mathring{\mathbf{N}} + \tilde{\mathbf{T}} \\ &= R\,\mathring{\mathbf{N}} + (F\,\mathring{\mathbf{N}} - R\,\mathring{\mathbf{N}}) = F\,\mathring{\mathbf{N}}.\end{aligned} \tag{5.2.46}$$

Since $[\mathbf{V}_1\ \mathbf{V}_2\ \mathring{\mathbf{N}}]$ is an orthogonal matrix, from (5.2.45) and (5.2.46) we get $R + \mathring{\mathbf{T}} \tilde{\mathbf{N}}^t = F$. Thus $(R, \mathring{\mathbf{T}}, \tilde{\mathbf{N}})$ is one solution of (5.2.44).

In the second solution, the sign of β is changed and α is unchanged. Equations (5.2.45) and (5.2.46) still hold for this second solution. Thus the second solution satisfies (5.2.44). □

5.2.5 Two Solutions Are Both Valid

Now we know that the two solutions both satisfy (5.2.44). One question arises as to whether they both are consistent with the observed images. As stated in the following theorem they correspond to two planes (under different motions) that give the same images at the two time instants. Therefore, two solutions are valid interpretations.

Theorem 5.3. Given n point correspondences represented by $(\mathbf{X}_i, \mathbf{X}'_i)$, $i = 1, 2, \cdots, n$. Let F be a matrix that satisfies (5.2.15) for every $(\mathbf{X}_i, \mathbf{X}'_i)$. If there are rotation matrices R_a, R_b, unit vectors $\mathring{\mathbf{T}}_a$, $\mathring{\mathbf{T}}_b$ and vectors $\tilde{\mathbf{N}}_a$ and $\tilde{\mathbf{N}}_b$, such that

$$F = R_a + \mathring{\mathbf{T}}_a \tilde{\mathbf{N}}_a^t = R_b + \mathring{\mathbf{T}}_b \tilde{\mathbf{N}}_b^t \tag{5.2.47}$$

and $\tilde{\mathbf{N}}_a \cdot \mathbf{X}_i > 0$, $\tilde{\mathbf{N}}_b \cdot \mathbf{X}_i > 0$, $i = 1, 2, \cdots, n$, hold true, then there exist two planes with $\tilde{\mathbf{N}}_a$ and $\tilde{\mathbf{N}}_b$ as normals, respectively, such that if they undergo motions represented by $(R_a, \mathring{\mathbf{T}}_a)$ and $(R_b, \mathring{\mathbf{T}}_b)$, respectively, they render the same pair of images (with image vectors $\mathbf{X}_i$ and $\mathbf{X}'_i$ at the two time instants) and the corresponding 3-D points are all located in the forward half space before and after the motion. (If $\tilde{\mathbf{N}}_a \cdot \mathbf{X}_i < 0$ or $\tilde{\mathbf{N}}_b \cdot \mathbf{X}_i < 0$ is true for a few points, then the conclusion still holds except that the corresponding point is located in the back half space.)

Proof. See Appendix 5.B.

Fig. 5.1 shows an example where two solutions render the same pair of images with the positive depths for all the points. Four corners of a planar square board are used as feature points. Before motion, the board is orthogonal to the optical axis of the camera with the optical axis passing through its center. The motion consists a rotation about the vertical axis by a small angle and a translation towards the camera. Fig. 5.1(a) shows the setup and motion of the board, where board A moves to board A′ (both boards are shown with solid bold lines). In addition to the veridical interpretation (that agrees with what actually happened), another illusive interpretation (that does not agree with what actually happened) exists that is consistent with these two images. This illusive interpretation corresponds to board positions shown in Fig. 5.1(b) with dashed bold lines. That is, in Fig. 5.1(b) if board B moves to board B′, we get the exactly

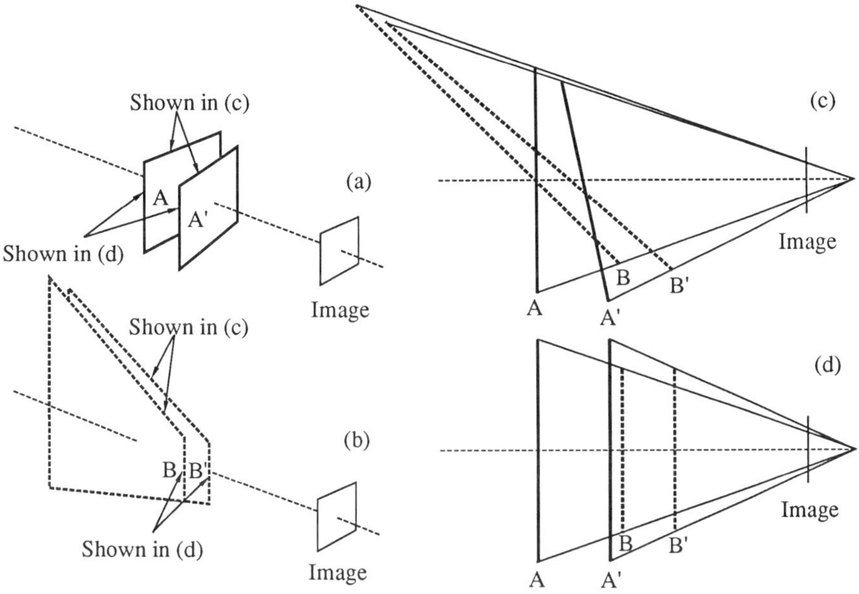

Fig. 5.1. Two solutions, veridical and illusive, render the same pair of images (see text). **(a)** Illustration of the setup and the veridical solution. **(b)** Illusive solution. **(c)** A combined top view. **(d)** A combined side view, with closer vertical edges shown only.

same images as in Fig. 5.1(a). To show the relationships between these two interpretations more precisely, the top edges of the boards in these two interpretations are shown in Fig. 5.1(c) as a top view combining Figs. 5.1(a) and (b). Fig. 5.1(d) is the corresponding side view, where for each board only the vertical edge that is closer to the reader is shown for clearity. As can be seen from the figure, the image projections of the corresponding corners of boards are the same for those two interpretations.

In some cases, we are able to reject the illusive solution on the ground that all the 3-D points should have positive depths. The illusive solution may have some recovered points in the back half space of the camera. In other words, one of the conditions $\tilde{\mathbf{N}}_a \cdot \mathbf{X} > 0$ and $\tilde{\mathbf{N}}_b \mathbf{X} > 0$ stated in Theorem 5.3 may be violated for some image points. However, it is proved in the Theorem 5.3 that once these conditions for positive depths are satisfied before motion, positive depths are guaranteed after motion for illusive solution as well. We have conducted a series of simulations, and we were often not be able to reject the illusive solution based on the condition of positive depths. The example shown in Fig. 5.1 has positive depths for both solutions and thus the illusive solution can not be rejected. If the plane of the illusive solution is so tilted that the intersections of the plane and the projection lines of some points are located in the back half space, as illustrated in Fig. 5.2, then the illusive solution can be rejected.

5.2.6 Special Cases

Now let us turn to the cases where F^tF has multiple eigenvalues. We first derive the necessary and sufficient conditions for F^tF to have multiple eigenvalues, starting from the following lemma.

Lemma 5.1. Let the eigenvalues of F^tF be $\lambda_1, \lambda_2, \lambda_3$, with $\lambda_1 \le \lambda_2 \le \lambda_3$. Then

(1) F^tF has multiple eigenvalues iff $\mathbf{T} \,//\, R\mathbf{N}$,
(2) $\lambda_1 = 1 < \lambda_3$ iff $\mathbf{T} \,//\, R\mathbf{N}$ and $2R\mathbf{N}\cdot\mathbf{T} > -\|\mathbf{T}\|^2\|\mathbf{N}\|^2$,
(3) $\lambda_1 < 1 = \lambda_3$ iff $\mathbf{T} \,//\, R\mathbf{N}$ and $2R\mathbf{N}\cdot\mathbf{T} < -\|\mathbf{T}\|^2\|\mathbf{N}\|^2$,
(4) $\lambda_1 = 1 = \lambda_3$ iff $\mathbf{T} \,//\, R\mathbf{N}$ and $2R\mathbf{N}\cdot\mathbf{T} = -\|\mathbf{T}\|^2\|\mathbf{N}\|^2$.

Proof. Notice that $\|\mathbf{N}\| \ne 0$, since $\mathbf{N}^t\mathbf{x} = 1$.

For (2): From (5.2.28), $\lambda_1 = 1 < \lambda_3$ iff $a^2 + b^2 = 0$ and $c > 1$. From (5.2.24), we have $a^2 + b^2 = 0$ and $c > 1$ iff $R\mathbf{V}_1\cdot\mathbf{T} = \mathbf{0}$ and $R\mathbf{V}_2\cdot\mathbf{T} = \mathbf{0}$ and $2R\mathbf{N}\cdot\mathbf{T} > -\|\mathbf{T}\|^2\|\mathbf{N}\|^2$. But from (5.2.40), $R\mathbf{V}_1\cdot\mathbf{T} = \mathbf{0}$ and $R\mathbf{V}_2\cdot\mathbf{T} = 0$ iff $\mathbf{T} \,//\, R\mathbf{N}$.

The proofs for cases (3) and (4) are similar to that of (2) and so are omitted. (1) immediately follows from (2), (3) and (4). □

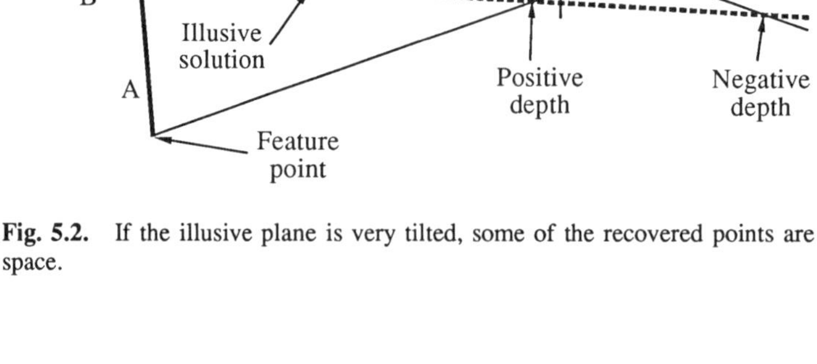

Fig. 5.2. If the illusive plane is very tilted, some of the recovered points are in the back half space.

Lemma 5.1 gives algebraic conditions for the multiple eigenvalues to occur. Further investigation of these conditions leads to the following new theorem.

Theorem 5.4. Let the eigenvalues of F^tF be $\lambda_1, \lambda_2, \lambda_3$, with $\lambda_1 \leq \lambda_2 \leq \lambda_3$. Then

(1) F^tF has multiple eigenvalues iff $\mathbf{T} // R\mathbf{N}$,

(2) $\lambda_1 = 1 < \lambda_3$ iff $\mathbf{T} // R\mathbf{N}$, and the absolute distance between the plane and the origin increases due to motion,

(3) $\lambda_1 < 1 = \lambda_3$ iff $\mathbf{T} // R\mathbf{N}$, and the absolute distance between the plane and the origin decreases due to motion,

(4) $\lambda_1 = 1 = \lambda_3$ iff $\mathbf{T} // R\mathbf{N}$, and the absolute distance between the plane and the origin dose not change.

Before proving this theorem, we first discuss the geometrical meanings of this theorem. Remember that we represent the motion of the plane by a rotation followed a translation. Fig. 5.3 shows the intermediate position of the plane that is rotated by the matrix R. Therefore, the normal of this rotated plane is $R\mathbf{N}$. The final position (shown with dashed line in Fig. 5.3) is determined by the translation vector $\mathbf{T}$. If and only if the translation is aligned with the normal of the rotated plane, multiple eigenvalues occur. The sign and the magnitude of the translation determine which case occurs. If the translation is such that the final absolute distance between the origin and the plane increases, Case (2) occurs. The absolute distance can be increased in two ways, as shown in Fig. 5.3. The first is to translate the rotated plane further away from the origin, in which the translation and the normal of the rotated plane have the same directions. The

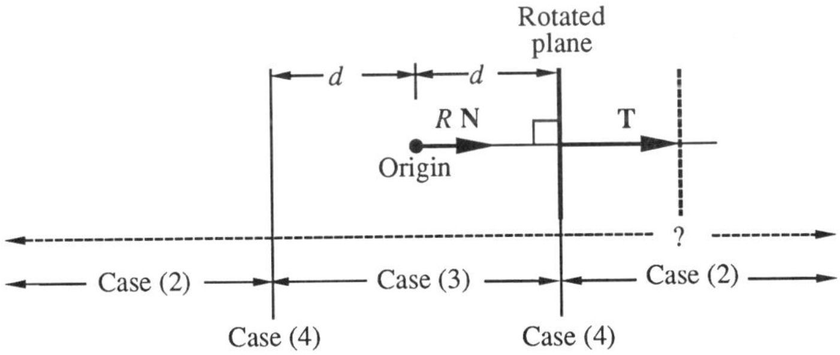

Fig. 5.3. Illustration of the necessary and sufficient conditions for the multiple eigenvalues to occur. Multiple eigenvalues occur if and only if $\mathbf{T}$ is parallel to $R\mathbf{N}$. The sign and the magnitude of $\mathbf{T}$ determine one of the Cases (2) to (4) in Theorem 5.4.

second way is to translate the plane to the back of the camera until the absolute distance exceeds the original distance. Case (3) occurs if the absolute distance decreases, this can be done only by translating the camera towards the origin as shown in Fig. 5.3. Although the magnitude of translation can be so large that the plane passes the origin, the final absolute distance is less than the original distance in Case (3). Finally, Case (4) occurs if the absolute distance does not change. Only two situations are associated with this case: the translation vanishes or the plane is translated to its mirror position with respect the origin.

Now we prove Theorem 5.4 using Lemma 5.1. According to the definition of $\mathbf{N}$ in (5.2.5) it follows that

$$\|\mathbf{N}\| = 1/d$$

where d is the absolute distance between the origin and the plane. Replacing $\|\mathbf{N}\|$ by $1/d$ in the inequalities in Lemma 15. and using the condition $\mathbf{T} \,//\, R\mathbf{N}$, we can see that those inequalities are in fact constraints on the magnitude of translation in terms of d. From Lemma 15., what we need to prove are Case (2) to Case (4).

First consider Case (2). $\mathbf{T} \,//\, R\mathbf{N}$ includes two situations: $\mathbf{T}$ and $R\mathbf{N}$ have either the same directions or the opposite directions. (a) Suppose the former is true, i.e., $R\mathbf{N}{\cdot}\mathbf{T} = \|\mathbf{N}\|\,\|\mathbf{T}\|$. The last inequality of Case (2) in Lemma 5.1 implies $2 > -\|\mathbf{T}\|\,\|\mathbf{N}\|$, which is always true. (b) Suppose $\mathbf{T}$ and $R\mathbf{N}$ have opposite directions, i.e., $R\mathbf{N}{\cdot}\mathbf{T} = -\|\mathbf{N}\|\,\|\mathbf{T}\|$. The last inequality of Case (2) in Lemma 5.1 implies $2 < \|\mathbf{T}\|\,\|\mathbf{N}\|$, or equivalently $\|\mathbf{T}\| > 2d$. This means that if translation is in the direction towards the origin (see Fig. 5.3), the magnitude of translation must be large enough so that the absolute distance is increased. Summarizing these two situations, the condition $\mathbf{T} \,//\, R\mathbf{N}$ and $2R\mathbf{N}{\cdot}\mathbf{T} > -\|\mathbf{T}\|^2\|\mathbf{N}\|^2$ in Lemma 5.1 is equivalent to the following condition: $\mathbf{T} \,//\, R\mathbf{N}$ and the absolute distance between the plane and the origin increases due to motion.

Case (3) can be proved in a similar way: In Case (3), $\mathbf{T}$ and $\mathbf{N}$ cannot have the same directions, otherwise the last inequality in Case (3) of Lemma 5.1 leads to $\|\mathbf{T}\| < -2d$ which is impossible. If $\mathbf{T}$ and $\mathbf{N}$ have the opposite directions, the last inequality means $\|\mathbf{T}\| < 2d$. That is, the absolute distance must decrease.

Case (4) occurs if and only if either $\mathbf{T}$ vanishes (then conditions in Lemma 5.1 is trivially satisfied) or $\mathbf{T}$ does not vanish and $\mathbf{T} \,//\, R\mathbf{N}$ and $2R\mathbf{N}{\cdot}\mathbf{T} = -\|\mathbf{T}\|^2\|\mathbf{N}\|^2$. In the latter situation with $\mathbf{T} \neq \mathbf{0}$, $\mathbf{T}$ and

$\mathbf{N}$ cannot have the same directions since this leads to impossible equation $\|\mathbf{T}\| = -2d$. Now suppose that $\mathbf{T}$ and $\mathbf{N}$ have the opposite directions. The corresponding necessary and sufficient condition in Lemma 5.1 is then equivalent to $\|\mathbf{T}\| = 2d$. That is, the absolute distance does not change (see Fig. 5.3). □

In Fig. 5.3, if the final position of the rotated plane is located to the left of the origin, then the other side of the plane faces the camera after motion, since rotation does not change the side. Then from Theorem 5.4, we get the following corollary immediately.

Corollary 5.2. Assuming that the translation does not vanish and the same side of the plane faces camera before and after motion, only two cases are associated with multiple eigenvalues:

(a) $\lambda_1 = 1 < \lambda_3$ iff $\mathbf{T} \mathbin{/\!/} R\mathbf{N}$, and $\mathbf{T}$ and $R\mathbf{N}$ have the same directions.

(b) $\lambda_1 < 1 = \lambda_3$ iff $\mathbf{T} \mathbin{/\!/} R\mathbf{N}$, and $\mathbf{T}$ and $R\mathbf{N}$ have the opposite directions. □

Now we turn to the solutions for these special cases. If F^tF has an eigenvalue with multiplicity of 2 ($\lambda_1 = 1$ or $\lambda_3 = 1$ but not both), the two planes in (5.2.36) degenerate into one plane. This gives steps for Cases (2) and Case (3), respectively, in the algorithm (Appendix 5.E).

If F^tF has eigenvalue of multiplicity of 3 ($\lambda_1 = \lambda_2 = \lambda_3$), Case (4) occurs (see Appendix 5.E). Then it is clear from (5.2.18) that F is orthogonal. For $\mathbf{V}_i \in \Pi$, we still have $F\mathbf{V}_i = R\mathbf{V}_i$, $i = 1,2$. Since F is orthogonal, we have

$$F\mathring{\mathbf{N}} = R\mathring{\mathbf{N}} + \mathbf{T}\|\mathbf{N}\| = \pm R\mathring{\mathbf{N}}. \tag{5.2.48}$$

If $\det(F) = 1$, $F\mathring{\mathbf{N}} = R\mathring{\mathbf{N}}$. Thus $R[\mathbf{V}_1\ \mathbf{V}_2\ \mathring{\mathbf{N}}] = F[\mathbf{V}_1\ \mathbf{V}_2\ \mathring{\mathbf{N}}]$. Therefore $R = F$. From $R\mathring{\mathbf{N}} = F\mathring{\mathbf{N}} = (R + \mathbf{T}\mathbf{N}^t)\mathring{\mathbf{N}} = R\mathring{\mathbf{N}} + \mathbf{T}\|\mathbf{N}\|$ and $\|\mathbf{N}\| \neq 0$, we get $\mathbf{T} = \mathbf{0}$. $\mathring{\mathbf{N}}$ cannot be determined.

If $\det(F) = -1$, $F\mathring{\mathbf{N}} = -R\mathring{\mathbf{N}}$. From (5.2.48) we have

$$\mathbf{T} = \frac{-2R\mathring{\mathbf{N}}}{\|\mathbf{N}\|}. \tag{5.2.49}$$

Then

$$F = R + \mathbf{T}\mathbf{N}^t = R - 2R\mathring{\mathbf{N}}\mathring{\mathbf{N}}^t = R(I_3 - 2\mathring{\mathbf{N}}\mathring{\mathbf{N}}^t).$$

Noticing $(I_3 - 2\mathring{\mathbf{N}}\mathring{\mathbf{N}}^t)(I_3 - 2\mathring{\mathbf{N}}\mathring{\mathbf{N}}^t) = I_3$ for any unit vector $\mathring{\mathbf{N}}$, we get (5.E.17). From (5.2.41) and (5.2.49) we get (5.E.18).

Finally, we need to show that given any $\mathring{\mathbf{N}}$, R in (5.E.17) and $\tilde{\mathbf{T}}$ in (5.E.18) give a solution for $R + \tilde{\mathbf{T}}\mathring{\mathbf{N}}^t = F$. First, R in (5.E.17) is a rotation matrix, since $I_3 - 2\mathring{\mathbf{N}}\mathring{\mathbf{N}}^t$ is a reflection with respect to a plane that has $\mathring{\mathbf{N}}$ as a normal and goes through the origin, and F is also a reflection. With R in (5.E.17) and $\tilde{\mathbf{T}}$ in (5.E.18), for any unit $\mathring{\mathbf{N}}$ we have

$$R + \mathbf{T}\mathbf{N}^t = R + \tilde{\mathbf{T}}\mathring{\mathbf{N}}^t = R - 2R\mathring{\mathbf{N}}\mathring{\mathbf{N}}^t = R(I_3 - 2\mathring{\mathbf{N}}\mathring{\mathbf{N}}^t)$$

$$= F(I_3 - 2\mathring{\mathbf{N}}\mathring{\mathbf{N}}^t)(I_3 - 2\mathring{\mathbf{N}}\mathring{\mathbf{N}}^t) = F.$$

To summarize, for motion of planar surface, there are generally two valid solutions, one veridical and one illusive. If the translation is aligned with the normal of the rotated plane, two solutions becomes one. If the translation is not vanished and the same side of the plane faces the camera before and after motion, these two cases are all that can occur.

If the translation vanishes, the motion parameters R and $\mathbf{T}$ can be determined but the structure $\mathbf{N}$ cannot be determined.

The complete algorithm is relegated to Appendix 5.E.

5.3 Inherent Uniqueness

It is mentioned in Chapter 3 that for the linear algorithm designed for general surfaces to reach a unique solution, the points must not lie in a quadric that passes the projection centers at which two images are taken. Obviously, this condition is just for the specific linear algorithm, and we called it *a*lgorithm-*d*ependent *c*ondition (ADC). The algorithm-independent condition (AIC), on the other hand, specifies the necessary and sufficient condition that uniquely determines the solution, independent of whether we have an algorithm to compute it or not. The AIC for the determination of motion from general surface is still unknown [5.1, 6], although it is clear that the AIC is weaker than the ADC of the existing linear algorithms. Symbolically, we express this by AIC < ADC, where < means "weaker". In other words, the linear algorithms leave a gap between what has been solved now and what may be solved. Some special point configurations with more than 7 point correspondences and most point configurations with 5 to 7 point correspondences are located in this gap. The situation is similar for the algorithms in Chapter 6 that use line correspondences. See Fig. 5.4(a).

However, quite unexpectedly, we will see that for determining motion and structure from a rigid planar surface, it turns out that AIC = ADC is true. That is, for planar surfaces, we have a linear algorithm which can guarantee a unique solution whenever the solution is uniquely determined by the input data. The linear algorithm goes all the way: no cap between AIC and ADC is left (see Fig. 5.4(b)). If there is a unique solution, the linear algorithm guarantees to obtain it. If in a condition the linear algorithm fails, so will any algorithm (provided with the same information). This means that the problem for planar case is now completely solved by the linear algorithm, and in this sense, the problem for planar case is closed.

5.3.1 Decomposability

Theorem 5.3 states that under two conditions: (I) if the matrix F that satisfies (5.2.15) for all the points, or equivalently satisfies (5.2.17), and (II) the matrix F is decomposable as in (5.2.47), then there existing corresponding plane and motion which rend the same images, aside from possibly a few negative depths. In solving for (5.2.17) the vector **h** is regarded as a free vector. Therefore, it is not clear up to now whether such a solution **h** will always result in an F that satisfies condition (II). This issue is closely related to the inherent uniqueness of the problem.

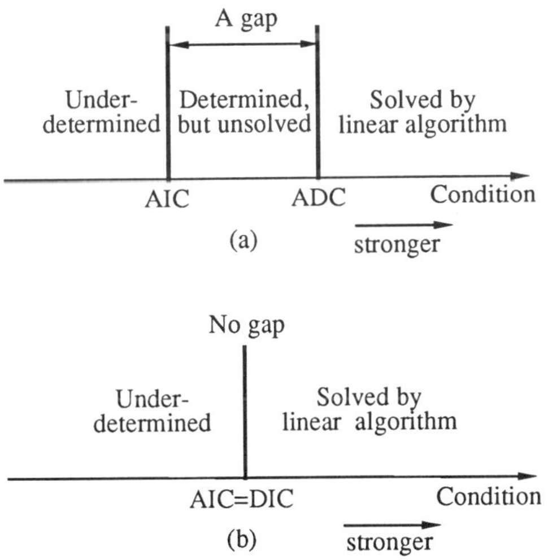

Fig. 5.4. (**a**) The linear algorithms for general surfaces cannot solve all the problems that are solvable. (**b**) For planar surfaces, the corresponding noniterative linear algorithm goes all the way: it solves all that are solvable.

Now, the question is whether any 3×3 matrix, say F_s, is always decomposable, i.e., there exists a scale factor k, a rotation matrix R, a vector $\tilde{\mathbf{T}}$ and a unit vector $\mathring{\mathbf{N}}$ such that

$$F_s = k(R + \tilde{\mathbf{T}}\mathring{\mathbf{N}}^t) \tag{5.3.1}$$

is exactly true (the matrix F_s can be considered as a matrix computed in (5.2.17)). Note that the form of decomposition (5.3.1) can always interchange with the decomposition

$$F_s = k(R + \mathring{\mathbf{T}}\tilde{\mathbf{N}}^t). \tag{5.3.2}$$

Since when $\tilde{\mathbf{T}} = 0$ in (5.3.1), $F = R$ and $\tilde{\mathbf{N}}$ is replaced by a zero vector and $\mathring{\mathbf{T}}$ can be any unit vector. It is also true from (5.3.2) to (5.3.1): a zero $\tilde{\mathbf{H}}$ in (5.3.2) leads to a zero $\tilde{\mathbf{T}}$ in (5.3.1).

It can be seen that there are exactly 9 degrees of freedom on the right-hand side of (5.3.1): one for k, three for R, three for $\tilde{\mathbf{T}}$ and two for $\mathring{\mathbf{N}}$. In other words, a decomposable matrix has the same number of degrees of freedom as any 3×3 matrix. This suggests that it might be true that any 3×3 matrix is decomposable. In fact, we have the following theorem:

Theorem 5.5. Any 3×3 matrix F_s that has a rank no less than 2 is decomposable, i.e., there exists a scale factor k, a 3×3 rotation matrix R, a 3-D vector $\tilde{\mathbf{T}}$ and a unit 3-D vector $\mathring{\mathbf{N}}$ such that (5.3.1) holds exactly (similar for (5.3.2)).

Proof. Suppose a matrix F_s is given whose rank is no less than two. Then the second smallest eigenvalue of $F_s^t F_s$, γ_2, is positive. Let $k = \sqrt{\gamma_2}$, we define $F = k^{-1} F_s$. Therefore, the second smallest eigenvalue, λ_2, of $F^t F$ is equal to 1. Determine H for F as in (5.2.18) and define two orthonormal vectors $\mathbf{V}_1$ and $\mathbf{V}_2$ as in (5.2.38), and a rotation matrix R as in (5.2.39). Is there still a unique rotation R that satisfies (5.2.39)? The answer is positive. To prove this, it is equivalent to prove that $F\mathbf{V}_1$ and $F\mathbf{V}_2$ are still orthonormal vectors. It follows from (5.2.18) and (5.2.38) that

$$\begin{aligned}(F\mathbf{V}_i)^t(F\mathbf{V}_j) &= \mathbf{U}_i^t H^t F^t F H \mathbf{U}_j \\ &= \mathbf{U}_i^t \mathrm{diag}(\lambda_1, \lambda_2, \lambda_3)\mathbf{U}_j = \delta(i - j)\end{aligned}$$

for $i = 1, 2$, where $\delta(k) = 1$ if $k=0$ and $\delta(k) = 0$ otherwise. Therefore, there is a unique rotation matrix that rotates these two orthonormal vectors exactly to the other two orthonormal vectors so that

$$R\mathbf{V}_i = F\mathbf{V}_i \tag{5.3.3}$$

exactly holds for $i = 1, 2$. (So R in Equation (5.E.11) of the algorithm presented in Appendix 5.E gives a rotation matrix even if the data are contaminated by noise.)

Then we define $\tilde{\mathbf{T}}$ and $\mathring{\mathbf{N}}$ using (5.2.42), and (5.2.40). Equations (5.2.38), (5.2.40) and (5.3.3) lead to

$$(R + \tilde{\mathbf{T}}\mathring{\mathbf{N}}^t)\mathbf{V}_i = R\mathbf{V}_i = F\mathbf{V}_i, \tag{5.3.4}$$

for $i = 1, 2$. Equation (5.2.42) gives

$$(R + \tilde{\mathbf{T}}\mathring{\mathbf{N}}^t)\mathring{\mathbf{N}} = R\mathring{\mathbf{N}} + \tilde{\mathbf{T}} = R\mathring{\mathbf{N}} + (F\mathring{\mathbf{N}} - R\mathring{\mathbf{N}}) = F\mathring{\mathbf{N}}. \tag{5.3.5}$$

Then,

$$F = R + \tilde{\mathbf{T}}\mathring{\mathbf{N}}^t$$

follows from (5.3.4) and (5.3.5). Therefore, F is decomposable and (5.3.1) is proved. □

5.3.2 Rank Condition Is Algorithm-Independent

Theorem 5.5 leads to a very important conclusion: The condition that $\text{rank}(A) = 8$ is not only a sufficient condition for our algorithm to reach two solutions (except the special cases in Theorem 5.4), it is in fact an algorithm-independent necessary condition!

Theorem 5.6. There exist the exact number of solutions as given by the linear algorithm, if and only if $\text{rank}(A) = 8$.

Proof. We have proved that $\text{rank}(A) = 8$ is a sufficient condition for our algorithm, and thus it is a sufficient condition for the problem itself. What we need to prove is that if $\text{rank}(A) \neq 8$ there exist infinitely many solutions that render the same images.

Certainly $\text{rank}(A) > 8$ is impossible for coplanar points undergoing a rigid motion since $\mathbf{h}$ in (5.2.17) is a nonzero vector. Suppose $\text{rank}(A) < 8$. Then in (5.2.17) there are at least a solution $\mathbf{h}$ that corresponds to the true motion, and another solution $\mathbf{h}'$, which is linearly independent of $\mathbf{h}$. For any real number k, $\mathbf{h}(k) \triangleq \mathbf{h} + k\mathbf{h}'$ is also a solution of (5.2.17). Let $F_s(k)$ be the matrix determined from

$\mathbf{h}(k)$, as in (5.E.4) of Appendix 5.E. $F_s(0)$ corresponds to the true solution and thus it is decomposable, and the rank of $F_s(0)$ is no less than two. Thus, $F_s(0)$ has at least one 2 by 2 minor determinant that is not equal to zero. Since a determinant is continuous with respect to its elements, there exists a sufficiently small positive number $\delta > 0$, such that for any k, $|k| < \delta$, the corresponding minor determinant of $F_s(k)$ is also non-zero. That is, for and k, $|k| < \delta$, the matrix $F_s(k)$ has a rank no less than two. Therefore, these infinitely many $F_s(k)$ are decomposable. On the other hand, there is no number $c(k)$ so that $F_s(k) = c(k)F_s(0)$ holds true, because $\mathbf{h}$ and $\mathbf{h}'$ are linearly independent. Therefore, there are infinitely many matrices $F(k)$, which are not scaled version of $F_s(k)$, so that the median eigenvalue of $F(k)^t F(k)$ is unit (see (5.E.8) in Appendix 5.E). Finally, from Theorem 5.3 we know that all these $F(k)$ give infinitely many solutions for motion and structure which render the same pair of images. □

5.3.3 The Fundamental Theorem

Now we have established that $rank(A) = 8$ is not only sufficiently to our algorithm, it is in fact a necessary condition to determine at most two solutions (except for the Case 4 in Theorem 5.4). We summarize our results in the following fundamental theorem:

The Fundamental Theorem. Given two perspective views of a rigidly moving planar surface, the minimum number of solutions of $(R, \mathring{\mathbf{T}}, \tilde{\mathbf{N}})$ falls into the following cases according to the true motion and true position of the surface:

Case (1) $\mathbf{T} \times R\,\mathbf{N} \neq \mathbf{0}$: the minimum number of solutions is 2;

Case (2) $\mathbf{T} \,//\, R\,\mathbf{N}$, and the absolute distance between the plane and the origin increases due to motion: the minimum number of solutions is 1;

Case (3) $\mathbf{T} \,//\, R\,\mathbf{N}$, and the absolute distance between the plane and the origin decreases due to motion: the minimum number of solutions is 1;

Case (4) $\mathbf{T} \,//\, R\,\mathbf{N}$, and the absolute distance between the plane and the origin dose not change, which includes two subcases:

Subcase (4a) $\mathbf{T} = 0$: the minimum number of solutions for $(R, \mathbf{T})$ is 1, but infinite for $\tilde{\mathbf{N}}$.

Subcase (4b) the origin is located half way between the rotated-only plane and the fully moved plane: the minimum number of solutions is infinite.

All the above-mentioned solutions render the same images. Some of the solutions may force some points to be reconstructed in the back half space, in which case one may identify that the corresponding solution is illusive. Whenever the number of existing solution reaches the above minimum, the linear algorithm presented in Appendix 5.E guarantees to provide exactly these solutions. The necessary and sufficient conditions for the number of solution to reach the above minimum can be given in the following three ways:

(A) In terms of the rank of A: $rank(A) = 8$.

(B) In terms of the n points in 2-D image plane: there exists a set of four object points such that no image projections of any three points in this set are collinear in any of the two images.

(C) In terms of the n points in 3-D surface: there exists a set of four points in the object plane such that no three points in this set are collinear in the object plane and if the object plane is extended, it does not go through the center of projection before and after motion.

Proof. Since our above discussion has established many related results, the proof of this Fundamental Theorem can be concise. The solutions were derived when we discuss the linear algorithm and we have proved that they all satisfy (5.2.15). Theorem 5.3 concludes that they all render the same images. Theorem 5.4 tells under what configurations the different cases occur. $rank(A) = 8$ is the sufficient condition for the linear algorithm to give these minimum numbers of solution but Theorem 5.6 states that this condition is necessary, for the problem itself, to restrict the number of existing solution to this minimum. The geometrical necessary and sufficient conditions given by Theorem 5.1 and Corollary 5.1 are then not only necessary and sufficient for $rank(A) = 8$ but also necessary and sufficient for the problem itself, independent of specific algorithm. □

The Fundamental Theorem not only answers the inherent uniqueness question of this problem, but also gives a complete list of all the cases, each is associated with the clear geometrical conditions. Best of all, we know that all can be solved by the linear algorithm presented. If the linear algorithm fails to give these solutions, so does any algorithm.

5.3.4 Plane-Perceivable Surfaces

In this subsection, we investigate the problem from a different perspective. Previously, we assumed that the points are coplanar and they undergo a rigid motion. Now, suppose that we do not know that the points are coplanar. The questions arise as to in which case one can detect, from image points, that the points are not coplanar, and in which case one cannot.

According to Theorems 5.3 and 5.5 we know that if $\text{rank}(A) \leq 8$, there exist solutions each corresponds to coplanar points undergoing a rigid motion, no matter whether the actual points are coplanar or not. (Sometimes the solution may result in some negative depths, but for the discussion in this subsection, we put this aside). Therefore, we do not have any solution only when $\text{rank}(A) = 9$. The following theorem gives a necessary and sufficient condition on the configuration of points to result in $\text{rank}(A) \leq 8$, or equivalently, according to the Fundamental Theorem, there exists an interpretation that a planar patch undergoes a rigid motion.

Theorem 5.7. From two views, there exist an interpretation that a planar patch undergoes a rigid motion, if and only if there exists a 3×3 nonzero matrix K such that all the observed points lie in the following intersection of two quadratic surfaces before motion:

$$(\mathbf{x} - \mathbf{O}')\times K\mathbf{x} = 0 \tag{5.3.6}$$

where $\mathbf{O}' = -R^t\mathbf{T}$.

Proof. See Appendix 5.C.

It is clear that the three scalar equations in (5.3.6) are quadratic in the components of $\mathbf{x}$. Thus, each equation determines a quadratic surface. It is easy to prove that in (5.3.6) there are just two scalar equations that are independent. Therefore (5.3.6) determines the intersection of two quadratic surfaces. Such an intersection gives a curve in general, with plane as a special case. But we will still use the more general term "surface". The origin and $\mathbf{O}'$ are in this intersection. If we consider the object as stationary and the motion is due to that of the camera, the origin and $\mathbf{O}'$ are the positions of the camera at time t_1 and t_2, respectively.

We know that there is a solution that gives coplanar 3-D points if and only if $\text{rank}(A) \leq 8$. According to the above theorem, if and only if the points lie on a surface in the form of (5.3.6), the images can be interpreted as the projections of a set of coplanar points undergoing a rigid motion. Therefore, we call the surfaces in (5.3.6) *plane-perceivable* surfaces. Given two images, if the 3-D points are from such a surface, one cannot tell they are coplanar or not and can always interpret them as coplanar. If and only if they do not lie on any plane-perceivable surface in (5.3.6), one can detect that the points are not coplanar by checking $\text{rank}(A) > 8$. Fig. 5.5 illustrates the relations among planes, plane-perceivable surfaces and the rest of detectable non-planar surfaces. We will see that plane-perceivable surfaces include many nonplanar surfaces. This is due to the fact that image projections do not provide complete information for what happened in 3-D.

Now we investigate how many points are needed so that it is possible for the points do not lie on any plane-perceivable surface. It is easy to see that the scalar equations in (5.3.6) are linear in the elements of K. Each point on the surface, other than the origin and $\mathbf{O}'$, gives at most two independent linear equations in the elements of K (since one scalar equation in (5.3.6) is redundant). Any set of 4 points on the surface gives 8 such linear equations for the elements in K, and a non-zero solution K exists. Therefore, four points always lie on some plane-perceivable surface in (5.3.6). Since four points are generally not coplanar, the set of plane-perceivable surfaces does include many non-coplanar surfaces (including curves)! Similarly, we conclude that 5 points generally do not lie on any surface in the form of (5.3.6). Thus, with 5 points it is possible to detect that they are not coplanar. (We have conducted a simulation in which 20 examples of 5-point set have been randomly generated in space. In these examples the smallest eigenvalue of $A^t A$ ranges from 10^{-5} to 10^{-3}.)

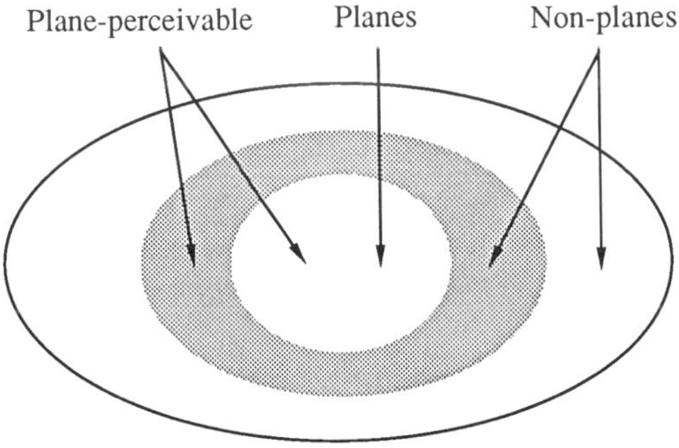

Fig. 5.5. The relations among planes, plane-perceivable surfaces and non-planar surfaces. Plane-perceivable surfaces include some of the nonplanar surfaces (including curves) and all the visible planes.

5.3.5 Optimality

Now we turns to the stability of the solution in the presence of noise. Theorem 5.5 is not only important to the inherent uniqueness, it is also important to understand the optimality of the linear algorithm. From Theorem 5.5, we know that from a noise-contaminated version of F_s is decomposable, as long as the noise is not so large and coincidental that the contaminated matrix happens to degenerate into a rank-one matrix. Then our algorithm computes the solutions that fit this F_s exactly. In other words, no constraint needs to be considered in the solution for F_s in (5.E.3) of Appendix 5.E, since the probability for the noise contaminated F_s to be a rank-one matrix is zero.

From (5.2.15)-(5.2.17), (5.E.3) and Theorem 5.5, it immediately follows that our algorithm determines a rotation matrix R, a unit 3-D vector $\mathring{\mathbf{T}}$, whose sign ensures that (5.E.14) of Appendix 5.E is false, and a 3-D vector $\tilde{\mathbf{N}}$ to minimize

$$\sum_{i=1}^{n} \|\mathbf{X}_i' \times (R + \mathring{\mathbf{T}}\tilde{\mathbf{N}}^t)\mathbf{X}_i\|_{z=0}^2. \tag{5.3.7}$$

where $\|\cdot\|_{z=0}$ denotes the Euclidean norm of the projection on to x-y plane: $\|(a, b, c)\|_{z=0}^2 = a^2 + b^2$. If all the three rows in (5.2.16) are included in A defined in (5.E.1), then $\|\cdot\|_{z=0}$ in (5.3.7) is replaced by the regular Euclidean norm $\|\cdot\|$. In the absence of noise, the minimum value in (5.3.7) is equal to zero. Theorem 5.3 implies that the images can be interpreted as a planar patch under going rigid motion if and only if

$$\mathbf{X}_i' \times (R + \mathring{\mathbf{T}}\tilde{\mathbf{N}}^t)\mathbf{X}_i = \mathbf{0} \tag{5.3.8}$$

is satisfied for every image point pair $(\mathbf{X}_i, \mathbf{X}_i')$. Therefore, the objective function (5.3.7) uses all the constraint in the problem, except the constraint on negative depth which is automatically satisfied for the veridical solution.

Since the variance of the equation residual of (5.3.8) varies from point to point. The minimum variance estimator requires that the objective function in (5.3.7) be weighted by the inverse of the residual variance. Let $E' = [e'_{ij}] \triangleq R + \mathring{\mathbf{T}}\tilde{\mathbf{N}}^t$, and $M = [m_{ij}]$ be a matrix constructed from E' so that $m_{11} = e'_{21}$, $m_{12} = -e'_{11}$, $m_{21} = e'_{22}$, $m_{22} = -e'_{12}$ and the remaining entries of M_i are zeros. Then the computed weight w_i^2 can be expressed concisely by

$$w_i^2 = \sigma^{-2}(\|E'\mathbf{X}_i\|_{z=0}^2 + \|M\mathbf{X}_i'\|^2)^{-1},$$

assuming the error in different components of image plane points is uncorrelated and has a zero mean and a constant variance σ^2. The objective function in (5.3.7) can be replaced by the optimal one (based on equation error):

$$\sum_{i=1}^{n} w_i^2 \|\mathbf{X}_i' \times (R + \mathring{\mathbf{T}}\tilde{\mathbf{N}}^t)\mathbf{X}_i\|^2. \tag{5.3.9}$$

However, the weight w_i^2 is a function of unknown motion parameters and structure. The solution of (5.3.9) cannot be solved in a closed-form and so an iterative process needs to be used to minimize the left-hand side of (5.3.9). Compared to (5.3.9), the objective function in (5.3.8) is simpler and can be directly computed suing the linear algorithm presented. Based the properties of the linear algorithm we discussed, as well as the result of our simulation, the minimizer of (5.3.7) is expected to be good enough for most applications.

5.3.6 Three-View Problem

Two views allow two interpretations in general: one is veridical, the other is illusive. In this section, we investigate how to reject the illusive interpretation.

Suppose that in addition we have observed an extra point that, we know a priori, does not belong to the plane. Then generally, we can reject the illusive solution as follows. Using the coplanar points, we determine two solutions. If the solution $(R, \mathring{\mathbf{T}})$ is veridical, from (5.2.2) we know that $\mathbf{X}'$, $R\mathbf{X}$ and $\mathring{\mathbf{T}}$ are coplanar for all the point pairs $(\mathbf{X}, \mathbf{X}')$ including the extra point that is not in the plane. However, if the solution is illusive, this will not, in general, be true for the extra point. This fact can be used to reject the illusive solution. Therefore, to reach one solution, this method just needs one extra non-coplanar points, instead of two in [5.2]. However, since it is unlikely for one to have a priori knowledge that a specified point is not coplanar with the rest, this method is not very useful in practice.

Fortunately, provided with three distinct images of coplanar points that undergo a rigid motion, in general we can uniquely determine the motion parameters and the normalized normal of the object plane. Suppose that we have three image frames taken at time instants t_1, t_2 and t_3, respectively. Consider the two motions: one from t_2 to t_1 and the other from t_2 to t_3. We solve for the motion parameters of these two motions and the unit normal $\mathring{\mathbf{N}}$ of the object plane at time t_2. The veridical solutions for these two motions should have the same answer for the unit normal $\mathring{\mathbf{N}}$ of the object plane at time t_2. According to the following theorem, the illusive solutions generally do not agree on the unit normal of the object plane. Therefore, generally we can determine which solution is veridical based on the agreement on the the unit normal at time t_2.

Theorem 5.8. Let F and F' correspond, respectively, to two motions from one position. Suppose there exist two solutions $(R_a, \tilde{\mathbf{T}}_a, \mathring{\mathbf{N}}_a)$ and $(R_b, \tilde{\mathbf{T}}_b, \mathring{\mathbf{N}}_b)$ such that

$$F = R_a + \tilde{\mathbf{T}}_a \mathring{\mathbf{N}}_a^t = R_b + \tilde{\mathbf{T}}_b \mathring{\mathbf{N}}_b^t$$

Let H be an orthogonal matrix with $\det(H) = 1$ such that

$$H^t F^t F H = \operatorname{diag}(\lambda_1, 1, \lambda_3)$$

and $\lambda_1 < 1 < \lambda_3$.

Similarly, for F' we have two solutions such that

$$F' = R'_a + \tilde{\mathbf{T}}'_a (\mathring{\mathbf{N}}'_a)^t = R'_b + \tilde{\mathbf{T}}'_b (\mathring{\mathbf{N}}'_b)^t .$$

Then, both solutions agree on the unit normal, i.e.,

$$\mathring{\mathbf{N}}_a = \mathring{\mathbf{N}}'_a, \quad \text{and} \quad \mathring{\mathbf{N}}_b = \mathring{\mathbf{N}}'_b,$$

if and only if the ambiguity condition is satisfied, i.e., there exists an orthogonal matrix Q and a positive number k such that

$$F' = Q \operatorname{diag}(\sqrt{1 - k(1 - \lambda_1)}, 1, \sqrt{1 + k(\lambda_3 - 1)}\,) H^t$$

with $1 - k(1 - \lambda_1) \geq 0$ or

$$F' = Q \operatorname{diag}(\sqrt{1 + k(1 - \lambda_1)}, 1, \sqrt{1 - k(\lambda_3 - 1)}\,) H^t$$

with $1 - k(\lambda_3 - 1) \geq 0$. □

Generally the ambiguity condition is not satisfied according to the following observation. We know that $F = R + \tilde{\mathbf{T}}\mathring{\mathbf{N}}^t$ has 8 degrees of freedom (3 in R, 3 in $\tilde{\mathbf{T}}$ and 2 in $\mathring{\mathbf{N}}$). With the fixed unit normal $\mathring{\mathbf{N}}$ the degree of freedom of F' is $8 - 2 = 6$. The ambiguity condition restricts the degree of freedom for F', with fixed unit normal $\mathring{\mathbf{N}}$, to 4 (3 in Q and one in k). In other words, the ambiguity condition reduces the degree of freedom for F' by two.

As an example, it is easy to see that the illusive solutions for the two motions agree on the unit normal if the motions correspond to $(R, \mathbf{T})$ and $(QR, Q\mathbf{T})$, respectively, where Q is any rotation matrix. That is, there are many cases where three-views do not give a unique solution. But if we consider all three possible interframe motions among three views and examine the agreement on object normal at three time instants, more cases can have a unique solution [5.8].

Given a sequence of images, as long as there exist three consecutive frames that do not satisfy the ambiguity condition, all the veridical interframe motions through the sequence can be determined. In fact, we know that veridical solution can be identified from two candidate solutions if either veridical motion is known or veridical unit normal of the object plane is known. Therefore, for example, the veridical unit normal at time t_1 can be used to identify the veridical motion from t_1 to t_2, which in turn can be used to identify the veridical unit normal at t_2, and so on. Such a propagation of the veridical solutions through the sequence determines the veridical solution of every interframe motion and the orientation of object plane at every time instant.

5.4 Examples

The experiments were designed to demonstrate the performance of the motion estimation algorithm and the associated error estimation. The experiments with real images are presented after the results of simulation.

5.4.1 Simulations

Object plane intersects the optical axis at $z = 11$. The image is a unit square with a unit focal length. The object feature points are generated randomly with a uniform distribution in each of the alternative four quadrants of the object plane. Therefore, every quadrant has roughly the same number of points so that the image plane is fully used. Only those points that are visible both before and after motion are used as feature points. The image coordinates of the points are digitized according to the resolution of the camera.

The principles of error estimation have been discussed in Section 3.3. The formulations for the algorithm here are presented in

Appendix 5.F. We assume that the digitization errors are uniformly distributed between plus half and minus half of the pixel size. Thus, the variance of the errors in the image coordinates is $\sigma^2 = s^2/12$, where s is the spacing between the quantization levels in the images. This variance is used in the error estimation. Both simulation and the formulation of error estimation indicate that the error in solution is roughly proportional to the variance of the noise. Therefore, we need only to show here the results with a typical noise level. We choose image resolution to be 256×256 pixels here. Fig. 5.6 shows the results of a sequence of trials with 5 point correspondences. The normal of the object plane is (0.5, 0.4, 0.7) before motion. The rotation is about an axis (1, 1, 1) by an angle 5°. The translation vector is (0.2, 0.3, −2). In Fig. 5.6, the actual relative errors and the estimated errors in 40 random trials (randomly generated points on the plane) are shown in the order of trials. As can be seen from these figures, the estimated errors are strongly correlated with the actual errors. As we discussed, the estimated errors are especially important to detect some relatively unreliable configurations.

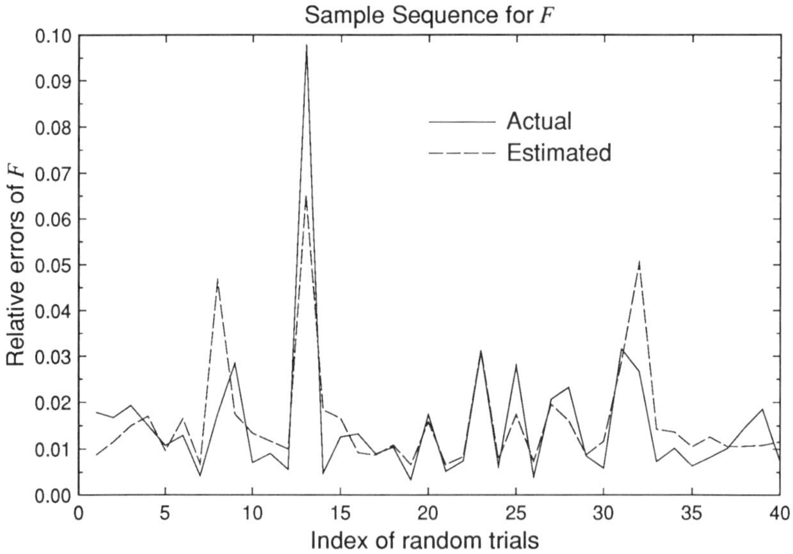

Fig. 5.6(a)

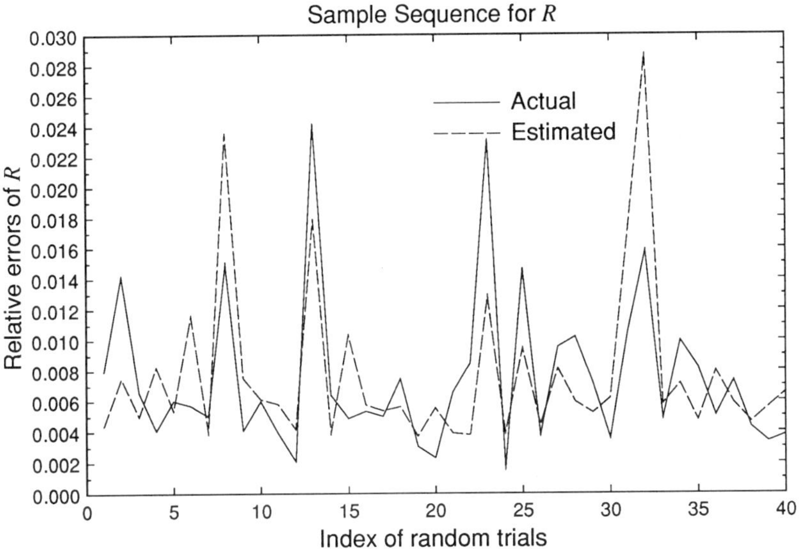

Fig. 5.6(b)

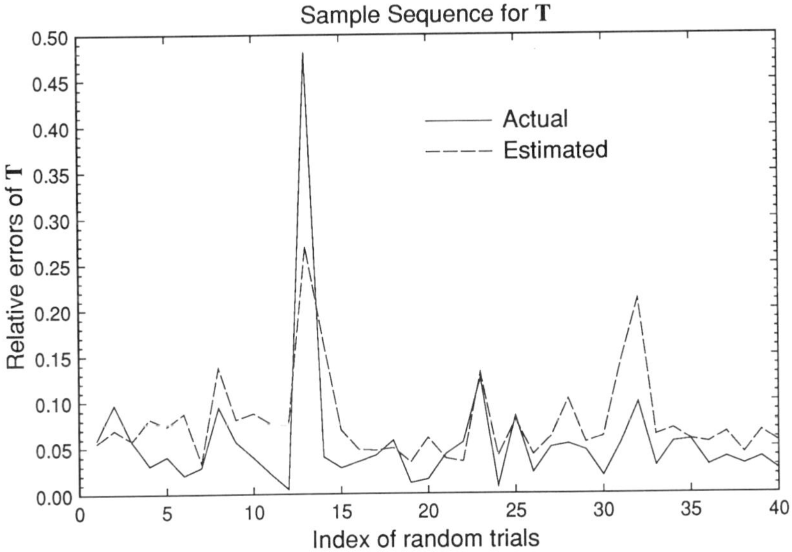

Fig. 5.6(c)

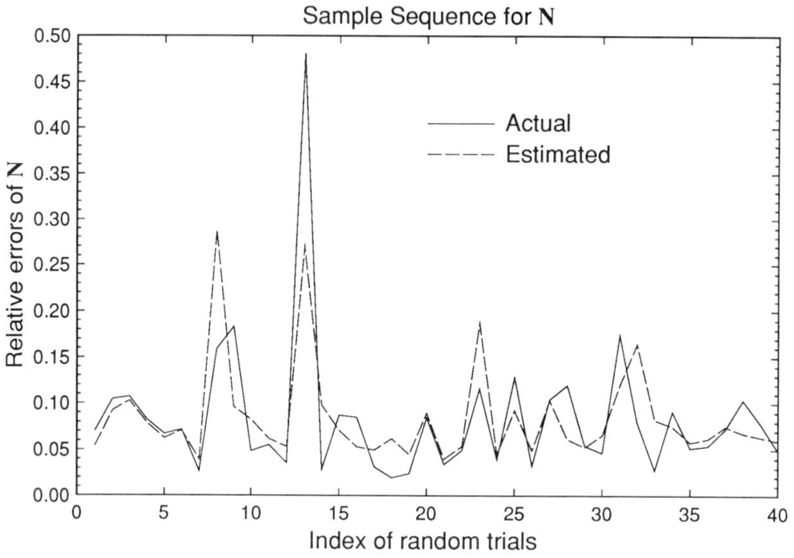

Fig. 5.6(d)

Fig. 5.6. The actual relative errors and the estimated relative errors of **(a)** F, **(b)** R, **(c)** **T**, and **(d)** **N**. The horizontal index is the order of random trials.

The average performance of the error estimation as well as that of the motion estimation algorithm is presented in Fig. 5.7. The data shown are based on 200 random trials. The deviation of error estimation is defined as the average of the absolute difference between the estimated relative error and the actual relative error. The bias of error estimation is defined as the difference between the mean of the estimated relative errors and the mean of the actual relative errors. As can been seen from Fig. 5.7, the actual mean relative errors decrease very fast when the number of points increases beyond the required minimum of 4. This indicates that it is very effective to reduce the error by using a few more points in addition to the minimally required 4. In Fig. 5.7, the mean deviation between the estimated error and the actual error is about a half of the actual error, except for the cases where the number of points is small (e.g., four points). When the number of point correspondences is equal to 4, there is a reasonably high probability for the randomly generated points to form a nearly degenerate configuration. When the point configuration are degenerate or nearly degenerate, the difference between the estimated error and the actual error is expected to be large. This is one of the reasons for the large deviations and bias in the 4-point case. Some individual simulations still show a good agreement between the estimated errors and the actual errors in the 4-point case.

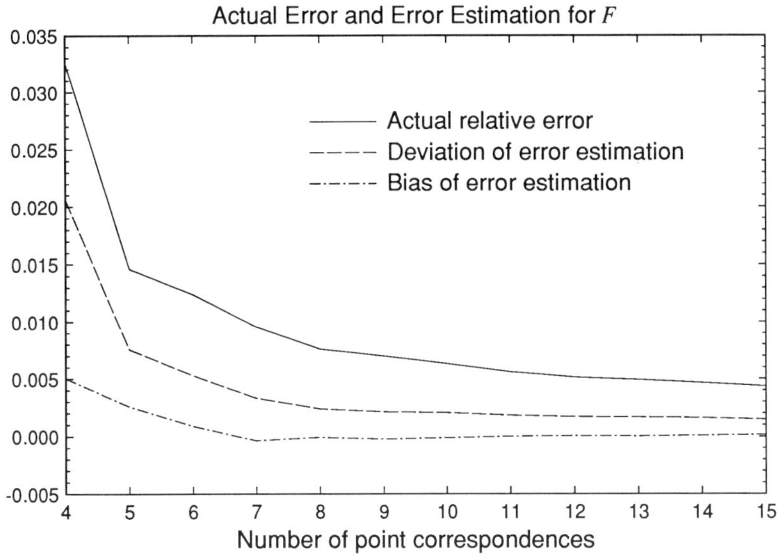

Fig. 5.7(a)

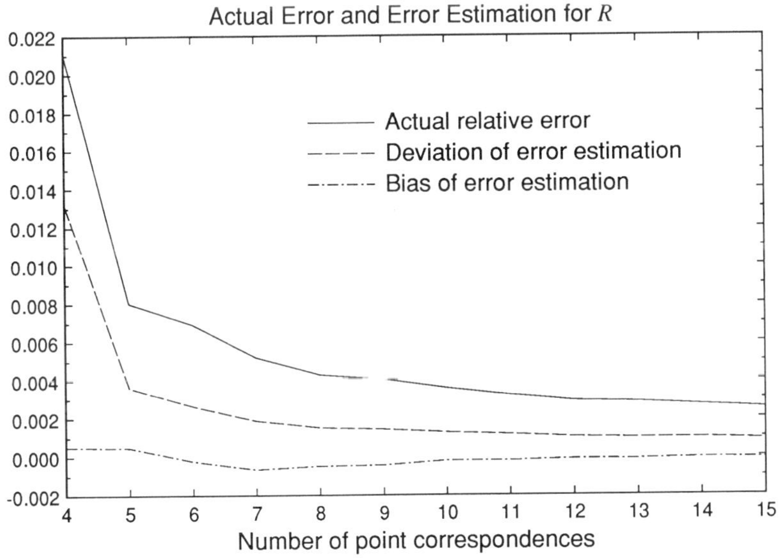

Fig. 5.7(b)

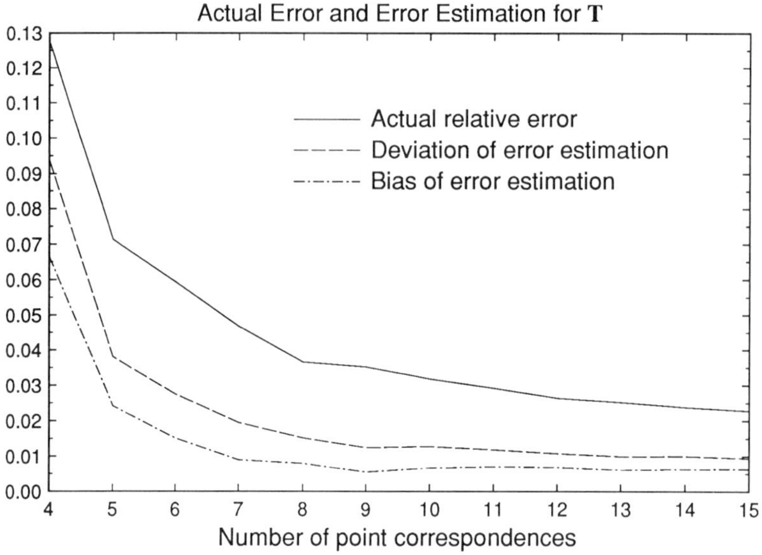

Fig. 5.7(c)

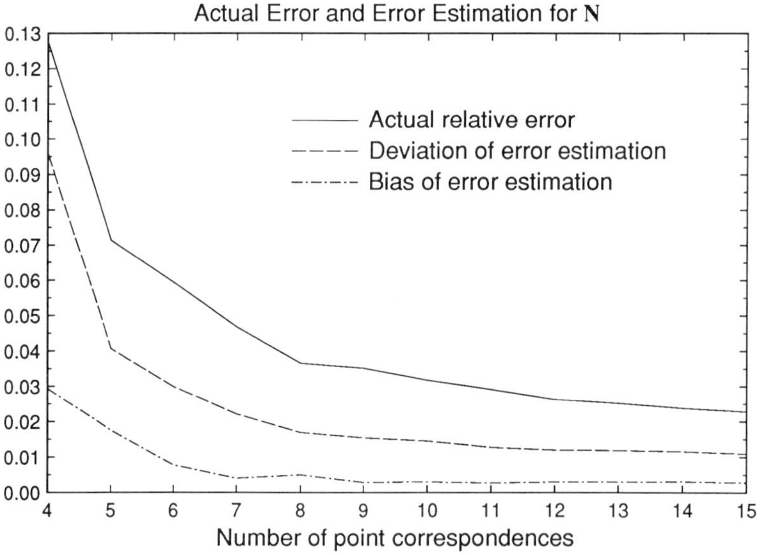

Fig. 5.7(d)

Fig. 5.7. The actual mean relative errors, deviation of error estimation and bias of error estimation for with trials versus the number of point correspondences for **(a)** F, **(b)** R, **(c)** $\mathbf{T}$, and **(d)** $\mathbf{N}$. Same motion as in Fig. 5.6.

5.4.2 Real Images

The setup is the same as that in Chapter 2. Fig. 5.8 shows three views of a bulletin board, called Image 1, Image 2 and Image 3, respectively. As we discussed, there are generally two solutions from two views, and adding a third view generally leads to a unique solution. The image matching algorithm discussed in Chapter 2 was used to establish the image plane displacement fields between two images (equivalent to point correspondences). Fig. 5.9 shows sampled displacement fields of the two motions: Motion 1-2 (from Image 1 to Image 2) and Motion 1-3 (from Image 1 to Image 3). Table 5.1 shows the estimated parameters and the corresponding estimated relative errors for Motion 1-2 and Table 5.2 shows those for Motion 1-3. For error estimation, the variance of the noise in image points is assumed to be equal to that of a uniformly distributed random noise ranging ±1 pixel.

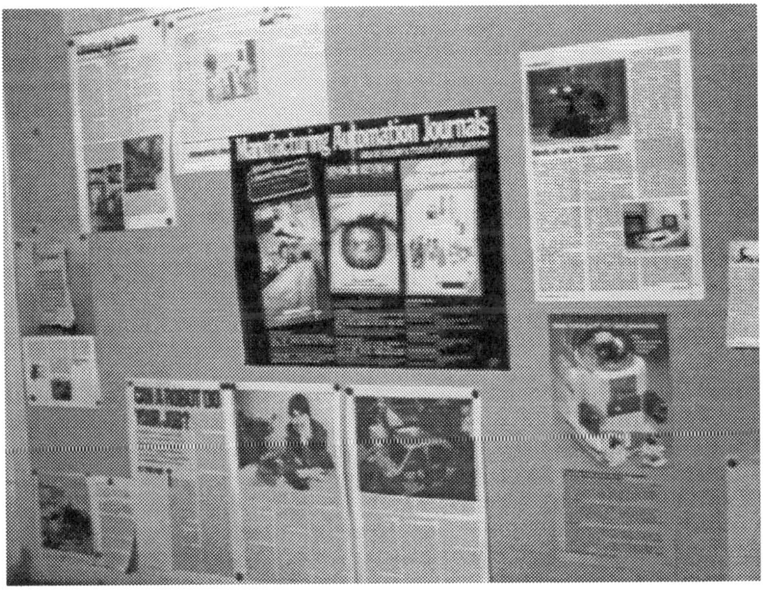

Fig. 5.8(a)

Fig. 5.8(b)

Fig. 5.8(c)
Fig. 5.8. Three views of a bulletin board. **(a)** Image 1; **(b)** Image 2; **(c)** Image 3.

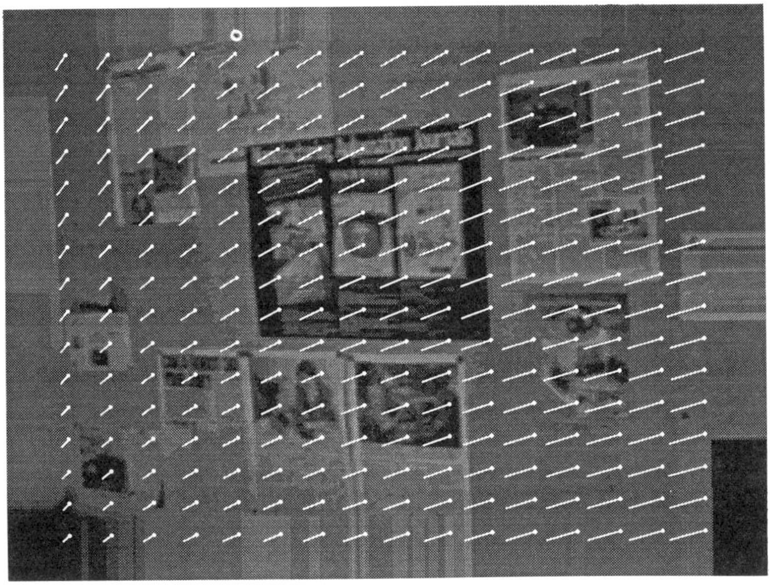

Fig. 5.9(a)

Fig. 5.9(b)

Fig. 5.9. Samples of the computed displacement fields for the bulletin scene, superimposed on the blurred extended intensity image. **(a)** Image 1 to Image 2 (1-2); **(b)** Image 1 to Image 3 (1-3).

Table 5.1. Results for Bulletin Board Images (1-2)

Solution 1 Parameters	x	y	z
Unit normal	-0.08	-0.966	0.24
Unit translation	0.04	0.58	0.81
Rotation axis	-0.95	0.29	0.08
Rotation angle		-1.8°	
E. R. error of normal		0.16	
E. R. error of translation		0.13	
E. R. error of rotation matrix		0.009	
Solution 2 Parameters	x	y	z
Unit normal	0.04	0.58	0.81
Unit translation	-0.08	-0.97	0.24
Rotation axis	0.986	0.13	0.10
Rotation angle		-1.8°	
E. R. error of normal		0.12	
E. R. error of translation		0.18	
E. R. error of rotation matrix		0.016	

Table 5.2. Results for Bulletin Board Images (1-3)

Solution 1 Parameters	x	y	z
Unit normal	0.12	0.35	0.930
Unit translation	0.28	-0.77	0.57
Rotation axis	0.61	0.79	0.02
Rotation angle		0.5°	
E. R. error of normal		0.13	
E. R. error of translation		0.22	
E. R. error of rotation matrix		0.013	
Solution 2 Parameters	x	y	z
Unit normal	0.27	-0.76	0.59
Unit translation	0.14	0.32	0.936
Rotation axis	-0.936	-0.30	0.17
Rotation angle		-2.7°	
E. R. error of normal		0.20	
E. R. error of translation		0.17	
E. R. error of rotation matrix		0.006	

Let $\mathbf{N}_{i-j,k}$ denote the k-th solution of the unit object normal for motion from Image i to Image j. The correct solution for each motion is determined by the consistency of object normal at Image 1:

$$\|\mathbf{N}_{1-2,1} - \mathbf{N}_{1-3,1}\| = 1.50,$$

$$\|\mathbf{N}_{1-2,1} - \mathbf{N}_{1-3,2}\| = 0.54,$$

$$\|\mathbf{N}_{1-2,2} - \mathbf{N}_{1-3,1}\| = 0.27,$$

$$\|\mathbf{N}_{1-2,2} - \mathbf{N}_{1-3,2}\| = 1.38.$$

Since the solution 2 of Motion 1-2 and solution 1 of Motion 1-3 give the most consistent object normals, the corresponding solutions are what we needed. The corresponding solutions roughly agree with visual inspection (since the rotation angles are very small, the rotation axes are not important), although lens distortion is visible from images.

5.5 Conclusions

The problem of determining motion and structure of a rigid planar patch from two monocular images has been completely solved. (1) We have established the minimum solution set: the number of solutions in this set is generally two (one veridical one illusive), rarely one (veridical) or infinitely many (only one veridical and all others are illusive), all rendering the same images. (2) We have established intuitive necessary and sufficient geometrical conditions that correspond to each of these cases. (3) The solution set is minimum in the sense that it cannot be reduced further no matter how many more point correspondences are used and what kind of algorithm is used. This conclusion holds true for any type of feature correspondences, since point correspondence can be used to specify any other types of feature correspondences. (4) We have a linear algorithm that guarantees to obtain these solutions, no more, no less.

Unlike the linear algorithm for general surfaces discussed in Chapter 3, the linear algorithm here solves the problem whenever it is solvable. The key to this is the establishment of AIC = ADC, the disappearance of the gap between the algorithm independent condition and the algorithm dependent one. Based on this, we have introduced and investigated plane-perceivable surfaces and established a new optimality of the linear algorithm.

5.6 Bibliographical Notes

Like the subject with curved surfaces, the analysis of motion and structure from planar surfaces has also received intensive attention during the last decade. After a few iterative algorithms, a linear algorithm has been developed by Tsai & Huang [5.7] using the technique of singular value decomposition. They have established that from two images, there are generally two candidate solutions. A sufficient condition to guarantee rank(A)=8 is given in [5.8] using a result of projective geometry: If four point correspondences are available, no three object points are collinear, then rank(A)=8. Weng, Huang & Ahuja [5.10] have established that condition as a necessary and sufficient one: the linear algorithm reaches the unique solution (the minimum solution set) if and only if, among n point correspondences, there exists a set of four points, no three are collinear in the plane, and the (infinite) object plane does not go through the projection center of the camera before and after motion. In addition to the true solution, Tsai & Huang's algorithm generally generates an extra candidate solution (the same is true in the case of optical flow [5.3]), which was proved in [5.10] to be a valid solution: Both the true and the extra ones rend the same images.

The inherent nature of the linear algorithm was reported in Weng, Ahuja & Huang [5.11] recently. The fact that the linear algorithm can solve all the cases that are solvable is somewhat surprising. Due to this fact, the uniqueness of the solution from the linear algorithm is not only for a specific linear algorithm, but much more fundamentally for the original problem itself.

Longuet-Higins [5.3] as well as Waxman & Wohn [5.9] studied the analysis of plane motion in terms of motion velocity (optical flow approach).

In some special cases, it is possible to determine motion parameters from only partial information about point matching, although the stability of the solution may degrade. For example, Negahdaripour & Horn [5.5] used a *direct method* to compute the small motion of a planar scene. By their method, the motion parameters of the plane are estimated directly from the spatial and temporal gradient of the image intensity, and the complete optical flow is not necessary.

Appendices

Appendix 5.A Conditions on the Rank of A

We first prove a lemma which is needed for the proofs of Theorem 5.1 and Corollary 5.1.

Lemma 5.2. Assume the object plane does not go through the center of projection (origin). The projection of a set of points in the object plane lie on a straight line in the image plane if and only if they lie on a straight line in the object plane.

Proof. The equation of the object plane that does not go through the origin can be written as $ax + by + cz = 1$, If the points lie on a straight line in image plane, they lie in the intersection of $ax + by + cz = 1$ and another plane $ex + fy + gz = 1$ that does not go through the origin either (such a plane exists since the origin is not in the line). For the intersection to be a line, (a, b, c) is not parallel to (e, f, g). These two equations give

$$(a - e)x + (b - f)y + (c - g)z = 0$$

or

$$(a - e)(x/z) + (b - f)(y/z) + c - g = 0.$$

That is, their images lie on a straight line

$$(a - e)u + (b - f)v + (c - g) = 0$$

in the image plane provided $(a - e)^2 + (b - f)^2 \neq 0$. In fact, if $(a - e)^2 + (b - f)^2 = 0$, we have $c = g$, and thus

$$(a, b, c) \,//\, (e, f, g),$$

which is a contradiction.

Conversely, assume the image of the points lie on a straight line $eu + fv + g = 0$ in the image plane with $e^2 + f^2 \neq 0$. Their 3-D coordinates must satisfy $e(x/z) + f(y/z) + g = 0$, or

$$ex + fy + gz = 0. \tag{5.A.1}$$

The intersection of the object plane $ax + by + cz = 1$ and the plane in (5.A.1) gives a straight line in 3-D provided (a, b, c) is not parallel to (e, f, g). In fact (a, b, c) is never parallel to (e, f, g). If they are parallel, $ax + by + cz = 1$ and $ex + fy + gz = 0$ have no intersection, which is impossible since the object points are in the intersection. □

Theorem 5.1. $\text{rank}(A) = 8$ if and only if there exists a set of four object points such that no image projections of any three points in this set are collinear in any of the two images.

Proof. For the sufficiency part, assume such a set of four points exists. We prove the corresponding 8 rows of A has a rank 8. We prove by contradiction. Suppose that these 8 rows are linearly dependent. We have two vectors $\mathbf{a} = (a_1, a_2, a_3, a_4)$ and $\mathbf{b} = (b_1, b_2, b_3, b_4)$ with $\mathbf{a} \neq \mathbf{0}$ or $\mathbf{b} \neq \mathbf{0}$, such that (see (5.2.16))

$$\sum_{i=1}^{4} a_i \mathbf{X}_i = \mathbf{0}, \tag{5.A.2}$$

$$\sum_{i=1}^{4} b_i \mathbf{X}_i = \mathbf{0}, \tag{5.A.3}$$

$$\sum_{i=1}^{4} a_i v'_i \mathbf{X}_i + \sum_{i=1}^{4} b_i u'_i \mathbf{X}_i = \mathbf{0}. \tag{5.A.4}$$

Let $X = [\mathbf{X}_1 \ \ \mathbf{X}_2 \ \ \mathbf{X}_3 \ \ \mathbf{X}_4]$ and $\mathbf{c} = (c_1, c_2, c_3, c_4)$, where

$$c_i = a_i u'_i + b_i v'_i.$$

Equations (5.A.2)-(5.A.4) give $X\mathbf{a} = \mathbf{0}$, $X\mathbf{b} = \mathbf{0}$ and $X\mathbf{c} = \mathbf{0}$. Without loss of generality assume $\mathbf{a} \neq \mathbf{0}$. Since no three points lie on a straight line, no a_i can be zero. We have $\text{rank}(X) \geq 3$. Therefore, $\mathbf{a} \,//\, \mathbf{b} \,//\, \mathbf{c}$. There exist constants k and c such that $b_i = ka_i$ and

$$ca_i = c_i = (a_i u'_i + b_i v'_i)(u'_i + kv'_i)a_i,$$

$1 \leq i \leq 4$. Since $a_i \neq 0$, $1 \leq i \leq 4$, we have $u'_i + kv'_i = c$, $1 \leq i \leq 4$. This means that all four points lie on a straight line after motion. A contradiction. (From Lemma 5.2 we see that if such a set of four points exists before motion, the only way for the condition to be violated after motion is that the object plane goes through the center of projection after motion.)

For the necessity part, first we prove that if all the image points are collinear in the image plane, $\text{rank}(A) \leq 5$. Assume all the points lie on a straight line (called common line). Let $\mathbf{v}$ be the unit normal of the plane that contains the common line and passes through the origin. We have $\mathbf{X}_i^t \mathbf{v} = 0$, $i=1, 2, \cdots, n$. Define four 9-dimensional vectors:

$$\mathbf{h}_1 = \begin{bmatrix} \mathbf{v} \\ 0 \\ 0 \end{bmatrix}, \quad \mathbf{h}_2 = \begin{bmatrix} 0 \\ \mathbf{v} \\ 0 \end{bmatrix}, \quad \mathbf{h}_3 = \begin{bmatrix} 0 \\ 0 \\ \mathbf{v} \end{bmatrix}, \quad \mathbf{h}_4 = \begin{bmatrix} \mathbf{e}_1 \\ \mathbf{e}_2 \\ \mathbf{e}_3 \end{bmatrix} \tag{5.A.5}$$

where $[\mathbf{e}_1 \ \mathbf{e}_2 \ \mathbf{e}_3] = F^t$. From (5.2.17) and the structure of A in (5.2.16), we have $A\,\mathbf{h}_i = 0$, $i=1, 2, 3, 4$. Furthermore we prove that the four vectors in (5.A.5) are linearly independent. Otherwise, $\mathbf{h}_4$ is a linear combination of $\mathbf{h}_1$, $\mathbf{h}_2$ and $\mathbf{h}_3$, since it is clear that $\mathbf{h}_1$, $\mathbf{h}_2$ and $\mathbf{h}_3$ are linearly independent. This means that the rank of F is at most one which is contradictory to (5.2.29). Since the equation $A\,\mathbf{h} = 0$ has at least four linearly independent solutions in (5.A.5),

$$\text{rank}(A) \leq 9 - 4 = 5.$$

Now we assume that the condition of the theorem is not satisfied and prove rank$(A) < 8$. If $n < 4$ the theorem is trivialy true. Now let $n \geq 4$. If any set of three points are collinear, it is clear all points lie on a single straight line. Then rank of A is no more than 5 as we proved earlier. Now assume there exists a set of three points that are not collinear. Without loss of generality, assume they are $\mathbf{X}_1$, $\mathbf{X}_2$, and $\mathbf{X}_3$. $\mathbf{X}_4$ must lie on a line that contains two of $\mathbf{X}_1$, $\mathbf{X}_2$ and $\mathbf{X}_3$ (otherwise the set of these four points violate our assumption). Without loss of generality, let $\mathbf{X}_2$, $\mathbf{X}_3$ and $\mathbf{X}_4$ lie on a straight line (call this common line, see Fig. 5.10).

We prove all the $\mathbf{X}_i$'s, $i > 4$ must lie on the common line. This is proved by the induction on i. Assume it is true for $i = j \geq 4$. Suppose $\mathbf{X}_{j+1}$ is not on the common line. We will get a

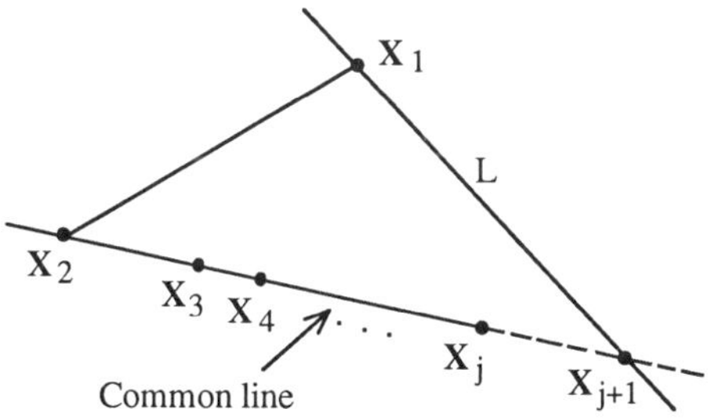

Fig. 5.10. Illustration for the proof of Theorem 5.1.

contradiction. The straight line l that passes through $\mathbf{X}_1$ and $\mathbf{X}_{j+1}$ can contain at most one point $\mathbf{X}_k$ on the common line, for some k, $2 \le k \le j$, (otherwise $\mathbf{X}_1$ would have been in the common line). Any two points other than $\mathbf{X}_k$ on the common line together with $\mathbf{X}_1$ and $\mathbf{X}_{j+1}$ form a set of four points that satisfy the condition. A contradiction. Therefore $\mathbf{X}_{j+1}$ is on the common line.

Since all the points lie on a single line except $\mathbf{X}_1$, the rank of A is no more than $5 + 2 = 7$. □

Corollary 5.1. rank(A) = 8 if and only if

(1) there exists a set of four points in the object plane such that no three points in this set are collinear in the object plane and

(2) if the object plane is extended, it does not go through the center of projection before and after motion.

Proof. If (1) and (2) are satisfied, by Lemma 5.2 the condition of Theorem 5.1 is satisfied. Thus rank(A) = 8. Conversely assume rank(A) = 8. (2) must be satisfied otherwise all the points lie on a straight line and, therefore, rank(A) ≤ 5 as proved in the proof of Theorem 5.1. (1) must be satisfied according to Lemma 2.

Appendix 5.B Two Solutions from F

Theorem 5.3. Given n point correspondences represented by $(\mathbf{X}_i, \mathbf{X}'_i)$, $i = 1, 2, \cdots, n$. Let F be a matrix that satisfies (5.2.15) for every $(\mathbf{X}_i, \mathbf{X}'_i)$. If there are rotation matrices R_a, R_b, unit vectors $\mathring{\mathbf{T}}_a$, $\mathring{\mathbf{T}}_b$ and vectors $\tilde{\mathbf{N}}_a$ and $\tilde{\mathbf{N}}_b$, such that

$$F = R_a + \mathring{\mathbf{T}}_a \tilde{\mathbf{N}}_a^t = R_b + \mathring{\mathbf{T}}_b \tilde{\mathbf{N}}_b^t \tag{5.B.1}$$

and $\tilde{\mathbf{N}}_a \cdot \mathbf{X}_i > 0$, $\tilde{\mathbf{N}}_b \cdot \mathbf{X}_i > 0$, $i = 1, 2, \cdots, n$, hold true. Then there exist two planes with $\tilde{\mathbf{N}}_a$ and $\tilde{\mathbf{N}}_b$ as normals, respectively, such that if they undergo motions represented by $(R_a, \mathring{\mathbf{T}}_a)$ and $(R_b, \mathring{\mathbf{T}}_b)$, respectively, they render the same pair of images (with image vectors $\mathbf{X}_i$ and $\mathbf{X}'_i$ at the two time instants) and the corresponding 3-D points are all located in the forward half space before and after the motion. (If $\tilde{\mathbf{N}}_a \cdot \mathbf{X}_i < 0$ or $\tilde{\mathbf{N}}_b \cdot \mathbf{X}_i < 0$ is true for a few points, then the conclusion still holds except that the corresponding point is located in the back half space.)

Proof. For any point pair $(\mathbf{X}, \mathbf{X}')$ in $\{ (\mathbf{X}_i, \mathbf{X}'_i) \mid i = 1, 2, \cdots, n \}$, give its two positive depths z_a and z_b before motion by

$$z_a = \frac{1}{\tilde{\mathbf{N}}_a^t \mathbf{X}}, \quad z_b = \frac{1}{\tilde{\mathbf{N}}_b^t \mathbf{X}}. \tag{5.B.2}$$

Their corresponding depths after motion, z'_a and z'_b, are determined by (5.2.14):

$$z'_a \mathbf{X}' = z_a F \mathbf{X}, \quad z'_b \mathbf{X}' = z_b F \mathbf{X}. \tag{5.B.3}$$

The equations (5.B.3) can be exactly satisfied because we have (5.2.15). Moreover, the solutions for depths in (5.B.3) are all positive since there exist true positive depths z and z' that satisfy (5.2.14) and positive depths before motion in (5.B.2) are given conditions. Specifically, positive depths after motion are guaranteed for the illusive solution as well. We get the 3-D object points as follows

$$\mathbf{x}_a = z_a \mathbf{X}, \quad \mathbf{x}'_a = z'_a \mathbf{X}', \tag{5.B.4.a}$$

$$\mathbf{x}_b = z_b \mathbf{X}, \quad \mathbf{x}'_b = z'_b \mathbf{X}'. \tag{5.B.4.b}$$

Using the equations (5.B.1)-(5.B.4) gives

$$\begin{aligned} \mathbf{x}'_a &= z'_a \mathbf{X}' = F z_a \mathbf{X} = (R_a + \mathring{\mathbf{T}}_a \tilde{\mathbf{N}}_a^t) z_a \mathbf{X} \\ &= R_a z_a \mathbf{X} + \mathring{\mathbf{T}}_a (\tilde{\mathbf{N}}_a^t \mathbf{X}) z_a = R_a \mathbf{x}_a + \mathring{\mathbf{T}}_a. \end{aligned} \tag{5.B.5.a}$$

Similarly we get

$$\mathbf{x}'_b = R_b \mathbf{x}_b + \mathring{\mathbf{T}}_b. \tag{5.B.5.b}$$

For the two solutions corresponding to subscripts a and b, respectively, (5.B.2) guarantees all the points lie on the plane before motion:

$$\tilde{\mathbf{N}}_a^t \mathbf{x}_a = 1, \quad \tilde{\mathbf{N}}_b^t \mathbf{x}_b = 1.$$

(5.B.5.a) and (5.B.5.b) are equations of the rigid motion, so the points after motion also lie on a plane. Furthermore, (5.B.5) states that the two solutions have the motions corresponding to R_a, $\mathring{\mathbf{T}}_a$ and R_b, $\mathring{\mathbf{T}}_b$, respectively. (5.B.4) ensures that both solutions render the same images. □

Appendix 5.C Plane-Perceivable Surfaces

Theorem 5.7. From two views, there exist an interpretation that a planar patch undergoes a rigid motion, if and only if there exists a

3×3 nonzero matrix K such that all the observed points lie in the following intersection of two quadratic surfaces before motion:

$$(\mathbf{x} - \mathbf{O}') \times K \mathbf{x} = \mathbf{0} \tag{5.C.1}$$

where $\mathbf{O}' = -R^t \mathbf{T}$.

Proof. According to Theorem 5.6, there exists such an interpretation if and only if $\mathrm{rank}(A) \le 8$. We investigate the rank of A. First assume that the rank of A is less than 9. Then there exists a nonzero vector $\mathbf{h}$ such that (5.2.17) and (5.2.16) hold for all the points. Since

$$[-v'\mathbf{X}^t \quad u'\mathbf{X}^t \quad \mathbf{0}] = -v\,[\mathbf{X}^t \quad \mathbf{0} \quad -u'\mathbf{X}^t] + u\,[\mathbf{0} \quad \mathbf{X}^t \quad -v'\mathbf{X}^t]$$

the third equation in (5.2.16) is redundant once the first two equations are given. Thus (5.2.16) gives (5.2.15). From (5.2.14) we obtain

$$\mathbf{x}' \times F \mathbf{x} = \mathbf{0}. \tag{5.C.2}$$

From (5.2.1) it follows that

$$(R\mathbf{x} + \mathbf{T}) \times F \mathbf{x} = \mathbf{0}, \tag{5.C.3}$$

or

$$(\mathbf{x} + R^t \mathbf{T}) \times R^t F \mathbf{x} = \mathbf{0}. \tag{5.C.4}$$

Letting $K = R^t F$ and $O' = -R^t \mathbf{T}$, we get (5.C.1).

Conversely assume (5.C.1) holds. Letting $F = RK$ $(F \ne O)$ we have (5.C.4). Thus (5.C.3) and (5.C.2) hold. From the fact that the depths of the points are not zeros, we have (5.2.15) and then (5.2.16) and (5.2.17). Since $\mathbf{h} \ne \mathbf{0}$, the rank of A is less than 9. □

Appendix 5.D Condition on Consistent Normals

Theorem 5.8. Let F and F' correspond, respectively, to two motions from one position. Suppose there exist two solutions $(R_a, \tilde{\mathbf{T}}_a, \mathring{\mathbf{N}}_a)$ and $(R_b, \tilde{\mathbf{T}}_b, \mathring{\mathbf{N}}_b)$ such that

$$F = R_a + \tilde{\mathbf{T}}_a \mathring{\mathbf{N}}_a^t = R_b + \tilde{\mathbf{T}}_b \mathring{\mathbf{N}}_b^t.$$

Let H be an orthogonal matrix with $\det(H) = 1$ such that

$$H^t F^t F H = \mathrm{diag}(\lambda_1, 1, \lambda_3) \tag{5.D.1}$$

and $\lambda_1 < 1 < \lambda_3$.

Similarly, for F' we have two solutions such that

$$F' = R'_a + \tilde{\mathbf{T}}'_a(\mathring{\mathbf{N}}'_a)^t = R'_b + \tilde{\mathbf{T}}'_b(\mathring{\mathbf{N}}'_b)^t.$$

Then, both solutions agree on the unit normal, i.e.,

$$\mathring{\mathbf{N}}_a = \mathring{\mathbf{N}}'_a, \quad \text{and} \quad \mathring{\mathbf{N}}_b = \mathring{\mathbf{N}}'_b, \tag{5.D.2}$$

if and only if the ambiguity condition is satisfied, i.e., there exists an orthogonal matrix Q and a positive number k such that

$$F' = Q \ \mathrm{diag}(\sqrt{1 - k(1 - \lambda_1)}, 1, \sqrt{1 + k(\lambda_3 - 1)}\,)H^t \tag{5.D.3.a}$$

with $1 - k(1 - \lambda_1) \geq 0$ or

$$F' = Q \ \mathrm{diag}(\sqrt{1 + k(1 - \lambda_1)}, 1, \sqrt{1 - k(\lambda_3 - 1)}\,)H^t \tag{5.D.3.b}$$

with $1 - k(\lambda_3 - 1) \geq 0$.

Proof. Specify the columns of H by $H = [\mathbf{h}'_1 \ \mathbf{h}'_2 \ \mathbf{h}'_3]$. Let $H' = [\mathbf{h}'_1 \ \mathbf{h}'_2 \ \mathbf{h}'_3]$ be orthogonal matrices such that

$$(H')^t(F')^tF'H' = \mathrm{diag}(\lambda'_1, 1, \lambda'_3). \tag{5.D.4}$$

If we change the signs of the columns of H or H', (5.D.1) and (5.D.4) still hold. Thus, we can assume $\det(H) = \det(H') = 1$. Then the column vectors of H and H' form a right-hand orthonormal basis, respectively.

We first prove the necessity part. Assume (5.D.2) holds. We apply the algorithm to F and F'. From (5.E.12) and the sign determined by (5.E.14), we have (using (5.E.10))

$$\pm \mathring{\mathbf{N}}_a = \mathbf{v}_1 \times \mathbf{v}_2 = (\alpha\mathbf{h}_1 + \beta\mathbf{h}_3) \times \mathbf{h}_2$$

$$= \alpha(\mathbf{h}_1 \times \mathbf{h}_2) + \beta(\mathbf{h}_3 \times \mathbf{h}_2) = \alpha\mathbf{h}_3 - \beta\mathbf{h}_1 \tag{5.D.5.a}$$

and for the second solution:

$$\pm \mathring{\mathbf{N}}_b = (\alpha\mathbf{h}_1 - \beta\mathbf{h}_3) \times \mathbf{h}_2$$

$$= \alpha(\mathbf{h}_1 \times \mathbf{h}_2) - \beta(\mathbf{h}_3 \times \mathbf{h}_2) = \alpha\mathbf{h}_3 + \beta\mathbf{h}_1. \tag{5.D.5.b}$$

Similarly, we obtain the equations corresponding to (5.D.5.a) and (5.D.5.b) for F':

$$\pm \mathring{\mathbf{N}}'_a = \mathbf{V}'_1 \times \mathbf{V}'_2 = \alpha' \mathbf{h}'_3 - \beta' \mathbf{h}'_1, \tag{5.D.6.a}$$

$$\pm \mathring{\mathbf{N}}'_b = \mathbf{V}'_1 \times \mathbf{V}'_2 = \alpha' \mathbf{h}'_3 + \beta' \mathbf{h}'_1. \tag{5.D.6.b}$$

If the first solution for F' gives $\mathring{\mathbf{N}}_a$ $(\mathring{\mathbf{N}}'_a = \mathring{\mathbf{N}}_a)$, we change, if necessary, the sign of $(\mathbf{h}'_1, \mathbf{h}'_3)$ without change $\det(H')$ to make the signs of (5.D.5.a) and (5.D.6.a) the same. Then (5.D.5) and (5.D.6) give

$$\alpha \mathbf{h}_3 - \beta \mathbf{h}_1 = \alpha' \mathbf{h}'_3 - \beta' \mathbf{h}'_1, \quad \alpha \mathbf{h}_3 + \beta \mathbf{h}_1 = s_1(\alpha' \mathbf{h}'_3 + \beta' \mathbf{h}'_1) \tag{5.D.7.I}$$

where $s_1 \in \{-1, 1\}$.

Otherwise the second solution for F' gives $\mathring{\mathbf{N}}_a$ $(\mathring{\mathbf{N}}'_b = \mathring{\mathbf{N}}_a)$. Then we change, if necessary, the sign of $(\mathbf{h}'_1, \mathbf{h}'_2)$ without changing $\det(H')$ to make the signs of (5.D.5.a) and (5.D.5.b) the same. Then (5.D.5) and (5.D.5) give

$$\alpha \mathbf{h}_3 - \beta \mathbf{h}_1 = \alpha' \mathbf{h}'_3 + \beta' \mathbf{h}'_1, \quad \alpha \mathbf{h}_3 + \beta \mathbf{h}_1 = s_2(\alpha' \mathbf{h}'_3 - \beta' \mathbf{h}'_1) \tag{5.D.7.II}$$

where $s_2 \in \{-1, 1\}$. Now consider (5.D.7.I). Suppose $s_1 = 1$. Summing and subtracting the corresponding sides of the two equations in (5.D.7.I) give

$$2\alpha \mathbf{h}_3 = 2\alpha' \mathbf{h}'_3, \qquad 2\beta \mathbf{h}_1 = 2\beta \mathbf{h}'_1.$$

Since α and β are positive and H and H' are orthogonal matrices with unit determinants, it follows that

$$\alpha' = \alpha, \qquad \beta' = \beta, \qquad H = H'.$$

Similarly we get the similar equations for $s_1 = -1$. We also consider the two cases in (5.D.7.II). Thus totally we get four cases:

$$\text{Ia: } \begin{cases} \alpha' = \alpha \\ \beta' = \beta \\ H' = H \end{cases}, \qquad \text{Ib: } \begin{cases} \alpha' = \beta \\ \beta' = \alpha \\ H' = H \ \mathrm{rdiag}(1, 1, -1) \end{cases}, \tag{5.D.8.I}$$

$$\text{IIa: }\begin{cases}\alpha' = \alpha\\ \beta' = \beta\\ H' = H\ \text{diag}(-1, -1, 1)\end{cases},\quad \text{IIb: }\begin{cases}\alpha' = \beta\\ \beta' = \alpha\\ H' = H\ \text{rdiag}(-1, 1, 1)\end{cases} \tag{5.D.8.II}$$

where rdiag$(\lambda_1, \lambda_2, \lambda_3)$ denotes the (unconventional) diagonal matrix where the diagonal line is defined from top-right to bottom-left. In fact, the cases Ia and IIa give (5.D.3.a), and Ib and IIb give (5.D.3.b). We prove that the case IIb gives (5.D.3.b). The proofs for the other cases are similar and therefore are omitted. From the case IIb in (5.D.8.II), using (5.E.9) gives

$$\frac{\lambda_3' - 1}{\lambda_3' - \lambda_1'} = \frac{1 - \lambda_1}{\lambda_3 - \lambda_1}, \qquad \frac{1 - \lambda_1'}{\lambda_3' - \lambda_1'} = \frac{\lambda_3 - 1}{\lambda_3 - \lambda_1}.$$

Letting $k = (\lambda_3' - \lambda_1')/(\lambda_3 - \lambda_1)$ it follows that

$$\lambda_1' = 1 - k(\lambda_3 - 1), \qquad \lambda_3' = 1 + k(1 - \lambda_1). \tag{5.D.9}$$

Since $k > 0$, and $\lambda_1 < 1 < \lambda_3$, we know $\lambda_1' < 1 < \lambda_3'$. Then (5.D.4), (5.D.9) and (5.D.8.II) give

$$\begin{aligned}(F')^t F' &= H'\ \text{diag}(\lambda_1', 1, \lambda_3')\ (H')^t\\ &= H\ \text{rdiag}(-1, 1, 1)\ \text{diag}(\lambda_1', 1, \lambda_3')\ \text{rdiag}(-1, 1, 1)^t\ H^t\\ &= H\ \text{diag}(\lambda_3', 1, \lambda_1')\ H^t\\ &= H\ \text{diag}(1 + k(1 - \lambda_1), 1, 1 - k(\lambda_3 - 1))\ H^t.\end{aligned} \tag{5.D.10}$$

Let $C = \text{diag}(\sqrt{1 + k(1 - \lambda_1)}, 1, \sqrt{1 - k(\lambda_3 - 1)}\,)H^t$. Noticing (5.D.4) and (5.D.10) we know that $F'H$ and CH are orthogonal matrices. From (5.D.9) we know that one can establish column correspondences, according to vector length, between $F'H$ and CH so that the corresponding column vectors have the same lengths. Therefore, there exists an orthonormal matrix (rotation or reflection) such that $F'H = QCH$, *i.e.*, $F' = QC$. This is (5.D.3.b).

Now we prove the sufficiency part. Assume (5.D.3.a) or (5.D.3.b) holds, we prove (5.D.2). We present the proof for the case where (5.D.3.b) holds. The proof corresponding to (5.D.3.a) is similar and is omited. Letting $H' = [\mathbf{h}_3\ \mathbf{h}_2\ \ \mathbf{h}_1]$, we get (5.D.4), where

$$\lambda_1' = 1 - k(\lambda_3 - 1), \quad \text{and} \quad \lambda_3' = 1 + k(1 - \lambda_1).$$

By the algorithm we have (we use prime to denote the corresponding variables for F')

$$\alpha' = \left[\frac{\lambda_3' - 1}{\lambda_3' - \lambda_1'}\right]^{1/2} = \left[\frac{1 - \lambda_1}{\lambda_3 - \lambda_1}\right]^{1/2} = \beta,$$

$$\beta' = \left[\frac{1 - \lambda_1'}{\lambda_3' - \lambda_1'}\right]^{1/2} = \left[\frac{\lambda_3 - 1}{\lambda_3 - \lambda_1}\right]^{1/2} = \alpha,$$

$$\mathbf{V}_1' = \alpha'\mathbf{h}_1' + \beta'\mathbf{h}_3' = \beta\mathbf{h}_3 + \alpha\mathbf{h}_1,$$

$$\mathbf{V}_2' = \mathbf{h}_2,$$

and

$$\pm \mathring{\mathbf{N}}_a' = \mathbf{V}_1' \times \mathbf{V}_2' = \beta\mathbf{h}_3 \times \mathbf{h}_2 + \alpha\mathbf{h}_1 \times \mathbf{h}_2$$

$$= -\beta\mathbf{h}_1 + \alpha\mathbf{h}_3 = (\alpha\mathbf{h}_1 + \beta\mathbf{h}_3) \times \mathbf{h}_2 = \pm \mathring{\mathbf{N}}_a.$$

From the definition of F in (5.2.13), we know that given an F the sign of the normal $\mathbf{N}$ can be coupled with the sign of $\mathbf{T}$. In other words, we can let

$$\mathring{\mathbf{N}}_a' = \mathring{\mathbf{N}}_a,$$

and according to Theorem 5.5, the algorithm determines R_a', $\tilde{\mathbf{T}}_a$ such that

$$F' = R'_a + \tilde{\mathbf{T}}_a \mathring{\mathbf{N}}_a^t.$$

Similarly we have $\mathring{\mathbf{N}}_b' = \mathring{\mathbf{N}}_b$ and

$$F' = R'_b + \tilde{\mathbf{T}}_b \mathring{\mathbf{N}}_b^t.$$

Appendix 5.E Two-View Algorithm

Step (i). *Solve for intermediate parameter matrix* $F = R + \mathbf{TN}^t$:

Given $\mathbf{X}_i = (u_i, v_i, 1)^t$, $\mathbf{X}_i' = (u_i', v_i', 1)^t$, $i = 1, 2, \cdots, n$, the corresponding image vectors of n ($n \geq 4$) points, let A be a $2n$ by 9 matrix such that

$$A = \begin{bmatrix} \mathbf{X}_1^t & \mathbf{0} & -u_1'\mathbf{X}_1^t \\ \mathbf{0} & \mathbf{X}_1^t & -v_1'\mathbf{X}_1^t \\ \mathbf{X}_2^t & \mathbf{0} & -u_2'\mathbf{X}_2^t \\ \mathbf{0} & \mathbf{X}_2^t & -v_2'\mathbf{X}_2^t \\ \cdot & \cdot & \cdot \\ \cdot & \cdot & \cdot \\ \cdot & \cdot & \cdot \\ \mathbf{X}_n^t & \mathbf{0} & -u_n'\mathbf{X}_n^t \\ \mathbf{0} & \mathbf{X}_n^t & -v_n'\mathbf{X}_n^t \end{bmatrix} \tag{5.E.1}$$

and $\mathbf{h}$ be a 9-dimensional vector

$$\mathbf{h} = (h_1, h_2, h_3, h_4, h_5, h_6, h_7, h_8, h_9)^t. \tag{5.E.2}$$

We solve for unit vector $\mathbf{h}$ in the following

$$\min_{\mathbf{h}} \|A\,\mathbf{h}\|, \qquad \textit{subject to}: \|\mathbf{h}\|=1. \tag{5.E.3}$$

The solution of $\mathbf{h}$ is a unit eigenvector of A^tA associated with the smallest eigenvalue. Then F_s is determined by

$$F_s = \begin{bmatrix} h_1 & h_2 & h_3 \\ h_4 & h_5 & h_6 \\ h_7 & h_8 & h_9 \end{bmatrix}. \tag{5.E.4}$$

Let $H = [\mathbf{h}_1\ \mathbf{h}_2\ \mathbf{h}_3]$ be a 3×3 orthogonal matrix such that

$$H^t F_s^t F_s H = \mathrm{diag}(\gamma_1, \gamma_2, \gamma_3) \tag{5.E.5}$$

with $\gamma_1 \le \gamma_2 \le \gamma_3$, where $\mathrm{diag}(\gamma_1, \gamma_2, \gamma_3)$ denotes the conventional diagonal matrix with the corresponding diagonal elements. Then

$$F = \frac{1}{\sqrt{\gamma_2}} F_s. \tag{5.E.6}$$

If

$$\sum_i \mathbf{X}_i' \cdot F \mathbf{X}_i < 0 \tag{5.E.7}$$

then $F \leftarrow -F$. The summation in (5.E.7) is over several values of i's to reduce the sensitivity to noise (usually three or four values of i will suffice).

Step (ii). *Solve for* R, $\mathring{\mathbf{T}}$ *and* $\tilde{\mathbf{N}}$ *from* F:

We have

$$H^t F^t F H = \mathrm{diag}(\gamma_1/\gamma_2,\ 1,\ \gamma_3/\gamma_2) \triangleq \mathrm{diag}(\lambda_1, \lambda_2, \lambda_3). \tag{5.E.8}$$

Case (1). $\lambda_1 < 1 < \lambda_3$ (iff $\mathbf{T}\times(R\,\mathbf{N}) \neq \mathbf{0}$):

There exist two solutions that give the same images. Let

$$\alpha = \left[\frac{\lambda_3 - 1}{\lambda_3 - \lambda_1}\right]^{1/2}, \qquad \beta = \left[\frac{1 - \lambda_1}{\lambda_3 - \lambda_1}\right]^{1/2}. \tag{5.E.9}$$

The first solution. Let

$$\mathbf{V}_1 = \alpha \mathbf{h}_1 + \beta \mathbf{h}_3, \qquad \mathbf{V}_2 = \mathbf{h}_2. \tag{5.E.10}$$

Then

$$R = [F\mathbf{V}_1 \;\; F\mathbf{V}_2 \;\; F\mathbf{V}_1 \times F\mathbf{V}_2][\mathbf{V}_1 \;\; \mathbf{V}_2 \;\; \mathbf{V}_1 \times \mathbf{V}_2]^t. \tag{5.E.11}$$

Let

$$\mathring{\mathbf{N}} = \mathbf{V}_1 \times \mathbf{V}_2, \tag{5.E.12}$$

and

$$\tilde{\mathbf{T}} = F\mathring{\mathbf{N}} - R\mathring{\mathbf{N}}. \tag{5.E.13}$$

If

$$\sum_i (\mathbf{X}_i' \times R\,\mathbf{X}_i) \cdot (\tilde{\mathbf{T}} \times \mathbf{X}_i') < 0, \tag{5.E.14}$$

then $\mathring{\mathbf{N}} \leftarrow -\mathring{\mathbf{N}}$ and re-evaluate (5.E.13) (change the sign of $\tilde{\mathbf{T}}$). As in (5.E.7), $\sum$ in (5.E.14) sums over several values of i to suppress noise. Finally, we get

$$\mathring{\mathbf{T}} = \frac{\tilde{\mathbf{T}}}{\|\tilde{\mathbf{T}}\|}, \tag{5.E.15}$$

and

$$\tilde{\mathbf{N}} = \|\tilde{\mathbf{T}}\|\mathring{\mathbf{N}}. \tag{5.E.16}$$

The second solution. Change the sign of the β we got in (5.E.9) (i.e., $\beta \leftarrow -\beta$) and keep the α unchanged. (5.E.10)-(5.E.16) give the second solution.

Case (2). $\lambda_1 = 1 < \lambda_3$ (iff $\mathbf{T} // R\mathbf{N}$ and the absolute distance between the plane and the origin increases due to motion):

In this case, $\alpha = 1$ and $\beta = 0$. (5.E.9)-(5.E.16) give the unique solution.

Case (3). $\lambda_1 < 1 = \lambda_3$ (iff $\mathbf{T} // R\mathbf{N}$ and the absolute distance between the plane and the origin decreases due to motion):

In this case, $\alpha = 0$ and $\beta = 1$. (5.E.9)-(5.E.16) give the unique solution.

Case (4). $\lambda_1 = 1 = \lambda_3$ (iff $\mathbf{T} // R\mathbf{N}$ and the absolute distance between the plane and the origin does not change):

If $\det(F) > 0$, report $\mathbf{T} = 0$. $R = F$. $\tilde{\mathbf{N}}$ can not be determined.

$\det(F) < 0$ occurs only if the back side of the plane faces the camera after motion, which is impossible for an opaque plane. If the plane is transparent and the points on the plane are visible on both sides, this case can happen. If so, the solutions are infinitely many. For any unit $\mathring{\mathbf{N}}$, the following is a solution

$$R = F(I_3 - 2\mathring{\mathbf{N}}\mathring{\mathbf{N}}^t) \tag{5.E.17}$$

$$\tilde{\mathbf{T}} = -2R\mathring{\mathbf{N}}. \tag{5.E.18}$$

(5.E.15) and (5.E.16) give $\mathring{\mathbf{T}}$ and $\tilde{\mathbf{N}}$, respectively.

Appendix 5.F Error Estimation

In steps (i), and (ii), the algorithm computes the eigenvalues and the eigenvectors of the corresponding matrices. What we need to do is to estimate the perturbation of the corresponding matrices from the perturbation in the image coordinates. We use the first order approximation to estimate these perturbations in the matrices.

For Step (i). Assume the components of the image vector

$\mathbf{X}_i = (u_i, v_i, 1)$ and $\mathbf{X}'_i = (u'_i, v'_i, 1)$ have errors. (The third component 1 in each image vector is accurate.) Let u_i, v_i, u'_i and v'_i have additive errors δ_{u_i}, δ_{v_i}, $\delta_{u'_i}$ and $\delta_{v'_i}$, respectively, for $1 \le i \le n$. From (5.E.1) we get:

$$\Delta_A^t = \begin{bmatrix} \delta_{u_1} & 0 & \cdot\cdot & 0 \\ \delta_{v_1} & 0 & \cdot\cdot & 0 \\ 0 & 0 & \cdot\cdot & 0 \\ 0 & \delta_{u_1} & \cdot\cdot & \delta_{u_n} \\ 0 & \delta_{v_1} & \cdot\cdot & \delta_{v_n} \\ 0 & 0 & \cdot\cdot & 0 \\ -\delta_{u_1}u'_1 - \delta_{u'_1}u_1 & -\delta_{u_1}v'_1 - \delta_{v'_1}u_1 & \cdot\cdot & -\delta_{u_n}v'_n - \delta_{v'_n}u_n \\ -\delta_{v_1}u'_1 - \delta_{u'_1}v_1 & -\delta_{v_1}v'_1 - \delta_{v'_1}v_1 & \cdot\cdot & -\delta_{v_n}v'_n - \delta_{v'_n}v_n \\ -\delta_{u'_1} & -\delta_{v'_1} & \cdot\cdot & -\delta_{v'_n} \end{bmatrix}. \quad (5.F.1)$$

Assuming that the errors in different points and different components of the image coordinates are uncorrelated and have the same variance σ^2, we get

$$\Gamma_{A^t} = \sigma^2 \operatorname{diag}(P_1, P_2, \cdots, P_n)$$

where P_i, $1 \le i \le n$, is a 18×18 sub-matrix:

$$P_i = \begin{bmatrix} J & O & -u'_iJ & O & J & -v'_iJ \\ O & O & O & O & O & O \\ -u'_iJ & O & M_{1i} & O & -u'_iJ & u'_iv'_iJ \\ O & O & O & O & O & O \\ J & O & -u'_iJ & O & J & -v'_iJ \\ -v'_iJ & O & u'_iv'_iJ & O & -v'_iJ & M_{2i} \end{bmatrix}$$

where O is a 3×3 zero matrix and

$$J = \begin{bmatrix} 1 & 0 & 0 \\ 0 & 1 & 0 \\ 0 & 0 & 0 \end{bmatrix},$$

$$M_{1i} = \mathbf{X}_i\mathbf{X}_i^t + (u'_i)^2 J, \qquad M_{2i} = \mathbf{X}_i\mathbf{X}_i^t + (v'_i)^2 J.$$

Now, consider the error of $\mathbf{h}$ in (5.E.3). From Theorem 3.1 and (5.E.2), we have (note that $\mathbf{h}$ is an eigenvector of A^tA instead of A):

$$\boldsymbol{\delta}_{\mathbf{h}} \cong H\,\Delta H^t\,\Delta_{A^tA}\,\mathbf{h}$$
$$= H\,\Delta H^t[h_1 I_9\ h_2 I_9\ \cdots\ h_9 I_9]\boldsymbol{\delta}_{A^tA} \triangleq G_{\mathbf{h}}\boldsymbol{\delta}_{A^tA}\,. \tag{5.F.2}$$

In the above equations, we have rewritten the matrices Δ_{A^tA} by $\boldsymbol{\delta}_{A^tA}$ and moved the perturbation to the right end of the expression. In this way, the perturbation of the eigenvector is then the linear transformation (by matrix $G_{\mathbf{h}}$) of the perturbation vector $\boldsymbol{\delta}_{A^tA}$. Since we have Γ_{A^t} $(=\Gamma_A^t)$ in (5.F.1), we need to relate $\boldsymbol{\delta}_{A^tA}$ in (5.F.2) to $\boldsymbol{\delta}_{A^t}$. Using first order approximation, we get

$$\Delta_{A^tA} \cong A^t\Delta_A + \Delta_A^t A. \tag{5.F.3}$$

Letting $A^t = [a_{ij}]^t \triangleq [\mathbf{A}_1\ \mathbf{A}_2\ \cdots\ \mathbf{A}_n]$ we write

$$\boldsymbol{\delta}_{A^tA} \cong G_{A^tA}\,\boldsymbol{\delta}_{A^t} \tag{5.F.4}$$

where G_{A^tA} can be easily determined from (5.F.3):

$$G_{A^tA} = [F_{ij}] + [G_{ij}]$$

where $[F_{ij}]$ and $[G_{ij}]$ are matrices with 9 by n sub-matrices F_{ij} and G_{ij}, respectively, $F_{ij} = a_{ji}I_9$, and G_{ij} is a 9×9 matrix with the i-th column being the column vector $\mathbf{A}_j$ and all other columns being zeros. From (5.F.2) and (5.F.4) we get

$$\boldsymbol{\delta}_{\mathbf{h}} \cong G_{\mathbf{h}}\boldsymbol{\delta}_{A^tA} \cong G_{\mathbf{h}}G_{A^tA}\,\boldsymbol{\delta}_{A^t} \triangleq D_{\mathbf{h}}\boldsymbol{\delta}_{A^t}.$$

Then $\Gamma_{\mathbf{h}} \cong D_{\mathbf{h}}\Gamma_{A^t}D_{\mathbf{h}}^t$. Since $\mathbf{F}^t = \pm\,\mathbf{h}/\sqrt{|\gamma_2|}$, there exists a permutation matrix M such that $\mathbf{F} = \pm\,M\mathbf{h}/\sqrt{|\gamma_2|}$. We have

$$\boldsymbol{\delta}_F \cong \pm\frac{1}{\sqrt{|\gamma_2|}}M\,\boldsymbol{\delta}_{\mathbf{h}} \cong \pm\frac{1}{\sqrt{|\gamma_2|}}MD_{\mathbf{h}}\boldsymbol{\delta}_{A^t} \triangleq \pm\,D_F\,\boldsymbol{\delta}_{A^t}$$

where the sign is negative iff (5.E.7) holds.

Starting from the covariance matrix of the perturbation in A^t, we got the covariance matrix of the perturbation in the eigenvector of A^tA. Finally we get the perturbation vector of F, $\boldsymbol{\delta}_F$. For the perturbation vectors of R, $\mathring{\mathbf{T}}$ and $\mathbf{N}$ we will get the linear expression in terms of $\boldsymbol{\delta}_F$. The corresponding covariance matrices then can be obtained. For example, if we get $D_{\mathring{\mathbf{T}}}$ such that $\boldsymbol{\delta}_{\mathring{\mathbf{T}}} \cong D_{\mathring{\mathbf{T}}}\boldsymbol{\delta}_F$, it follows that $\Gamma_{\mathring{\mathbf{T}}} \cong D_{\mathring{\mathbf{T}}}\Gamma_F D_{\mathring{\mathbf{T}}}^t$.

The solution of Step (i) needs the eigenvector of A^tA associated with the smallest eigenvalue. The smallest eigenvalue is a simple zero eigenvalue when rank$(A) = 8$ (nondegenerate configuration), When rank$(A) < 8$ (i.e., when degenerate configurations occur), the solution $\mathbf{h}$ in Step (i) is very sensitive to noise. As can be seen from (5.F.6), the second diagonal element of Δ is infinite when $\lambda_1 = \lambda_2$, which makes the estimated errors infinite.

However, in most real applications, we do not know the noise-free A. We only know the noise-corrupted A: $A(\varepsilon)$. Therefore, $A(\varepsilon)$ is used to estimate A. In the presence of noise, generally, the rank of $A(\varepsilon)$ is full mathematically and the smallest eigenvalue of $A(\varepsilon)^t A(\varepsilon)$ is a small positive number. If noise is reasonably small, when rank$(A) < 8$ we have $\lambda_1 \approx \lambda_2$. Then large estimates of errors are still generated. From a slightly different point of view, we can regard A as a "noise-corrupted" matrix by adding $-\Delta_A$ to the matrix $A(\varepsilon)$. Now the error is the deviation of the true solution from the noise-corrupted solution. This observation justify our use of the noise-corrupted A to estimate errors.

For Step (ii). To obtain the perturbation of R, $\mathring{\mathbf{T}}$, and $\tilde{\mathbf{N}}$, we need the perturbation of H in (5.E.5) and that of α and β in (5.E.9), which in turn need the perturbation of the eigenvalues and the eigenvectors of F^tF. Theorem 3.1 can be used again to give a first order perturbation of the eigenvalues and eigenvectors (column vectors in H) of F^tF. According to Theorem 3.1, for i=1, 2, 3, we have

$$\delta_{\gamma_i} \cong \mathbf{h}_i^t \Delta_{F_s^tF_s} \mathbf{h}_i = \mathbf{d}_i^t \delta_{F_s^tF_s}$$

where

$$\mathbf{d}_i = (h_{1i}h_{1i},\, h_{2i}h_{1i},\, h_{3i}h_{1i},\, h_{1i}h_{2i},\, \cdots,\, h_{3i}h_{3i})^t$$

and $H = [h_{ij}]$. Therefore,

$$\delta_\gamma \triangleq \begin{bmatrix} \delta_{\gamma_1} \\ \delta_{\gamma_2} \\ \delta_{\gamma_3} \end{bmatrix} \cong \begin{bmatrix} \mathbf{d}_1^t \\ \mathbf{d}_2^t \\ \mathbf{d}_3^t \end{bmatrix} \delta_{F_s^tF_s} \triangleq G_\gamma \delta_{F_s^tF_s} = G_\gamma \gamma_2 \delta_{F^tF}, \tag{5.F.5}$$

since $F_s^tF_s = \gamma_2 F^tF$. We need δ_{F^tF}, which can be derived in terms of δ_F.

$$\Delta_{F^tF} \cong F^t \Delta_F + \Delta_F^t F. \tag{5.F.6}$$

Letting $F^t = [e_{ij}]^t \triangleq [\mathbf{F}_1\ \mathbf{F}_2\ \mathbf{F}_3]$ we write

$$\delta_{F^tF} \cong D_{F^tF}\, \delta_F \tag{5.F.7}$$

where D_{F^tF} can be easily determined from (5.F.6):

$$D_{F^tF} = \text{diag}(F^t, F^t, F^t) + [G_{ij}]$$

where diag(F^t, F^t, F^t) is a diagonal matrix whose elements are 3×3 submatrices. $[G_{ij}]$ is a matrix with 3×3 sub-matrices G_{ij}'s. G_{ij} is a 3×3 matrix with the j-th row vector being the column vector $\mathbf{F}_i$ and all other rows being zeros. From (5.F.5) and (5.F.7) we get

$$\delta_\gamma \cong G_{\gamma\gamma_2} \delta_{F^tF} \cong G_{\gamma\gamma_2} D_{F^tF}\, \delta_F \triangleq D_\gamma \delta_F .$$

Now we discuss the perturbation of the eigenvectors of F^tF, i.e., the column vectors in H. Using Theorem 3.1 gives

$$\delta_H \triangleq \begin{bmatrix} \delta_{\mathbf{h}_1} \\ \delta_{\mathbf{h}_2} \\ \delta_{\mathbf{h}_3} \end{bmatrix} \cong \begin{bmatrix} H\Delta_1 H^t [h_{11} I_3 \;\; h_{21} I_3 \;\; h_{31} I] \\ H\Delta_2 H^t [h_{12} I_3 \;\; h_{22} I_3 \;\; h_{32} I] \\ H\Delta_3 H^t [h_{13} I_3 \;\; h_{23} I_3 \;\; h_{33} I] \end{bmatrix} \delta_{F^tF}$$

$$\triangleq G_H\, \delta_{F^tF} \cong G_H D_{F^tF}\, \delta_F \triangleq D_H\, \delta_F$$

where

$$\Delta_1 = \text{diag}(0,\, (\lambda_1 - \lambda_2)^{-1},\, (\lambda_1 - \lambda_3)^{-1}),$$

$$\Delta_2 = \text{diag}((\lambda_2 - \lambda_1)^{-1},\, 0,\, (\lambda_2 - \lambda_3)^{-1}),$$

$$\Delta_3 = \text{diag}((\lambda_3 - \lambda_1)^{-1},\, (\lambda_3 - \lambda_2)^{-1},\, 0\ ).$$

From (5.E.11) we know that the perturbations in R depend on those in $\mathbf{V}_1$, $\mathbf{V}_2$ and F. From (5.E.10), the perturbations of $\mathbf{V}_1$ and $\mathbf{V}_2$ depend on those in α, β and H. Now we consider the first order perturbation in α and β. From (5.E.8) and (5.E.9) it follows that

$$\alpha = \left\{ \frac{\lambda_3 - 1}{\lambda_3 - \lambda_1} \right\}^{1/2} = \left\{ \frac{\gamma_3 - \gamma_2}{\gamma_3 - \gamma_1} \right\}^{1/2},$$

$$\beta = \left[\frac{1-\lambda_1}{\lambda_3-\lambda_1}\right]^{1/2} = \left[\frac{\gamma_2-\gamma_1}{\gamma_3-\gamma_1}\right]^{1/2}.$$

The noise corrupted α can be written as $\alpha(\delta_{\gamma_1}, \delta_{\gamma_2}, \delta_{\gamma_3})$, with the noise-free α being $\alpha(0, 0, 0)$. Expending $\alpha(\delta_{\gamma_1}, \delta_{\gamma_2}, \delta_{\gamma_3})$ at point (0, 0, 0) using Taylor series for multiple variables, it follows that

$$\alpha(\delta_{\gamma_1}, \delta_{\gamma_2}, \delta_{\gamma_3}) \cong \alpha + \frac{\sqrt{\gamma_3-\gamma_2}}{2\sqrt{(\gamma_3-\gamma_1)^3}}\delta_{\gamma_1}$$

$$-\frac{1}{2\sqrt{(\gamma_3-\gamma_1)(\gamma_3-\gamma_2)}}\delta_{\gamma_2} + \frac{\gamma_2-\gamma_1}{2\sqrt{(\gamma_3-\gamma_1)^3(\gamma_3-\gamma_2)}}\delta_{\gamma_3}.$$

Similarly we get the first order approximation of $\beta(\delta_{\gamma_1}, \delta_{\gamma_2}, \delta_{\gamma_3})$. Writing the the first order perturbation of α and β in matrix form, we have:

$$\begin{bmatrix}\delta_\alpha\\ \delta_\beta\end{bmatrix} \cong \frac{1}{\sqrt{(\gamma_3-\gamma_1)^3(\gamma_2-\gamma_1)(\gamma_3-\gamma_2)}}$$

$$\begin{bmatrix}(\gamma_3-\gamma_2)\sqrt{\gamma_2-\gamma_1} & -(\gamma_3-\gamma_1)\sqrt{\gamma_2-\gamma_1} & \sqrt{(\gamma_2-\gamma_1)^3}\\ -\sqrt{(\gamma_3-\gamma_2)^3} & -(\gamma_3-\gamma_1)\sqrt{\gamma_3-\gamma_2} & (\gamma_2-\gamma_1)\sqrt{\gamma_3-\gamma_2}\end{bmatrix}\begin{bmatrix}\delta_{\gamma_1}\\ \delta_{\gamma_2}\\ \delta_{\gamma_3}\end{bmatrix}$$

$$\triangleq G_{\alpha\beta}\delta_\gamma \cong G_{\alpha\beta}D_\gamma\delta_F \triangleq D_{\alpha\beta}\delta_F.$$

Let $\mathbf{G}_\alpha^t$ and $\mathbf{G}_\beta^t$ be the row vectors of $D_{\alpha\beta}$, *i.e.*, $G_{\alpha\beta} = [\mathbf{G}_\alpha \;\; \mathbf{G}_\beta]^t$. From (5.E.10) we have the first order perturbation of $\mathbf{V}_1$ and $\mathbf{V}_2$:

$$\delta_{\mathbf{V}_1} \cong \delta_\alpha\mathbf{h}_1 + \alpha\delta_{\mathbf{h}_1} + \delta_\beta\mathbf{h}_3 + \beta\delta_{\mathbf{h}_3}$$

$$= (\mathbf{h}_1\mathbf{G}_\alpha^t + \mathbf{h}_3\mathbf{G}_\beta^t + [\alpha I_3 \;\; O \;\; \beta I_3]D_H)\delta_F \triangleq D_{\mathbf{V}_1}\delta_F, \tag{5.F.8}$$

$$\delta_{\mathbf{V}_2} = \delta_{\mathbf{h}_2} \cong [O \;\; I_3 \;\; O]D_H\delta_F \triangleq D_{\mathbf{V}_2}\delta_F. \tag{5.F.9}$$

Now we find D_R such that $\delta_R = D_R\delta_F$. First, we consider the perturbation of $\mathbf{W}_i \triangleq F\mathbf{V}_i$, i=1, 2. Letting $\mathbf{V}_i = (v_{1i}, v_{2i}, v_{3i})$, we obtain

$$\delta_{\mathbf{W}_i} \cong \Delta_F\mathbf{V}_i + F\delta_{\mathbf{V}_i}$$

$$= ([v_{1i}I_3 \;\; v_{2i}I_3 \;\; v_{3i}I_3] + FD_{\mathbf{V}_i})\delta_F \triangleq D_{\mathbf{W}_i}\delta_F .$$

For the conciseness of notations, we define a new vector **K**.

$$\delta_{\mathbf{K}} \triangleq \begin{bmatrix} \delta_{\mathbf{V}_1} \\ \delta_{\mathbf{V}_2} \\ \delta_{\mathbf{W}_1} \\ \delta_{\mathbf{W}_2} \end{bmatrix} \cong \begin{bmatrix} D_{\mathbf{V}_1} \\ D_{\mathbf{V}_2} \\ D_{\mathbf{W}_1} \\ D_{\mathbf{W}_2} \end{bmatrix} \delta_F \triangleq D_{\mathbf{K}}\delta_F . \tag{5.F.10}$$

Since $R = [\mathbf{W}_1, \mathbf{W}_2, \mathbf{W}_1 \times \mathbf{W}_2][\mathbf{V}_1, \mathbf{V}_2, \mathbf{V}_1 \times \mathbf{V}_2]^t$, evaluating the first order perturbation, one can construct a matrix G_R such that

$$\delta_R \cong G_R \, \delta_{\mathbf{K}}. \tag{5.F.11}$$

Then, (5.F.10) and (5.F.11) give

$$\delta_R \cong G_R \, \delta_{\mathbf{K}} \cong G_R D_{\mathbf{K}} \delta_R \triangleq D_R \, \delta_F . \tag{5.F.12}$$

Now we consider the perturbation of $\tilde{\mathbf{N}}$. From (5.E.12), (5.F.8), (5.F.9) and the possible sign change according to (5.E.14) we get

$$\pm \, \delta_{\mathring{\mathbf{N}}} \cong \mathbf{V}_1 \times \delta_{\mathbf{V}_2} + \delta_{\mathbf{V}_1} \times \mathbf{V}_2 = [\mathbf{V}_1]_\times \delta_{\mathbf{V}_2} - [\mathbf{V}_2]_\times \delta_{\mathbf{V}_1}$$

$$= ([\mathbf{V}_1]_\times D_{\mathbf{V}_2} - [\mathbf{V}_1]_\times D_{\mathbf{V}_2})\delta_F \triangleq \pm \, D_{\mathring{\mathbf{N}}}\delta_F . \tag{5.F.13}$$

From (5.E.16), we get the perturbation of $\tilde{\mathbf{N}}$

$$\delta_{\tilde{\mathbf{N}}} \cong \|\tilde{\mathbf{T}}\| \delta_{\mathring{\mathbf{N}}} \cong \|\tilde{\mathbf{T}}\| D_{\mathring{\mathbf{N}}} \delta_F .$$

From (5.E.13), we obtain the first order perturbation of $\tilde{\mathbf{T}}$:

$$\delta_{\tilde{\mathbf{T}}} \cong (F - R)\delta_{\mathring{\mathbf{N}}} + (\Delta_F - \Delta_R)\mathring{\mathbf{N}}. \tag{5.F.14}$$

Letting $\mathring{\mathbf{N}} = (n_1, n_2, n_3)$, and $M_{\mathring{\mathbf{N}}} \triangleq [n_1 I_3 \;\; n_2 I_3 \;\; n_3 I_3]$ we have $\Delta_F \mathring{\mathbf{N}} = M_{\mathring{\mathbf{N}}}\delta_F$ and $\Delta_R \mathring{\mathbf{N}} = M_{\mathring{\mathbf{N}}}\delta_R$. Using (5.F.12), (5.F.13) and (5.F.14) gives

$$\delta_{\tilde{\mathbf{T}}} = (F - R)D_{\mathring{\mathbf{N}}}\delta_F + M_{\mathring{\mathbf{N}}}\delta_F - M_{\mathring{\mathbf{N}}}\delta_R$$

$$\cong ((F - R)D_{\mathring{\mathbf{N}}} + M_{\mathring{\mathbf{N}}} - M_{\mathring{\mathbf{N}}}D_R)\delta_F \triangleq D_{\tilde{\mathbf{T}}}\delta_F .$$

Then $\delta_{\hat{\mathbf{T}}} \cong \|\tilde{\mathbf{T}}\|^{-1}\delta_{\tilde{\mathbf{T}}}$.

As in Step (i), in Step (ii) we estimate the errors by using the perturbed F, R and $\|\tilde{\mathbf{T}}\|$ *etc*., to substitute the noise-free ones.

In summary, the perturbation vectors of R, $\mathring{\mathbf{T}}$ and $\tilde{\mathbf{N}}$ are expressed in terms of linear transformation of perturbation in F.

6. From Line Correspondences

Using lines as the features for analysis has some advantages. This chapter addresses the subject of using lines to estimate the motion and structure parameters. A closed-form solution to this problem is derived. The corresponding linear algorithm makes use of redundancy in the data to improve the accuracy of the solution in the presence of noise. We also study the uniqueness of the solution as well as the necessary and sufficient conditions for the algorithm to give the unique solution. The optimization and performance bound are also investigated.

6.1 Lines as Features

The choice of types of features depends on the availability of the features in the images and the reliability of their measurement. When points are not available in large quantities, other features such as lines or contours can be used. Since higher level features like lines and contours are determined by a set of pixels, the redundancy in the edge pixels make it possible to locate those features accurately in image plane. In this chapter, we discuss motion estimation based on lines. In practice, one certainly may use different types of features as many as possible to obtain robust solutions. However, the study of the use of single type of features like lines is very important, theoretically and practically, to the general use of multiple types of features. It provides insights into the roles of this type of features in the solutions to the problem.

Because the end points of an extracted line are very unstable [6.2, 3, 6], the lines we will deal with are straight lines without known end points. (Using end points becomes a point-based problem and it is discussed in the previous chapters.) For example, the end points often do not correspond to physical points and move as the view point changes. Many factors such as lighting and surface reflection often changes the position of the end points when view point changes. However, the location and orientation of the line can generally be determined reliably by a line fitting along a sequence of edge points. In other words, long lines are preferred since they provide more edge points to allow a more accurate measurement of the line position.

We first present a guideline of our approach to those problems. As we mentioned above, a common characteristic of linear algorithms is solving for a matrix of intermediate unknowns through linear equations. Those intermediate unknowns are not independent. In other words, there exist constraints on the variables of intermediate unknowns, and there are more intermediate unknown variables than the "independent" unknowns. The linear equations are solved without using those constraints (otherwise, we are forced to solve nonlinear equations). The resulting intermediate unknowns contain redundant information. One of the objective of our linear algorithm is to make good use of such a redundancy to improve the accuracy of the solutions in the presence of noise. On the other hand, the problem to be investigated here involves three image frames with line features and thus, it is significantly more complicated than two-frame point-based problem. We derive compact computational steps to avoid, as much as possible, possible degenerate cases and spurious solutions that may otherwise be generated. As a result, we are able to investigate the problem of uniqueness. Although our linear algorithm is designed to well utilize the redundancy in the data, the solutions are not overall optimal. However, those solutions can be used as an initial guess for further improvement through optimization. Since the optimization is nonlinear, the good initial guess provided by the solution of the linear algorithm is very crucial to the correct convergence of the optimization. We will compare the error variance of our optimal solution with that of the Cramér-Rao lower error bound.

6.2 Solution and Algorithm

This section presents a linear algorithm for the motion and structure estimation. We first show that, inherently, motion cannot be determined from lines in just two images.

6.2.1 Why Two Views Are Not Sufficient

It is more convenient to investigate this issue from the view-point of the world coordinate system. In this system, the camera is moving and the scene is stationary. Let a camera system consist of a center of projection and an image plane both of which are rigidly connected. At each time instant, define a camera system at the corresponding position along the trajectory of motion associated with the corresponding image. The problem to be investigated is equivalent to the following: Fixing the first camera system at a known position and orientation, we want to determine the position and orientation of the second camera system and the 3-D positions of the lines, from the projections of lines in the two image planes (see Fig. 6.1).

We show that this is an underdetermined problem and it allows arbitrary motion interpretations. For each line in an image plane, we define a 3-D plane, called the *projection plane* of the line, which passes through the center of projection and the line. For each line correspondence, two camera systems determine two corresponding projection planes, whose intersection gives the line in 3-D. Suppose that the second camera system is *arbitrarily* perturbed away from the correct position, with the projection planes fixed with the camera system. Since any two non-parallel 3-D planes intersect and the corresponding intersection yields a line, every pair of projection planes still intersects as long as the perturbation is not so large and accidental that two corresponding projection planes become parallel (see Fig. 6.1). In other words, the arbitrarily perturbed position of the

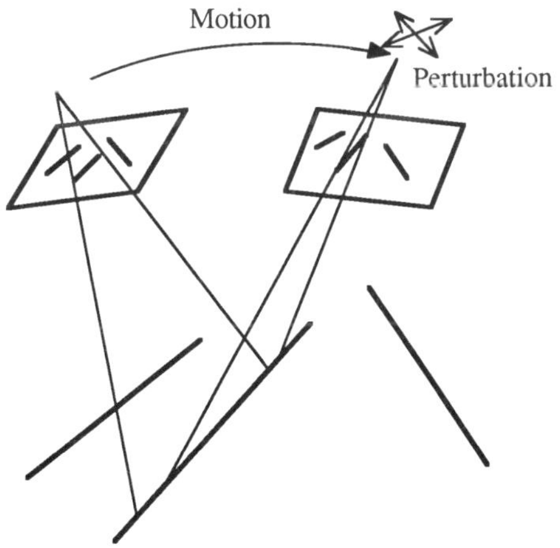

Fig. 6.1. Motion and structure can not be determined from lines in two images.

second camera system still gives a 3-D line configuration that is consistent with the two images observed. Therefore, the solution to the position and orientation of the camera is arbitrary (at least in an open set including the correct one), as is the corresponding 3-D line structure.

If a third image is added, it is possible to determine the position of the second and the third camera systems as well as the 3-D position of the lines, because the intersection of three projection planes is generally not a line. In the following, we discuss determining motion and 3-D line structure using line correspondences through three images.

6.2.2 From Three Views

Let the camera-centered coordinate system be fixed on the camera with the origin coinciding with the center of projection of the camera, and the z-axis coinciding with the optical axis and pointing to front as shown in Fig. 3.1. Thus, in this coordinate system, camera is fixed and the scene is moving. Visible objects are always located in front of the camera, i.e., $z > 0$.

In the camera-centered coordinate system, a line passing through a point $\mathbf{x}_p$ (to be specific, let $\mathbf{x}_p$ be the point on the line that is the closest to the origin) with direction $\mathbf{l}$ at time t_0 can be expressed in the following parametrical form

$$\mathbf{x}_0 = \mathbf{x}_p + k\mathbf{l} \tag{6.2.1}$$

where the subscript in $\mathbf{x}_0$ means time t_0, and k is parameter. At another time instant, t_1, the line is moved from t_0 by a rotation, represented by a rotation matrix R, and then a translation, represented by a translation vector $\mathbf{T}$. That is, any point at position $\mathbf{x}_1$ at time t_1 is related to its position $\mathbf{x}_0$ at times t_0 by

$$\mathbf{x}_1 = R\,\mathbf{x}_0 + \mathbf{T}. \tag{6.2.2}$$

The line equation at time t_1 is

$$t_1\colon \quad \mathbf{x}_1 = R\,\mathbf{x}_0 + \mathbf{T} = (R\,\mathbf{x}_p + \mathbf{T}) + kR\,\mathbf{l}. \tag{6.2.3}$$

It is easy to see that after motion, the line at time t_1 passes through point $R\,\mathbf{x}_p + \mathbf{T}$ with a direction $R\,\mathbf{l}$. Similarly at another time instant t_2, the line is rotated by a rotation matrix S and then translated by a vector $\mathbf{U}$ from time t_0. The line equation at t_2 is

$$t_2: \quad \mathbf{x}_2 = S\,\mathbf{x}_0 + \mathbf{U} = (S\,\mathbf{x}_p + \mathbf{U}) + kS\,\mathbf{l}. \tag{6.2.4}$$

Notice that it is not necessary that $t_0 \le t_1 \le t_2$ holds. The order of the three time instants t_0, t_1 and t_2 can be arbitrary (see Fig. 6.2).

6.2.3 Two Important Equations

We define the *projection normal* of a line as the normal of the projection plane (plane that passes through the line and the center of projection — origin). Since the projection normal is orthogonal to the line and the position vector of a point on the line, it is easy to get the projection normals at the three time instants from (6.2.2)-(6.2.4)

$$t_0: \quad \mathbf{n}_0 = \mathbf{x}_p \times \mathbf{l}; \tag{6.2.5}$$

$$t_1: \quad \mathbf{n}_1 = (R\,\mathbf{x}_p + \mathbf{T}) \times R\,\mathbf{l} = R\,((\mathbf{x}_p + R^{-1}\mathbf{T}) \times \mathbf{l}) = R\,(\mathbf{n}_0 + R^{-1}\mathbf{T} \times \mathbf{l}); \tag{6.2.6}$$

$$t_2: \quad \mathbf{n}_2 = (S\,\mathbf{x}_p + \mathbf{U}) \times S\,\mathbf{l} = S\,((\mathbf{x}_p + S^{-1}\mathbf{U}) \times \mathbf{l}) = S\,(\mathbf{n}_0 + S^{-1}\mathbf{U} \times \mathbf{l}). \tag{6.2.7}$$

Equation (6.2.6) gives

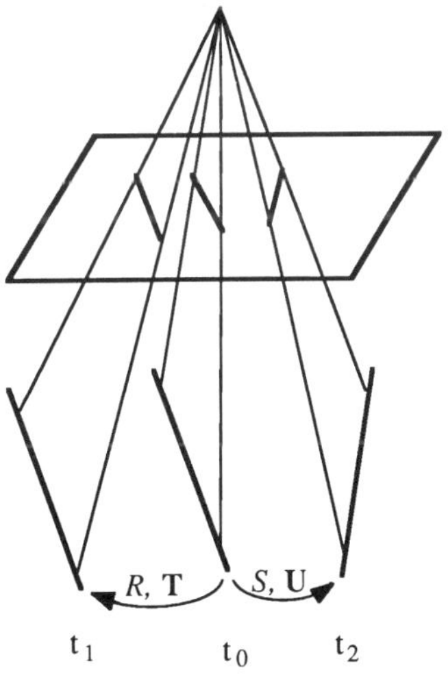

Fig. 6.2. Motion and structure from lines in three images.

$$R^{-1}\mathbf{n}_1 = \mathbf{n}_0 + R^{-1}\mathbf{T}\times\mathbf{l}. \tag{6.2.8}$$

Using the vector identity $\mathbf{a}\times(\mathbf{b}\times\mathbf{c}) = (\mathbf{a}\cdot\mathbf{c})\mathbf{b} - (\mathbf{a}\cdot\mathbf{b})\mathbf{c}$ and (6.2.8) yields

$$\mathbf{n}_0\times R^{-1}\mathbf{n}_1 = \mathbf{n}_0\times(R^{-1}\mathbf{T}\times\mathbf{l}) = (\mathbf{n}_0\cdot\mathbf{l})R^{-1}\mathbf{T} - (\mathbf{n}_0\cdot R^{-1}\mathbf{T})\mathbf{l} = -(\mathbf{n}_0\cdot R^{-1}\mathbf{T})\mathbf{l}. \tag{6.2.9}$$

The last equation follows from the fact that $\mathbf{n}_0\cdot\mathbf{l} = 0$. Using $\mathbf{n}_0$ solved from (6.2.8) gives

$$\mathbf{n}_0\cdot R^{-1}\mathbf{T} = (R^{-1}\mathbf{n}_1 - R^{-1}\mathbf{T}\times\mathbf{l})\cdot R^{-1}\mathbf{T} = R^{-1}\mathbf{n}_1\cdot R^{-1}\mathbf{T} = \mathbf{n}_1\cdot\mathbf{T}. \tag{6.2.10}$$

Equations (6.2.9) and (6.2.10) yield

$$\mathbf{n}_0\times R^{-1}\mathbf{n}_1 = -(\mathbf{n}_1\cdot\mathbf{T})\mathbf{l}. \tag{6.2.11}$$

Similarly we get

$$\mathbf{n}_0\times S^{-1}\mathbf{n}_2 = -(\mathbf{n}_2\cdot\mathbf{U})\mathbf{l}. \tag{6.2.12}$$

6.2.4 A Geometrical View

These two equations, (3.11) and (6.2.12), can also be proved geometrically as shown in the following:

For each line we arbitrarily choose its direction from two possible alternative ones and thus represent it by a vector $\mathbf{l}$. Viewed along this direction, the configuration can be shown in Fig. 6.3, where the line vector $\mathbf{l}$ points towards the paper (a cross marks the tail of the vector), and the point "O" denotes the center of projection. We first assume that the line does not go through the center of projection "O". From the original motion equation (6.2.2) it follows that

$$R^{-1}\mathbf{x}_1 = \mathbf{x}_0 + R^{-1}\mathbf{T}. \tag{6.2.13}$$

This means that, if the moved line is rotated back by R^{-1}, the resulting composite motion is a pure translation represented by vector $R^{-1}\mathbf{T}$. This composite motion is shown in Fig. 6.3 (only the projection of $R^{-1}\mathbf{T}$ onto the viewing plane can be shown in the figure). Viewed from the direction employed for Fig. 6.3, the vector $R^{-1}\mathbf{T}$ can lie on either side of the projection plane of the line $\mathbf{l}$ (the plane that passes "O" and $\mathbf{l}$, visible as a line in Fig. 6.3). In order to show both cases, we let the corresponding vectors $R^{-1}\mathbf{T}$ (first motion) and $S^{-1}\mathbf{U}$

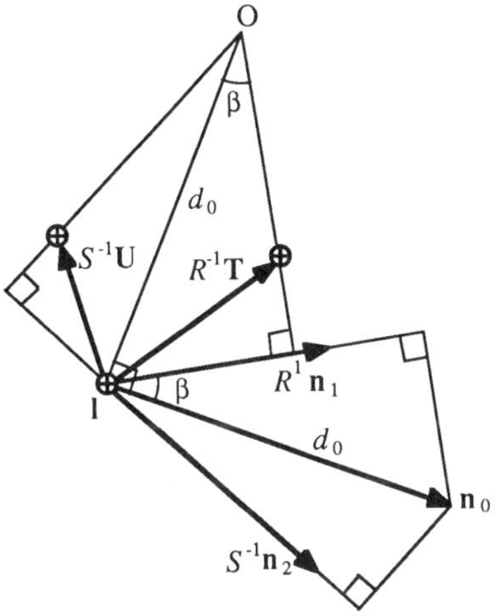

Fig. 6.3. Geometrical illustration of equations (6.2.11) and (6.2.12).

(second motion) lie on different sides in Fig. 6.3. The vector $R^{-1}\mathbf{n}_1$ is the projection normal of the plane after the pure translation $R^{-1}\mathbf{T}$, and thus, it is orthogonal to the line that is translated by vector $R^{-1}\mathbf{T}$ from **l**. Because the composite motion is a translation, the vector $R^{-1}\mathbf{n}_1$ is also orthogonal to **l** as shown in Fig. 6.3. Therefore, both $R^{-1}\mathbf{n}_1$ and $\mathbf{n}_0$ are orthogonal to **l**: $\mathbf{l} \,//\, (\mathbf{n}_0 \times R^{-1}\mathbf{n}_1)$, which gives the "alignment part" of (6.2.11).

By definition in (6.2.5), the length of $\mathbf{n}_0$, d_0, is equal to the distance between the line and the origin. In Fig. 6.3, there exist two congruent right triangles, determined by two equal angles β's and the two equal hypotenuses with length d_0. The corresponding sides opposite to β's, respectively, should be equal: One side is equal to $\|\mathbf{n}_0 \times R^{-1}\mathbf{n}_1\| / \|\mathbf{n}_1\|$, and the other is equal to

$$R^{-1}\mathbf{n}_1 \cdot R^{-1}\mathbf{T} / \|\mathbf{n}_1\| = \mathbf{n}_1 \cdot \mathbf{T} / \|\mathbf{n}_1\|. \tag{6.2.14}$$

This proves the "length part" of (6.2.11).

What remains to be established is the "sign part". As we mentioned above, the vector $R^{-1}\mathbf{T}$ can lie on either side of the projection plane of **l**: the side as is in Fig. 6.3 or the side of $S^{-1}\mathbf{U}$ shown in Fig. 6.3. In the former case, $\mathbf{n}_0 \times R^{-1}\mathbf{n}_1$ has a direction opposite to **l**, and we have $\mathbf{n}_1 \cdot \mathbf{T} \geq 0$ because $R^{-1}\mathbf{n}_1$ and $R^{-1}\mathbf{T}$ lie on the same side of the projection plane of **l** and the angle between them is an internal angle of a right triangle. $\mathbf{n}_1 \cdot \mathbf{T} = 0$ holds true if and only if $\mathbf{T} = \mathbf{0}$, and so does $\mathbf{n}_0 \times R^{-1}\mathbf{n}_1 = \mathbf{0}$. This concludes the "sign part" of (6.2.11) for the former case. For the latter case, the vector $\mathbf{n}_0 \times R^{-1}\mathbf{n}_1$ gets the same direction as $\mathbf{l}$, and $\mathbf{n}_1 \cdot \mathbf{T} \le 0$ because $R^{-1}\mathbf{n}_1$ and $R^{-1}\mathbf{T}$ are located on the different sides of the projection plane of $\mathbf{l}$ and the angle between them is an external angle of a right triangle. Therefore, the "sign part" of (6.2.11) is always true for both cases. Suppose the line $\mathbf{l}$ does go through the senter of projection "O". Then $\mathbf{n}_0 = \mathbf{0}$ according to the definition of (6.2.5) and $R^{-1}\mathbf{T}$ is orthogonal to $R^{-1}\mathbf{n}_1$ as can be seen from Fig. 6.3. Thus, equation (6.2.11) still holds true since both sides vanish. This completes the proof for (6.2.11). The proof for (6.2.12) is analogous.

Compared with the geometrical proof, the algebraic derivation discussed earlier appears to be more rigorous, but less intuitive. From the geometrical proof, one can see what properties are used to determine the solution.

6.2.5 Intermediate Parameters

Multiplying both sides of (6.2.11) by $\mathbf{n}_2 \cdot \mathbf{U}$ and those of (6.2.12) by $\mathbf{n}_1 \cdot \mathbf{T}$ yields

$$(\mathbf{n}_2 \cdot \mathbf{U})(\mathbf{n}_0 \times R^{-1}\mathbf{n}_1) = (\mathbf{n}_1 \cdot \mathbf{T})(\mathbf{n}_0 \times S^{-1}\mathbf{n}_2). \tag{6.2.15}$$

Or,

$$[\mathbf{n}_0]_\times \mathbf{B} = 0 \tag{6.2.16}$$

where $\mathbf{B} = (\mathbf{n}_2 \cdot \mathbf{U})R^{-1}\mathbf{n}_1 - (\mathbf{n}_1 \cdot \mathbf{T})S^{-1}\mathbf{n}_2$. Letting $R = [\mathbf{R}_1\ \mathbf{R}_2\ \mathbf{R}_3]$ and $S = [\mathbf{S}_1\ \mathbf{S}_2\ \mathbf{S}_3]$, $\mathbf{B}$ can be expressed as

$$\mathbf{B} = \begin{bmatrix} \mathbf{n}_1^t(\mathbf{R}_1\mathbf{U}^t - \mathbf{T}\mathbf{S}_1^t)\mathbf{n}_2 \\ \mathbf{n}_1^t(\mathbf{R}_2\mathbf{U}^t - \mathbf{T}\mathbf{S}_2^t)\mathbf{n}_2 \\ \mathbf{n}_1^t(\mathbf{R}_3\mathbf{U}^t - \mathbf{T}\mathbf{S}_3^t)\mathbf{n}_2 \end{bmatrix} \triangleq \begin{bmatrix} \mathbf{n}_1^t E\, \mathbf{n}_2 \\ \mathbf{n}_1^t F\, \mathbf{n}_2 \\ \mathbf{n}_1^t G\, \mathbf{n}_2 \end{bmatrix} \tag{6.2.17}$$

where we define the *intermediate parameters* (E, F, G):

$$E = \mathbf{R}_1\mathbf{U}^t - \mathbf{T}\mathbf{S}_1^t, \quad F = \mathbf{R}_2\mathbf{U}^t - \mathbf{T}\mathbf{S}_2^t, \quad G = \mathbf{R}_3\mathbf{U}^t - \mathbf{T}\mathbf{S}_3^t. \tag{6.2.18}$$

We have

$$[\mathbf{n}_0]_\times \begin{bmatrix} \mathbf{n}_1^t E\, \mathbf{n}_2 \\ \mathbf{n}_1^t F\, \mathbf{n}_2 \\ \mathbf{n}_1^t G\, \mathbf{n}_2 \end{bmatrix} = \mathbf{0}. \tag{6.2.19}$$

Equation (6.2.19) is a vector equation involving motion parameters R, $\mathbf{T}$, S, $\mathbf{U}$ and observables $\mathbf{n}_0$, $\mathbf{n}_1$, and $\mathbf{n}_2$. The norm of $\mathbf{n}_0$, $\mathbf{n}_1$ and $\mathbf{n}_2$ are arbitrary in (6.2.19). Therefore, unit vectors can be used. The three scalar equations in (6.2.19) are linear in the 9×3 = 27 components of the intermediate parameters (E, F, G). Since $\text{rank}([\mathbf{n}_0]_\times) = 2$ for $\mathbf{n}_0 \neq \mathbf{0}$, (6.2.19) has at most two independent scalar equations. From each line correspondence through three perspective views, we get a set of corresponding projection normals: $\mathbf{n}_0$, $\mathbf{n}_1$ and $\mathbf{n}_2$. If we have at least 13 line correspondences through three views, we might have 26 independent scalar equations. If so, we can solve for the intermediate parameters (E, F, G) up to a scale factor based on (6.2.19). When a matrix is determined up to a scale factor, we say that it is *essentially* determined. The condition to have 26 independent scalar equations is discussed in the next section. Now it is assumed that the intermediate parameters (E, F, G) are essentially determined. For convenience, we solve for the normalized intermediate parameters (E_s, F_s, G_s) with $\|E_s\|^2 + \|F_s\|^2 + \|G_s\|^2 = 1$, such that

$$(E_s, F_s, G_s) = \alpha(E, F, G) \tag{6.2.20}$$

where α is an unknown scale factor. The motion parameters are to be determined from the normalized intermediate parameters.

It is easy to see from (6.2.18) that $\|\mathbf{T}\|^2 + \|\mathbf{U}\|^2$ is proportional to $\|E\|^2 + \|F\|^2 + \|G\|^2$. If the scene is scaled with respect to the origin by a positive factor of k and the translations $\mathbf{T}$ and $\mathbf{U}$ are also scaled by k, we get the same images. Therefore, $\|\mathbf{T}\|^2 + \|\mathbf{U}\|^2$ cannot be determined from the monocular images. For simplicity of notation, we drop the subscript s and let $\|E\|^2 + \|F\|^2 + \|G\|^2 = 1$, with the understanding that (E, F, G) are known only up to a scale factor. As shown later, the rotation matrices are independent of this scale factor.

6.2.6 Motion from Intermediate Parameters

Let $\mathbf{V}_i = \mathbf{T}\times\mathbf{R}_i$, $i = 1, 2, 3$. From (6.2.18) we have $E^t\mathbf{V}_1 = \mathbf{0}$, $F^t\mathbf{V}_2 = \mathbf{0}$ and $G^t\mathbf{V}_3 = \mathbf{0}$. If the ranks of E, F, G are all equal to two, $\mathbf{V}_i$ can be essentially determined from (E, F, G). Then the translation vector $\mathbf{T}$ can be essentially determined by $\mathbf{T}\cdot\mathbf{V}_i = 0$, $i = 1, 2, 3$. However the ranks of E, F, G are not always equal to 2. The following theorem enumerates all the possible cases.

Theorem 6.1. Assume $\mathbf{T} \neq \mathbf{0}$ and $\mathbf{U} \neq \mathbf{0}$. Then there exist unit vectors $\mathbf{V}_1$, $\mathbf{V}_2$, and $\mathbf{V}_3$ such that

$$E^t\mathbf{V}_1 = \mathbf{0}, \tag{6.2.21}$$

$$F^t\mathbf{V}_2 = \mathbf{0}, \tag{6.2.22}$$

$$G^t\mathbf{V}_3 = \mathbf{0}. \tag{6.2.23}$$

and the ranks of E, F and G fall into three cases.

Case 1. All of E, F, G have rank two. $\mathbf{V}_i$ is then essentially determined. Let $A = [\mathbf{V}_1\ \mathbf{V}_2\ \mathbf{V}_3]$. Then $\text{rank}(A) = 2$ and $\mathbf{T}$ is essentially determined by $A^t\mathbf{T} = \mathbf{0}$.

Case 2. Two of E, F, G have rank two, and the third has a rank one. Without loss of generality, let $\text{rank}(E) = 1$. Let $A = [\mathbf{V}_2\ \mathbf{V}_3]$. If $\text{rank}(A) = 2$, $\mathbf{T}$ is still essentially determined by $A^t\mathbf{T} = \mathbf{0}$. Otherwise, $\mathbf{T}$ is essentially determined by $\mathbf{T} \,//\, (\mathbf{E}_i\times\mathbf{V}_2)\times\mathbf{V}_2$, where $\mathbf{E}_i$ is any non-zero column vector of E. $(\mathbf{E}_i\times\mathbf{V}_2)\times\mathbf{V}_2 \neq \mathbf{0}$ is guaranteed.

Case 3. Only one of E, F, G has a rank two, the other two matrices have a rank one. Without loss of generality let $\text{rank}(G) = 2$. Then there are two orthogonal solutions in (6.2.21) and (6.2.22), respectively:

$$E^t\mathbf{V}_{1a} = \mathbf{0}, \quad E^t\mathbf{V}_{1b} = \mathbf{0},$$

$$F^t\mathbf{V}_{2a} = \mathbf{0}, \quad F^t\mathbf{V}_{2b} = \mathbf{0}.$$

where $\mathbf{V}_{1a}\cdot\mathbf{V}_{1b} = 0$ and $\mathbf{V}_{2a}\cdot\mathbf{V}_{2b} = 0$. One and only one of the two equations

$$\mathbf{V}_3\cdot(\mathbf{V}_{1a}\times\mathbf{V}_{1b}) = 0 \tag{6.2.24}$$

and

$$\mathbf{V}_3 \cdot (\mathbf{V}_{2a} \times \mathbf{V}_{2b}) = 0 \tag{6.2.25}$$

holds. $\mathbf{T} \,//\, \mathbf{V}_{1a} \times \mathbf{V}_{1b}$ if (6.2.24) holds. $\mathbf{T} \,//\, \mathbf{V}_{2a} \times \mathbf{V}_{2b}$ if (6.2.25) holds.

Proof. See Appendix 6.A.

From Theorem 6.1, we know that $\mathbf{T}$ can be essentially determined. Similarly if we apply E^t, F^t, G^t to Theorem 6.1, we know that $\mathbf{U}$ can also be essentially determined. In a word, we can determine unit vectors $\mathring{\mathbf{T}}_s$ and $\mathring{\mathbf{U}}_s$ such that $\mathring{\mathbf{T}}_s \times \mathbf{T} = \mathbf{0}$ and $\mathring{\mathbf{U}}_s \times \mathbf{U} = \mathbf{0}$.

The following theorem states the uniqueness of the solution for motion parameters from the intermediate parameters. The condition $\mathbf{T} \neq \mathbf{0}$, $\mathbf{U} \neq \mathbf{0}$ and $R^t\mathbf{T} \neq S^t\mathbf{U}$ used in the theorem is called *distinct locations condition.* In section 5 we will see that this condition turns out to be a necessary condition for essentially determining intermediate parameters by (6.2.19). It is a sufficient condition in the following theorem.

Theorem 6.2. Given (E, F, G), the solution for $R, \mathbf{T}, S, \mathbf{U}$ is unique provided $\mathbf{T} \neq \mathbf{0}$, $\mathbf{U} \neq \mathbf{0}$ and $R^t\mathbf{T} \neq S^t\mathbf{U}$.

Proof. From Theorem 6.1, we can determine $\mathring{\mathbf{T}}_s$ and $\mathring{\mathbf{U}}_s$, such that $\mathbf{T} = s_1 \|\mathbf{T}\| \mathring{\mathbf{T}}_s$ and $\mathbf{U} = s_2 \|\mathbf{U}\| \mathring{\mathbf{U}}_s$, where $s_1, s_2 \in \{-1, 1\}$. For four combinations of the values of s_1 and s_2, we have four sets of equations:

$$(s_1, s_2): \quad \begin{cases} E = s_1 \|\mathbf{U}\| \mathbf{R}_1 \mathring{\mathbf{U}}_s^t - s_2 \|\mathbf{T}\| \mathring{\mathbf{T}}_s \mathbf{S}_1^t \\ F = s_1 \|\mathbf{U}\| \mathbf{R}_2 \mathring{\mathbf{U}}_s^t - s_2 \|\mathbf{T}\| \mathring{\mathbf{T}}_s \mathbf{S}_2^t . \\ G = s_1 \|\mathbf{U}\| \mathbf{R}_3 \mathring{\mathbf{U}}_s^t - s_2 \|\mathbf{T}\| \mathring{\mathbf{T}}_s \mathbf{S}_2^t \end{cases} \tag{6.2.26}$$

Premultiplying both sides of the first equation in (6.2.26) yields

$$[\mathring{\mathbf{T}}_s]_\times E = s_1 \|\mathbf{U}\| [\mathring{\mathbf{T}}_s]_\times \mathbf{R}_1 \mathring{\mathbf{U}}_s^t . \tag{6.2.27}$$

Post-multiplying both sides by $\mathring{\mathbf{U}}_s$ gives

$$[\mathring{\mathbf{T}}_s]_\times E \mathring{\mathbf{U}}_s = s_1 \|\mathbf{U}\| [\mathring{\mathbf{T}}_s]_\times \mathbf{R}_1 . \tag{6.2.28}$$

Applying the same operations to the second and the third equations in (6.2.26) gives the other two equations similar to (6.2.28). Combining these three equations yields

$$[\mathring{\mathbf{T}}_s]_\times [E\mathring{\mathbf{U}}_s \quad F\mathring{\mathbf{U}}_s \quad G\mathring{\mathbf{U}}_s] = s_1 \|\mathbf{U}\| [\mathring{\mathbf{T}}_s]_\times R. \tag{6.2.29}$$

Since R is a rotation matrix, $\|R\mathbf{x}\| = \|\mathbf{x}\|$. We get $\|\mathbf{U}\|$ from (6.2.29):

$$\|\mathbf{U}\| = \|[\mathring{\mathbf{T}}_s]_\times\|^{-1} \|[\mathring{\mathbf{T}}_s]_\times [E\mathring{\mathbf{U}}_s \quad F\mathring{\mathbf{U}}_s \quad G\mathring{\mathbf{U}}_s]\|. \tag{6.2.30}$$

Considering the transposed version of E, F, G, similarly we get

$$[\mathring{\mathbf{U}}_s]_\times [E^t\mathring{\mathbf{T}}_s \quad F^t\mathring{\mathbf{T}}_s \quad G^t\mathring{\mathbf{T}}_s] = -s_2 \|\mathbf{T}\| [\mathring{\mathbf{U}}_s]_\times S. \tag{6.2.31}$$

$\|\mathbf{T}\|$ is determined by an equation similar to (6.2.30). The equations (6.2.29) and (6.2.31) both have the form $A = BR$, with $\text{rank}(B) = 2$ ($[\mathbf{x}]_\times$ has a rank 2 if $\mathbf{x} \neq \mathbf{0}$, which is shown in Appendix 6.A). Therefore, rotation matrices R and S are uniquely determined in (6.2.29) and (6.2.31), respectively. In the presence of noise, we solve for a rotation matrix R in the following

$$\min_R \|A - BR\|, \text{ subject to: } R \text{ is a rotation matrix.} \tag{6.2.32}$$

The solution of this problem is discussed in Appendix 3.B in Chapter 3, and is presented later with the algorithm.

However, there exist four combinations for all the possible signs of (s_1, s_2) in (6.2.26). The following lemma states that only one combination has a solution for rotation matrices R and S from (6.2.26).

Lemma 6.1. Assume $\mathbf{T} \neq \mathbf{0}$, $\mathbf{U} \neq \mathbf{0}$, and $R^t\mathbf{T} \neq S^t\mathbf{U}$. Only one assignment for (s_1, s_2), $s_1, s_2 \in \{-1, 1\}$, has a solution for rotation matrices R and S from (6.2.26).

Proof. See Appendix 6.B.

By substituting into (6.2.26) the four assignments for (s_1, s_2), we get a unique solution R and S and the assignment of (s_1, s_2) which is the one that satisfies (6.2.26). □

On the other hand, (E, F, G) can only be essentially determined, i.e., up to a scale factor. From (6.2.29)-(6.2.31), it is easy to see that the scale factor does not affect the solution of the rotation matrices R and S. However the translation vector pair $(\mathbf{T}, \mathbf{U})$ is essentially determined, i.e., the ratio between $\|\mathbf{T}\|$ and $\|\mathbf{U}\|$ is still determined. We can choose any sign for (E, F, G) and solve for the translation vector pair to get $\mathbf{T}_s$ and $\mathbf{U}_s$ such that $(\mathbf{T}, \mathbf{U}) = \alpha(\mathbf{T}_s, \mathbf{U}_s)$ with unknown α. The absolute value of α can not determined from monocular images. The sign of α can be determined in the following.

6.2.7 Structure and Sign of Translation Vectors

From (6.2.5)-(6.2.7) we get $\mathbf{l}\cdot\mathbf{n}_0 = 0$, $\mathbf{l}\cdot R^{-1}\mathbf{n}_1 = 0$ and $\mathbf{l}\cdot S^{-1}\mathbf{n}_2 = 0$. For each line we solve for a unit vector $\mathring{\mathbf{l}}$ such that $\mathbf{l} \,//\, \mathring{\mathbf{l}}$ in the following

$$\min_{\mathring{\mathbf{l}}} \|[\mathbf{n}_0 \;\; R^{-1}\mathbf{n}_1 \;\; S^{-1}\mathbf{n}_2]^t \, \mathring{\mathbf{l}} \,\|, \text{ subject to: } \|\mathring{\mathbf{l}}\,\| = 1. \qquad (6.2.33)$$

If the rank of $[\mathbf{n}_0 \;\; R^{-1}\mathbf{n}_1 \;\; S^{-1}\mathbf{n}_2]$ is no more than one, the line position cannot be recovered.

For each line, let $\mathbf{x}_p$ be a point on the line that is the closest to the origin. $d_0 \triangleq \|\mathbf{x}_p\,\|$ is the positive distance of the line to the origin. Since $\mathbf{x}_p \cdot \mathbf{l} = 0$, from (6.2.5) it follows that

$$\|\mathbf{n}_0\| = \|\mathbf{x}_p \times \mathbf{l}\| = \|\mathbf{x}_p\,\| \, \|\mathbf{l}\|. \qquad (6.2.34)$$

Although we use (6.2.5)-(6.2.7) to define the projection normals, the scale factor of those normals is immaterial since it will be canceled out later in (6.2.36). Using (6.2.11) and (6.2.34) yields

$$|\mathbf{T}\cdot\mathbf{n}_1| = \|\mathbf{l}\|^{-1} \|\mathbf{n}_0 \times R^{-1}\mathbf{n}_1\| = \|\mathbf{x}_p\,\| \, \|\mathbf{n}_0\|^{-1} \|\mathbf{n}_0 \times R^{-1}\mathbf{n}_1\|. \qquad (6.2.35)$$

Dividing both sides by $\|\mathbf{n}_1\|$ gives the distance of the line to the origin

$$d_0 = \|\mathbf{x}_p\,\| = \|\mathring{\mathbf{n}}_0 \times R^{-1}\mathring{\mathbf{n}}_1\|^{-1} \, |\mathbf{T}\cdot\mathring{\mathbf{n}}_1|. \qquad (6.2.36)$$

Notice that d_0 is proportional to $\|\mathbf{T}\|$. When $\mathbf{T}_s$ replaces for $\mathbf{T}$ in (6.2.36), $\|\mathbf{x}_p\,\|$ is essentially determined.

Let $\mathring{\mathbf{v}}$ be a unit vector that is parallel to $\mathbf{x}_p$ and always points to the positive z-direction. That is, $\mathring{\mathbf{v}} = \pm\|\mathbf{n}_0 \times \mathbf{l}\|^{-1}\mathbf{n}_0 \times \mathbf{l}$, such that $\mathbf{Z}\cdot\mathring{\mathbf{v}} > 0$. Then $\mathbf{x}_p = \pm d_0 \mathring{\mathbf{v}}$.

To determine the sign for the translation vectors, first consider the motion equation (6.2.2).

$$\mathbf{x}_1 = R\,\mathbf{x}_0 + \mathbf{T}. \qquad (6.2.37)$$

Multiplying both sides by -1 we get

$$-\mathbf{x}_1 = R\,(-\mathbf{x}_0) + (-\mathbf{T}). \qquad (6.2.38)$$

Equation (6.2.38) implies that when a point $\mathbf{x}_0$ is rotated by R and translated by $\mathbf{T}$ to a point $\mathbf{x}_1$, its mirror image (with respect to the origin) $-\mathbf{x}_0$ is rotated by R and translated by $-\mathbf{T}$. The original line and the mirror image line produce the same images through projection plane. Therefore if $\mathbf{T}$ and $\mathbf{U}$ are the true translation vectors, $-\mathbf{T}$ and $-\mathbf{U}$ are the translation vectors of the mirror image. Obviously, if a point is located in front of the camera, its mirror image is at the back of the camera. Therefore if we assume all the visible scene is in front of the camera, the translation that corresponds to the structure located at back of the camera is not the correct one. As to the lines, what is the criterion for a line to be in front of the camera? We can assume that most lines in front of the camera have $\mathbf{x}_p$, the closest to the origin on the line, with a positive (depth) z-component. This assumption is called *majority positive depths assumption* (see Fig. 6.4). The assumption holds for majority of the lines, because those lines whose closest point, $\mathbf{x}_p$, has a negative depth are not visible through the camera lens. A few visible lines may violate this positive depth assumption. For example, when we look horizontally in a car that is running down hill, the two curb lines are visible, but $\mathbf{x}_p$ of these two lines have negative z-components since $\mathbf{x}_p$ is at back of the eyes. However, those lines with negative z-components generally are in the minority among all visible lines.

Let $\mathbf{T}_s = s\mathbf{T}$, where $s \in \{1, -1\}$ is unknown. Assume a line at time t_0 with $\mathbf{x}_p$, the point closest to the origin. At time t_1 its new position vector should be orthogonal to the new projection normal

$$\mathbf{n}_1 \cdot (R\mathbf{x}_p + \mathbf{T}) = \mathbf{n}_1 \cdot (R\mathbf{x}_p + s\mathbf{T}_s) = 0. \qquad (6.2.39)$$

Assume that $\mathbf{x}_p$ has a positive z-component at time $\mathbf{t}_0$, $\mathbf{x}_p = \|\mathbf{x}_p\|\mathring{\mathbf{v}}$. From (6.2.39) it follows that

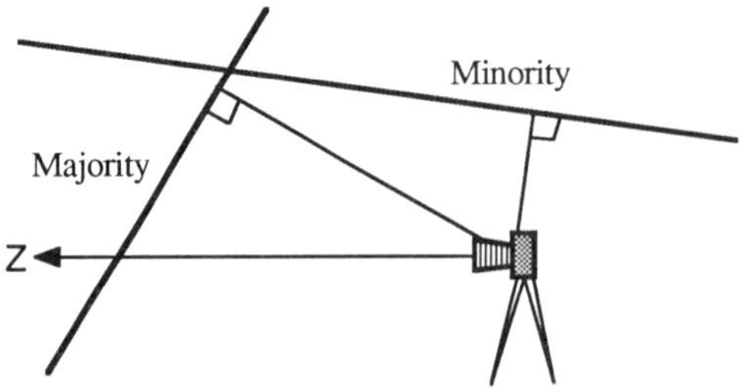

Fig. 6.4. Majority positive depths assumption.

$$\mathbf{n}_1 \cdot (\|\mathbf{x}_p\| R \mathring{\mathbf{v}} + s\mathbf{T}_s) = 0. \tag{6.2.40}$$

Note that (6.2.40) cannot hold for both $s = 1$ and $s = -1$ unless $\mathbf{n}_1 \cdot \mathbf{T}_s = 0$, which is rare. In a word, if $\mathbf{x}_p$ has a positive z-component, the correct sign for s satisfies (6.2.40) and the incorrect one generally does not. If $\mathbf{x}_p$ has a negative z-component, we have $\mathbf{x}_p = -\|\mathbf{x}_p\| \mathring{\mathbf{v}}$ which yields

$$\mathbf{n}_1 \cdot (-\|\mathbf{x}_p\| R \mathring{\mathbf{v}} + s\mathbf{T}) = 0. \tag{6.2.41}$$

In other words, a point with a negative z-component satisfies (6.2.40) if the sign for s is reversed.

We determine the sign s in the following way. For each line, if $s = 1$ satisfies (6.2.40) a vote is added to the set POS, and if $s = -1$ satisfies (6.2.40) a vote is added to the set NEG. After voting by all lines, if POS has more votes than NEG, $s = 1$ supports the majority positive depths assumption. Thus, we let $s = 1$ and $\mathbf{T} = +\mathbf{T}_s$. All the lines that voted for POS should have $\mathbf{x}_p = +\|\mathbf{x}_p\| \mathring{\mathbf{v}}$ since they are majority and the results are consistent with (6.2.40). For all the lines that voted for NEG should have $\mathbf{x}_p = -\|\mathbf{x}_p\| \mathring{\mathbf{v}}$ since they are minority and the results are consistent with (6.2.41).

Otherwise POS has less votes. This means that $s = -1$ supports the majority positive depths assumption. Therefore, $s = -1$ and $\mathbf{T} = -\mathbf{T}_s$. All the lines that voted for POS should have $\mathbf{x}_p = -\|\mathbf{x}_p\| \mathring{\mathbf{v}}$ since they are minority and the results are consistent with (6.2.41). All the lines that voted for NEG should have $\mathbf{x}_p = \|\mathbf{x}_p\| \mathring{\mathbf{v}}$ since they are majority and again the results are consistent with (6.2.40). Thus, the sign of $\mathbf{T}$ and the locations of the lines can be determined based on the majority positive depths assumption.

The motion from t_0 to t_2 can be analyzed in a similar way.

6.2.8 In the Presence of Noise

In this section, we consider a few schemes to combat noise, while still keeping the algorithm linear.

Since short lines in the images are generally not as reliably determined as long lines, less weight should be assigned to the short lines in (6.2.19) when solving for E, F, G. Let the length of the lines in the images at time t_i be l_i, $i = 0, 1, 2$. A simple weight for the line can be

$$w = (l_0^{-1} + l_1^{-1} + l_2^{-1})^{-1}. \tag{6.2.42}$$

Since only two of the three scalar equations in (6.2.19) are independent, any row of $[\mathbf{n}_0]_\times$ could be zero vector which results in a trivial $0 = 0$ equation. To ensure that none of the two independent equations is missed, we can use all three equations of (6.2.19). (This is useful in the presence of noise.) These three equations are scaled by the weight in (6.2.42) in the system of linear equations formed by combining (6.2.19) for all the lines.

Since a noise corrupted matrix almost always has a full rank, the conditions on the rank of the matrices should then be modified accordingly in the presence of noise. A discussion of the sensitivity of the eigenvectors to the perturbation of the matrix can be found in Appendix 3.A in Chapter 3. A rough measurement for the error of the eigenvector associated with the smallest eigenvalue λ_1 is $(\lambda_1 - \lambda_2)^{-1}$ where λ_2 is the second smallest eigenvalue. The solution of $\mathbf{V}_i$ $i = 1, 2, 3$ in (6.2.21)-(6.2.23) is the eigenvector of EE^t, FF^t, GG^t, respectively, associated with the smallest eigenvalue. The reliability of those solutions is roughly proportional to the difference between the smallest and the second smallest eigenvalues. Case 1 and Case 2 in Theorem 6.1 can be combined by using a weighted A. Let the i-th column vectors of A be weighted by the difference of the two smallest eigenvalues of EE^t, FF^t and GG^t, for $i = 1, 2, 3$, respectively. For example, if rank(E) is close to one, the corresponding weight in A is close to zero, which is Case 2.

We have observed considerable improvements in simulations by using the weighted schemes discussed above.

In determining the distance d_0, the motion from t_0 to t_1 and that from t_0 to t_2 can both be used to enchance the robustness.

6.2.9 Algorithm

Now, we are ready to present the algorithm. In the algorithm, ε denotes a small positive threshold to accommodate noise. Without noise, ε should be zero. With noise, ε can be estimated by the error estimation approach in Section 3.3 of Chapter 3 or determined empirically. Though ε should be different in different parts of the algorithm, a single ε will be used for the simplicity of notation.

Step (i): Solving for (E, F, G) up to a scale factor. Given n line correspondences through three views. Let the unit projection normal at time t_i be $\mathbf{n}_i$, $i = 0, 1, 2$. Solve for (E, F, G) in the following

$$\min_{(E,F,G)} \sum_{\text{lines}} w_i^2 \left\| [\mathbf{n}_0]\times \begin{bmatrix} \mathbf{n}_1^t E\, \mathbf{n}_2 \\ \mathbf{n}_1^t F\, \mathbf{n}_2 \\ \mathbf{n}_1^t G\, \mathbf{n}_2 \end{bmatrix} \right\|^2, \tag{6.2.43}$$

subject to $\|E\|^2 + \|F\|^2 + \|G\|^2 = 1$, where the weight w_i for i-th line is given in (6.2.42). (6.2.43) can be written in the form

$$\min_{\mathbf{y}} \|D\mathbf{y}\|^2, \text{ subject to: } \|\mathbf{y}\| = 1, \tag{6.2.44}$$

where D is a $3n$ by 27 matrix determined from projection normals, and $\mathbf{y}$ is a 27-dimensional unit vector. The solution for unit vector $\mathbf{y}$ is the unit eigenvector of D^tD associated with the smallest eigenvalue.

Step (ii): Determining $\mathring{\mathbf{T}}_s$ and $\mathring{\mathbf{U}}_s$ such that $\mathbf{T} \,//\, \mathring{\mathbf{T}}_s$ and $\mathbf{U} \,//\, \mathring{\mathbf{U}}_s$. Let

$$H_e \triangleq [\mathbf{h}_{e1} \;\; \mathbf{h}_{e2} \;\; \mathbf{h}_{e3}],$$

$$F_f \triangleq [\mathbf{h}_{f1} \;\; \mathbf{h}_{f2} \;\; \mathbf{h}_{f3}],$$

and

$$H_g \triangleq [\mathbf{h}_{g1} \;\; \mathbf{h}_{g2} \;\; \mathbf{h}_{g3}]$$

be orthogonal matrices such that

$$H_e^t E E^t H_e = \text{diag}\{\lambda_{e1}, \lambda_{e2}, \lambda_{e3}\}, \quad \lambda_{e1} \le \lambda_{e2} \le \lambda_{e3}; \tag{6.2.45}$$

$$H_f^t F F^t H_f = \text{diag}\{\lambda_{f1}, \lambda_{f2}, \lambda_{f3}\}, \quad \lambda_{f1} \le \lambda_{f2} \le \lambda_{f3}; \tag{6.2.46}$$

$$H_g^t G G^t H_g = \text{diag}\{\lambda_{g1}, \lambda_{g2}, \lambda_{g3}\}, \quad \lambda_{g1} \le \lambda_{g2} \le \lambda_{g3}. \tag{6.2.47}$$

Case 1. The median of the set $C = \{\lambda_{e2}, \lambda_{f2}, \lambda_{g2}\}$ is larger than ε.

Let $A = [(\lambda_{e2} - \lambda_{e1})\mathbf{h}_{e1} \;\; (\lambda_{f2} - \lambda_{f1})\mathbf{h}_{f1} \;\; (\lambda_{g2} - \lambda_{g1})\mathbf{h}_{g1}]$.

a) If the second smallest eigenvalue of A^tA is larger than ε (rank$(A) \ge 2$), $\mathring{\mathbf{T}}_s$ is determined up to a scale factor by

$$\min_{\mathring{\mathbf{T}}_s} \|A^t \mathring{\mathbf{T}}_s\|, \text{ subject to: } \|\mathring{\mathbf{T}}_s\| = 1. \tag{6.2.48}$$

b) Otherwise (rank(A) $\approx$ 1), determine the smallest number in set C. If λ_{e2} is the smallest in set C, then $\mathring{\mathbf{T}}_s$ // $(\mathbf{E}_i \times \mathbf{h}_{f1}) \times \mathbf{h}_{f1}$, where $\mathbf{E}_i$ is a nonzero column vector of E. If λ_{f2} or λ_{g2} is the smallest in C, $\mathring{\mathbf{T}}_s$ is determined by a similar equation (circularly rotating e, f, g and E, F, G).

Case 2. The median of the set $C = \{\lambda_{e2}, \lambda_{f2}, \lambda_{g2}\}$ is not larger than ε. Determine the maximum of the set C. Without loss of generality, assume λ_{g2} is the maximum.

$$\mathring{\mathbf{T}}_s = \begin{cases} \mathbf{h}_{e1} \times \mathbf{h}_{e2} & \text{if } |\mathbf{h}_{g1} \cdot (\mathbf{h}_{e1} \times \mathbf{h}_{e2})| < |\mathbf{h}_{g1} \cdot (\mathbf{h}_{f1} \times \mathbf{h}_{f2})| \\ \mathbf{h}_{f1} \times \mathbf{h}_{f2} & \text{otherwise.} \end{cases} \tag{6.2.49}$$

Replacing E, F, G by E^t, F^t, G^t, similarly determine $\mathring{\mathbf{U}}_s$.

Step (iii): Determining R and S. Let

$$G_R = [\mathring{\mathbf{T}}_s]_\times [E\mathring{\mathbf{U}}_s \;\; F\mathring{\mathbf{U}}_s \;\; G\mathring{\mathbf{U}}_s],$$

$$G_S = -[\mathring{\mathbf{U}}_s]_\times [E^t\mathring{\mathbf{T}}_s \;\; F^t\mathring{\mathbf{T}}_s \;\; G^t\mathring{\mathbf{T}}_s], \tag{6.2.50}$$

and $\|\mathbf{U}\| = \|G_R\|/\sqrt{2}$, $\|\mathbf{T}\| = \|G_S\|/\sqrt{2}$. Then let $G_R \leftarrow \|\mathbf{U}\|^{-1} G_R$, and $G_S \leftarrow \|\mathbf{T}\|^{-1} G_S$. Solve for R_p, R_n, S_p and S_n in the following

$$\min_{R_p} \|G_R - [\mathring{\mathbf{T}}_s]_\times R_p]\|, \quad \min_{R_n} \|-G_R - [\mathring{\mathbf{T}}_s]_\times R_n]\|, \tag{6.2.51}$$

$$\min_{S_p} \|G_S - [\mathring{\mathbf{U}}_s]_\times S_p]\|, \quad \min_{S_n} \|-G_S - [\mathring{\mathbf{U}}_s]_\times S_n]\|, \tag{6.2.52}$$

subject to that R_p, R_n, S_p and S_n are rotation matrices. Both (6.2.51) and (6.2.52) have the form (noticing $\|D - CR\| = \|R^t C^t - D^t\|$):

$$\min_R \|RC - D\|, \text{ subject to: } R \text{ is a rotation matrix.} \tag{6.2.53}$$

Where $C = [\mathbf{C}_1\ \mathbf{C}_2\ \mathbf{C}_3]$, $D = [\mathbf{D}_1\ \mathbf{D}_2\ \mathbf{D}_3]$. The solution of (6.2.53) is as follows.

Define a 4×4 matrix B by

$$B = \sum_{i=1}^{3} B_i^t B_i$$

where

$$B_i = \begin{bmatrix} 0 & (\mathbf{C}_i - \mathbf{D}_i)^t \\ \mathbf{D}_i - \mathbf{C}_i & [\mathbf{D}_i + \mathbf{C}_i]_\times \end{bmatrix}.$$

Let $\mathbf{q} = (q_0, q_1, q_2, q_3)^t$ be the unit eigenvector of B associated with the smallest eigenvalue. The solution of rotation matrix R in (6.2.53) is

$$R = \begin{bmatrix} q_0^2+q_1^2-q_2^2-q_3^2 & 2(q_1q_2 - q_0q_3) & 2(q_1q_3 + q_0q_2) \\ 2(q_2q_1 + q_0q_3) & q_0^2-q_1^2+q_2^2-q_3^2 & 2(q_2q_3 - q_0q_1) \\ 2(q_3q_1 - q_0q_2) & 2(q_3q_2 + q_0q_1) & q_0^2-q_1^2-q_2^2+q_3^2 \end{bmatrix}.$$

Substituting four assignments

$$(1, 1, R_p, S_p), \quad (-1, 1, R_n, S_p),$$
$$(1, 1, R_p, S_n), \quad (-1, -1, R_n, S_n),$$

for (s_t, s_u, R, S) in

$$\begin{cases} E = s_t \|\mathbf{U}\|\mathbf{R}_1\mathring{\mathbf{U}}_s^t - s_u \|\mathbf{T}\|\mathring{\mathbf{T}}_s\mathbf{S}_1^t, \\ F = s_t \|\mathbf{U}\|\mathbf{R}_2\mathring{\mathbf{U}}_s^t - s_u \|\mathbf{T}\|\mathring{\mathbf{T}}_s\mathbf{S}_2^t, \\ G = s_t \|\mathbf{U}\|\mathbf{R}_2\mathring{\mathbf{U}}_s^t - s_u \|\mathbf{T}\|\mathring{\mathbf{T}}_s\mathbf{S}_2^t, \end{cases} \tag{6.2.54}$$

the assignment best satisfying (6.2.54), in the Euclidean norm sense, gives the correct assignment for (s_t, s_u, R, S). Then

$$\mathbf{T} = s_t \|\mathbf{T}\|\mathring{\mathbf{T}}_s, \quad \mathbf{U} = s_u \|\mathbf{U}\|\mathring{\mathbf{U}}_s.$$

Step (iv): Determining structure $\mathring{\mathbf{l}}$ and $\mathbf{x}_p$. For each line, solve for the direction of the line represented by a vector $\mathring{\mathbf{l}}$ in the following

$$\min_{\mathring{\mathbf{l}}} \|[\mathbf{n}_0 \;\; R^{-1}\mathbf{n}_1 \;\; S^{-1}\mathbf{n}_2]^t \, \mathring{\mathbf{l}}\|, \text{ subject to: } \|\mathring{\mathbf{l}}\| = 1.$$

Then,

$$d_0 = \frac{|\mathbf{T}\cdot\mathring{\mathbf{n}}_1|}{2\|\mathring{\mathbf{n}}_0 \times R^{-1}\mathring{\mathbf{n}}_1\|} + \frac{|\mathbf{U}\cdot\mathring{\mathbf{n}}_2|}{2\|\mathring{\mathbf{n}}_0 \times S^{-1}\mathring{\mathbf{n}}_2\|},$$

$$\mathring{\mathbf{v}} = \pm \frac{\mathring{\mathbf{n}}_0 \times \mathring{\mathbf{l}}}{\|\mathring{\mathbf{n}}_0 \times \mathring{\mathbf{l}}\|},$$

where the sign is such that the third component of $\mathring{\mathbf{v}}$ is non-negative.

Let POS and NEG be empty sets. For each line i do the following: If

$$|\mathbf{n}_1\cdot(d_0 R\,\mathring{\mathbf{v}} + \mathbf{T})| + |\mathbf{n}_2\cdot(d_0 S\,\mathring{\mathbf{v}} + \mathbf{U})|$$

$$< |\mathbf{n}_1\cdot(d_0 R\,\mathring{\mathbf{v}} - \mathbf{T})| + |\mathbf{n}_2\cdot(d_0 S\,\mathring{\mathbf{v}} - \mathbf{U})|,$$

i is added into the set POS. Otherwise, i is added to the set NEG.

Finally, if $\|\text{POS}\| > \|\text{NEG}\|$, for each line i,

$$\mathbf{x}_p = \begin{cases} d_0\mathring{\mathbf{v}} & \text{if i} \in \text{POS} \\ -d_0\mathring{\mathbf{v}} & \text{otherwise.} \end{cases}$$

Otherwise, if $\|\text{POS}\| < \|\text{NEG}\|$,

$$\mathbf{T} \leftarrow -\mathbf{T}, \quad \mathbf{U} \leftarrow -\mathbf{U}.$$

And for each line i,

$$\mathbf{x}_p = \begin{cases} -d_0\mathring{\mathbf{v}} & \text{if i} \in \text{POS} \\ d_0\mathring{\mathbf{v}} & \text{otherwise.} \end{cases}$$

6.3 Degeneracy

In the last section, it is established by Theorem 6.2 that as long as $\mathbf{T} \neq \mathbf{0}$, $\mathbf{U} \neq \mathbf{0}$ and $\mathbf{T}^t R \neq \mathbf{U}^t S$ the solution of motion parameters from the intermediate parameters (E, F, G) is unique. The fact that (E, F, G) can only be determined up to a scale factor does not affect the solution of the rotation matrices R and S. The direction of the translation and the structure of the lines can be determined based on the majority positive depths assumption. Therefore, the translation and the closest points on the lines are determined up to a positive scale factor.

First, let us see what the condition

$$\mathbf{T} \neq \mathbf{0}, \quad \mathbf{U} \neq \mathbf{0}, \quad \mathbf{T}^t R \neq \mathbf{U}^t S \tag{6.3.1}$$

means. For a more intuitive interpretation, we consider the case where the scene is stationary and the camera is moving. $\mathbf{x}_1$ in (6.2.2) is the position of the point at time t_1 in a coordinate system fixed on the camera. If the scene is stationary the point $\mathbf{x}_0$ is fixed, and the transformation that transform $\mathbf{x}_1$ to $\mathbf{x}_0$ corresponds to the motion of the camera. From (6.2.2) it follows that

$$\mathbf{x}_0 = R^{-1}\mathbf{x}_1 - R^t\mathbf{T}. \tag{6.3.2}$$

Therefore, the motion of the camera is a rotation R^{-1} followed by a translation $-R^{-1}\mathbf{t}$. Since the senter of projection at time t_0 is at the origin and the rotation is about the origin, the senter of projection at time t_1 is at $\mathbf{O}_1 = -R^t\mathbf{T}$. Similarly the senter of projection at time t_2 is at $\mathbf{O}_2 = -S^t\mathbf{U}$. Thus the condition in (6.3.1) is equivalent to the condition

$$\mathbf{O}_1 \neq \mathbf{0}, \quad \mathbf{O}_2 \neq \mathbf{0}, \quad \mathbf{O}_1 \neq \mathbf{O}_2. \tag{6.3.3}$$

That is to say that no positions of the senter of projection of the camera coincide, or in other words, the translation between any two views does not vanish. So the condition in (6.3.1) or (6.3.3) is called *distinct locations condition.*

The intermediate parameters (E, F, G) are essentially determined by (6.2.44), or equivalently by (6.2.19), if and only if the rank of D is not under 26. If (E, F, G) are not essentially determined by (6.2.19) we say that degeneracy occurs. The degeneracy condition can be tested by calculating the rank of D. However in the presence of noise, the rank of D is mostly full. The method of error estimation in Section 3.3 of Chapter 3 can be used to access the accuracy of the intermediate parameters and the final motion parameters. If the estimated error in the solution is large, we say that degeneracy or near degeneracy occurs.

The following theorem gives the necessary and sufficient conditions for degeneracy in terms of 3-D line configurations at time t_0 and the motion parameters.

Theorem 6.3. (E, F, G) is not essentially determined by (6.2.19) or equivalently, $\text{rank}(D) < 26$ in (6.2.44), if and only if there exist no trivial parameters $(\tilde{E}, \tilde{F}, \tilde{G})$ such that

$$[\mathbf{n}]_\times \begin{bmatrix} ((\mathbf{x}_p - \mathbf{O}_1)\times\mathbf{l})^t \tilde{E}((\mathbf{x}_p - \mathbf{O}_2)\times\mathbf{l}) \\ ((\mathbf{x}_p - \mathbf{O}_1)\times\mathbf{l})^t \tilde{F}((\mathbf{x}_p - \mathbf{O}_2)\times\mathbf{l}) \\ ((\mathbf{x}_p - \mathbf{O}_1)\times\mathbf{l})^t \tilde{G}((\mathbf{x}_p - \mathbf{O}_2)\times\mathbf{l}) \end{bmatrix} = \mathbf{0} \tag{6.3.4}$$

is satisfied for all lines $\mathbf{x} = \mathbf{x}_p + k\mathbf{l}$ at time t_0. $(\tilde{E}, \tilde{F}, \tilde{G})$ is trivial if and only if

$$(\tilde{E}, \tilde{F}, \tilde{G}) = \alpha(P_1 - Q_1^t, P_2 - Q_2^t, P_3 - Q_3^t)$$

for some real number α, where P_i and Q_i are matrices with the i-th column being $\mathbf{O}_1$ and $\mathbf{O}_2$, respectively, and the other columns are zero vectors.

Proof. See Appendix 6.C.

It can be seen that the degeneracy depends on two factors: one is the motion parameters, the other is the configuration of 3D lines. We have the following corollary:

Corollary 6.1. If the distinct locations condition is not satisfied, the intermediate parameters is not essentially determined by (6.2.19).

Proof. See Appendix 6.D.

Therefore, if the distinct locations condition is not satisfied, (E, F, G) cannot be essentially determined by (6.2.19) regardless of the structure of the lines. If the distinct locations condition is satisfied and (6.2.19) is still degenerate, we say the line configuration is degenerate. The following corollary gives an example of degenerate line configurations.

Corollary 6.2. If the directions of lines are coplanar the line configuration is degenerate.

Proof. Let all the lines be orthogonal to a vector $\mathbf{v}$. First, we want to prove $[\mathbf{l}]_\times[\mathbf{v}]_\times[\mathbf{l}]_\times = \mathbf{0}$. In fact, it is easy to verify the identity

$$[\mathbf{x}]_\times[\mathbf{y}]_\times = (\mathbf{x}\cdot\mathbf{y})I - \mathbf{y}\mathbf{x}^t.$$

Then it follows that

$$[\mathbf{l}]_\times[\mathbf{v}]_\times[\mathbf{l}]_\times = 0I - \mathbf{v}\mathbf{l}^t[\mathbf{l}]_\times = -\mathbf{v}\mathbf{0}^t = \mathbf{0}.$$

Then, letting $\tilde{E} = \tilde{F} = \tilde{G} = [\mathbf{v}]_\times$, the second column vector on the left hand side of (6.3.4) vanishes. In fact,

$$((\mathbf{x}_p - \mathbf{O}_1)\times\mathbf{l})^t[\mathbf{v}]_\times((\mathbf{x}_p - \mathbf{O}_2)\times\mathbf{l})$$

$$= (\mathbf{x}_p - \mathbf{O}_1)^t[\mathbf{l}]_\times[\mathbf{v}]_\times[\mathbf{l}]_\times(\mathbf{O}_2 - \mathbf{x}_p) = \mathbf{0}.$$

In other words, (6.3.4) is satisfied for nontrivial $(\tilde{E}, \tilde{F}, \tilde{G})$. Therefore, the configuration is degenerate. □

If all the lines lie in a plane, the direction of lines must be coplanar (they are orthogonal to the normal of the plane). Therefore, planar scene is a degenerate case. Obviously, the case of coplanar directions is more general, since lines do not have to be in a plane.

For lines whose matrix $[\mathbf{n}_0 \; R^{-1}\mathbf{n}_1 \; S^{-1}\mathbf{n}_2]$ in (6.2.33) has a rank no more than one, the line position cannot be recovered. We neglect this line. This happens if and only if the the line lies in the plane determined by the origin and $\mathbf{O}_1$ and $\mathbf{O}_2$.

From Theorems 6.2 and 6.3 and the majority positive depths assumption, we come to the conclusion that if the distinct locations condition is satisfied and the line structure is not degenerate, the motion parameters and the structure of the line can be uniquely determined (up to a scale factor for translation and line distances).

Another form of necessary and sufficient condition is presented in the following theorem.

Theorem 6.4. Assume 13 line correspondences at time t_0, with the i-th line being represented by $\mathbf{x} = \mathbf{x}_{pi} + \mathbf{l}_i$ ($\mathbf{x}_{pi}$ is the point closest to the origin) and the normal for the plane that passes though the line and $\mathbf{O}_k$ being $\tilde{\mathbf{n}}_{ki}$, $k = 1, 2$, $i = 1, 2 \cdots 13$. Then (E, F, G) is not essentially determined by (6.2.19), or equivalently, rank$(D) < 26$ in (6.2.44), if and only if there exist no a_i, b_i, $i = 1, 2, 3, \cdots, 13$, not all of which are zeros, such that

$$\sum_{i=1}^{13}(a_i \mathbf{x}_{pi} + b_i \mathbf{l}_i)\tilde{\mathbf{n}}_{2i}\tilde{\mathbf{n}}_{3i} = \mathbf{0}$$

where **ab** denotes tensor out product of **a** and **b**. For m-dimensional vector **a** and n-dimensional vector **b**, **ab** has mn components which are products of all the possible combinations between one element in **a** and one in **b**.

Proof. See Appendix 6.E.

Before we end this section, two points should be emphasized here. (1) The fact that the linear algorithm can not give unique solution under degenerate line configurations dose not mean that no algorithms can solve the problem under these configurations. Other algorithms, e.g., some nonlinear algorithms, can still give a correct solution under these configurations. However, as we mentioned, the uniqueness of solution by a nonlinear algorithm is unknown and the correct convergence of a nonlinear algorithm is not guaranteed. (2) We do not assume any a priori knowledge of the scene other than rigidity. However, if we know some additional information, some of the degenerate configurations can be solved. For example, if we know that the scene is planar, the intersection of those coplanar lines gives feature points (no matter whether the lines meet at those points or not) and planar point based algorithms can be used to solve the problem.

6.4 Optimization

The closed-form solution discussed above make use of least-squares criterion in several steps. However, the solution is not overall optimum. The reason is that we have not considered noise distributions or different amount of noise in different components of the variables. Our objective in this section is to obtain "best possible" estimates from the noise corrupted data.

6.4.1 Lines from Pixels

An image line consists of a sequence of edge points. The image position of the line is typically determined from those edge points by a line fitting algorithm. Since lines are used as primitives, it is natural and efficient to consider the parameters of lines as the variables of observations. A line in (u, v) image plane can be represented by

$$v = au + b \tag{6.4.1}$$

or

$$u = av + b \tag{6.4.2}$$

where a and b are parameters of the line. In order to avoid the case of vertical line in (6.4.1) or horizontal line in (6.4.2) and to improve the numerical stability, both representations are used. For each line, the use of (6.4.1) or (6.4.2) is determined such that $|a| \le 1$ (a one-bit flag for each line indicates which is used). The variable u in (6.4.1) and v in (6.4.2) can be considered as the image coordinates of edge pixels (see Fig. 6.5). In any case, a line parameter vector $\mathbf{a} \triangleq (a, b)^t$ is observed. Certainly, it is not appropriate to assume that the parameters here are equally reliable. We need to evaluate the covariance matrix of the error in the line parameter vectors. Suppose that the edge point (u_i, v_i) in a line is corrupted by additive, zero mean, uncorrelated, equal variance noise in equations (6.4.1) and (6.4.2):

$$v_i = au_i + b + \delta_{v_i}, \tag{6.4.3}$$

$$u_i = av_i + b + \delta_{u_i}. \tag{6.4.4}$$

This agrees with our method to determine a line: An edge point on a line is determined along one pixel row or column (see Fig. 6.5). If a least squares fitting is employed to edge points, for (6.4.3) we get

$$A\mathbf{a} = \mathbf{V} + \boldsymbol{\delta} \tag{6.4.5}$$

where

$$A = \begin{bmatrix} u_1 & u_2 & \cdot & \cdot & \cdot & u_n \\ 1 & 1 & \cdot & \cdot & \cdot & 1 \end{bmatrix}^t, \tag{6.4.6}$$

$\mathbf{V} = (v_1, v_2, \cdots, v_n)$ and $\boldsymbol{\delta}$ is the corresponding noise vector. Similar results can be obtained for (6.4.4). In order to use the Gauss-Markov Theorem presented in Section 4.5, we need to know the covariance

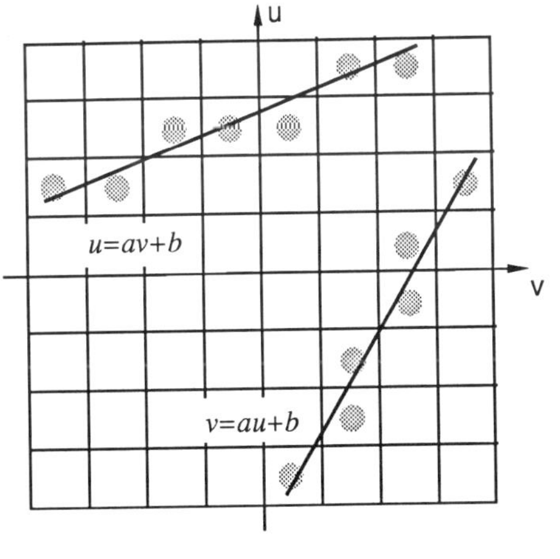

Fig. 6.5. Line fitting and line equations.

matrix of $\boldsymbol{\delta}$. The component of $\boldsymbol{\delta}$ is the error of the detected edge pixel location in the direction orthogonal to that of the index variable. As an approximation, we assume that this type of error is unbiased and uncorrelated. Although this assumption is not exactly true even in the case of digitization error, we expect the correlation is not so significant that the accuracy of estimates derived from this assumption is not considerably degraded. We will return to this point in the discussion of our simulation, in which we will see that the accuracy of the estimates does not change much if the actual (correlated) digitization noise is replaced by uncorrelated noise. Using the condition $\Gamma_{\mathbf{y}} = \sigma^2 I$ and the Gauss-Markov Theorem, the minimum variance estimator of $\mathbf{a}$ is

$$\mathbf{a}^* = (A^t A)^{-1} A^t \mathbf{V}. \tag{6.4.7}$$

The covariance matrix of the error of $\mathbf{a}^*$ is given by

$$\Gamma_{\mathbf{a}^*} = \sigma^2 (A^t A)^{-1} = \sigma^2 \begin{bmatrix} \sum_{i=1}^{n} u_i^2 & \sum_{i=1}^{n} u_i \\ \sum_{i=1}^{n} u_i & n \end{bmatrix}^{-1}. \tag{6.4.8}$$

6.4.2 Optimal Solution

Now we discuss how to determine the unknown motion parameter vector $\mathbf{m}$ from the lines that are determined by the least-squares fitting. Let $\mathbf{a}^*_{ij}$ be the observed line parameter vector of the ith line in jth image frame, and $\mathbf{a}_{ij}(\mathbf{m})$ be the corresponding computed line parameter vector from the motion parameter $\mathbf{m}$ (the structure $\mathbf{x}$ can be optimally determined from $\mathbf{m}$ and be suppressed from $\mathbf{a}^*_{ij}(\mathbf{m}, \mathbf{x})$). A nonlinear extension of the Gauss-Markov theorem leads to minimizing

$$\sum_{i=1}^{n} \sum_{j=1}^{3} (\mathbf{a}^*_{ij} - \mathbf{a}_{ij}(\mathbf{m})) \Gamma_{\mathbf{a}^*}^{-1} (\mathbf{a}^*_{ij} - \mathbf{a}_{ij}(\mathbf{m})). \tag{6.4.9}$$

In other words, we want to minimize the discrepancies between the measured observables and the inferred observables. Such an objective function can be directly extend to the use of other types of features. According to the Gauss-Markov theorem, the discrepancies to be minimized should be weighted by the inverse of the covariance matrix of the expected errors in the corresponding measurements. For example, if lines and points are used as matching features, the objective function to be minimized is the sum of the weighted line discrepancies shown in (6.4.9) and the weighted point discrepancies discussed in Chapter 4.

The minimization of nonlinear function (6.4.9) can be solved iteratively from a good initial guess. The closed-form solution of the linear algorithm discussed above can be used as an initial guess solution for the nonlinear optimization here.

6.4.3 How Accurate It Can Be

With a solution based on noise corrupted data, an important question to ask is how close the error is from the minimum error allowed by the information in the data. Since the noise in the data is random, this question should be investigated in terms of statistics. As discussed in Subsection 4.6.2, there exist theoretical bounds for the covariance matrix of any estimator. Cramér-Rao bound is one of them. We can compare the expected errors with that of a "best possible" unbiased estimator using Cramér-Rao bound. The evaluation of Cramér-Rao bound requires noise distribution. With Gaussian noise, the Cramér-Rao bound takes a simpler form.

For the problem investigated here, $\mathbf{m}$ consists of the independent motion parameters (three in R, three in S, five in normalized (unit) version of $(\mathbf{T}, \mathbf{U})$. The structure $\mathbf{x}$ can be estimated from the motion parameter vector $\mathbf{m}$ by the methods in step (iv) of the algorithm presented in the previous section. The observation vector $\mathbf{y}$ is a vector which consists of vectors $\mathbf{a}$'s in (6.4.5) for every line in every image. Therefore $\mathbf{y}$ is a $6n$-dimensional vector for n line correspondences (n lines in each image). Given any motion parameter vector $\mathbf{m}$, we can compute its corresponding structure $\mathbf{x}$ and the vector $\mathbf{y}_0(\mathbf{m})$. Suppose the observation vector is given by additive Gaussian noise: $\mathbf{y} = \mathbf{y}_0(\mathbf{m}) + \boldsymbol{\delta}$, where $\boldsymbol{\delta}$ is a Gaussian noise vector with a zero mean and a block diagonal covariance matrix C with each 2 by 2 block given by (6.4.8). Then $p(\mathbf{y}, \mathbf{m})$ is a 6n-dimensional distribution with mean $\mathbf{y}_0(\mathbf{m})$ and the same block diagonal covariance matrix C. The Cramér-Rao bound for any unbiased estimator gives

$$\mathbb{E}(\hat{\mathbf{m}} - \mathbf{m})(\hat{\mathbf{m}} - \mathbf{m})^t \geq \left\{ \left[\frac{\partial \mathbf{y}_0(\mathbf{m})}{\partial \mathbf{m}} \right]^t C^{-1} \left[\frac{\partial \mathbf{y}_0(\mathbf{m})}{\partial \mathbf{m}} \right] \right\}^{-1}.$$

6.5 Simulations

Simulations have been performed to demonstrate correctness of the algorithm as well as sensitivity of the solution to noise.

6.5.1 Setup

The focal length of the camera is one unit and the image is a $s \times s$ square. Lines are generated randomly for time t_0. The centers of the lines are uniformly distributed between depth 5 and depth 15. The orientation of the lines are uniformly distributed over all directions. The length of the line are uniformly distributed in $[4s, 8s]$ and only the visible part of the lines are used for line fitting. (Usually, a line following procedure will detect connected edge points along the line. However, a small portion of missing edge points does not significantly degrade the accuracy of the measured line.) The lines are moved to another position at time t_1 by a rotation represented by a matrix R and then a translation represented by translation vector $\mathbf{T}$. Similarly at time t_2 the lines are moved from t_0 by a rotation and then a translation represented by S and $\mathbf{U}$, respectively.

The lines at each time are projected onto the image plane and the edge points are corrupted by digitization noise: For a 256×256 image, there are 256×256 pixels and the image coordinates of edge points have 256 evenly spaced levels for u and v coordinates, respectively. The real image coordinates of edge points are round off to the closest levels. To simulate common line finding algorithms the line position is determined by a least square fitting to the visible edge points (pixel centers) of the line. As we discussed, the relatively higher accuracy of the line position obtained by such a line fitting is one of the main advantages for using lines as matching primitives.

The errors shown in the following are all relative errors. As we defined before, the relative error of a matrix (vector is a column matrix) is defined by the Euclidean norm of the error matrix (difference between the estimated one and the true one) divided by the Euclidean norm of the true matrix.

6.5.2 Linear Algorithm

In noise-free cases, the errors in the solution given by the algorithm are mainly caused by computer round off errors. With a double precision representation of 64 bits on a SUN workstation, the relative errors in the solution were of the order of 10^{-10} if edge points were not rounded off to pixel centers.

In the presence of noise, the errors in solution depend on the configuration of the lines. To show the general sensitivity to the noise, the errors are averaged through 100 random trials, each trial has the same motion but has a different set of randomly generated lines. Different amount of noise can be simulated by image digitization noise at different image resolutions. Simulations of our line-based algorithm and the results of error estimation discussed in Chapter 4 and Chapter 5 showed that the errors in the solutions are roughly proportional to the amount of noise in image plane. For example, if the noise in image plane is doubled, the errors in solutions are also doubled. For the data shown here, the image resolution used is 256×256.

Since the errors in the solutions are very sensitive to the field of view, we show the results using two image sizes: $s = 1$ (equivalent to a f=35mm lens of a 35mm camera) and $s = 0.7$ (equivalent to a f=50mm lens of a 35mm camera). The results of $s = 1$ are shown in Fig. 6.6. The motion parameters are as follows: R corresponds to a rotation about axis (1, 1, 1) by an angle of 6°, and S corresponds to a rotation about axis (0, 1, −1) by an angle of 5°. $\mathbf{T} = (2, -2, 2)$. $\mathbf{U} = (-1, 2, -2)$. Fig. 6.6(a) shows the average relative errors of the intermediate parameters (E, F, G), R and $\mathbf{T}$, versus the number of line correspondences. Fig. 6.6(b) shows the average relative errors of S and $\mathbf{U}$. Those two figures show that with a minimal 13 lines the errors are relatively large. With a minimal number of line correspondences no overdetermination is available in the set of equations of (6.2.19). Therefore the intermediate parameters cannot be determined accurately. Furthermore, some short lines in the images may give very unreliable projection normals. After a few lines are added to the minimally required 13, the errors decreases significantly. (It seems that far more than 13 pairs of lines can be matched from real world images [6.3, 6]. We may expect overdetermination over 13 line correspondences is generally available in practice.) The location of a line can be specified by the direction of the line ($\mathring{\mathbf{l}}$), the projection plane passing through the line and the origin (seen by the camera as an observable) and the distance from the line to the origin ($\|\mathbf{x}_p\|$).

Fig. 6.6(c) shows the average errors of the recovered direction of lines $\mathring{\mathbf{l}}$ and those of recovered relative errors of the distance from the line to the origin.

Fig. 6.7 shows the relative errors for $s = 0.7$. The rotation parameters are the same as those for $s = 1$ in Fig. 6.6. The translations are modified accordingly to account for the reduced field of view: $\mathbf{T} = (1.5, -1.5, 2)$ and $\mathbf{U} = (-1, 1.5, -2)$. As can be seen that the errors in this case are considerably larger than the corresponding errors for a larger field of view (Fig. 6.6).

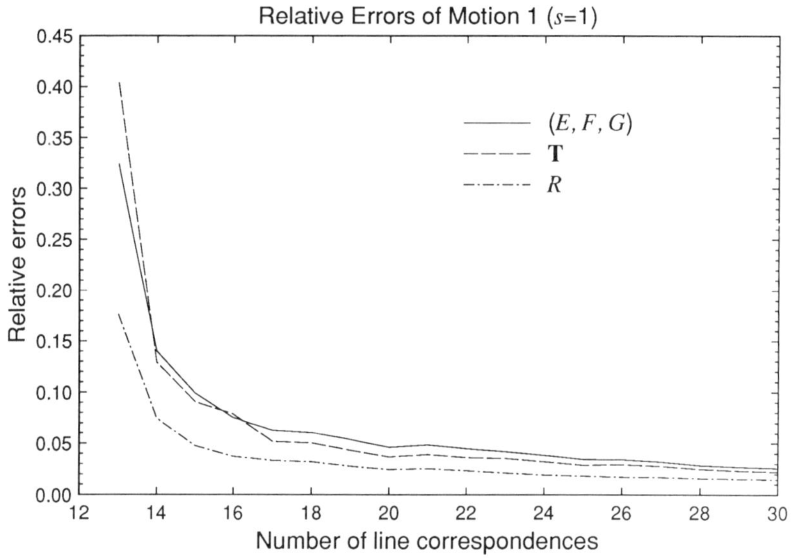

Fig. 6.6(a)

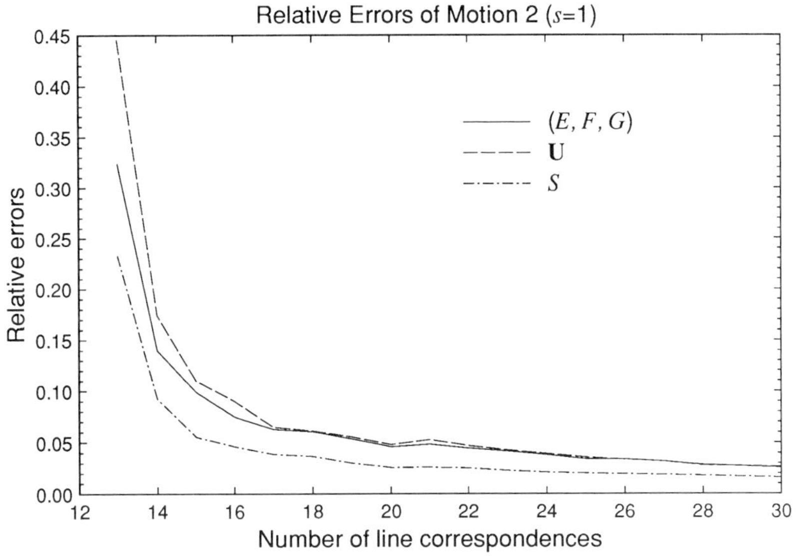

Fig. 6.6(b)

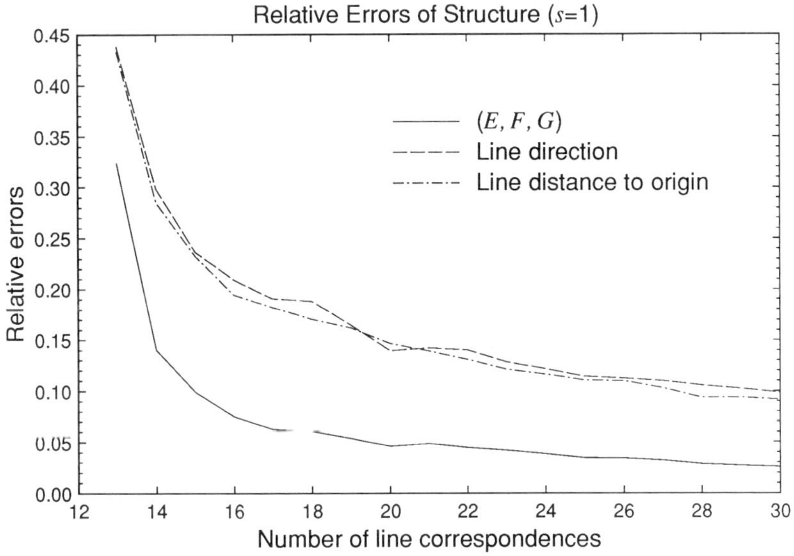

Fig. 6.6(c)

Fig. 6.6. Average relative errors of for $s = 1$ (image size: $s \times s$) through 100 random trials. **(a)** Motion parameters of motion 1. **(b)** Motion parameters of motion 2. **(c)** Structure.

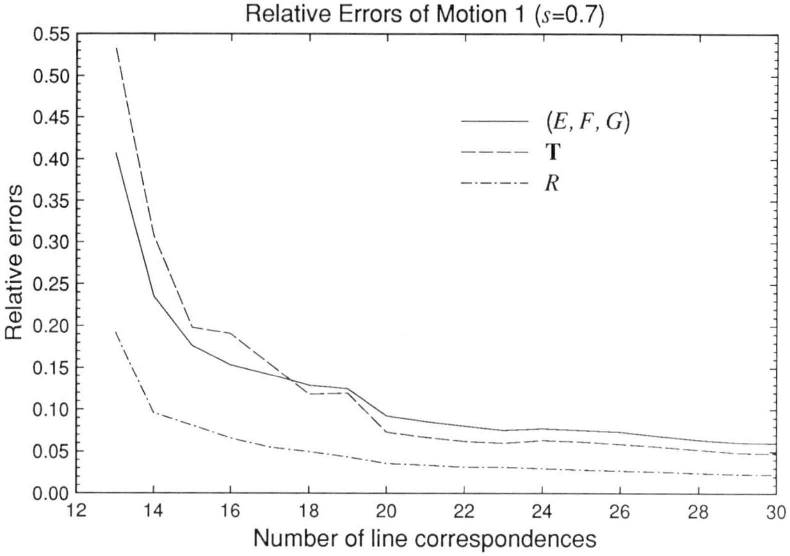

Fig. 6.7(a)

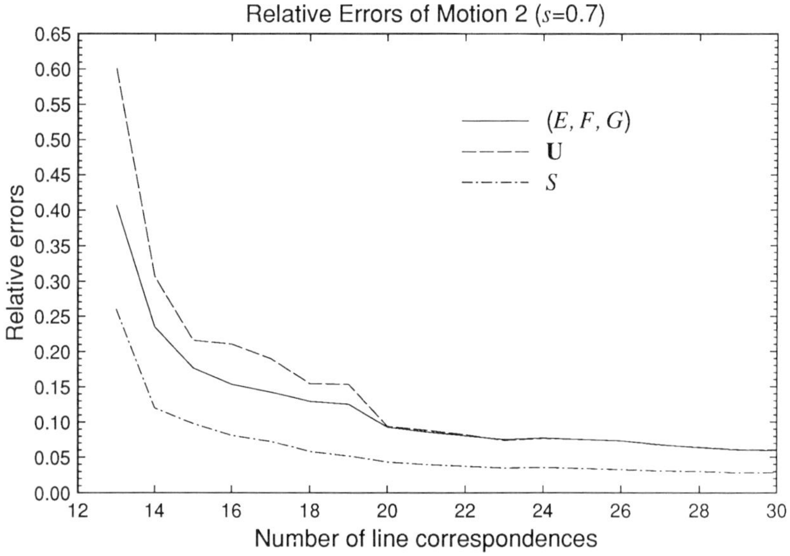

Fig. 6.7(b)

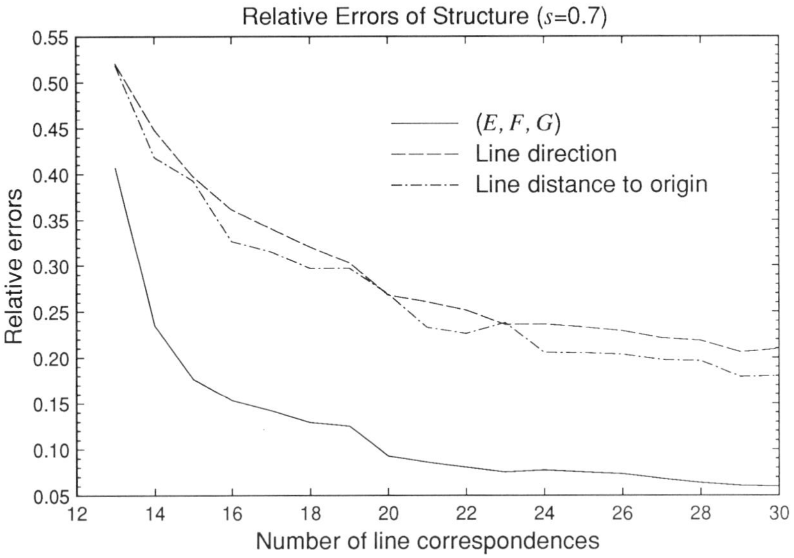

Fig. 6.7(c)

Fig. 6.7. Average relative errors of for $s = 0.7$ (image size: $s \times s$) through 100 random trials. **(a)** Motion parameters of motion 1. **(b)** Motion parameters of motion 2. **(c)** Structure.

6.5.3 Algorithm with Optimization

We then present the results with optimization. In actual implementation, to obtain a better solution before evaluate (6.4.9), the minimization consists of two substeps. First, the equation error of (6.2.43) is minimized. Then the resulting parameters are used as an initial guess solution to minimize (6.4.9). The errors of the final optimal solutions are shown in Fig. 6.8, together with those of the initial guesses shown in Fig. 6.6. Very significant improvements over those of the linear algorithm are achieved. The results of minimizing equation errors in (6.2.43) and those of minimizing (6.4.9) are shown in Fig. 6.9. As can be seen, although minimizing equation errors of (6.2.43) is not as good as minimizing (6.4.9) the difference is not very large. Therefore, when an exact optimal solution is not necessary, minimizing the simpler expression (6.2.43) may suffice.

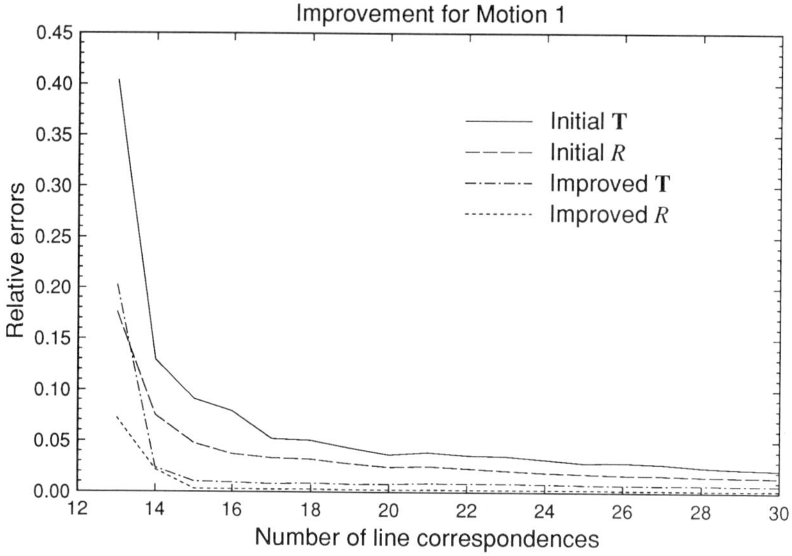

Fig. 6.8(a)

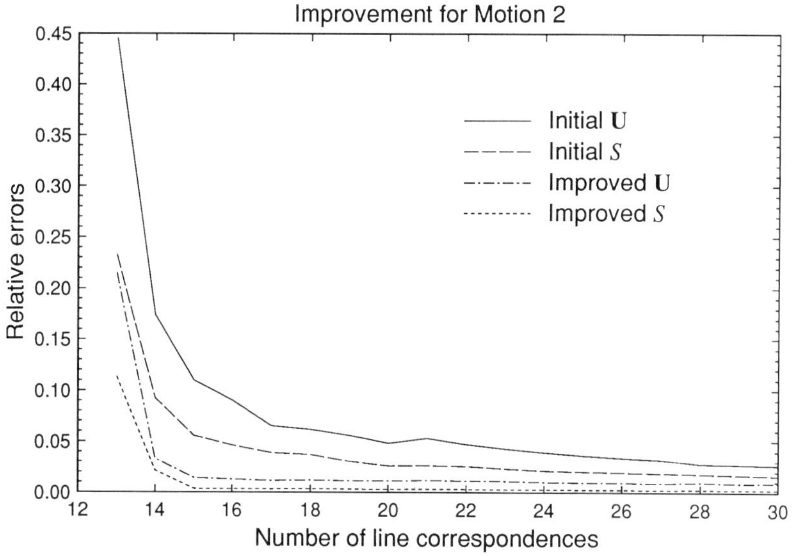

Fig. 6.8(b)

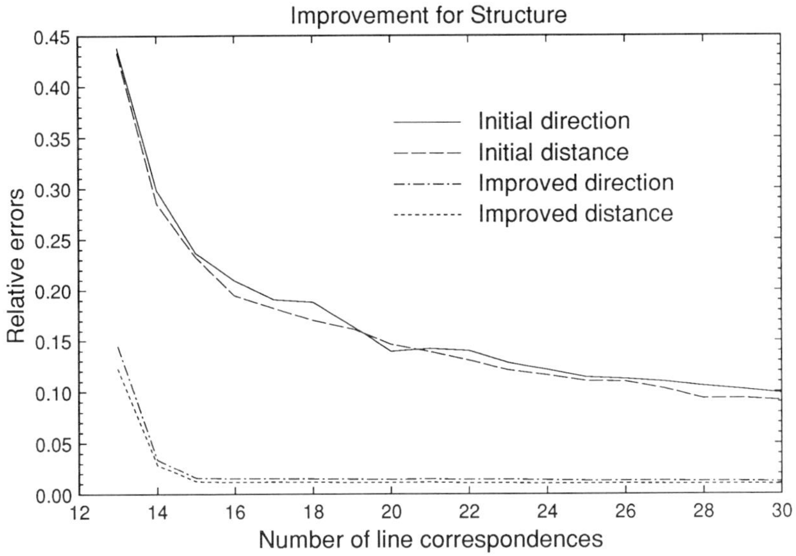

Fig. 6.8(c)

Fig. 6.8. Improvement of optimization over the linear algorithm versus number of line correspondences. 100 random trials. Image size 1×1. (**a**) Motion parameters of motion 1. (**b**) Motion parameters of motion 2. (**c**) Structure.

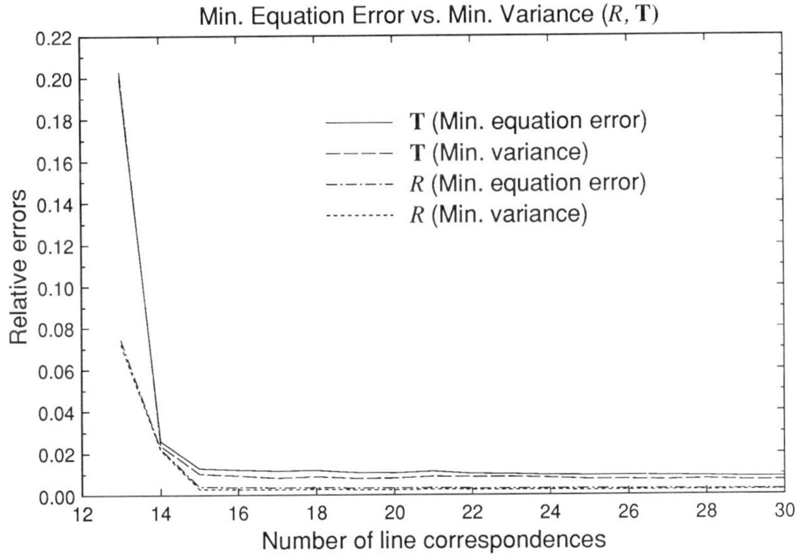

Fig. 6.9(a)

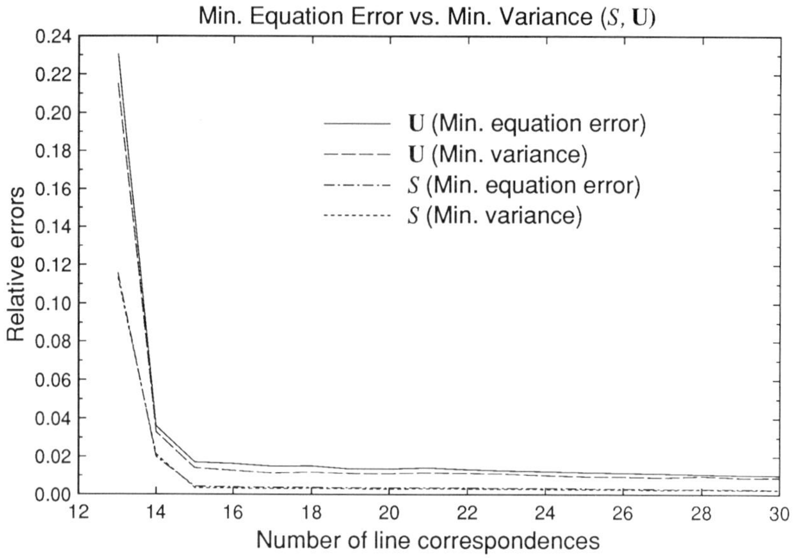

Fig. 6.9(b)

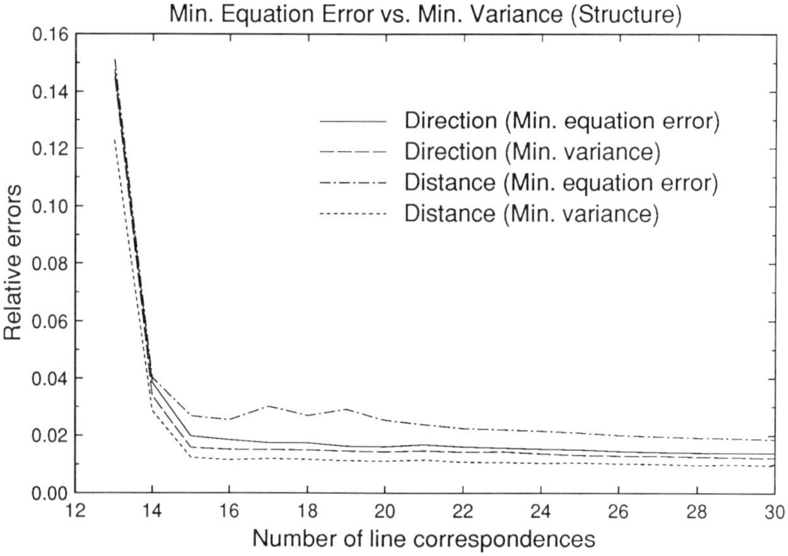

Fig. 6.9(c)

Fig. 6.9. Minimizing equation error (6.2.43) versus minimizing variance, (6.4.9), versus number of line correspondences. 100 random trials. Image size 1×1. **(a)** Motion parameters of motion 1. **(b)** Motion parameters of motion 2. **(c)** Structure.

6.5.4 Comparison with the Bound

In Fig. 6.10, the Cramér-Rao lower bound is shown together with the actual errors. The setup and the motion parameters are the same as those in Fig. 6.6, and white Gaussian noise with a variance equal to that of digitization noise of 256×256 images is added to edge points. The error bounds are very close to the actual errors except for small number of line correspondences. With a small number of lines, the errors in the initial guesses are relatively large (see Figs. 6.5 and 6.6) and thus, the true optimal solutions cannot always be reliably obtained. However, as shown in Fig. 6.10, with a few extra lines in addition to a minimal 13, linear algorithm followed by optimization essentially reached the Cramér-Rao lower bound, the lower error bound of a best possible unbiased estimator.

It should be mentioned that the noise added in Fig. 6.10 is uncorrelated Gaussian noise, while in Figs. 6.6 through 6.9 it is actual spatial digitization noise, which is not exactly uncorrelated. In (6.4.7) we have assumed that the digitization noise is uncorrelated, which will potentially degrade the performance. However, a comparison between Figs. 6.8 and 6.10 for the improved results shows that the corresponding errors are similar. It seems to suggest that the assumption that the digitization noise is uncorrelated does not affect the performance significantly.

6.5.5 Comparison with the Point-Based Algorithm

A question one may raise here is the following: Can a line-based algorithm reach an accuracy, in the presence of noise, which is close to that of a point-based algorithm? Let us compare Fig. 6.10 with Fig. 4.18 in Chapter 4. Both line-based algorithm and point-based algorithm have essentially reached the respective Cramér-Rao bounds. When the number of lines is sufficiently large, for example, larger than 18 in Fig. 6.10, the line-based algorithm is stable to be close to the bound. If we compare the two algorithms on the basis that a point in the point-based algorithm is regarded as a line in the line-based algorithm, we can see that the corresponding errors in Fig. 6.10 are close to those in Fig. 4.18 if the number of feature correspondences is larger than 18. For instance, with 20 feature point correspondences, $\mathbf{T}$ and R in Fig. 4.18 have relative errors 0.014 and 0.0032, respectively, while $\mathbf{T}$ and R in Fig. 6.10 have relative errors 0.010 and 0.0025, respectively.

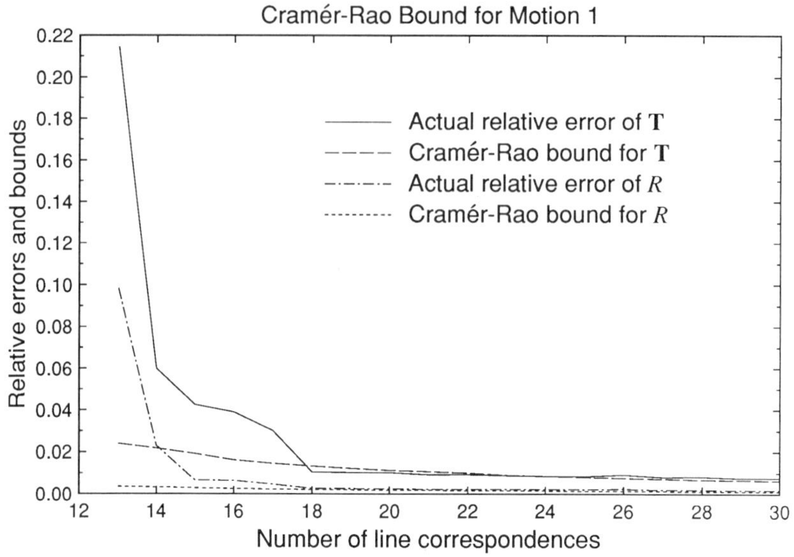

Fig. 6.10(a)

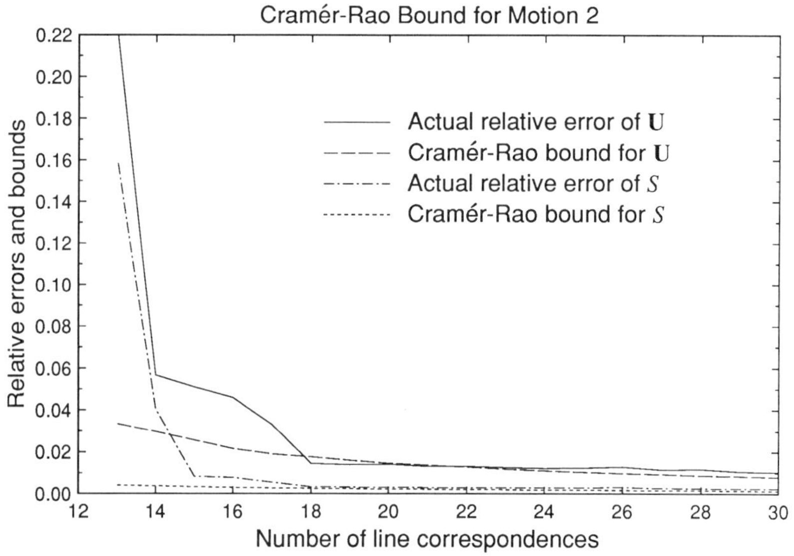

Fig. 6.10(b)

Fig. 6.10. Actual errors, Cramér-Rao bound for Gaussian noise, vs. number of line correspondences. 100 random trials. Image size 0.7×0.7. **(a)** Motion parameters of motion 1. **(b)** Motion parameters of motion 2.

Therefore, the result of simulations seems to suggest that the performance of a line-based algorithm can be as good as a point-based algorithm, if the discussed optimization process is used starting with a good initial guess provided by the linear algorithm.

6.6 Conclusions and Discussions

A linear algorithm is presented for estimating motion and structure parameters from line correspondences. The uniqueness of the solution has been established. As long as the coefficient matrix of the linear equations (6.2.44) is not degenerate, the algorithm gives a unique solution to the motion parameters. Some necessary and sufficient conditions for the linear algorithm to successfully reach a unique solution are presented.

The method of optimal estimation is applied to the problem here. The reliability of each measured line is represented by an error covariance matrix. Those covariance matrices are utilized to reach an optimal solution with a goal of minimizing the expected errors in motion and structure parameters. In order to reliably reach the minimal point of the nonlinear objective function, the closed-form solution is used as an initial guess solution.

From the results of our simulations, it appears that the accuracy of the solutions by our optimal line-based algorithm is close to that of the corresponding optimal point-based algorithm presented in Chapter 4 (with the same amount of image plane noise and the same number of line or point correspondences). A line-to-line correspondence provides only one component of the image plane displacement (along the normal to the line), while a point-to-point correspondence provides both components. Therefore, in some sense, a line-to-line correspondence contains less information than a point-to-point correspondence. How can one expect the line-based algorithm to perform as well as the point-based algorithm? The key is that a line-based algorithm can use the redundancy in the edge points to obtain more accurate measurement of line positions. In other words, a line fitting step used our algorithm is very important for line-based algorithms.

Since the accuracy of the optimal solutions is close to the Cramér-Rao lower error bound for any unbiased estimator, the obtained performance appear to leave little improvement beyond.

6.7 Bibliographical Notes

In the literature, line finding is often treated in the context of edge detection. In some of the reports, straight line detection is emphasized (e.g., Nevatia & Babu [6.8] and Burns, Hanson &

Riseman [6.2]). Ayache & Faverjon [6.1], and McIntosh & Mutch have reported their approaches to matching straight line segments between stereo images.

Yen & Huang [6.12] and Liu & Huang [6.4] proposed iterative nonlinear algorithms to solve the motion parameters from line correspondences. A different approach was reported by Mitiche, Seida & Aggarwal [6.7] where the property of angular invariance between lines is used. Faugeras, Lustman & Toscani [6.3] approximated the nonlinear equations by some linear equations, and based on which the iterated extended Kalman filter was employed to estimate the motion parameters. All these algorithms do not give a closed-form solution to the problem.

Spetsakis & Aloimonos [6.9] and Liu & Huang [6.5] recently proposed linear algorithms for estimating motion and structure parameters from line correspondences. Noting a very high sensitivity to noise, only results with noise-free data were reported [6.9, 5]. A more compact algorithm that uses redundancy in the intermediate parameters was proposed by Weng et al [6.10], in which the uniqueness of the solution by the linear algorithm has been established. Although the latter linear algorithm gives some results in the presence of small noise, as shown in this chapter, its robustness is still not satisfactory. The use of optimization for line-based algorithm [6.11] has so drastically improved the robustness of the line-based approach that it can rival the point-based counterpart.

Appendices

Appendix 6.A Ranks

First we prove the following lemma which will be used for the proof of Theorem 6.1.

Lemma 6.2. Let $\mathbf{R}_1$, $\mathbf{S}_1$, $\mathbf{T}$ and $\mathbf{U}$ be nonzero vectors, and

$$E = \mathbf{R}_1\mathbf{U}^t - \mathbf{T}\mathbf{S}_1^t.$$

Then

1) If rank$(E) = 0$, then $\mathbf{R}_1 \mathbin{//} \mathbf{T}$ and $\mathbf{S}_1 \mathbin{//} \mathbf{U}$.

2) If rank$(E) = 1$, then $\mathbf{R}_1 \mathbin{//} \mathbf{T}$ or $\mathbf{S}_1 \mathbin{//} \mathbf{U}$.

3) If $\mathbf{R}_1 \mathbin{//} \mathbf{T}$, or $\mathbf{U} \mathbin{//} S_1$, then rank$(E) \le 1$.

Proof. 1) rank$(E) = 0$ implies $\mathbf{R}_1\mathbf{U}^t = \mathbf{T}\mathbf{S}_1^t$, the conclusion of 1) immediately follows.

2) $\text{rank}(E) = 1$ implies there exist two nonzero vectors $\mathbf{a}$ and $\mathbf{b}$ such that

$$E = \mathbf{R}_1\mathbf{U}^t - \mathbf{T}\mathbf{S}_1^t = \mathbf{a}\mathbf{b}^t.$$

Let $\mathbf{b}$, $\mathbf{b}_1$, $\mathbf{b}_2$ be nonzero vectors and they are mutually orthogonal. Post-multiplying both sides of the above equation by $\mathbf{b}_1$ and $\mathbf{b}_2$ yields

$$\mathbf{R}_1(\mathbf{U}\cdot\mathbf{b}_1) - \mathbf{T}(\mathbf{S}_1\cdot\mathbf{b}_1) = 0,$$

$$\mathbf{R}_1(\mathbf{U}\cdot\mathbf{b}_2) - \mathbf{T}(\mathbf{S}_1\cdot\mathbf{b}_2) = 0.$$

If $\mathbf{R}_1 \times \mathbf{T} \neq 0$, we have

$$\mathbf{U}\cdot\mathbf{b}_1 = \mathbf{U}\cdot\mathbf{b}_2 = \mathbf{S}_1\cdot\mathbf{b}_1 = \mathbf{S}_2\cdot\mathbf{b}_2 = 0$$

which implying $\mathbf{U} \,//\, \mathbf{b}$ and $\mathbf{S}_1 \,//\, \mathbf{b}$ and therefore, $\mathbf{U} \,//\, \mathbf{S}_1$.

3) Let $\mathbf{R}_1 \,//\, \mathbf{T}$. There exists a number k such that $\mathbf{T} = k\mathbf{R}_1$. Thus $E = \mathbf{R}_1(\mathbf{U} - k\mathbf{S}_1)^t$, which implies $\text{rank}(E) \leq 1$. Similarly the case $\mathbf{S}_1 \,//\, \mathbf{U}$ can be proved. □

Theorem 6.1. Assume $\mathbf{T} \neq \mathbf{0}$ and $\mathbf{U} \neq \mathbf{0}$. E, F and G are always singular matrices. Therefore there exist unit vectors $\mathbf{V}_1$, $\mathbf{V}_2$, and $\mathbf{V}_3$ such that

$$E^t\mathbf{V}_1 = \mathbf{0}, \tag{6.A.1}$$

$$F^t\mathbf{V}_2 = \mathbf{0}, \tag{6.A.2}$$

$$G^t\mathbf{V}_3 = \mathbf{0}. \tag{6.A.3}$$

The ranks of E, F and G fall into three cases.

Case 1. All of E, F, G have a rank two. $\mathbf{V}_i$ is unique up to a scale factor. Let $A = [\mathbf{V}_1\ \mathbf{V}_2\ \mathbf{V}_3]$. Then $\text{rank}(A) = 2$ and $\mathbf{T}$ is essentially determined by $A^t\mathbf{T} = \mathbf{0}$.

Case 2. Two of E, F, G have a rank two, and the rest one has a rank one. Without loss of generality, let $\text{rank}(F) = \text{rank}(G) = 2$. Let $A = [\mathbf{V}_2\ \mathbf{V}_3]$. If $\text{rank}(A) = 2$, $\mathbf{T}$ is still essentially determined by $A^t\mathbf{T} = 0$. Otherwise, $\mathbf{T}$ is essentially determined by $\mathbf{T} \,//\, (\mathbf{E}_i \times \mathbf{V}_2) \times \mathbf{V}_2$, where $\mathbf{E}_i$ is any non-zero column vector of E. $(\mathbf{E}_i \times \mathbf{V}_2) \times \mathbf{V}_2 \neq \mathbf{0}$ is guaranteed.

Case 3. Only one of E, F, G has a rank two, the other two matrices have a rank one. Without loss of generality let rank(G) = 2. Then there are two orthogonal solutions in (6.A.1) and (6.A.2), respectively:

$$E^t \mathbf{V}_{1a} = \mathbf{0}, \quad E^t \mathbf{V}_{1b} = \mathbf{0}, \tag{6.A.4}$$

$$F^t \mathbf{V}_{2a} = \mathbf{0}, \quad F^t \mathbf{V}_{2b} = \mathbf{0}, \tag{6.A.5}$$

where $\mathbf{V}_{1a} \cdot \mathbf{V}_{1b} = 0$ and $\mathbf{V}_{2a} \cdot \mathbf{V}_{2b} = 0$. One and only one of two equations

$$\mathbf{V}_3 \cdot (\mathbf{V}_{1a} \times \mathbf{V}_{1b}) = 0 \tag{6.A.6}$$

and

$$\mathbf{V}_3 \cdot (\mathbf{V}_{2a} \times \mathbf{V}_{2b}) = 0 \tag{6.A.7}$$

holds true. $\mathbf{T} \mathbin{//} \mathbf{V}_{1a} \times \mathbf{V}_{1b}$ if (6.A.6) holds. $\mathbf{T} \mathbin{//} \mathbf{V}_{2a} \times \mathbf{V}_{2b}$ if (6.A.7) holds.

Proof. If $\mathbf{R}_1 \mathbin{//} \mathbf{T}$, Lemma 6.2 concludes that rank(E) ≤ 1. Otherwise $E^t \mathbf{V}_1 = \mathbf{0}$ where $\mathbf{V}_1 = \mathbf{R}_1 \times \mathbf{T} \ne \mathbf{0}$. Thus rank($E$) ≤ 2. Similarly rank(F) ≤ 2 and rank(G) ≤ 2.

Case 1. Since the ranks of E, F, G are all equal to 2, from Lemma 6.2, $\mathbf{T}$ is not parallel to any column vectors of R, and $\mathbf{U}$ is not parallel to any column vectors of S. Thus $\mathbf{V}_i \mathbin{//} \mathbf{T} \times \mathbf{R}_i$, $i = 1, 2, 3$. rank(A) = 2 if rank(M) = 2 where

$$M = [\mathring{\mathbf{T}} \times \mathbf{R}_1 \ \mathring{\mathbf{T}} \times \mathbf{R}_2 \ \mathring{\mathbf{T}} \times \mathbf{R}_3] = [\mathring{\mathbf{T}}]_\times R$$

and $\mathring{\mathbf{T}} = \|\mathbf{T}\|^{-1}\mathbf{T}$. We prove that rank($M$) = 2: In fact, let the unit vectors $\mathring{\mathbf{T}}$, $\mathring{\mathbf{T}}_2$ and $\mathring{\mathbf{T}}_3$ be such that $Q = [\mathring{\mathbf{T}} \ \mathring{\mathbf{Q}}_2 \ \mathring{\mathbf{Q}}_3]$ is an orthonormal 3×3 matrix. $G = R^t Q$ is also orthonormal. Post-multiplying the two sides of (6.A.13) by G, we get

$$MG = [\mathring{\mathbf{T}}]_\times RG = [\mathring{\mathbf{T}}]_\times Q = [\ \mathbf{0} \quad \mathring{\mathbf{T}} \times \mathring{\mathbf{Q}}_2 \ \mathring{\mathbf{T}} \times \mathring{\mathbf{Q}}_3].$$

We see that the second and the third columns of MG are orthonormal from the definition of Q. Thus, rank(M) = rank(MG) = 2.

Case 2. We need to prove that if rank(A) ≤ 1, $\mathbf{T}$ is essentially determined by $\mathbf{T} \mathbin{//} (\mathbf{E}_i \times \mathbf{V}_2) \times \mathbf{V}_2$ and it is true that $(\mathbf{E}_i \times \mathbf{V}_2) \times \mathbf{V}_2 \ne \mathbf{0}$. Since $\mathbf{V}_i \mathbin{//} \mathbf{T} \times \mathbf{R}_i$, $i = 2, 3$, rank(A) ≤ 1 implies

$$\mathbf{0} = (\mathbf{T}\times\mathbf{R}_2)\times(\mathbf{T}\times\mathbf{R}_3) = ((\mathbf{T}\times\mathbf{R}_2)\cdot\mathbf{R}_3)\mathbf{T} - 0\mathbf{R}_3$$

where the last equation follows by using the vector identity

$$\mathbf{a}\times(\mathbf{b}\times\mathbf{c}) = (\mathbf{a}\cdot\mathbf{c})\mathbf{b} - (\mathbf{a}\cdot\mathbf{b})\mathbf{c}.$$

Thus $(\mathbf{T}\times\mathbf{R}_2)\cdot\mathbf{R}_3 = 0$, i.e., $\mathbf{T}$, $\mathbf{R}_2$, $\mathbf{R}_3$ are coplanar. Therefore

$$\mathbf{T}\cdot\mathbf{R}_1 = 0 \tag{6.A.8}$$

and $\mathbf{V}_2 \mathbin{//} \mathbf{T}\times\mathbf{R}_2 \mathbin{//} \mathbf{R}_1$. Thus $\mathbf{V}_2 = \pm\mathbf{R}_1$. Since $\mathrm{rank}(E) \le 1$ and $\mathbf{T}\cdot\mathbf{R}_1 = 0$, we get $\mathbf{S}_1 \mathbin{//} \mathbf{U}$ from Lemma 6.2. Let $\mathbf{U} = k\mathbf{S}_1$ for some real number k. We have $E = (k\mathbf{R}_1 - \mathbf{T})\mathbf{S}_1^t$ and $\mathbf{E}_i = l(k\mathbf{R}_1 - \mathbf{T})$ for some real number $l \neq 0$ since $\mathbf{S}_1 \neq \mathbf{0}$. Therefore, using (6.A.8) yields

$$(\mathbf{E}_i\times\mathbf{V}_2)\times\mathbf{V}_2 = l((k\mathbf{R}_1 - \mathbf{T})\times\mathbf{R}_1)\times\mathbf{R}_1 = -l(\mathbf{T}\times\mathbf{R}_1)\times\mathbf{R}_1 = l\mathbf{T}.$$

Then $\mathbf{T} \neq \mathbf{0}$ and $l \neq 0$ give $(\mathbf{E}_i\times\mathbf{V}_2)\times\mathbf{V}_2 \neq \mathbf{0}$.

Case 3. From Lemma 6.2, it follows that

$$\mathbf{T} \mathbin{//} \mathbf{R}_1 \quad \text{or} \quad \mathbf{T} \mathbin{//} \mathbf{R}_2, \tag{6.A.9}$$

and

$$\mathbf{U} \mathbin{//} \mathbf{S}_1 \quad \text{or} \quad \mathbf{U} \mathbin{//} \mathbf{S}_2. \tag{6.A.10}$$

Since $\mathbf{R}_1$ and $\mathbf{R}_2$ are orthogonal vectors, and so are $\mathbf{S}_1$ and $\mathbf{S}_2$, (6.A.9) and (6.A.10) give only two possible combinations: (i) $\mathbf{T} \mathbin{//} \mathbf{R}_1$ and $\mathbf{U} \mathbin{//} \mathbf{S}_2$. (ii) $\mathbf{T} \mathbin{//} \mathbf{R}_2$ and $\mathbf{U} \mathbin{//} \mathbf{S}_2$.

For (i): Letting $\mathbf{T} = k\mathbf{R}_1$, $(k \neq 0)$, and $\mathbf{U} = l\mathbf{S}_2$, $(l \neq 0)$ yields

$$E = \mathbf{R}_1(\mathbf{U} - k\mathbf{S}_1)^t,$$

$$F = (l\mathbf{R}_2 - \mathbf{T})\mathbf{S}_2^t = (l\mathbf{R}_2 - k\mathbf{R}_1)\mathbf{S}_2^t,$$

which gives

$$\mathbf{V}_{1a}\times\mathbf{V}_{1b} \mathbin{//} \mathbf{R}_1, \tag{6.A.11}$$

$$\mathbf{V}_{2a}\times\mathbf{V}_{2b} \mathbin{//} (l\mathbf{R}_2 - k\mathbf{R}_1). \tag{6.A.12}$$

On the other hand, $\mathbf{V}_3 \mathbin{//} \mathbf{T}\times\mathbf{R}_3 = k\mathbf{R}_1\times\mathbf{R}_3 = k\mathbf{R}_2$. Therefore

$$\mathbf{V}_3 \cdot (\mathbf{V}_{1a} \times \mathbf{V}_{1b}) = 0, \tag{6.A.13}$$

$$\mathbf{V}_3 \cdot (\mathbf{V}_{2a} \times \mathbf{V}_{2b}) \neq 0.$$

Then (6.A.11) leads to

$$\mathbf{T} \mathbin{//} \mathbf{R}_1 \mathbin{//} \mathbf{V}_{1a} \times \mathbf{V}_{1b} .$$

For (ii): Similarly we have

$$\mathbf{V}_3 \cdot (\mathbf{V}_{1a} \times \mathbf{V}_{1b}) \neq 0,$$

$$\mathbf{V}_3 \cdot (\mathbf{V}_{2a} \times \mathbf{V}_{2b}) = 0,$$

$$\mathbf{T} \mathbin{//} \mathbf{R}_2 \mathbin{//} \mathbf{V}_{2a} \times \mathbf{V}_{2b} .$$

□

Appendix 6.B Unique Consistent Assignment

We first need to prove the following lemma.

Lemma 6.3. Suppose $\mathbf{T} \neq \mathbf{0}$. Then

$$[\mathbf{T}]_\times R_p = -[\mathbf{T}]_\times R_n \tag{6.B.1}$$

yields the relation

$$R_n = R(\mathbf{T}, \pi) R_p \tag{6.B.2}$$

where $R(\mathbf{T},\pi)$ is the rotation matrix representing the rotation about vector $\mathbf{T}$ by angle π.

Proof. Let $\mathring{\mathbf{T}} = \|\mathbf{T}\|^{-1}\mathbf{T}$, and the matrix $[\mathring{\mathbf{T}}\ \mathring{\mathbf{T}}_1\ \mathring{\mathbf{T}}_2]$ form a right-handed orthonormal matrix. (6.B.1) gives

$$R_p^t [\mathring{\mathbf{T}}]_\times = -R_n^t [\mathring{\mathbf{T}}]_\times . \tag{6.B.3}$$

Since $\operatorname{rank}([\mathring{\mathbf{T}}]_\times) = 2$, and $[\mathring{\mathbf{T}}]_\times \mathring{\mathbf{T}} = 0$, the columns of $[\mathring{\mathbf{T}}]$ and $\mathring{\mathbf{T}}$ span the 3-dimensional space R^3. Since $\mathring{\mathbf{T}}_1$ and $\mathring{\mathbf{T}}_2$ are all orthogonal to $\mathring{\mathbf{T}}$, they can be represented by the linear combination of the columns of $[\mathring{\mathbf{T}}]_\times$: $\mathbf{T}_i = [\mathring{\mathbf{T}}]_\times \mathbf{Y}_i$, $i = 1, 2$. From (6.B.3) it follows that

$$R_p^t \mathring{\mathbf{T}}_i = R_p^t [\mathring{\mathbf{T}}]_\times \mathbf{Y}_i = -R_n^t [\mathring{\mathbf{T}}]_\times \mathbf{Y}_i = -R_n^t \mathring{\mathbf{T}}_i ,$$

$i = 1, 2$. Therefore

$$R_n R_p^t [\mathring{\mathbf{T}}\ \mathring{\mathbf{T}}_1\ \mathring{\mathbf{T}}_2] = [\mathbf{X}\ -\mathring{\mathbf{T}}_1\ -\mathring{\mathbf{T}}_2] \tag{6.B.4}$$

where $\mathbf{X}$ must be equal to $\mathring{\mathbf{T}}$ since the left-hand side is a rotation matrix. Since

$$[\mathbf{T}\ -\mathring{\mathbf{T}}_1\ -\mathring{\mathbf{T}}_2] = R(\mathring{\mathbf{T}}, \pi)[\mathbf{T}\ \mathring{\mathbf{T}}_1\ \mathring{\mathbf{T}}_2]. \tag{6.B.5}$$

(6.B.4) and (6.B.5) give

$$R_n R_p^t = R(\mathring{\mathbf{T}}, \pi).$$

This leads to (6.B.2) directly. □

Lemma 6.1. Assume $\mathbf{T} \neq \mathbf{0}$, $\mathbf{U} \neq \mathbf{0}$, and $R^t\mathbf{T} \neq S^t\mathbf{U}$. Only one assignment for (s_1, s_2), $s_1, s_2 \in \{-1, 1\}$, has a solution in (6.2.26) for rotation matrices R and S.

Proof. Since there exists at least one set of solution: the true one, we prove that solutions for two combinations of (s_1, s_2) yield contradiction. Assume $(s_1, s_2) = (\dot{s}_1, \dot{s}_2)$ is the correct assignment, which gives the solutions

$$(R, S) = (R_p, S_p). \tag{6.B.6}$$

That is,

$$\mathbf{T} = \dot{s}_1 \|\mathbf{T}\| \mathring{\mathbf{T}}_s, \qquad \mathbf{U} = \dot{s}_2 \|\mathbf{U}\| \mathring{\mathbf{U}}_s. \tag{6.B.7}$$

Reversing the assign for s_1 yields R_n in (6.2.29) and reversing the sign for s_2 yields S_n in (6.2.31). From (6.2.29) we get

$$[\mathring{\mathbf{T}}_s]_\times R_p = -[\mathring{\mathbf{T}}_s]_\times R_n.$$

From Lemma 3, it follows that

$$R_n = R(\mathbf{T},\pi) R_p \tag{6.B.8}$$

where $R(\mathbf{T},\pi)$ is the rotation matrix representing the rotation about vector $\mathbf{T}$ by an angle π. Similarly we obtain

$$S_n = S(\mathbf{U},\pi) S_p. \tag{6.B.9}$$

We first prove that it is impossible that for both $(\pm\dot{s}_1, \dot{s}_2)$ there exist solutions. Otherwise, from the first equation in (6.2.26) we have (the subscripts denote the corresponding columns of the rotation matrix),

$$E = \dot{s}_1 \|\mathbf{U}\| \mathbf{R}_{p1} \mathring{\mathbf{U}}_s^t - \dot{s}_2 \|\mathbf{T}\| \mathring{\mathbf{T}}_s \mathbf{S}_{p1}^t,$$

$$E = -\dot{s}_1 \|\mathbf{U}\| \mathbf{R}_{n1} \mathring{\mathbf{U}}_s^t - \dot{s}_2 \|\mathbf{T}\| \mathring{\mathbf{T}}_s \mathbf{S}_{p1}^t.$$

Subtracting both sides yields

$$\|\mathbf{U}\| \mathbf{R}_{p1} \mathring{\mathbf{U}}_s^t = -\|\mathbf{U}\| \mathbf{R}_{n1} \mathring{\mathbf{U}}_s^t.$$

Since $\mathbf{U} \neq \mathbf{0}$ we get

$$\mathbf{R}_{p1} = -\mathbf{R}_{n1}.$$

From the rest two equations of (6.2.26) we get the similar results for the other two columns of the rotation matrices R_p and S_n. Therefore

$$R_p = -R_n.$$

This is a contradiction because R_p and R_n are both rotation matrices whose determinants are equal to one.

Similarly $(\dot{s}_1, \pm\dot{s}_2)$ cannot both have solutions.

Now only possible another assignment that has the solution is $(-\dot{s}_1, -\dot{s}_2)$. Suppose it has a solution. For $i = 1, 2, 3$, the equations of (6.2.26) give

$$E = \dot{s}_1 \|\mathbf{U}\| \mathbf{R}_{pi} \mathring{\mathbf{U}}_s^t - \dot{s}_2 \|\mathbf{T}\| \mathring{\mathbf{T}}_s \mathbf{S}_{pi}^t,$$

$$E = -\dot{s}_1 \|\mathbf{U}\| \mathbf{R}_{ni} \mathring{\mathbf{U}}_s^t + \dot{s}_2 \|\mathbf{T}\| \mathring{\mathbf{T}}_s \mathbf{S}_{ni}^t.$$

These two equations yield

$$\dot{s}_1 \|\mathbf{U}\| (\mathbf{R}_{pi} + \mathbf{R}_{ni}) \mathring{\mathbf{U}}_s^t = \dot{s}_2 \|\mathbf{T}\| \mathring{\mathbf{T}}_s (\mathbf{S}_{pi}^t + \mathbf{S}_{ni}^t). \tag{6.B.10}$$

From (6.B.8), (6.B.9) and (6.B.10) we get

$$\|\mathbf{U}\| (R(\mathring{\mathbf{T}}, \pi) + I) \mathbf{R}_{pi}) \mathring{\mathbf{U}}_s^t = \|\mathbf{T}\| \mathring{\mathbf{T}}_s (S(\mathring{\mathbf{U}}, \pi) + I) \mathbf{S}_{pi})^t. \tag{6.B.11}$$

Since

$$\mathring{\mathbf{T}}_s^t R(\mathring{\mathbf{T}}, \pi) = \mathring{\mathbf{T}}_s,$$

premultiplying both sides of (6.B.11) by $\mathring{\mathbf{T}}_s^t$ and post-multiplying the result by $\mathring{\mathbf{U}}_s$ give

$$2\dot{s}_1 \|\mathbf{U}\| \mathring{\mathbf{T}}_s^t \mathbf{R}_{pi} = 2\dot{s}_2 \|\mathbf{T}\| \mathbf{S}_{pi}^t \mathring{\mathbf{U}}_s$$

where $i = 1, 2, 3$. Therefore

$$2\dot{s}_1 \|\mathbf{U}\| \mathring{\mathbf{T}}_s^t R_p = 2\dot{s}_2 \|\mathbf{T}\| \mathring{\mathbf{U}}_s^t S_p . \tag{6.B.12}$$

Since R and S are rotation matrices, from (6.B.12) we have $\|\mathbf{T}\| = \|\mathbf{U}\|$. Then (6.B.12), (6.B.6) and (6.B.7) yield

$$\mathbf{T}^t R = \mathbf{U}^t S.$$

This is a contradiction to $R^t \mathbf{T} \neq S^t \mathbf{U}$. □

Appendix 6.C Degeneracy

Theorem 6.3. (E, F, G) is not essentially determined by (6.2.19) or equivalently, rank$(D) < 26$ in (6.2.44), if and only if there exist no trivial parameters $(\tilde{E}, \tilde{F}, \tilde{G})$ such that

$$[\mathbf{n}]_\times \begin{bmatrix} ((\mathbf{x}_p - \mathbf{O}_1)\times\mathbf{l})^t \tilde{E}((\mathbf{x}_p - \mathbf{O}_2)\times\mathbf{l}) \\ ((\mathbf{x}_p - \mathbf{O}_1)\times\mathbf{l})^t \tilde{F}((\mathbf{x}_p - \mathbf{O}_2)\times\mathbf{l}) \\ ((\mathbf{x}_p - \mathbf{O}_1)\times\mathbf{l})^t \tilde{G}((\mathbf{x}_p - \mathbf{O}_2)\times\mathbf{l}) \end{bmatrix} = \mathbf{0} \tag{6.C.1}$$

is satisfied for all lines $\mathbf{x} = \mathbf{x}_p + k\mathbf{l}$ at time t_0. $(\tilde{E}, \tilde{F}, \tilde{G})$ is trivial if and only if

$$(\tilde{E}, \tilde{F}, \tilde{G}) = \alpha(P_1 - Q_1^t, P_2 - Q_2^t, P_3 - Q_3^t)$$

for some real number α, where P_i and Q_i are matrices with the ith column being $\mathbf{O}_1$ and $\mathbf{O}_2$, respectively, and the other columns are zero vectors.

Proof. From (6.2.6) and (6.2.7) it follows that

$$\mathbf{n}_1 = R(\mathbf{n}_0 - \mathbf{O}_1\times\mathbf{l}), \tag{6.C.2}$$

$$\mathbf{n}_2 = S(\mathbf{n}_0 - \mathbf{O}_2\times\mathbf{l}). \tag{6.C.3}$$

Substituting (6.C.2) and (6.C.3) into (6.2.19) yields (6.C.1), where we define

$$(\tilde{E}, \tilde{F}, \tilde{G}) = (R^t ES, R^t FS, R^t GS)$$

$$= (P_1 - Q_1^t, P_2 - Q_2^t, P_3 - Q_3^t). \tag{6.C.4}$$

The last equation in (6.C.4) follows from (6.2.18) using the fact that R and S are orthonormal.

(E, F, G) in (6.2.18) satisfies (6.2.19). From (6.C.4), trivial $(\tilde{E}, \tilde{F}, \tilde{G})$ satisfies (6.C.1). From (6.C.4), (6.C.1) has only trivial solution if and only if (6.2.19) essentially determines (E, F, G). □

Appendix 6.D Distinct Locations Are Necessary

Corollary 6.1. If the distinct locations condition is not satisfied, the intermediate parameters are not essentially determined by (6.2.19).

Proof. If $\mathbf{O}_1 = \mathbf{0}$ let $(\tilde{E}, \tilde{F}, \tilde{G}) = (V_1^t, V_2^t, V_3^t)$, where V_i is a matrix with the ith column being an arbitrary vector $\mathbf{v}_i$ and the other columns are zeros. (6.C.1) becomes $[\mathbf{n}]_\times \mathbf{n}[\mathbf{v}_1\ \mathbf{v}_2\ \mathbf{v}_3]^t \mathbf{n}_2 = \mathbf{0}$, which holds since $[\mathbf{n}]_\times \mathbf{n} = \mathbf{0}$. This means that (6.C.1) has nontrivial solutions. Similarly, if $\mathbf{O}_2 = \mathbf{0}$, (6.C.1) has nontrivial solutions. Finally, if $\mathbf{O}_1 = \mathbf{O}_2$, then $\tilde{\mathbf{n}}_1 = \tilde{\mathbf{n}}_2$. As long as $\tilde{E}, \tilde{F}, \tilde{G}$ are antisymmetrical matrices, (6.C.1) holds, since $(\mathbf{v})^t M \mathbf{v} = 0$ holds for any symmetrical matrix M. Therefore (6.C.1) has non-trivial solutions. □

Appendix 6.E Alternative Degeneracy Condition

Theorem 6.4. Given 13 line correspondences at time t_0, with the ith line being represented by $\mathbf{x} = \mathbf{x}_{pi} + \mathbf{l}_i$ ($\mathbf{x}_{pi}$ is the point closest to the origin) and the normal for the plane that passes though the line and $\mathbf{O}_k$ is $\tilde{\mathbf{n}}_{ki}$, $k = 1, 2$, $i = 1, 2 \cdots 13$. (E, F, G) is not essentially determined by (6.2.19) or equivalently, rank$(D) < 26$ in (6.2.44), if and only if there exist no a_i, b_i, $i = 1, 2, \cdots, 13$, not all of which are zeros, such that

$$\sum_{i=1}^{13} (a_i \mathbf{x}_{pi} + b_i \mathbf{l}_i) \tilde{\mathbf{n}}_{2i} \tilde{\mathbf{n}}_{3i} = \mathbf{0}. \tag{6.E.1}$$

Proof. Obviously, the normal of plane passing though the line and $\mathbf{O}_k$ is equal to $(\mathbf{x}_p - \mathbf{O}_1) \times \mathbf{l}$, $k = 1, 2$. (6.C.1) can be rewritten as

$$[\mathbf{n}]_\times \begin{bmatrix} (\tilde{\mathbf{n}}_1)^t \tilde{E} \tilde{\mathbf{n}}_2 \\ (\tilde{\mathbf{n}}_1)^t \tilde{F} \tilde{\mathbf{n}}_2 \\ (\tilde{\mathbf{n}}_1)^t \tilde{G} \tilde{\mathbf{n}}_2 \end{bmatrix} = \mathbf{0}. \tag{6.E.2}$$

(6.E.2) has at most two independent scalar equations since $\text{rank}([\mathbf{n}]_\times) = 2$ for $\mathbf{n} \neq \mathbf{0}$. We need to exclude one equation that is a linear combination of other two. Let $\mathbf{a} = (a_1, a_2, a_3)^t$ and $a_1 a_2 a_3 = 0$. The condition $a_1 a_2 a_3 = 0$ implies at least one of the elements of $\mathbf{a}$ is zero. Assuming $\mathbf{n} = (n_1, n_2, n_3)^t$. Multiplying a_i to the ith scalar equation of (6.E.2) gives

$$\begin{aligned} -n_3 a_1 (\tilde{\mathbf{n}}_1)^t \tilde{F} \tilde{\mathbf{n}}_2 + n_2 a_1 (\tilde{\mathbf{n}}_1)^t \tilde{G} \tilde{\mathbf{n}}_2 &= 0, \\ n_3 a_2 (\tilde{\mathbf{n}}_1)^t \tilde{E} \tilde{\mathbf{n}}_2 - n_1 a_2 (\tilde{\mathbf{n}}_1)^t \tilde{G} \tilde{\mathbf{n}}_2 &= 0, \\ -n_2 a_3 (\tilde{\mathbf{n}}_1)^t \tilde{E} \tilde{\mathbf{n}}_2 + n_1 a_3 (\tilde{\mathbf{n}}_1)^t \tilde{F} \tilde{\mathbf{n}}_2 &= 0. \end{aligned} \tag{6.E.3}$$

The above equation hold for every line. Now we append one more subscript i to denote the corresponding values for the ith line. $\text{rank}(D) < 26$ if and only if the 26 rows of D is linearly independent. Considering the coefficients of elements of $\tilde{E}, \tilde{F}, \tilde{G}$ in (6.E.3), $\text{rank}(D) < 26$ if and only if

$$\begin{aligned} \sum_{i=1}^{13} (n_{3i} a_{2i} - n_{2i} a_{3i}) \tilde{\mathbf{n}}_{1i} \tilde{\mathbf{n}}_{2i} &= \mathbf{0}, \\ \sum_{i=1}^{13} (n_{1i} a_{3i} - n_{3i} a_{1i}) \tilde{\mathbf{n}}_{1i} \tilde{\mathbf{n}}_{2i} &= \mathbf{0}, \end{aligned} \tag{6.E.4}$$

and

$$\sum_{i=1}^{13} (n_{2i} a_{1i} - n_{1i} a_{2i}) \tilde{\mathbf{n}}_{1i} \tilde{\mathbf{n}}_{2i} = \mathbf{0},$$

where $a_{1i} a_{2i} a_{3i} = 0$ to make sure at most two equations of (6.E.2) are used. (6.E.4) can be rewritten using tensor notation

$$\sum_{i=1}^{13} (\mathbf{a}_i \times \mathbf{n}_i) \mathbf{n}_{1i} \mathbf{n}_{2i} = \mathbf{0}. \tag{6.E.5}$$

Since $\mathbf{n}_i = \mathbf{x}_{p_i} \times \mathbf{l}_i$, $i = 1, 2, \cdots, 13$, we will prove that for any projection normal $\mathbf{n}$ the following holds

$$\{\mathbf{a}\times\mathbf{n} \mid \mathbf{a} = (a_1, a_2, a_3)^t,\ a_1 a_2 a_3 = 0\}$$

$$= \{\mathbf{v} \mid \mathbf{v}\cdot\mathbf{n} = 0\} = \{a\mathbf{x}_p + b\mathbf{l} \mid \mathbf{n} = \mathbf{x}_p\times\mathbf{l},\ \text{any } a, b\} \tag{6.E.6}$$

where $\mathbf{x}_p$ is the position vector of the point on the line that is the closest to the origin and $\mathbf{l}$ is the direction of the line. Then (6.E.1) is equivalent to (6.E.5) which is equivalent to (6.C.1). That is, (6.C.1) has only trivial solutions if and only if (6.E.1) holds.

What remains to be proved is (6.E.6). The last equation is trivial since $\mathbf{n}$, $\mathbf{x}_p$ and $\mathbf{l}$ are orthogonal to one another. It is clear that $\{\mathbf{a}\times\mathbf{n} \mid \mathbf{a} = (a_1, a_2, a_3)^t,\ a_1 a_2 a_3 = 0\}$ is a subset of $\{\mathbf{v} \mid \mathbf{v}\cdot\mathbf{n} = 0\}$. Now we prove the reverse. Let $\mathbf{s}$ be orthogonal to $\mathbf{n}$. Define $\mathbf{b} = \mathbf{s}\times\mathbf{n}$. Obviously $\mathbf{b}\times\mathbf{n}$ is aligned with $\mathbf{s}$. If $\mathbf{b}$ has a zero component, $\mathbf{a} = k\mathbf{b}$ gives $\mathbf{a}\times\mathbf{n} = \mathbf{s}$ for some real number k. Otherwise, $\mathbf{b} = (b_1, b_2, b_3)^t$ has no zero components. Noticing that $\mathbf{n}$ is not a zero vector, without loss of generality, assume its jth element is not zero $n_j \neq 0$. Then let $\mathbf{a} = n_j\mathbf{b} - b_j\mathbf{n}$. $\mathbf{a}$ is a vector whose jth component is zero and $\mathbf{a}$ is not aligned with $\mathbf{n}$ (since $b_j \neq 0$ and $n_j \neq 0$). $\mathbf{a}$ is a linear combination of $\mathbf{b}$ and $\mathbf{n}$, so $\mathbf{a}$ is orthogonal to $\mathbf{s}$. $k\mathbf{a}\times\mathbf{n} = \mathbf{s}$ for some real number k. Therefore, $\mathbf{s}$ is in the subset of the left-hand side of (6.E.6). □

7. Stereo

From stereo image sequences, the 3-D positions of the object points can be determined by stereo triangulation. Therefore, motion estimation can deal with 3-D coordinates directly. In this sense, one faces a less difficult problem here than in the monocular case we discussed before. However, stereo motion and structure analysis has its own characteristics, especially in dealing with uncertainty of the 3-D points. In this chapter, we will discuss stereo motion estimation using different weighting schemes, including unweighted, scalar-weighted, and matrix-weighted objective functions. We will investigate the performance limit in terms of the theoretical error bound. We will also address problems caused by outliers or gross errors in input data, and the use of robust statistics to cope with these problems.

7.1 Stereo Camera Systems

Many of the previously published works use a simplified stereo camera system, in which two optical axes are parallel, and so are coordinate systems of two image planes. In reality, however, it is difficult to align two cameras physically so that thc intcrnal optical gcomctry of the cameras satisfies the above requirements. As a matter of fact, such an alignment is often not desirable. For example, the common field of view of the two cameras is larger if the two cameras gaze at the scene of interest with a convergence angle. We use a general stereo setup in our investigation.

Let the camera-centered coordinate system be fixed with the camera, the origin of the system coinciding with the optical center of the camera and the z-axis being aligned with the optical axis and pointing towards the scene, as shown in Fig. 7.1. As our convention, the coordinate system centered at the left camera is called the camera-centered

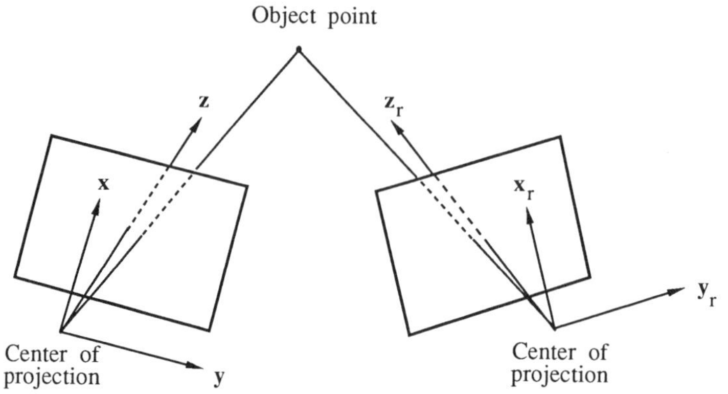

Fig. 7.1. Camera coordinate systems. The camera-centered coordinate system is centered at the left camera.

coordinate system and the coordinate system centered at the right camera is called the right-camera-centered coordinate system. In other words, we use the coordinate system centered at the left camera as the coordinate system of the entire stereo system. The stereo configuration, i.e., the orientation and position of the right camera, with respect to the left camera, is specified by a rotation matrix M, and a translation vector $\mathbf{B}$. Vector $\mathbf{x}_r = (x_r, y_r, z_r)^t$ represented in the right-camera-centered system is related to $\mathbf{x} = \mathbf{x}_l = (x_l, y_l, z_l)^t$ in the (left) camera-centered system by

$$\mathbf{x} = \mathbf{x}_l = M\mathbf{x}_r + \mathbf{B}, \tag{7.1.1}$$

where M and $\mathbf{B}$ are determined through camera calibration. In a pin-hole camera model, the image position of a point determines a projection line that passes through the optical center and the 3-D point. What is important is the direction of this projection line in the camera-centered system, and the focal length is immaterial. We can define a normalize the pin-hole camera model in which the focal length is equal to 1 and the image plane is at z=1. For a point $\mathbf{x} = (x, y, z)^t$, represented in the coordinate system centered at either camera, the corresponding image vector of the point $\mathbf{x}$ is:

$$\mathbf{X} = 1/z \ \mathbf{x} \tag{7.1.2}$$

which represents a vector starting from the origin and ending at the image point in the normalized camera model. The first two components $\mathbf{u} = (x/z, y/z)^t$ in $\mathbf{X}$ are defined as the image coordinates of the point. Thus, the depths z_l and z_r of the point, in the camera-centered and the right-camera-centered systems, respectively, can be determined from (7.1.1):

$$\mathbf{X}_l \; z_l = M \mathbf{X}_r \; z_r + \mathbf{B}. \tag{7.1.3}$$

This is a vector equation with three scalar equations and two unknowns. In the absence of noise, the three scalar equations are always consistent since the true depths give the solution. From (7.1.3) we can immediately obtain so called *epipolar constraint* illustrated in Fig. 7.2: $\mathbf{X}_l$, $M\mathbf{X}_r$ and $\mathbf{B}$ are coplanar (linearly dependent). Geometrically, the constraint means that two projection lines intersect in 3-D space. In the presence of noise, the epipolar constraint may be violated. This is caused by the error component orthogonal to the epipolar line in images (see Fig. 7.2).

7.2 Stereo Triangulation

We need to estimate the optimal 3-D position of the point from the noise contaminated observations, and assess the amount of error in the

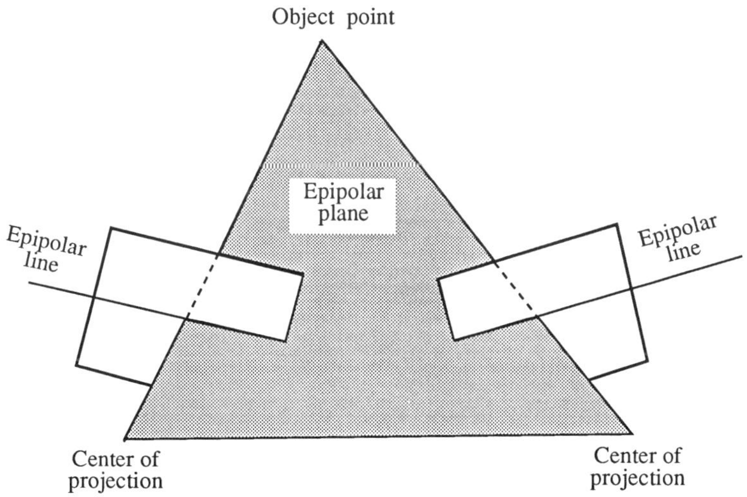

Fig. 7.2. Epipolar constraint.

estimates. Let the noise-contaminated projections in the left and right images be $\hat{\mathbf{u}}_l$ and $\hat{\mathbf{u}}_r$, respectively, and $\mathbf{u}_l$ and $\mathbf{u}_r$ be the corresponding noise-free ones. We have

$$\hat{\mathbf{u}}_r = \mathbf{u}_r + \boldsymbol{\delta}_r, \qquad \hat{\mathbf{u}}_l = \mathbf{u}_l + \boldsymbol{\delta}_l, \tag{7.2.1}$$

where $\boldsymbol{\delta}_r$ and $\boldsymbol{\delta}_l$ are additive noise vectors in the image plane. This noise accounts for image quantization noise, edge detecting error, feature matching error etc.

Suppose the correlation between components of the errors is negligible. Without loss of generality, we also assume the same error variance in the different components of image points (This is not true for CCD arrays with rectangular sensing cells, but an extension for this case is straight-forward). Let the estimated 3-D point $\hat{\mathbf{x}}$ have its projections $\mathbf{u}_l(\hat{\mathbf{x}})$ and $\mathbf{u}_r(\hat{\mathbf{x}})$ in left and right images, respectively. The observation vector in this problem consists of $\hat{\mathbf{u}}_l$ and $\hat{\mathbf{u}}_r$. Using the minimum variance estimation discussion in Section 4.5, the 3-D point $\hat{\mathbf{x}}$ should minimize

$$\|\hat{\mathbf{u}}_l - \mathbf{u}_l(\hat{\mathbf{x}})\|^2 + \|\hat{\mathbf{u}}_r - \mathbf{u}_r(\hat{\mathbf{x}})\|^2. \tag{7.2.2}$$

This is a nonlinear minimization problem. We first give an approximate solution. According to (7.1.3), solve for z_l and z_r which minimize

$$\|M\hat{\mathbf{X}}_r\, z_r + \mathbf{B} - \hat{\mathbf{X}}_l\, z_l\|. \tag{7.2.3}$$

This is a linear least-squares problem which has a closed-form solution. The estimated $\hat{\mathbf{x}}$ is then determined by

$$\hat{\mathbf{x}} = (M\hat{\mathbf{X}}_r z_r + \mathbf{B} + \hat{\mathbf{X}}_l z_l)/2. \tag{7.2.4}$$

The geometrical interpretation of this approximate solution is illustrated in Fig. 7.3. Due to noise, the two projection lines of the point from left and right cameras, respectively, do not intersect in space. The solution in (7.2.4) is the midpoint of the shortest line segment that connects these two projection lines. From this approximate solution, a few iterations can be performed to minimize (7.2.2). It has been observed that the solution in (7.2.4) is very close to the optimal solution that minimizes (7.2.2), unless the stereo configuration is very unusual (e.g., one camera is far behind the other camera). Therefore, in most cases the use of the approximate solution will not cause

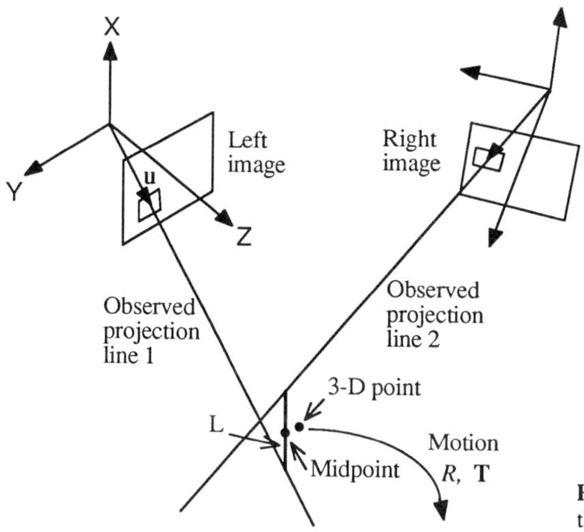

Fig. 7.3. Stereo triangulation in the presence of noise.

significant performance degradation. If the stereo correspondence satisfies the epipolar constraint, (7.1.3) is consistent, and thus, the depth z_l of a point can be directly determined based on the first two scalar equations of (7.1.3).

In summary, from a pair of stereo projections $\mathbf{u}_l$ and $\mathbf{u}_r$, we have constructed a function $\mathbf{c}(\mathbf{u}_l, \mathbf{u}_r)$ that gives the estimated 3-D position of a point:

$$\hat{\mathbf{x}} = \mathbf{c}(\hat{\mathbf{u}}_l, \hat{\mathbf{u}}_r). \tag{7.2.5}$$

This process is called stereo triangulation.

We take a look at the geometrical meaning of triangulation.

The stereo triangulation is very much like the motion case shown in Fig. 4.7. In motion analysis, the relative configuration between two cameras is the motion to be estimated. In stereo case, this relative configuration is called stereo configuration and is known. The observed position of the 3-D point in each image determine a projection line. Due to error in the image position of each point, the projection lines have certain uncertainty. The intersection of these two uncertain projection lines determines the uncertainty shape of the 3-D position of the point. In Fig. 4.7(b), the darker intensity indicates high probability density. In a typical stereo setup, the ratio of baseline length (the distance between the projection centers of two stereo cameras) to the depth of the point is around 10:1. This implies that the uncertainty shape is much more elongated than that shown in Fig. 4.7(b). The depth component can be 10 times less reliable than other two components.

Analytically, the uncertainty shape of the 3-D position of a point can be characterized by a covariance matrix. From (7.2.5), we have

$$\hat{\mathbf{x}} - \mathbf{x} \cong \frac{\partial \mathbf{c}(\hat{\mathbf{u}}_l, \hat{\mathbf{u}}_r)}{\partial \hat{\mathbf{u}}_l}(\hat{\mathbf{u}}_l - \mathbf{u}_l) + \frac{\partial \mathbf{c}(\hat{\mathbf{u}}_l, \hat{\mathbf{u}}_r)}{\partial \hat{\mathbf{u}}_r}(\hat{\mathbf{u}}_r - \mathbf{u}_r), \tag{7.2.6}$$

or,

$$\boldsymbol{\delta}_{\mathbf{x}} \cong \frac{\partial \mathbf{c}(\hat{\mathbf{u}}_l, \hat{\mathbf{u}}_r)}{\partial \hat{\mathbf{u}}_l}\boldsymbol{\delta}_{\mathbf{u}_l} + \frac{\partial \mathbf{c}(\hat{\mathbf{u}}_l, \hat{\mathbf{u}}_r)}{\partial \hat{\mathbf{u}}_r}\boldsymbol{\delta}_{\mathbf{u}_r}. \tag{7.2.7}$$

Denote the error covariance matrix of a vector $\mathbf{y}$ by $\Gamma_{\mathbf{y}}$. From (7.2.7) we have

$$\Gamma_{\mathbf{x}} \cong \frac{\partial \mathbf{c}(\hat{\mathbf{u}}_l, \hat{\mathbf{u}}_r)}{\partial \hat{\mathbf{u}}_l}\Gamma_{\mathbf{u}_l}\frac{\partial \mathbf{c}(\hat{\mathbf{u}}_l, \hat{\mathbf{u}}_r)^t}{\partial \hat{\mathbf{u}}_l} + \frac{\partial \mathbf{c}(\hat{\mathbf{u}}_l, \hat{\mathbf{u}}_r)}{\partial \hat{\mathbf{u}}_r}\Gamma_{\mathbf{u}_r}\frac{\partial \mathbf{c}(\hat{\mathbf{u}}_l, \hat{\mathbf{u}}_r)^t}{\partial \hat{\mathbf{u}}_r} \tag{7.2.8}$$

assuming $\boldsymbol{\delta}_{\mathbf{u}_l}$ and $\boldsymbol{\delta}_{\mathbf{u}_r}$ are uncorrelated.

According to the definition of covariance matrix, it is easy to see that a large uncertainty in depth component of $\mathbf{x}$ is indicated by a large element at the third row and third column in $\Gamma_{\mathbf{x}}$. If the uncertainty shape is titled, $\Gamma_{\mathbf{x}}$ will have large off-diagonal elements.

7.3 Closed-Form Solution

Due to the fact that the uncertainty shape of the 3-D position of a point is very elongated, proper weighting of each component is very important to motion estimation. In this section, we first discuss methods that give closed-form solution to motion parameters.

Suppose that the stereo camera system is moving in a static surrounding and a sequence of stereo images is taken. We need to estimate the motion of the camera system from two consecutive image pairs. In a camera-centered coordinate system, the scene is moving.

Equivalently, we determine the motion parameters of the scene in the camera-centered system. In this system, let a point $\mathbf{x}$ at time t_0 be moved to $\mathbf{x}'$ at time t_1. They are related by

$$\mathbf{x}' = R\mathbf{x} + \mathbf{T}, \tag{7.3.1}$$

where R is a rotation matrix and $\mathbf{T}$ is a translational vector. The objective is to determine R and $\mathbf{T}$ from a sequence of point correspondences: $\{(\mathbf{x}_i, \mathbf{x}_i')\}$.

7.3.1 A Matrix-Weighted Objective Function

In the presence of noise, we need to take into account the uncertainties in the points that are constructed by stereo triangulation. Using the estimated 3-D positions $\hat{\mathbf{x}}' = \mathbf{x}' + \boldsymbol{\delta}_{\mathbf{x}'}$ and $\hat{\mathbf{x}} = \mathbf{x} + \boldsymbol{\delta}_{\mathbf{x}}$, equation (7.3.1) gives

$$\hat{\mathbf{x}}' = R\hat{\mathbf{x}} + \mathbf{T} + \boldsymbol{\delta}, \tag{7.3.2}$$

where

$$\boldsymbol{\delta} = \boldsymbol{\delta}_{\mathbf{x}'} - R\,\boldsymbol{\delta}_{\mathbf{x}}. \tag{7.3.3}$$

Suppose that the errors in the observed points are uncorrelated between time t_0 and t_1. It follows from (7.3.3) that the residual vector $\boldsymbol{\delta}$ has a covariance matrix

$$\Gamma \triangleq \mathbb{E}\boldsymbol{\delta}\boldsymbol{\delta}^t = \Gamma_{\mathbf{x}'} + R\,\Gamma_{\mathbf{x}}R^{-1}. \tag{7.3.4}$$

We now suppose that the observation vector consists of a sequence of 3-D points at two time instants and that the errors in these observations are uncorrelated between different points and between different time instants. From (7.2.12) the motion parameters should minimize

$$\sum_{i=1}^{n} (R\hat{\mathbf{x}}_i + \mathbf{T} - \hat{\mathbf{x}}_i')^t\Gamma_i^{-1}(R\hat{\mathbf{x}}_i + \mathbf{T} - \hat{\mathbf{x}}_i'), \tag{7.3.5}$$

where n is the number of point correspondences and

$$\Gamma_i \triangleq \Gamma_{\mathbf{x}_i'} + R\,\Gamma_{\mathbf{x}_i}R^{-1}. \tag{7.3.6}$$

Remember that since the matrix R is a rotation matrix (orthonormal and right-handed), there are only three degrees of freedom in R.

Letting $\mathbf{a}$ denote a three dimensional vector consisting of the three independent parameters of the rotation matrix R, the expression (7.3.5) is a nonlinear function of a six-dimensional parameter vector

$$\mathbf{m} \triangleq (\mathbf{a}^t, \mathbf{T}^t)^t. \tag{7.3.7}$$

Thus, the objective is to determine $\mathbf{m}$ which minimizes the matrix weighted objective function (7.3.5).

7.3.2 Unweighted and Scalar-Weighted Versions

A special case of (7.3.5) is

$$\sum_{i=1}^{n} \|R\hat{\mathbf{x}}_i + \mathbf{T} - \hat{\mathbf{x}}_i'\|^2 \tag{7.3.8}$$

in that the weighting matrix Γ_i^{-1} is replaced by an identity matrix I. This objective function leads to an unweighted least-squares solution. Another special case of (7.3.5) is

$$\sum_{i=1}^{n} w_i^2 \|R\hat{\mathbf{x}}_i + \mathbf{T} - \hat{\mathbf{x}}_i'\|^2, \tag{7.3.9}$$

in that the weighting matrix Γ_i^{-1} is replaced by a scalar weight $w_i^2 I$. This leads to what is called a scalar-weighted least-squares solution.

However, as we discussed, since the depth component of a point is significantly less reliable than other two components, and the errors in the three components have considerable correlations (the uncertainty volume is elongated and tilted), an unweighed or even a scalar-weighted objective function cannot properly treat those uncertainties.

7.3.3 Matrix-Weighted Least-Squares Solution

In this subsection, we derive a closed-form solution for the matrix-weighted objective function (7.3.5). To do so, we need to simplify the weighting matrix in (7.3.6) so that it does not depend on the rotation matrix being computed. With a small rotation, the rotation matrix is roughly equal to an identity matrix, $R \approx I$, and the weighting matrix in (7.3.4) does not depend very much on R. The weighting matrix can be approximated by

$$\Gamma_i \triangleq \Gamma_{\mathbf{x}_i'} + \Gamma_{\mathbf{x}_i}. \tag{7.3.10}$$

If the rotation is so large that the simplification in (7.3.10) is not allowed, we can use the weighting matrix in (7.3.4) but R in (7.3.4) is replaced by a fixed rotation matrix which is estimated, e.g., by a closed-form scalar-weighted least-squares solution to be discussed in this subsection.

We first state a so called matrix-weighted centroid-coincidence theorem (MWCC theorem for short), whose proof is relegated to Appendix 7.A. The unweighted version of the centroid-coincidence theorem was proved in [7.9]. Here we extend the result to the more general matrix-weighted case.

MWCC Theorem. If R^* and $\mathbf{T}^*$ minimize (7.3.5) with the weighting matrix Γ_i^{-1} not depending on either R or $\mathbf{T}$, then the matrix-weighted centroids of $\{\hat{\mathbf{x}}_i'\}$ and $\{R^*\hat{\mathbf{x}}_i + \mathbf{T}^*\}$ must coincide:

$$\sum_{i=1}^{n} \Gamma_i^{-1}(R^*\hat{\mathbf{x}}_i + \mathbf{T}^*) = \sum_{i=1}^{n} \Gamma_i^{-1}\hat{\mathbf{x}}_i'. \tag{7.3.11}$$

Before looking for a closed-form solution to the matrix-weighted problem, we first consider the scalar-weighted least-squares solution that minimize (7.3.9). Replacing Γ_i^{-1} by matrix $(1/n)I$, the MWCC theorem gives the unweighted centroid-coincidence theorem presented in [7.9],

$$\frac{1}{n}\sum_{i=1}^{n} (R^*\hat{\mathbf{x}}_i + \mathbf{T}^*) = \frac{1}{n}\sum_{i=1}^{n} \hat{\mathbf{x}}_i'.$$

Replacing Γ_i^{-1} by a scalar matrix $w_i^2 I$, the MWCC theorem takes the form

$$\sum_{i=1}^{n} w_i^2(R^*\hat{\mathbf{x}}_i + \mathbf{T}^*) = \sum_{i=1}^{n} w_i^2\,\hat{\mathbf{x}}_i'$$

which can be rewritten as

$$R^*\sum_{i=1}^{n} w_i^2\hat{\mathbf{x}}_i + \sum_{i=1}^{n} w_i^2\mathbf{T}^* = \sum_{i=1}^{n} w_i^2\,\hat{\mathbf{x}}_i'. \tag{7.3.12}$$

Letting $\bar{\mathbf{x}} \triangleq \sum_{i=1}^{n} w_i^2\hat{\mathbf{x}}_i / \sum_{i=1}^{n} w_i^2$ and $\bar{\mathbf{x}}' \triangleq \sum_{i=1}^{n} w_i^2\hat{\mathbf{x}}_i' / \sum_{i=1}^{n} w_i^2$, it follows from (7.3.12) that

$$\mathbf{T}^* = \overline{\mathbf{x}}' - R^* \overline{\mathbf{x}}. \tag{7.3.13}$$

Then,

$$\begin{aligned} w_i(R^* \hat{\mathbf{x}}_i + \mathbf{T}^* - \hat{\mathbf{x}}_i') &= w_i(R^* \hat{\mathbf{x}}_i + \overline{\mathbf{x}}' - R^* \overline{\mathbf{x}} - \hat{\mathbf{x}}_i') \\ &= R^* w_i(\hat{\mathbf{x}}_i - \overline{\mathbf{x}}) - w_i(\hat{\mathbf{x}}_i' - \overline{\mathbf{x}}'). \end{aligned} \tag{7.3.14}$$

If R^* and $\mathbf{T}^*$ minimize the scalar-weighted objective function (7.3.9), we conclude from (7.3.14) that R^* must minimize

$$\sum_{i=1}^{n} \| R w_i(\hat{\mathbf{x}}_i - \overline{\mathbf{x}}) - w_i(\hat{\mathbf{x}}_i' - \overline{\mathbf{x}}') \|^2. \tag{7.3.15}$$

Noticing that the term under the summation has a form $\|R\mathbf{x} - \mathbf{y}\|^2$, we have

$$\begin{aligned} \|R\mathbf{x} - \mathbf{y}\|^2 &= (R\mathbf{x} - \mathbf{y})^t (R\mathbf{x} - \mathbf{y}) \\ &= \mathbf{x}^t R^t R \mathbf{x} - 2\mathbf{x}^t R^t \mathbf{y} + \|\mathbf{y}\|^2. \end{aligned} \tag{7.3.16}$$

Because R is orthonormal, $R^t R = I$, (7.3.16) is a linear function in the elements of R. The rotation matrix R^* that minimizes (7.3.15) can be solved for in a closed-form by the method presented in Appendix 7.B. An alternative way to solve for R that minimizes (7.3.15) uses singular value decomposition, which is presented in [7.1]. Once R^* is determined, $\mathbf{T}^*$ is determined based on (7.3.13).

We now consider the matrix-weighted least-squares solution. Since Γ_i^{-1} is a positive definite matrix, there is a matrix W_i such that

$$\Gamma_i^{-1} = W_i^t W_i.$$

Because Γ_i^{-1} is just a 3×3 matrix, such a W_i can be computed by a non-iterative algorithm. The objective function (7.3.5) can then be rewritten as

$$\sum_{i=1}^{n} \| W_i(R\hat{\mathbf{x}}_i + \mathbf{T} - \hat{\mathbf{x}}_i') \|^2. \tag{7.3.17}$$

From (7.3.11) of the MWCC theorem, we have

$$\sum_{i=1}^{n} \Gamma_i^{-1} R^* \hat{\mathbf{x}}_i + \sum_{i=1}^{n} \Gamma_i^{-1} \mathbf{T}^* = \sum_{i=1}^{n} \Gamma_i^{-1} \hat{\mathbf{x}}_i'.$$

It follows that

$$\mathbf{T}^* = (\sum_{i=1}^{n} \Gamma_i^{-1})^{-1} \sum_{i=1}^{n} \Gamma_i^{-1} \hat{\mathbf{x}}_i' - (\sum_{i=1}^{n} \Gamma_i^{-1})^{-1} \sum_{i=1}^{n} \Gamma_i^{-1} R^* \hat{\mathbf{x}}_i . \quad (7.3.18)$$

Substituting $\mathbf{T}$ in (7.3.17) by the right-hand side of (7.3.18) we get an expression which is a quadratic in the elements of R. This implies that when using general matrix weights, we cannot simplify the objective function to a linear expression in the elements of R as we did for the unweighted or the scalar-weighted cases in (7.3.16). This is due to the fact that matrix multiplication is generally not commutative. To give a concise form of the quadratic expression, we represent a rotation matrix by the corresponding vector as follows. For any matrix $R = [\mathbf{R}_1 \ \mathbf{R}_2 \ \mathbf{R}_3]^t$, we define

$$\mathbf{R} \triangleq \begin{bmatrix} \mathbf{R}_1 \\ \mathbf{R}_2 \\ \mathbf{R}_3 \end{bmatrix}.$$

Using this notation, we have

$$R\mathbf{x} = C(\mathbf{x})\mathbf{R}$$

we define a mapping from a three-dimensional vector $\mathbf{x}$ to a 3×9 matrix $C(\mathbf{x})$ as

$$C(\mathbf{x}) = \begin{bmatrix} \mathbf{x} & 0 & 0 \\ 0 & \mathbf{x} & 0 \\ 0 & 0 & \mathbf{x} \end{bmatrix}^t .$$

From (7.3.18), it follows that

$$\mathbf{T}^* = (\sum_{i=1}^{n} \Gamma_i^{-1})^{-1} \sum_{i=1}^{n} \hat{\mathbf{x}}_i' - (\sum_{i=1}^{n} \Gamma_i^{-1})^{-1} \sum_{i=1}^{n} \Gamma_i^{-1} C(\hat{\mathbf{x}}_i) \mathbf{R}^*$$

$$\triangleq \mathbf{d} - D\mathbf{R}^* . \quad (7.3.19)$$

where we define

$$\mathbf{d} = (\sum_{i=1}^{n} \Gamma_i^{-1})^{-1} \sum_{i=1}^{n} \hat{\mathbf{x}}_i'$$

and

$$D = (\sum_{i=1}^{n} \Gamma_i^{-1})^{-1} \sum_{i=1}^{n} \Gamma_i^{-1} C(\hat{\mathbf{x}}_i).$$

Then

$$W_i(R^* \hat{\mathbf{x}}_i + \mathbf{T}^* - \hat{\mathbf{x}}_i') = W_i(C(\hat{\mathbf{x}}_i)\mathbf{R}^* + \mathbf{d} - D\mathbf{R}^* - \hat{\mathbf{x}}_i')$$

$$= W_i(C(\hat{\mathbf{x}}_i) - D)\mathbf{R}^* - W_i(\hat{\mathbf{x}}_i' - \mathbf{d}) \triangleq A_i\mathbf{R}^* - \mathbf{b}_i. \tag{7.3.20}$$

with our definitions $A_i = W_i(C(\hat{\mathbf{x}}_i) - D)$ and $\mathbf{b}_i = W_i(\hat{\mathbf{x}}_i' - \mathbf{d})$. Given n point correspondences, define a new matrix A and a new vector $\mathbf{b}$ by

$$A \triangleq \begin{bmatrix} A_1 \\ A_2 \\ \vdots \\ A_n \end{bmatrix} \quad \text{and} \quad \mathbf{b} \triangleq \begin{bmatrix} \mathbf{b}_1 \\ \mathbf{b}_2 \\ \vdots \\ \mathbf{b}_n \end{bmatrix}. \tag{7.3.21}$$

According to (7.3.20) and (7.3.21), if we substitute R^* and $\mathbf{T}^*$ into the objective function in (7.3.17), we get

$$\|A\mathbf{R}^* - \mathbf{b}\|^2. \tag{7.3.22}$$

Therefore, the solution to the rotation matrix R^* must be such that (7.3.22) is minimized. The 9-dimensional vector $\mathbf{R}^*$ in (7.3.22) is subject to the constraint that it represents a rotation matrix. An iterative algorithm is required to search for a $\mathbf{R}^*$ that satisfies the constraint and minimizes (7.3.22). To avoid iteration, we first solve for an intermediate $\hat{\mathbf{R}}$ that minimizes (7.3.22) without the constraint:

$$\hat{\mathbf{R}} = (A^tA)^{-1}A^t\mathbf{b}$$

$$= \left\{\sum_{i=1}^{n} [C(\hat{\mathbf{x}}_i) - D)]^t \Gamma_i^{-1} [C(\hat{\mathbf{x}}_i) - D)]\right\}^{-1} \left\{\sum_{i=1}^{n} [C(\hat{\mathbf{x}}_i) - D)]^t \Gamma_i^{-1} [\hat{\mathbf{x}}_i' - \mathbf{d}]\right\}.$$

Then the rotation matrix R^* is determined by minimizing

$$\|R^* - \hat{R}\|^2 \tag{7.3.23}$$

using the method presented in Appendix 7.B. Finally, the translation is determined according to (7.3.19):

$$\mathbf{T}^* = \mathbf{d} - D\,\mathbf{R}^* . \tag{7.3.24}$$

Since the constraint in the corresponding vector of the rotation matrix is not considered in minimizing (7.3.22) and, instead, it is compensated later in minimizing (7.3.23), the performance is penalized. However, such a penalty is expected to be much less significant than the penalty otherwise caused by improper weighting. The simulation results presented in Section 7.6 have showed that this closed-form solution is significantly more reliable than both unweighted and scalar-weighted least-squares solutions.

7.3.4 Uniqueness

Because noise may cause a degenerate matrix to become nondegenerate, the uniqueness question should be studied in the absence of noise. A rigid motion can be uniquely determined from the correspondences of three noncollinear points. However, the above algorithm for the matrix-weighted solution requires at least 4 point correspondences as we will see in the following.

In (7.3.22), $\mathbf{R}^*$ is a 9-dimensional vector. Each point correspondence gives 3 scalar equations in the corresponding noise-free equation:

$$A\,\mathbf{R}^* = \mathbf{b}. \tag{7.3.25}$$

It seems that 3 point correspondences might be enough to uniquely determine $\mathbf{R}^*$. However, this is not the case, since the matrix A may not have a full rank. We consider a noise-free case, for which we can replace Γ_i^{-1} by $(1/n)I$ for all i. Based on the derivation of the closed-form solution of the matrix-weighted objective function, it can be seen that (7.3.25) is equivalent to the noise-free equation that corresponds to (7.3.15):

$$R\,(\mathbf{x}_i - \overline{\mathbf{x}}) = (\mathbf{x}_i' - \overline{\mathbf{x}}'). \tag{7.3.26}$$

Three points are always coplanar, and so are these three points plus their centroid $\overline{\mathbf{x}}$. Therefore, the three vectors $\mathbf{x}_i - \overline{\mathbf{x}}$, i=1, 2, 3, are coplanar. Consequently, R cannot be uniquely determined by three point correspondences based on (7.3.26) without imposing a constraint on R. With four point correspondences, and if the four vectors $\mathbf{x}_i - \overline{\mathbf{x}}$, i=1, 2, 3, 4, are not coplanar, R is uniquely determined by

(7.3.26). Therefore, four point correspondences are enough in general to uniquely determine the intermediate matrix $\hat{R}$. The fitting of a rotation matrix to the intermediate matrix $\hat{R}$ will improve the estimate of the rotation matrix, but will not affect the uniqueness. Once the rotation matrix is determined, the translation vector is also determined by (7.3.13), or (7.3.24) for the matrix-weighted solution.

In summary, while 3 is the minimum number of point correspondences needed for the unweighted or scalar-weighted solutions discussed in Subsection 7.3.3, the closed-form matrix-weighted least-squares solution requires at least 4 point correspondences. This is due to the fact that the constraint on the rotation matrix R is not considered in solving (7.3.22).

7.4 Iterative Optimal Solution

The solutions discussed in Section 7.3 are based on equation (7.3.2) and the observation is directly defined as the left-hand side of the equation. An observation vector defined directly from a simple basic equation results in a simpler relation between the parameters to be estimated and the observation vectors. This simplicity may be helpful to find a closed-form solution. However, the observation vectors defined this way often have complicated noise terms (for example, in (7.3.3) the noise term is related to the unknown parameters.) This causes difficulties in estimating the statistics of the noise term. Often, some simplifications have to be made in estimating those statistics, and consequently, the corresponding solution to the problem is not exactly optimal.

7.4.1 A Basic Objective Function

In this section, we define the observation vectors directly as the measured image coordinates of points. Since the function that maps the unknown parameters to the observations is nonlinear, some iterations have to be used by the algorithm. As a good initial guess is provided by a method discussed in Section 7.3, the convergence of the iterative algorithm is generally fairly reliable and relatively fast.

From two pairs of stereo images with correspondences, the parameters to be estimated are the structure of the points

$$\mathbf{x} \triangleq (\mathbf{x}_1, \mathbf{x}_2, \cdots, \mathbf{x}_n) \tag{7.4.1}$$

and the motion parameter vector **m** as defined in (7.3.7). Let **t** denote all the parameters to be estimated from two stereo pairs,

$$\mathbf{t} \triangleq (\mathbf{x}, \mathbf{m}), \tag{7.4.2}$$

and the two-dimensional image coordinate vector of the i-th point in image j (j=1 for left image and j=2 for right image) at time t_k be $\mathbf{u}_{i,j,k}$. Suppose observation vector **u** consists of all image vectors at time t_0 and t_1. Given **t**, the (noise-free) observation vector **u** can be directly determined by

$$\mathbf{u} = \mathbf{f}(\mathbf{t}). \tag{7.4.3}$$

In the presence of noise, the contaminated observation vector $\hat{\mathbf{u}}$ is given by

$$\hat{\mathbf{u}} = \mathbf{f}(\mathbf{t}) + \boldsymbol{\delta}_{\mathbf{u}}. \tag{7.4.4}$$

The noise term $\boldsymbol{\delta}_{\mathbf{u}}$ accounts for measurement noise in the image plane. Supposing $\boldsymbol{\delta}_{\mathbf{u}}$ has an approximately zero mean, and a covariance matrix $\sigma^2 I$, as discussed in Section 7.2 the optimal **t** minimizes

$$\|\mathbf{u} - \mathbf{f}(\mathbf{t})\|^2. \tag{7.4.5}$$

This objective function is based on the 2-D image plane noise model (7.4.4) and the complicated 3-D uncertainties in the measured 3-D points will be implicitly taken into account. The matrix-weighted or the scalar-weighted least-squares solution is used as an initial guess for an iterative algorithm (e.g., the Levenberg-Marquardt method or conjugate gradient method) that improves the initial guess by minimizing (7.4.5). The expected error in the parameter **t** is provided by a covariance matrix similar to (7.2.14).

However, the following points must be considered.

(1) The above method is computationally expensive, mainly due to the large dimension of the parameter space. For two pairs of stereo images with n point correspondences, the parameter space is $(3n+6)$-dimensional. For example, with 20 point correspondences, the iterative algorithm has to search in a 66-dimensional space!

(2) The direct extension to long image sequences is more computationally prohibitive. If many images are used, **u** includes all image points that have ever appeared. The number of such points may be

extremely large in the case of extended navigation. Due to occlusions or other reasons, a point can disappear and reappear many times in an image sequence. In other words, there is only a moderately large number of points that are currently visible at each time instant.

(3) The model is not suited for recursive computation. In extended navigation, it is impossible to store all the data. The old information should be stored in a concise manner and used efficiently. For example, the structure estimated based on the motion from t_0 to t_1 needs to used for the motion from t_1 to t_2.

We present in the next subsection a modified method, in which the structure of 3-D points is not included in the search space, and the corresponding model can be directly extended to recursive computation for long image sequences.

7.4.2 Optimization Using Space Decomposition

Let $\mathbf{x}_{i,k}$ denote the true 3-D position of the i-th point at time t_k, in the camera-centered coordinate system, and let $\mathbf{x}_{\bullet,k}$ denote the collection of all such points at time t_k. Given two stereo image pairs, the parameter vector $\mathbf{t}$ to be estimated consists of the interframe motion parameter vector $\mathbf{m}$, and the structure of the points $\mathbf{x}_{\bullet,0}$ at time t_0: $\mathbf{t} = (\mathbf{m}, \mathbf{x}_{\bullet,0})$ (or equivalently, we can consider the structure at time t_1). Let $\mathbf{u}_{i,j,k}$ denote the image plane vector of i-th point on j-side (j=1 for left image and j=2 for right image) at time t_k. The set of direct observation vectors consists of all noise corrupted version $\hat{\mathbf{u}}_{i,j,k}$ of $\mathbf{u}_{i,j,k}$ with k from 0 to 1. If the noise in the components of those observations is uncorrelated and has the same variance σ^2, according to minimum variance estimation, the objective function to minimize is

$$f(\mathbf{m}, \mathbf{x}_{\bullet,0}) \triangleq \sum_{i=1}^{n}\sum_{j=1}^{2}\sum_{k=0}^{1}\sigma^{-2}\|\mathbf{u}_{i,j,k}(\mathbf{m}, \mathbf{x}_{i,0}) - \hat{\mathbf{u}}_{i,j,k}\|^2 \tag{7.4.6}$$

where $\mathbf{u}_{i,j,k}(\mathbf{m}, \mathbf{x}_{i,0})$ is the noise-free projection computed from $\mathbf{m}$ and $\mathbf{x}_{i,0}$.

The above model is a natural model with two stereo pairs. Another alternative model, although less natural, is useful for recursive estimation from long image sequences. In this latter model, instead of treating the two stereo image pairs symmetrically, the observations at time t_0 are represented in the estimated structure $\hat{\mathbf{x}}_{i,0}$. The estimate $\hat{\mathbf{x}}_{i,0}$ and its estimated covariance matrix $\Gamma_{\mathbf{x}_{i,0}}$ can be determined by the methods discussed in Section 7.2. If $\hat{\mathbf{x}}_{i,0}$ is considered as an "observation" at time t_0, according to the minimum variance estimation, the objective function to be minimized is

$$f(\mathbf{m}, \mathbf{x}_{\bullet,0}) \triangleq \sum_{i=1}^{n} \{(\mathbf{x}_{i,0} - \hat{\mathbf{x}}_{i,0})^t \Gamma_{\mathbf{x}_{i,0}}^{-1} (\mathbf{x}_{i,0} - \hat{\mathbf{x}}_{i,0})$$

$$+ \sum_{j=1}^{2} \sigma^{-2} \| \mathbf{u}_{i,j,1}(\mathbf{m}, \mathbf{x}_{i,0}) - \hat{\mathbf{u}}_{i,j,1} \|^2\} \tag{7.4.7}$$

where we assume that (a) errors in $\hat{\mathbf{x}}_{\bullet,0}$ estimated from images before t_1 are uncorrelated between different points, (b) errors in the measured image coordinates of points $\{\hat{\mathbf{u}}_{i,j,1}\}$ are all uncorrelated and have the same variance σ, and (c) the errors in the previously estimated $\hat{\mathbf{x}}_{\bullet,0}$ are uncorrelated with those in the currently measured image coordinates. Due to the non-symmetrical nature of the distribution of errors in 3-D coordinates of a point constructed from triangulation, the objective function in (7.4.7) is not as good as that in (7.4.6). But it can be directly extended to highly efficient recursive estimation for long image sequences, because every interframe motion can be recursively regarded as a minimization problem with an objective function similar to (7.4.7).

The objective functions in (7.4.6) and (7.4.7) are neither linear nor quadratic in $\mathbf{m}$ and $\mathbf{x}_{\bullet,0}$, and therefore, an iterative algorithm is required to compute the solution: $\mathbf{m}$ and $\mathbf{x}_{\bullet,0}$. Instead of performing a computationally expensive direct optimization, we reduce the dimension of parameter search space first. Since the objective functions are continuous, we have

$$\min_{\mathbf{m}, \mathbf{x}_{\bullet,0}} f(\mathbf{m}, \mathbf{x}_{\bullet,0}) = \min_{\mathbf{m}} \{\min_{\mathbf{x}_{\bullet,0}} f(\mathbf{m}, \mathbf{x}_{\bullet,0})\} \triangleq \min_{\mathbf{m}} g(\mathbf{m}) \tag{7.4.8}$$

where

$$g(\mathbf{m}) \triangleq \min_{\mathbf{x}_{\bullet,0}} f(\mathbf{m}, \mathbf{x}_{\bullet,0}) \tag{7.4.9}$$

is the smallest "cost", computed by choosing the "best" structure $\mathbf{x}_{\bullet,0}$, with a given motion parameter vector $\mathbf{m}$. This is the space decomposition technique we used in Chapter 4. Fig. 4.8 illustrates this space decomposition. The space $(\mathbf{m}, \mathbf{x}_{\bullet,0})$ can be decomposed into two subspaces, corresponding to $\mathbf{m}$ and $\mathbf{x}_{\bullet,0}$, respectively. In the subspace of $\mathbf{m}$, an iterative algorithm (e.g, the Levenberg-Marquardt method or the conjugate gradient method) is used. In the subspace of $\mathbf{x}_{\bullet,0}$, a non-iterative method is used that gives the best $\mathbf{x}_{\bullet,0}$ for any given $\mathbf{m}$. According to the decomposition shown in (7.4.8), the search space in $\min_{\mathbf{m}} g(\mathbf{m})$ is just the 6-dimensional. With a good initial guess of $\mathbf{m}$ provided by the matrix-weighted solution discussed in Section 7.3, very few iterations are needed to reach the optimal solution.

Now we consider how to compute the best $\mathbf{x}_{\bullet,0}$ in (7.4.9) without resorting to iterations. In (7.4.7), there are two terms for each point, one is a matrix-weighted discrepancy of $\mathbf{x}_{i,0} - \hat{\mathbf{x}}_{i,0}$

$$(\mathbf{x}_{i,0} - \hat{\mathbf{x}}_{i,0})^t \, \Gamma_{\mathbf{x}_{i,0}}^{-1}(\mathbf{x}_{i,0} - \hat{\mathbf{x}}_{i,0}), \tag{7.4.10}$$

the other is

$$\sum_{j=1}^{2} \sigma^{-2} \|\mathbf{u}_{i,j,1}(\mathbf{m}, \mathbf{x}_{i,0}) - \hat{\mathbf{u}}_{i,j,1}\|^2. \tag{7.4.11}$$

The latter term corresponds to the triangulation of the point at time t_1. We have derived a method in Subsection 7.2.3 to give an estimate $\hat{\mathbf{x}}_{i,1}$ that minimizes (7.4.11). The corresponding error covariance matrix $\Gamma_{\mathbf{x}_{i,1}}$ of the point $\hat{\mathbf{x}}_{i,1}$ can also be estimated. In other words, we have two sample data for the same parameter vector $\mathbf{x}_{i,0}$. One is $\mathbf{p} \triangleq \hat{\mathbf{x}}_{i,0}$ with error covariance matrix $\Gamma_{\mathbf{p}} \triangleq \Gamma_{\mathbf{x}_{i,0}}$ and the other is the point moved back from $\hat{\mathbf{x}}_{i,1}$, $\mathbf{q} \triangleq R^t(\hat{\mathbf{x}}_{i,1} - \mathbf{T})$, with the error covariance matrix $\Gamma_{\mathbf{q}} = R^t \Gamma_{\mathbf{x}_{i,1}} R$. According to (7.2.6), the optimal $\mathbf{x}_{i,0}$ should minimize

$$(\mathbf{x}_{i,0} - \mathbf{p})^t \Gamma_{\mathbf{p}}^{-1}(\mathbf{x}_{i,0} - \mathbf{p}) + (\mathbf{x}_{i,0} - \mathbf{q})^t \Gamma_{\mathbf{q}}^{-1}(\mathbf{x}_{i,0} - \mathbf{q}). \tag{7.4.12}$$

The first term in (7.4.12) is exactly (7.4.10) and it can be proved that the second term in (7.4.12) is an approximation of (7.4.11). While minimizing the sum of (7.4.10) and (7.4.11) requires an iterative algorithm, according to (4.5.3) the optimal $\mathbf{x}_{i,0}^*$ that minimizes (7.4.12) is directly given by

$$\begin{aligned}
\mathbf{x}_{i,0}^* &= \Gamma_{\mathbf{q}}(\Gamma_{\mathbf{p}} + \Gamma_{\mathbf{q}})^{-1}\mathbf{p} + \Gamma_{\mathbf{p}}(\Gamma_{\mathbf{p}} + \Gamma_{\mathbf{q}})^{-1}\mathbf{q} \\
&= (\Gamma_{\mathbf{p}} + \Gamma_{\mathbf{q}} - \Gamma_{\mathbf{p}})(\Gamma_{\mathbf{p}} + \Gamma_{\mathbf{q}})^{-1}\mathbf{p} + \Gamma_{\mathbf{p}}(\Gamma_{\mathbf{p}} + \Gamma_{\mathbf{q}})^{-1}\mathbf{q} \\
&= \mathbf{p} + \Gamma_{\mathbf{p}}(\Gamma_{\mathbf{p}} + \Gamma_{\mathbf{q}})^{-1}(\mathbf{q} - \mathbf{p}).
\end{aligned} \tag{7.4.13}$$

Thus, given any motion parameter vector $\mathbf{m}$, a sequence of best points is determined based on (7.4.13) and the corresponding residual in (7.4.7) can be computed. When the best $\mathbf{m}$ is determined by an iterative algorithm to minimize the residual in (7.4.7), the corresponding set of points is the best solution for the structure. This algorithm will be used in Section 7.8 where long image sequences are utilized.

The space decomposition for the objective function in (7.4.6) is analogous, where both terms in the form of (7.4.11) lead to (7.4.12), and the estimated structure is given by (7.4.13). It can be seen that, given motion parameter vector $\mathbf{m}$, the structure is determined based on a 3-D symmetrical noise model (represented by covariance matrices) in order to eliminate iterations in the subspace of $\mathbf{x}_{i,0}$. Nevertheless, the motion parameter vector $\mathbf{m}$ is determined based on a non-symmetrical model (non-symmetrical in 3-D) in (7.4.6) or (7.4.7), since $\mathbf{m}$ is determined by minimizing f in (7.4.8). The non-symmetrical model has been used for both time instants in (7.4.6) and for the current time instant in (7.4.7). An iterative procedure is used to determine the best $\mathbf{m}$ that minimizes (7.4.6) with the corresponding $\mathbf{x}$ computed from $\mathbf{m}$ in a closed-form according to the space decomposition. This algorithm is called iterative optimization algorithm.

In summary, through our investigation of the nature of the objective functions (7.4.6) and (7.4.7), we have explored the relationships between $\mathbf{m}$ and $\mathbf{x}_{\bullet,0}$ so that the constraint between them is utilized. The computationally almost intractable optimization problem is then solved by a low dimensional iterative algorithm.

7.4.3 Estimating Errors

Since the actual errors in the solutions depend on random noise, it is reasonable to estimate the expected errors. More specifically, we estimate the covariance matrix of the estimated parameters. These covariance matrices not only give the expected accuracy of the estimates, they are also important for further estimation using the obtained estimates.

In the objective function in (7.4.7), each point has three "observations": $\hat{\mathbf{x}}_{i,0}$, $\hat{\mathbf{u}}_{i,1,1}$ and $\hat{\mathbf{u}}_{i,2,1}$. We define a 7-dimensional observation vector $\mathbf{u}_i$ which consists of these three observations. Then, $\hat{\mathbf{u}}$ in (7.4.4) consists of $\mathbf{u}_i$, $i=1, 2, \cdots, n$. Suppressing $\mathbf{x}_{\bullet,0}$ in (7.4.4) we have

$$\hat{\mathbf{u}} = \mathbf{f}(\mathbf{m}) + \boldsymbol{\delta}_{\mathbf{u}} \tag{7.4.14}$$

where $\mathbf{f}(\mathbf{m})$ is the computed observation from $\mathbf{m}$ (and $\mathbf{x}_{\bullet,0}$ which is computed from $\mathbf{m}$ as discussed in Subsection 7.4.2).

In the objective function in (7.4.6), each point has four "observations". With $\mathbf{u}_i$ corresponding to an 8-dimensional vector, (7.4.14) still applies to (7.4.6).

The estimated error covariance matrix $\Gamma_{\mathbf{m}}$ of motion parameter vector $\mathbf{m}$ is given in (4.6.1), where J is replaced by

$$\frac{\partial \mathbf{f}(\hat{\mathbf{m}})}{\partial \hat{\mathbf{m}}}$$

which is evaluated with the optimal estimates of $\mathbf{m}$ and $\mathbf{x}_{\bullet,0}$. According to the assumption that the image plane noise in $\boldsymbol{\delta}_{\mathbf{u}}$ are uncorrelated between different points, the covariance matrix of $\boldsymbol{\delta}_{\mathbf{u}}$ is a diagonal matrix.

For efficiency, the error covariance matrix of a point $\mathbf{x}_{i,0}$ is estimated based on the space-decomposition method. In Subsection 7.4.3, the error covariance matrix of $\mathbf{q}$ was given by $\Gamma_{\mathbf{q}} = R^t \Gamma_{\mathbf{x}_{i,1}} R$, which is a conditional covariance matrix conditioned on the given motion parameters. For error estimation here, the covariance matrix of $\mathbf{q}$ is unconditional and should take into account the errors in the estimated motion parameters. From the definition of vector $\mathbf{q}(\mathbf{m}) = R^{-1}(\hat{\mathbf{x}}_{i,0} - \mathbf{T})$, the covariance matrix of $\mathbf{q}$ should be

$$\Gamma_{\mathbf{q}} = R^{-1}\Gamma_{\mathbf{x}_{i,1}} R + \frac{\partial \mathbf{q}(\mathbf{m})}{\partial \mathbf{m}} \Gamma_{\mathbf{m}} \frac{\partial \mathbf{q}(\mathbf{m})^t}{\partial \mathbf{m}}.$$

Using (4.5.4), the error covariance matrix of the estimated 3-D position of the point $\mathbf{x}^*_{i,0}$ in (7.4.13) is estimated by

$$\begin{aligned}
\Gamma_{\mathbf{x}^*_{i,0}} &= \Gamma_{\mathbf{p}}(\Gamma_{\mathbf{p}} + \Gamma_{\mathbf{q}})^{-1}\Gamma_{\mathbf{q}} \\
&= \Gamma_{\mathbf{p}}(\Gamma_{\mathbf{p}} + \Gamma_{\mathbf{q}})^{-1}(\Gamma_{\mathbf{p}} + \Gamma_{\mathbf{q}} - \Gamma_{\mathbf{p}}) \\
&= \Gamma_{\mathbf{p}} - \Gamma_{\mathbf{p}}(\Gamma_{\mathbf{p}} + \Gamma_{\mathbf{q}})^{-1}\Gamma_{\mathbf{p}}.
\end{aligned}$$

The diagonal elements of this covariance matrix give the expected error variances of the corresponding components of the estimated vector.

7.5 Outliers and Robust Estimators

If the data are contaminated by well-behaved noise, the accuracy of the minimum variance optimal solution, discussed in the above section, is very good. What we mean by "well-behaved" is that the noise is limited to a small range and it is highly random and uncorrelated (or the correlation is known so that noise can be whitened). As simulations show, the error has essentially reached the Cramér-Rao lower error bound for any unbiased estimator under white Gaussian noise. We have also observed that its performance is not very sensitive to the distribution of noise, as long as the noise is well-behaved. For example, the average error of the solution almost remain the same, as shown in our experiments, when noise distribution is changed from Gaussian distribution to a uniform distribution with the same variance. However, if the data contain gross errors, the above solution is significantly affected.

As an example, let us suppose that we need to estimate the rigid motion from a set of 5 noisy coplanar points to another set of 5 points, as shown in Fig. 7.4. If the noise in the positions of the points is well-behaved, the least-squares fitting is generally good. However, if one of the point contains gross error, the least-squares fitting leads to very erroneous motion parameters as shown in Fig. 7.4. This implies that a conventional least-squares solution is very sensitive to gross errors.

Most structure and motion estimation methods presented in the literature are sensitive to gross errors or outliers, including all algorithms we discussed so far. Such a high sensitivity to gross errors is due to the use of conventional least-squares estimators. While a least-squares estimator is quite stable in the presence of well-behaved noise, it gives disastrous results when the input data are contaminated with a few outliers. In fact, when the gross error goes to infinity, so

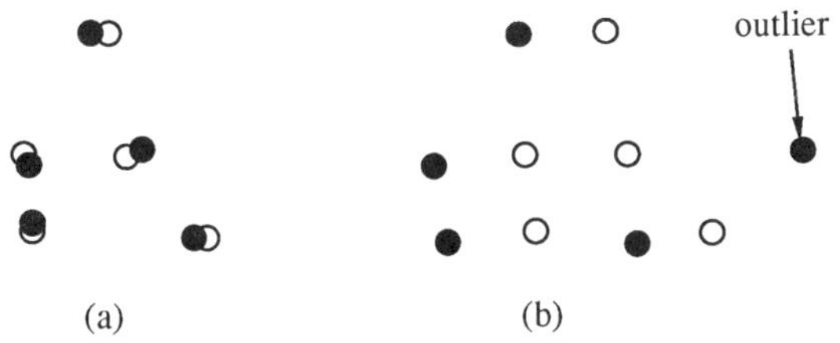

Fig. 7.4. Estimating rigid motion from one point set (white) to another point set (black) by least-squares fitting. **(a)** Good fit and accurate motion estimation with well-behaved noise. **(b)** Bad fit and inaccurate motion estimation with outliers.

does the error of the least-squares solution. In structure and motion analysis, outliers arise from feature mismatching, occlusion and other inadvertent events. Due to difficulties in stereo and temporal image matching, such outliers cannot always be easily eliminated within the feature matching stage. Therefore, immunity to outliers is essential to motion and structure estimation algorithms.

What one can do in the presence of gross errors is to detect the very bad data and appropriately reduce the contribution of these data to the solution. This is the basic idea behind the robust statistics. The term "robust" is used in this book in its general meaning, but the terms "robust estimator", "robust statistics" and "robust method" are reserved for those that are designed for handling gross errors or outliers.

7.5.1 Some Basic Concepts of Robust Statistics

Let us first consider the weighted least-squares estimate $\hat{x}$ from a series of samples $\{x_i\}$. The estimate $\hat{x}$ minimizes

$$\sum_i \rho(\hat{x} - x_i) = \sum_i w_i^2 (\hat{x} - x_i)^2 \tag{7.5.1}$$

where w_i is the weight for x_i to be determined, and $\rho(x) = w_i^2 x^2$. Taking the derivative with respect to $\hat{x}$ and setting the derivative to zero, we get

$$\sum_i \psi(\hat{x} - x_i) = 2\sum_i w_i^2 (\hat{x} - x_i) = 0 \tag{7.5.2}$$

where $\psi(x)$ is the derivative of $\rho(x)$:

$$\psi(x) = \frac{d\rho(x)}{dx}. \tag{7.5.3}$$

In robust statistics, $\psi(x)$ has several slightly different forms [7.10]. A typical one is the Tukey's biweight:

$$\psi_{tuk}(s) = \begin{cases} s(1 - s^2)^2 & \text{if } |s| \le 1 \\ 0 & \text{otherwise.} \end{cases} \tag{7.5.4}$$

Tukey's biweight $\psi(x)$ is normalized so that $s \ge 1$ corresponds to outliers. Letting c be the desired threshold for outliers and

$$s_i = (\hat{x} - x_i)/c,$$

one can substitute $\psi(\hat{x} - x_i)$ in (7.5.2) by the normalized Tukey's biweight $\psi_{tuk}(s_i)$ and thus get

$$\sum_i \psi_{tuk}(s_i) = 0. \tag{7.5.5}$$

The last equation in (7.5.2) can be rewritten as

$$\sum_i w_i^2 s_i = 0. \tag{7.5.6}$$

Comparing (7.5.5) and (7.5.6), and the form of the Tukey's biweight in (7.5.4), we know that the weight w_i^2 in (7.5.1) should be

$$w_i^2 = \begin{cases} (1 - s_i^2)^2 & \text{if } |s_i| \le 1 \\ 0 & \text{otherwise.} \end{cases} \tag{7.5.7}$$

This determines the robust weights w_i^2 in (7.5.1).

The weight in (7.5.7) suppresses residual $\hat{x} - x_i$ whose absolute value is large, relative to c. Particularly, if the residual exceeds the threshold c, the weight becomes zero. This means that any residual whose absolute value is equal to or large than c is completely discarded. This is how outliers are detected and suppressed.

7.5.2 A Robust Method for Motion Estimation

The robust estimator has a weighting profile that reduces or completely eliminates the influence of outliers. Although the robust estimator is theoretically immunc to a few gross errors, our experience indicates that the convergence problem of the robust estimator needs special attention. Due to varying weighting in robust methods, the convergence behavior of a robust estimator is more complicated than the least-squares counterpart. The flat-out shape of the weighting function implies that the objective function is relatively flat around the solution. This is a factor that may cause divergence. Another factor is that the initial guess provided is typically far away from the true value due to the outliers. With such a poor initial guess, it is difficult to converge to the global optimum solution. The detection of outliers also poses a challenging problem: A large residual error of certain equation does not necessarily correspond to an outlier. Incorrectly suppressing a large equation residual that does not correspond to an outlier leads to poor performance or divergence.

In order to successfully identify the outliers and improve the convergence, we formulate the robust estimation in the following way. Instead of concentrating on the residual errors of some motion equations, we investigate the residual errors directly on image plane, i.e., the place where measurement of the point data is taken. A large residual in image plane indicates the inconsistency between the measured data and the computed projection. An outlier in the measured image point accompanies a large inconsistency. This inconsistency in image plane bears a more direct link to outliers than the equation residuals of some other motion equations.

From (7.4.14), we define the residual vector, corresponding to a parameter vector $\mathbf{m}$, as

$$\mathbf{r}(\mathbf{m}) \triangleq \hat{\mathbf{u}} - \mathbf{f}(\mathbf{m}).$$

Notice that $\mathbf{r}(\mathbf{m})$ is not equal to $\boldsymbol{\delta}_{\mathbf{u}}$ in general, since the former is the residual corresponding to the parameter vector $\mathbf{m}$ that is not necessarily correct, and the latter is the actual noise. They are equal when $\mathbf{m}$ is the true parameter vector. In a least-squares solution as indicated by (7.4.5), one determines $\mathbf{m}$ which minimizes

$$\|\mathbf{r}(\mathbf{m})\|^2 = \sum_i (r_i(\mathbf{m}))^2$$

where $r_i(\mathbf{m})$ is the i-th component of $\mathbf{r}(\mathbf{m})$. Using robust weighting, we minimize instead

$$\sum_i w_i^2 (r_i(\mathbf{m}))^2 \tag{7.5.8}$$

where the robust weights w_i^2 are determined by (7.5.7) and $s_i(\mathbf{m}) = r_i(\mathbf{m})/c$. The value of c should be related to the scale (or standard deviation) of the well-behaved portion of the noise.

Since the median m of the absolute residual $\{|r_i|\}$ is a robust estimator of the standard deviation [7.10], we let $c = km$, where k is a positive value which is selected to include most useful data. For example, if the well-behaved portion of the noise can be well approximated by a Gaussian distribution, k can be selected based on χ^2 test so that the probability for the well-behaved residuals to fall out of km is sufficiently small. An overly small k will cause a waste of many useful data and an overly large k makes the estimator more sensitive to gross errors. Typically k may range from 3 to 5.

7.5.3 Using Closed-Form Solution

Another problem with the robust estimator is the convergence. Our experience indicates that the solution provided by a non-robust method (the one that does not take outliers into account) cannot lead to the correct convergence in minimizing (7.5.8) even if outliers are removed in latter iterations. This is of course attributed to the nonlinearity of the problem. But bad initial **m** provided an algorithm that cannot handle outliers is also a major reason. Therefore, a closed-form solution must be embedded into the iterative computation of robust estimates.

To improve the convergence of the iterative algorithm, we compute the initial **m** using a non-iterative algorithm that also uses robust weights. That is, the closed-form solution is computed to minimize (7.3.5) except that the weighting matrix Γ_i^{-1} in (7.3.5) is replaced by $w_i' \Gamma_i^{-1}$ where w_i' is the minimum of the weights w_i^2 in (7.5.7), that correspond to four projections of the i-th point (left, right, before and after motion), respectively. The weight w_i' is taken as the minimum because the point is an outlier whenever one of its projection is an outlier. With such a modification of weighting matrices, the closed-form matrix-weighted least-squares solution discussed in this chapter can still be directly used. The solution of this step can beforehand suppress the outliers and provide a good initial guess for the minimization of (7.5.8).

7.6 Examples

The performance of the above algorithm can be demonstrated through simulations, where ground truth and the amount of noise can be well controlled and the errors in the estimates can be accurately measured. Experiments with real images, on the other hand, intuitively indicate the performance of the algorithms, and test the complete approach in realistic situations.

7.6.1 Simulations

As our convention, the error in the rotation is measured as the relative error. The error measure under this definition gives a stable and meaningful measure over a wide range of rotation angles and therefore, can be used as a normalized measure to compare the accuracy between different rotations. But it is also useful to look into the accuracy of different components in a rotation. Here, we represent rotation by three angles: roll (γ), pitch (β) and yaw (α), which represent rotation angles about z, y, and x axis, respectively. The corresponding rotation matrix R is related to those three angles by the following equation,

$$R(\alpha, \beta, \gamma) = \begin{bmatrix} \cos\gamma & -\sin\gamma & 0 \\ \sin\gamma & \cos\gamma & 0 \\ 0 & 0 & 1 \end{bmatrix} \begin{bmatrix} \cos\beta & 0 & \sin\beta \\ 0 & 1 & 0 \\ -\sin\beta & 0 & \cos\beta \end{bmatrix} \begin{bmatrix} 1 & 0 & 0 \\ 0 & \cos\alpha & -\sin\alpha \\ 0 & \sin\alpha & \cos\alpha \end{bmatrix}. \tag{7.6.1}$$

The vector $\mathbf{a} = (\alpha, \beta, \gamma)$ was also used as the rotation parameter vector for iterative optimization (Section 7.4). The error in translation vector is defined as the norm of the difference vector between the estimated and true vectors. We do not use relative error for translation vector because this measure is very sensitive to the actual magnitude of translation. Unless stated otherwise, the unit used in this section is meter. The error in the image projections of points is simulated by additive zero-mean independent Gaussian noise. For the iterative optimization, the ''dunlsf'' subroutine in IMSL library [7.17, 18, 5] was used.

The 3-D points are generated randomly for each trial, between depth 2 and depth 15, with a uniform distribution. The field of view (side-to-side) of the cameras with a square image plane is about 53°. The two stereo cameras are arranged in such a way that the optical axes of two cameras intersect at the center of the block where random points are generated. They are separated along y-axis with a distance specified as baseline length (i.e., $\mathbf{B}$ in (7.1.1) is aligned with y-axis and the norm of $\mathbf{B}$ is called the baseline length). Only those points that fall into the field of view of both stereo cameras at t_0 and t_1 are used for motion analysis. The variance of Gaussian noise added to the image points is equal to that of the uniform digitization noise with a 256×256-pixel image. The average errors we will show are obtained through 1000 random trials each with a different set of 3-D points. The motion parameters are specified as the motion of points in the camera-centered coordinate system. In the data shown in the following Figs. 7.5 to 7.7, the motion is a rotation about a rotation axis (1, 0.2, 0.1) by an angle of 8°, followed by a translation of (–0.139, 1.352, –0.918). For the matrix-weighted least-squares solution, the matrix weight in (7.3.10) is used.

Unweighted, scalar-weighted, matrix-weighted, and optimal solutions. Fig. 7.5 shows the average error in motion parameters versus the number of point correspondences (equal to the number of points in each image). As show in Figs. 7.5(a) and (b), with a minimum of 4 point correspondences, the error in matrix-weighted least-squares solution is the largest, even larger than the unweighted least-squares solution. This is due to the fact that given 4 point correspondences, the closed-form matrix-weighted solution has no redundancy for computing the intermediated rotation matrix, while unweighted and scalar-weighted methods do (see Subsection 7.3.4). However, the closed-form matrix-weighted solution out-performs both unweighted and scalar-weighted solutions if the number of point correspondences is beyond 6. To show the data more clearly, Figs. 7.5(c) and (d) present the results starting with 6 point correspondences.

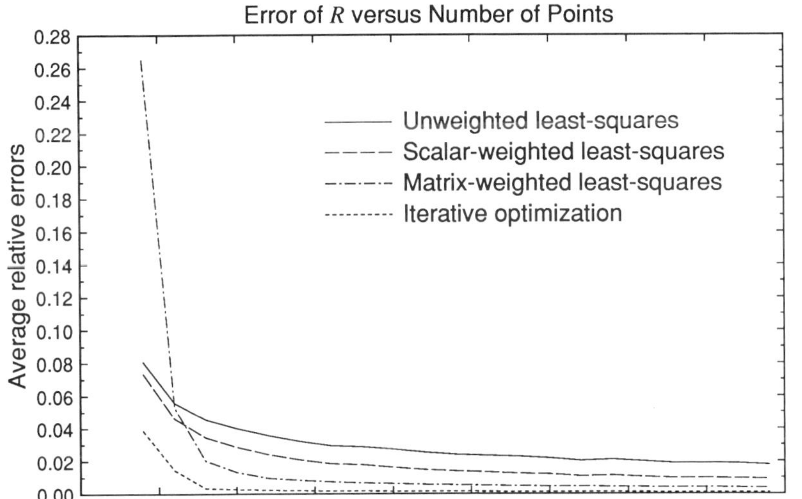

Fig. 7.5(a)

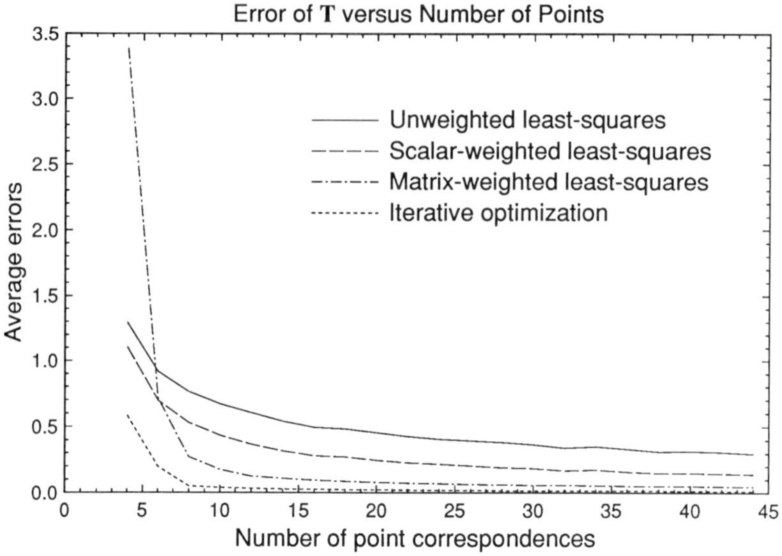

Fig. 7.5(b)

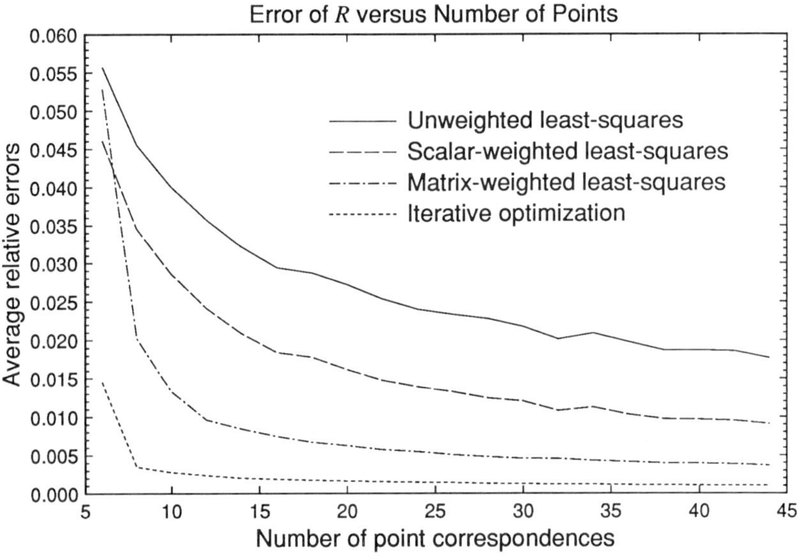

Fig. 7.5(c)

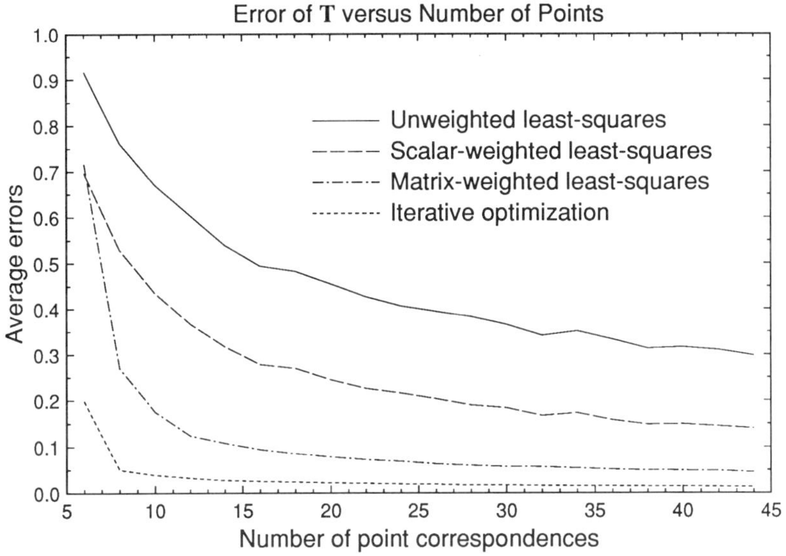

Fig. 7.5(d)

Fig. 7.5. Performance of four different solutions: unweighted least-squares, scalar-weighted least-squares, matrix-weighted least-squares and iterative optimization, versus the number of point correspondences. **(a)** Error of rotation matrix R. **(b)** Error of translation vector **T**. **(c)** Error of R starting from 6 point correspondences. **(d)** Error of **T** starting from 6 point correspondences. Baseline length = 0.5.

It can also been seen from Fig. 7.5 that the scalar-weighted solution, weighted by the reciprocal of the depth of each point after motion in the camera-centered coordinate system, is considerably better than the unweighted least-squares solution. With sufficiently many point correspondences, the new closed-form matrix-weighted solutions reduce the errors, on average, by about 80% compared with the unweighted solutions and by about 65% compared with the scalar-weighted solutions. The amount of improvement is remarkable. Therefore, the matrix-weighted method presented in this chapter is by far the best non-iterative method. At the cost of more computation, the results can be further improved by the iterative optimization algorithm presented in Section 7.4, as shown in Fig. 7.5.

Performance bounds. With errors in observations, it is impossible (zero probability) to compute the true parameters. Cramér-Rao bound gives a lower bound on the expected error in any estimated parameters. With Gaussian noise in observation, we evaluated the Cramér-Rao bound for any unbiased estimator as plotted in Fig. 7.6. The bound on the norm of the vector is equal to the trace of the bound matrix. The bound on the rotation matrix error is computed from the bound on the three Euler angles of 3-D rotation. No unbiased estimator can have expected parameter errors lower than that given by the bound. In Fig. 7.6, the average Cramér-Rao bound is compared with average errors based on 1000 random trials. It can be seen that the average errors of the iterative optimization solutions are very close to the theoretical bound. The matrix-weighted solutions still leave some gap to the bound but not large.

Baseline length. The effects of the baseline length are shown in Fig. 7.7. It can be seen that while the baseline is extended, the errors are consistently reduced. The improvement ratio of the matrix-weighted solutions, as well as the iterative optimization solutions, over unweighted solutions keep roughly constant when the baseline is extended to over 2 meters. Therefore, it is expected that a similar amount of improvement will be observed for typical navigation setups, where maximum object-to-camera distances are always considerably larger than the baseline length. It is obvious that a longer baseline will reduce the uncertainty of points in stereo triangulation. However, stereo matching will be more difficult with a wide baseline.

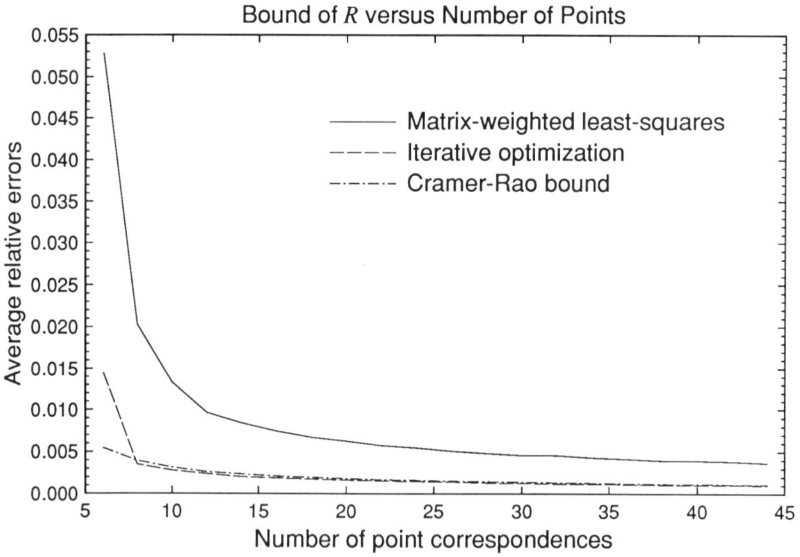

Fig. 7.6(a)

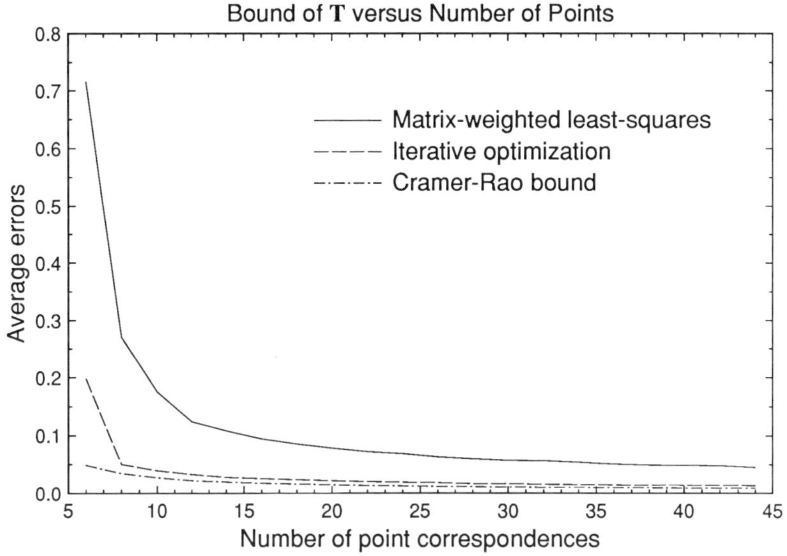

Fig. 7.6(b)

Fig. 7.6. Cramér-Rao bound and the performance achieved. (**a**) Error of rotation matrix R. (**b**) Error of translation vector **T**. Baseline length = 0.5.

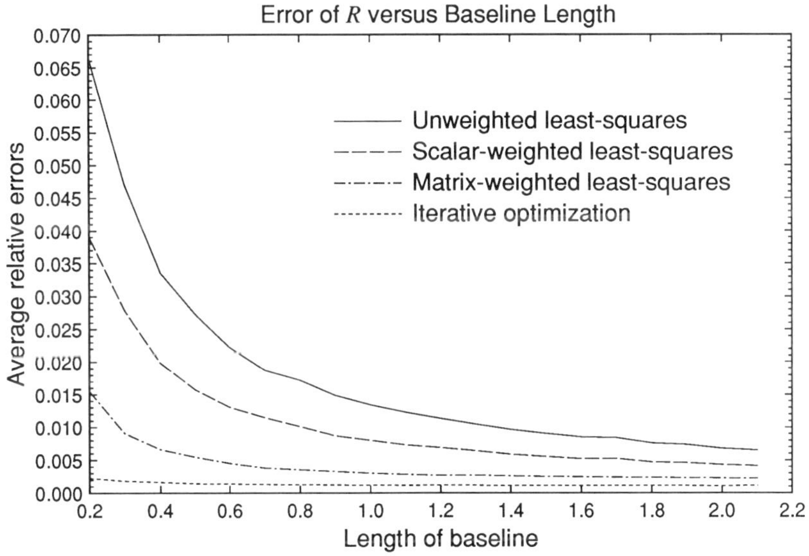

Fig. 7.7(a)

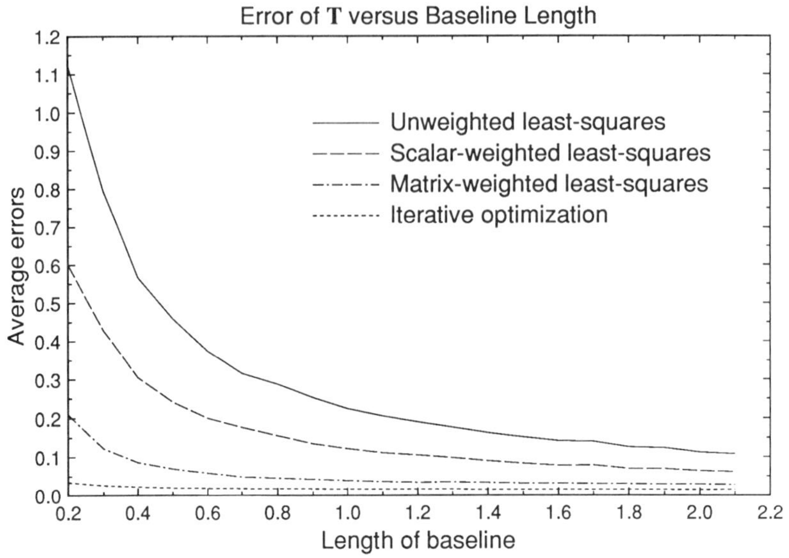

Fig. 7.7(b)

Fig. 7.7. Errors versus the length of the stereo baseline. **(a)** Error of rotation matrix R. **(b)** Error of translation vector $\mathbf{T}$. Number of point correspondences = 40.

The errors in the components of motion parameters are shown in Fig. 7.8, where the iterative optimization method is used. A somewhat surprising phenomenon in Fig. 7.8 is that the z-component in translation is the most reliable, except for very short baselines. This is contrary to the observation that the z-component of a point is the most unreliable one, and this may contribute to an unreliable z-component in translation. In fact, this phenomenon is caused by the correlation between rotation and translation: The images generated by a rotation about x-axis is similar to those generated by a translation in the direction of the y-axis. In other words, the error in yaw is strongly correlated with that in the y-component of translation. Such a strong correlation is confirmed by the covariance matrix of the estimated motion parameters. This accounts for the fact that the corresponding components are the most unreliable in Figs. 7.8(a) and (b), respectively. A pitch (rotation about the y-axis) and a translation in the x-direction in the camera-centered coordinate system (centered at the left camera) will result in similar images in the right camera, but with less similarity compared with the above case, since (a) the baseline is along the y-axis which allows the second camera to look from the side of the motion direction, and (b) the two cameras have a non-zero convergence angle (optical axes pointing to the center of points instead of being parallel), which makes the difference of two motions more observable from the second camera. Consequently, the pitch component of rotation and the corresponding x-component of translation are the second unreliable component in Figs. 7.8(a) and (b), respectively. The roll component of a rotation is more or less directly observable from images, it is the most reliable component in Fig. 7.8(a). Based on the above discussion, the z-component of translation is relatively reliable unless the baseline is so short that errors in the z-component of points significantly undermine the z-component of the estimated translation (e.g., when baseline has a length of 0.2 in Fig. 7.8).

Robust method. Simulation has confirmed that although the optimal solution has essentially reached the theoretical bound with a Gaussian distribution, it cannot handle gross errors. In the presence of outliers, the errors of the non-robust optimal solution are very large and the iterations often cannot converge. Even in those cases where the iteration converges, the error in the solution are mostly larger than that of the closed-form unweighted least-squares solution. In Fig. 7.9, the closed-form least-squares solution is compared with the robust solution. Fig. 7.9 shows the average errors of the motion parameters with the number of point correspondences from 20 to 50. To simulate the situations of gross error, 30% of the points after motion are converted

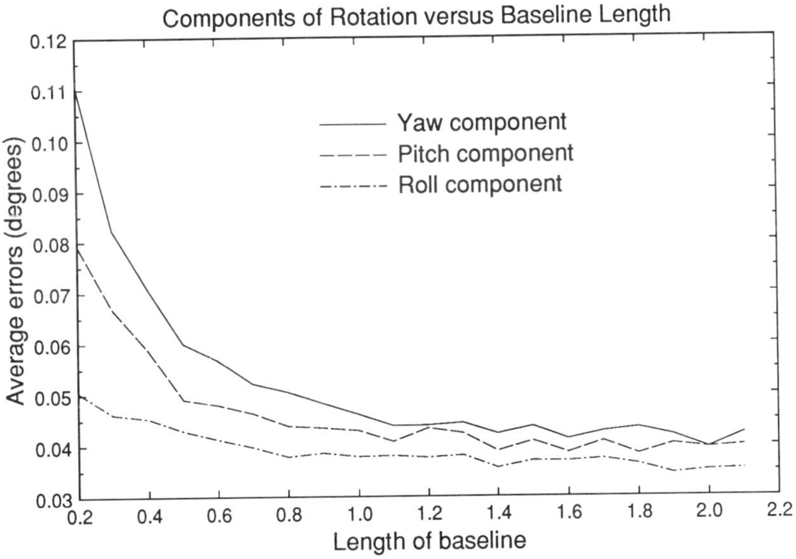

Fig. 7.8(a)

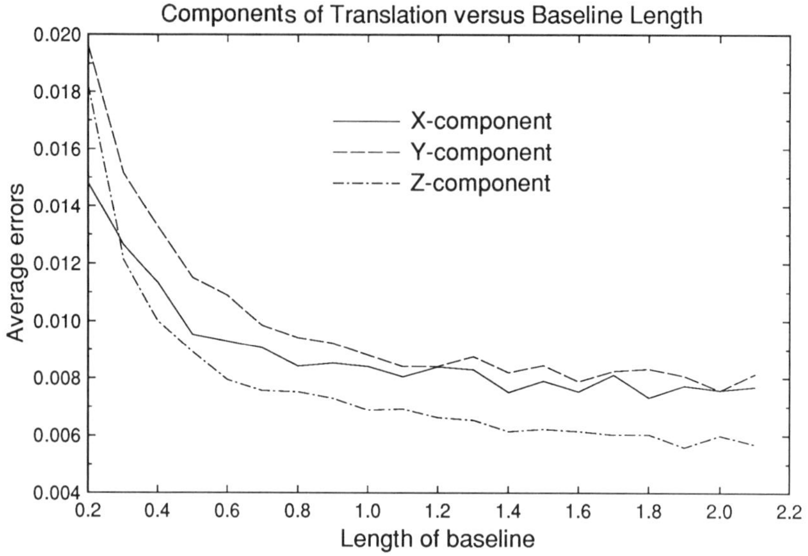

Fig. 7.8(b)

Fig. 7.8. Errors in the components of motion parameters estimated by the iterative optimization versus the length of the stereo baseline. **(a)** Error of rotation represented by roll-pitch-yaw. **(b)** Error of translation vector **T**. Number of point correspondences = 40.

to outliers, which is done by randomly moving their coordinates by a distance of a quarter of the image. For each number of point correspondences, 1000 trials were recorded to compute the mean errors. The results of the robust method were computed by our matrix-weighted algorithm with a robust weighting as discussed in Section 7.5, and then an iterative procedure (''dunlsf'' non-linear least-squares subroutine in IMSL) is called to minimize (7.5.8). From Fig. 7.9 we can see that the least-squares solution is extremely bad in the presence of gross errors but the robust solution remain relatively stable. The matrix-weighted least-squares solution is much less sensitive to outliers than the unweighted and scalar-weighted solutions. The errors of the matrix-weighted solution in the presence of outliers are about twice as large as those of the robust method.

7.6.2 Experiments with a Real Stereo Setup

Before computing depth map and motion parameters, we must first calibrate stereo cameras. We first briefly describe our calibration. Then the calibrated stereo setup will be used for motion and structure estimation.

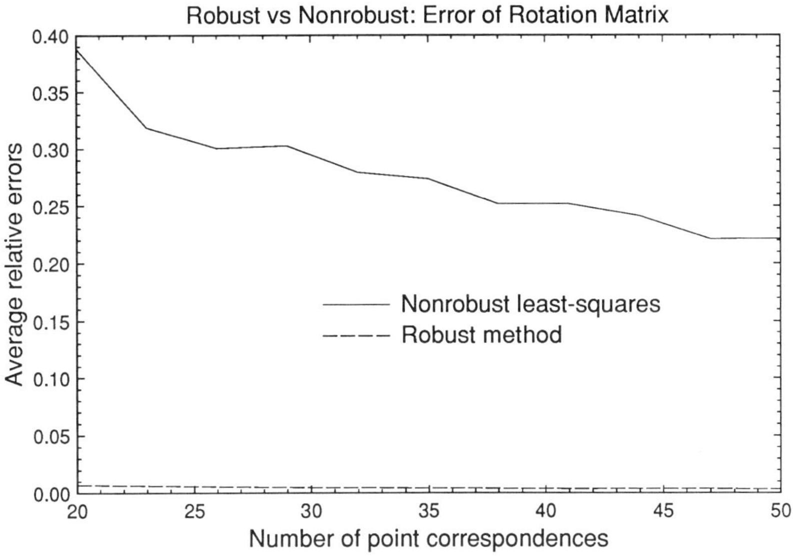

Fig. 7.9(a)

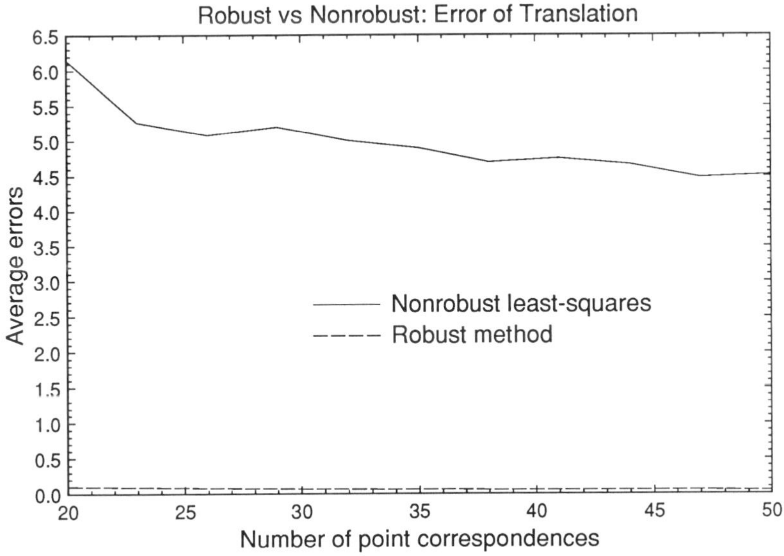

Fig. 7.9(b)

Fig. 7.9. Errors of the motion parameters estimated by the least-squares method and the robust method, versus the number of point correspondences. Rotation axis: (1, 0.2, 0.1). Rotation angle 8°. Translation: (–0.138770, 2.052886, –0.918066). **(a)** Error of rotation matrix R. **(b)** Error of translation vector $\mathbf{T}$.

Calibration. The setup of our experiment is as follows. Two COSMICA f=8.5mm CCD cameras were mounted on a high-precision six-joint IBM robot arm to form a stereo configuration. Each digital image grabbed from each camera has 480×512 pixels. The motion of the cameras can be controlled by the robot with precision. The two cameras are calibrated using a calibration method [7.26]. Each camera has four basic internal parameters (focal lengths in row and column numbers, and the coordinates of the principal point), and five distorsion parameters to compensate for radial and tangential distortions. The distortion compensation is as much as 23 pixels at the border of the image plane. To determine the 3-D position and orientation of the optical system in a world coordinate system, each camera has six independent external parameters, which are represented by a rotation axis, a rotation angle and a translation vector. The relative configuration between two cameras, M and $\mathbf{B}$ in (7.1.1), was directly computed from the external parameters of the two cameras. The baseline length is about 100mm.

Motion estimation with real world images. Two stereo pairs are shown in Fig. 7.10. After taking the first pair of stereo images, the robot arm was controlled to rotate about a vertical axis by an angle of –2.25° with first joint. But due to the fact that the arm was raised upward, the rotation axis in the camera was significantly away from vertical. If one views at the camera-centered coordinate system, the scene rotates about this axis, by an angle of 2.25° and then translates roughly to the left. The rotation axis and translation vector were not known precisely. After the motion, a second pair of stereo images was taken.

With two pairs of stereo images taken at two time instances t_0 and t_1, respectively, four-way matching is possible: two stereo matchings and two temporal matchings, as shown in Fig. 7.11. The two stereo matchings are: stereo matching before motion, and stereo matching after motion. The two temporal matchings are: left temporal matching, and right temporal matching. Of course, three-way matching is enough for motion analysis. For example, two stereo matchings give 3-D positions of a point at two time instances, and one temporal matching, either left or right, links these 3-D positions. For the sake of symmetry and robustness, we compute four-way matching instead, which provides certain degree of overdetermination. Obviously, due to noise, four-way matching may result in matching inconsistency. In our algorithm, inconsistent matches are averaged to give consistent

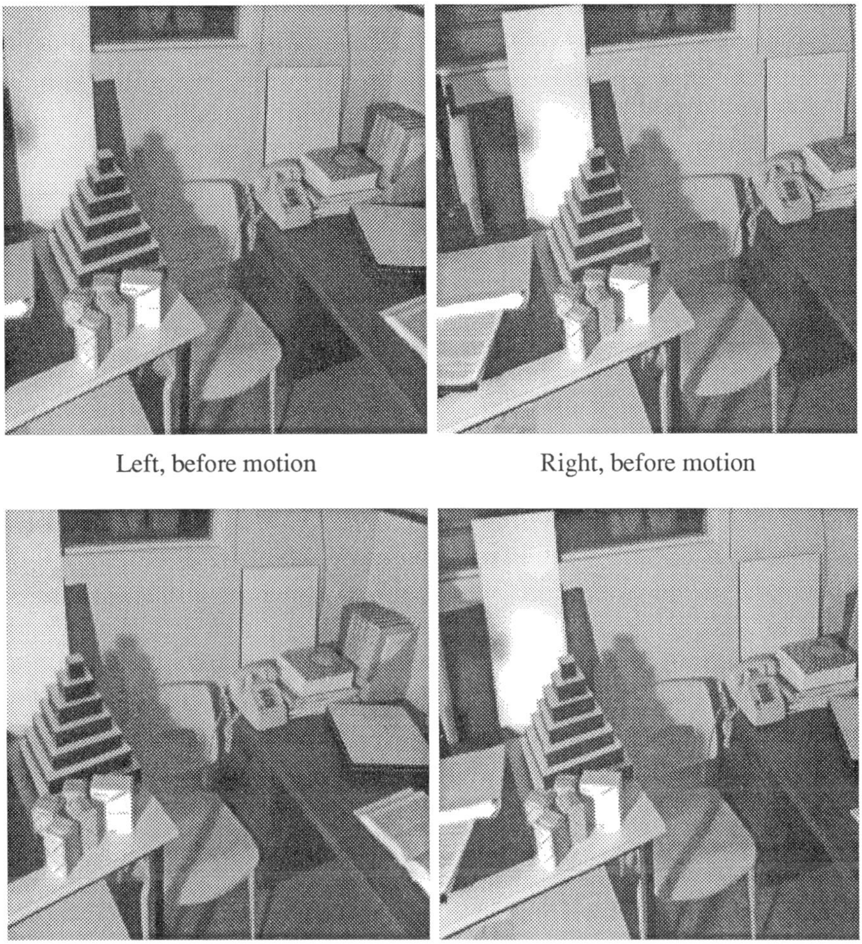

Fig. 7.10. Two stereo pairs. Upper row: stereo images before motion. Lower row: stereo images after motion.

matches.

The image matching algorithm presented in Chapter 2 was used to compute this four-way matching. It establishes correspondence for every pixel. To speed up the motion estimation, 151 equally sampled pixel points were automatically selected, out of 480×512 points, for motion analysis. In our experiment, five solutions were computed, three non-iterative ones and two iterative ones. Three non-iterative solutions are the unweighted, scaler weighted and matrix-weighted solutions. Two iterative solutions are the iterative optimal solution and the robust solution. Table 7.1 lists the the estimated angle by

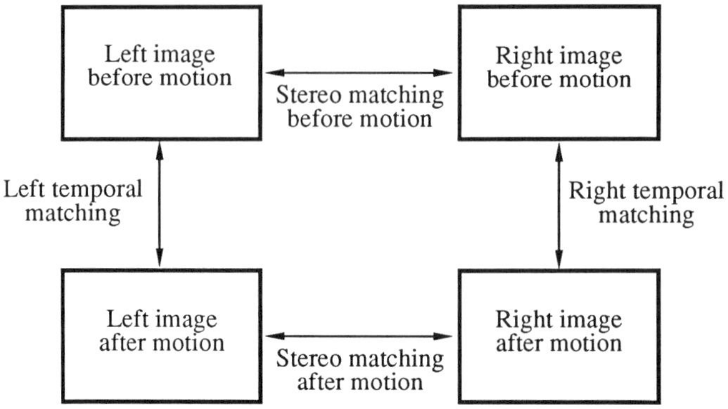

Fig. 7.11. Stereo matching and temporal matching.

each algorithm and the error. The unweighted and scalar weighted methods result in over 7° and 10°, respectively. This indicates that the solutions of unweighted and scalar-weighted solutions are hardly usable. The errors in rotation angles by other three solutions are of the similar magnitude, around one third of a degree. Table 7.2 shows complete motion parameters estimated by each algorithm. We can see that the translation vectors estimated by unweighted and scalar weighted methods are way off the expected values (the second component should be dominant and negative). Although the translation vectors estimated by matrix-weighted method, iterative optimization and robust method all show a dominant negative y-component, they are not very close. Based on our simulation with outliers, we believe that the estimates of matrix-weighted and robust methods are more credible. In fact, these two solutions are quite close to each other, but not to that of the iterative optimization. It should be noticed that in our setup, a short 100mm baseline, similar to that of human stereo vision, was used. This makes the ratio of average depth to baseline length about 12:1, a very large ratio which makes the uncertainty shape of each 3-D point be very elongated along depth direction. This explains why unweighted and scalar-weighted methods failed completely.

Table 7.2 also shows the computer time (user time + system time) used by each algorithm on a SUN-4 workstation. Although the matrix-weighted algorithm took longer than other two algorithms (0.35 sec versus 0.06 sec), it seems fast enough for real-time applications. Due to need of iteration, the iterative optimization and robust method

Table 7.1. Estimated rotation angles and errors

Methods	Estimated	Error
Unweighted LS	9.79°	7.54°
Scalar-weighted LS	12.78°	10.53°
Matrix-weighted LS	1.98°	-0.27°
Optimization	2.58°	0.33°
Robust	1.96°	-0.29°

Table 7.2. Motion parameters estimated and computer time consumed

Methods	Unweighted	Scalar-weighted	Matrix-weighted
Rotation axis	0.949 0.292 0.121	0.933 0.291 0.211	0.836 0.303 -0.457
Rotation angle	9.79°	12.78°	2.98°
Translation (mm)	-78.9 145.3 11.0	-106.9 201.2 17.5	-11.7 -51.6 -5.5
Computer time (sec)	0.06	0.06	0.35
Methods	Optimization	Robust	
Rotation axis	0.894 0.036 0.446	0.819 0.176 0.545	
Rotation angle	2.58°	1.96°	
Translation (mm)	5.9 -34.3 -15.7	-5.5 -54.3 -6.5	
Computer time (sec)	40.94	13.92	

took much longer 40.94 seconds and 13.92 seconds, respectively. Note that these times do not include image matching, which we think should be computed by a special purpose real-time hardware, instead of a general purpose computer.

Depth map from stereo images. In the above experiment, we have computed stereo matching using the pair of calibrated cameras. The stereo matching is represented by the displacement field which defines a 2-D disparity vector for every pixel. For each pixel, we can compute the 3-D position of the corresponding point through stereo triangulation. We computed a dense depth map which is defined at a density of 480×512. For display purpose, it is more appropriate to show the depth map with value $1/z$ where z is the depth between the point and the camera. This renders more range to near objects we are interested in. Fig. 7.12(a) shows the depth map in $1/z$ at time t_0 as an intensity image. The brighter the point is, the closer the point is located to the camera.

It is very desirable to test the accuracy of this depth map. To do this, we have selected a few lines in the scene which are representative and are easy to measure length. Those lines are shown in Fig. 7.12(b) as bright lines. A simple algorithm accepts the end points of each line in the image and computes the length of the line using the computed depth map. The absolute difference between the correct length and that computed from the depth map is defined as error for this line. Table 7.3 lists two types of statistical measurement of errors of lines: maximum error, root mean squared error (rms error) as well

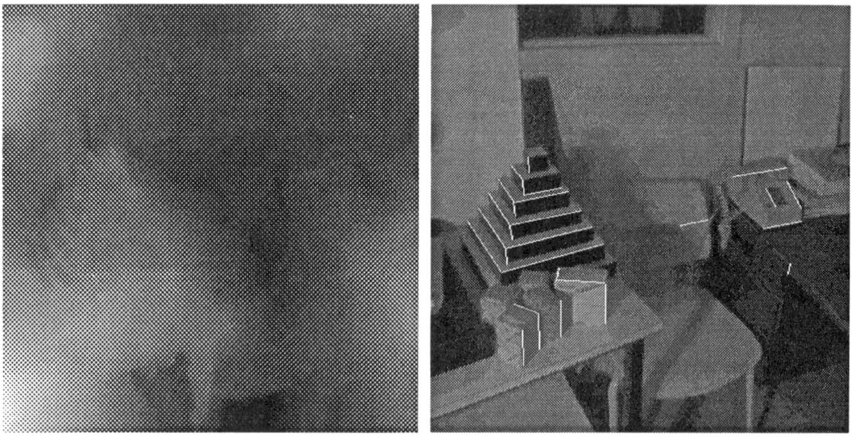

(a) (b)

Fig. 7.12. Computed depth map and test lines. **(a)** Computed depth map shown as intensity image with value $1/z$. **(b)** The test lines with known lengths, which are used to check the accuracy of the depth map. The test lines are highlighted in the image.

Table 7.3. Depth map accuracy test

Total Lines	Mean length	RMS error (rmse)
28	105.1mm	4.2mm
Maximum error	Mean depth $\overline{z}$	rmse / $\overline{z}$
8.3mm	1256.3mm	0.0033

as other data. As we can see from the table, on average, the depth is about 1256 mm. Out of this, the error is roughly 4 mm, a 0.3% error.

A point which needs to mention is that, in this particular experiment, we used wide-angle lenses (f=8mm) and a short baseline (100 mm). It is expected that a normal lens f=25mm will reduce the error by a fact of about 3, since the same pixel grid is focused on smaller region. A longer baseline will also significantly increase the accuracy. Under a long baseline, stereo disparity can be kept small by adjusting camera convergence.

7.7 Without Stereo Correspondences

The approach discussed in the previous sections makes use of stereo correspondences to obtain 3-D positions of the object points first, then estimates motion parameters from 3-D points.

7.7.1 Motivations

From a monocular image sequence, one can only determine the motion and structure up to a scale factor, while from a stereo sequence, one can determine them completely. The key to the complete determination in stereo case is that one can determine 3-D position from the stereo matching. If we do not try to establish stereo correspondences and establish only temporal correspondences for each camera, can we determine the motion and structure parameters of the scene completely? The answer is positive, provided that some conditions are met.

This approach has some advantages. One advantage arises from the fact that stereo matching is not necessary. Therefore, the two cameras can be directed in very different directions, say, one forward and the other sideward. An example is shown in Fig. 7.13. Since two cameras view in different directions, one of them may view a stable motion and the corresponding estimate is reliable. In figure 7.13, when the cameras move together with the mobile platform, the type of motion determines which camera can get accurate motion estimates. The final estimate can be a weighted sum of the estimates from the two cameras, with the weights determined by the reliability of each estimate. Thus, the result from multiple cameras renders accurate estimates for different motions. In fact, if we have three cameras

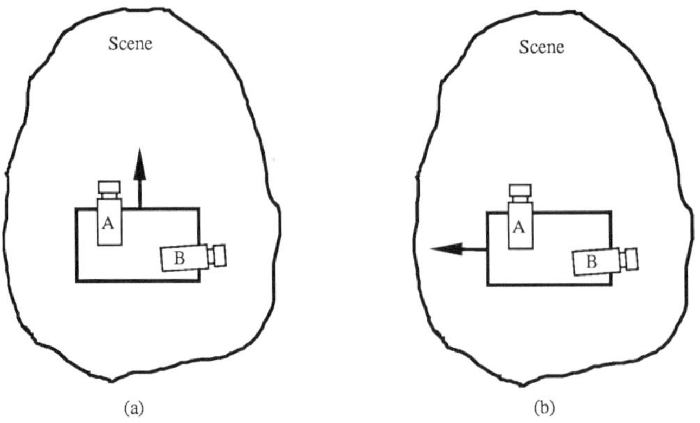

Fig. 7.13. One camera gives reliable motion estimate and the other gives unreliable estimate. The cameras are fixed on the mobile platform. The scene is stationary. **(a)** The motion estimate from camera A is more reliable than that from camera B. **(b)** The motion estimate from camera B is more reliable than that from camera A.

directed in three mutually orthogonal directions, accurate estimates are always available because at least one camera does not see lateral motion and thus its estimate is accurate.

Furthermore, this scheme can avoid stereo matching which has to deal with large disparities and occlusions. Temporal matching needs only to deal with small disparities. Matching difficulties arising from large disparities can be less critical here.

Another advantage of this scheme concerns the extent of recoverable structure of the scene. As shown in Fig. 7.14, with a conventional stereo setup, the recoverable surface lies in the intersection of the fields of the view of two cameras. But with the new scheme, the

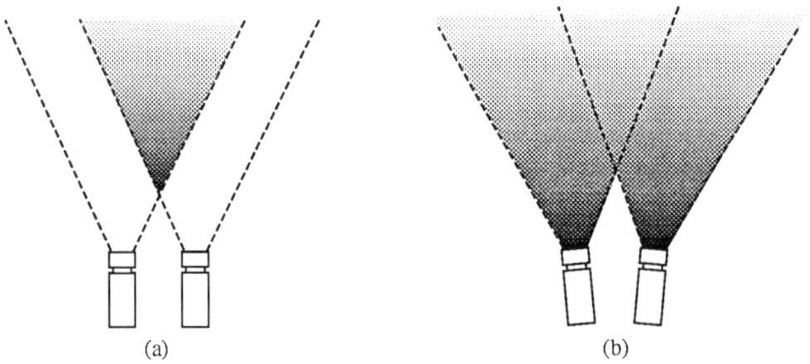

Fig. 7.14. **(a)** Conventional stereo can only determine depth for the intersection of two fields of view. **(b)** The new scheme can determine depth for the union of the two fields of view.

recoverable scene structure is the union of them. We can direct two cameras as needed to increase the recoverable scene.

7.7.2 Motions Seen from Two Cameras

To be more general, we assume that the camera setup shown in Fig. 7.1 is time varying. Let (7.1.1) holds for the system at time before motion. After motion we have a new stereo configuration such that

$$\mathbf{x}' = M'\mathbf{x}_r' + \mathbf{B}'. \tag{7.7.1}$$

Under the motion shown in (7.3.1), we need the representation of the motion seen by the right camera. Substituting $\mathbf{x}$ in (7.3.1) from (7.7.1) and $\mathbf{x}'$ in (7.3.1) from (7.1.1) yields

$$M'\mathbf{x}_r' + \mathbf{B}' = R(M\mathbf{x}_r + \mathbf{B}) + \mathbf{T}. \tag{7.7.2}$$

Rewriting the above equation gives

$$\mathbf{x}_r' = (M')^t RM\mathbf{x}_r + (M')^t(R\mathbf{B} - \mathbf{B}' + \mathbf{T}).$$

Letting

$$R_r = (M')^t RM \tag{7.7.3}$$

and

$$\mathbf{T}_r = (M')^t(R\mathbf{B} - \mathbf{B}' + \mathbf{T}) \tag{7.7.4}$$

(7.7.2) gives

$$\mathbf{x}_r' = R_r\mathbf{x}_r + \mathbf{T}_r.$$

This means that in the right-camera-centered coordinate system, the motion is a rotation represented by R_r followed by a translation represented by $\mathbf{T}_r$.

7.7.3 Determining Complete Motion and Structure

The known system configurations before and after motion provide M, M', $\mathbf{B}$ and $\mathbf{B}'$. Suppose $\alpha = \|\mathbf{T}\|$ and $\beta = \|\mathbf{T}_r\|$. If temporal matching is available to each camera, using monocular motion analysis we can compute R and $\mathring{\mathbf{T}} = \alpha^{-1}\mathbf{T}$ for the left camera and R_r and $\mathring{\mathbf{T}}_r = \beta^{-1}\mathbf{T}_r$ for the right camera. Our objective is to compute the scale factors α and β so that the motion and structure can be determined completely.

Define $\mathbf{K} \triangleq R\mathbf{B} - \mathbf{B}'$ and

$$\mathbf{V} \triangleq M'\mathbf{T}_r. \tag{7.7.5}$$

Then,

$$\mathring{\mathbf{V}} \triangleq \|\mathbf{V}\|^{-1}\mathbf{V} = M'\mathring{\mathbf{T}}_r$$

by virtue of (7.7.5). We can compute $\mathbf{K}$ and $\mathring{\mathbf{V}}$ from the motion parameters computed based on monocular images. From (7.7.4) we have

$$\beta\mathring{\mathbf{V}} - \alpha\mathring{\mathbf{T}} = \mathbf{K}. \tag{7.7.6}$$

This vector equation consists of three scalar equations in two unknowns α and β. To get a linear least-squares solution, we get the normal equation of (7.7.6):

$$\begin{bmatrix} 1 & -\mathring{\mathbf{T}}\cdot\mathring{\mathbf{V}} \\ -\mathring{\mathbf{T}}\cdot\mathring{\mathbf{V}} & 1 \end{bmatrix} \begin{bmatrix} \alpha \\ \beta \end{bmatrix} = \begin{bmatrix} -\mathring{\mathbf{T}}\cdot\mathbf{K} \\ \mathring{\mathbf{V}}\cdot\mathbf{K} \end{bmatrix}.$$

This equation fails to give a unique solution for (α, β), iff $\mathring{\mathbf{T}} \,//\, \mathring{\mathbf{V}}$, iff $\mathbf{K} \,//\, \mathbf{T}$, iff

$$(R\mathbf{B} - \mathbf{B}')\times\mathbf{T} = \mathbf{0}. \tag{7.7.7}$$

If the configuration is fixed, $\mathbf{B} = \mathbf{B}'$, the method fails if the motion is a pure translation $R = I$. We know that in the case of passive navigation, the motion is due to the motion of the camera system itself. Therefore degeneracy can always be avoided by a time-varying stereo system which is controlled such that (7.7.7) is not satisfied.

Once α and β are solved for, the translations $\mathbf{T}$ and $\mathbf{T}_r$ are completely determined. The 3-D positions of the object points can be completely determined.

Therefore, the scale factor that can not be determined by one monocular system can be determined by two monocular systems. The key is that the motions observed from the different camera coordinate systems are different. We relate this difference with the known relative configurations between two cameras and solve the scale factors. This approach can be easily extended to systems with more than two cameras. With more cameras, it is easier to avoid the degeneracy shown in (7.7.7).

In the absence of noise, from (7.7.3) we have

$$R = M'R_r M^t, \tag{7.7.8}$$

where the left side and right side are computed from the monocular motion analysis of the left camera and right camera, respectively. In the presence of noise, the two sides of (7.7.8) is not necessarily equal. In fact, this gives redundant information to improve R. For example, denoting $R_1(\alpha_1, \beta_1, \gamma_1) = R$ and $R_2(\alpha_2, \beta_2, \gamma_2) = M'R_r M^t$ (see (7.6.1)), the new rotation matrix can be re-estimated by

$$R((\alpha_1 + \alpha_2)/2, (\beta_1 + \beta_2)/2, (\gamma_1 + \gamma_2)/2).$$

Similarly, from (7.7.4) we have

$$\mathbf{T} = M'\mathbf{T}_r - R\mathbf{B} + \mathbf{B}'. \tag{7.7.9}$$

In the presence of noise, we can average two sides of (7.7.9) to get a new estimate of $\mathbf{T}$. If the error covariance matrices are available for R, R_r, $\mathring{\mathbf{T}}$ and $\mathring{\mathbf{T}}_r$, such simple averages should be replaced by a weighted averages.

7.7.4 Computational Aspect of the Degeneracy

Our simulations have shown an interesting phenomenon: In the presence of a moderate amount of noise, the degenerate cases almost never occur.

In the simulation, the 3-D points are generated randomly for each trial, between depth 2 and depth 15, with a uniform distribution. The baseline vector is given by $\mathbf{B} = (0, 0.5, 0)^t$. The field of view (side-to-side) of the cameras is about 53°. The variance of Gaussian noise added to the image points is equal to that of uniform digitization noise with a 256×256-pixel image. The temporal motion parameters were computed by the iterative optimization method with an objective function presented in Subsection 4.3.1. In Fig. 7.15, the rotation is about the axis (1, 0.2, 0.1), and the translation vector is (0.1, 0.2, 0), a lateral motion to the left-camera. The configuration of two cameras is known and fixed. According to our discussion, a zero rotation angle corresponds to degeneracy. To exam the degeneracy, we vary the rotation angle from 0° to 0.1°. Surprisingly, even with a zero rotation

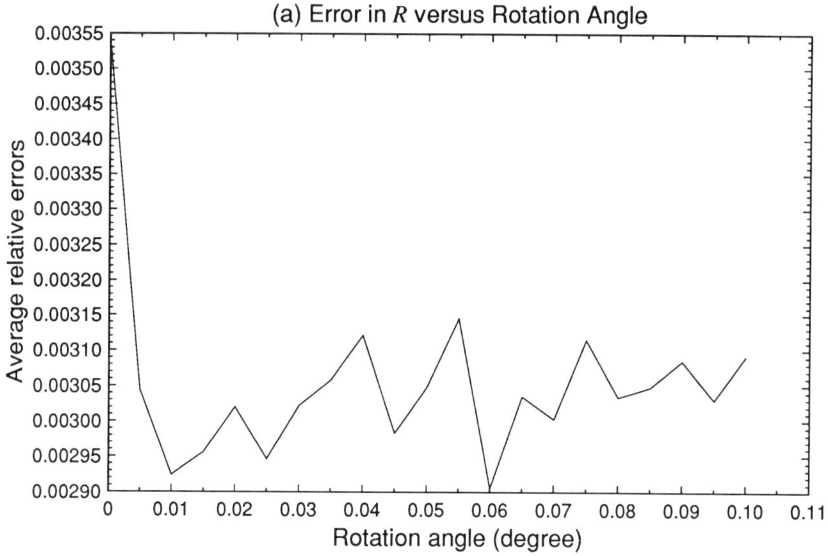

Fig. 7.15(a)

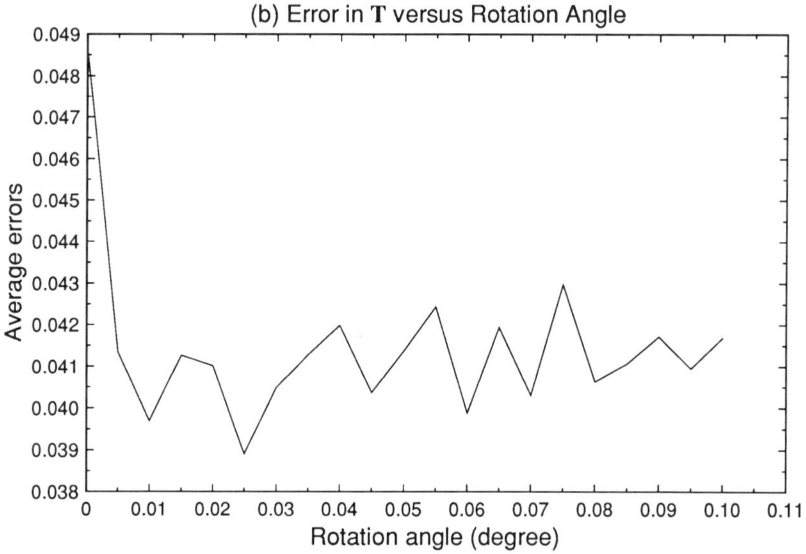

Fig. 7.15(b)

Fig. 7.15. Error of estimated motion without using stereo matching. 1000 trials. **(a)** Relative error of rotation matrix R. **(b)** Relative error of translation vector $\mathbf{T}$.

angle, which get rise to degeneracy according to (7.7.7) the estimated results are quite accurate as shown in Fig. 7.15. The error does not vary much with the amount of rotation. If we do not add noise to the image data, however, zero rotations gave over 100% errors in estimates. This can be explained as follows. With image noise, the observed rotation is not zero, although the true rotation is zero. This is indicated by a typical non-zero estimates of about 4° for rotation. Therefore, the amount of noise in fact avoids degeneracy. We have reduced the amount of noise to that of 4048×4048-pixels, and the effect of degeneracy only showed slightly by causing a slightly larger error in zero rotation than that in non-zero rotations. Therefore, it appears that in real applications, the degenerate configurations will almost never occur.

7.8 Long Image Sequences

With a long stereo image sequence, the amount of data to be processed is drastically larger than that for two stereo image pairs. Kalman filtering discussed in Section 4.7 is a method for computing weighted least-squares solution for a dynamic system. It is a sequential method in the sense that the observation data are sequentially fed into algorithm and new estimates are recursively computed from the previous estimates and the current observation. In contrast, a batch method is such that all observations are processed together in a batch fashion and estimates are determined directly from all observations.

A sequential method has some desirable properties. First, old observations can be discarded once they have been used for estimation. This is a must if the series of data is virtually infinite (e.g., in an extended navigation) or if the whole data set is so large that it exceeds the capacity of memory. Second, the method is efficient since relatively small amount of computation is required for updating estimates with each observation. Third, since a new estimate is computed at every time of observation, one does not need to wait for a complete set of observations to be collected before receiving estimates.

However, the sequential methods have relatively poor performance for nonlinear problems. As we discussed in Section 4.7, the estimated system matrices (i.e., Jacobian matrices) used for earlier observations are not updated after new observations are collected. Because the system matrices of a nonlinear system depend on the estimated system parameters, the system matrices for earlier observations are evaluated with parameters values close to the initial guess, and therefore, are very inaccurate. Those inaccurate system matrices almost always cause an early divergence, which requires many image frames to gradually "pull" the state trajectory back. The linear-model based covariance matrices, which determine the amount of parameter updating, are unable to reflect such a divergence and underestimate the errors in the current estimates. Consequently, these covariance matrices impede the correction of divergence and parameter updating. This accounts for why Kalman filtering requires many image frames (typically 20 to 40 frames) to converge to an acceptable solution. Such a slow convergence also implies a slow response to system changes, such as those caused by change of motion direction.

In this section, we employ a so called recursive-batch approach to estimate motion and structure from long image sequences. By this approach, observation data are grouped. Estimation is done in a sequential fashion among groups of data. Within each data group, estimation is done in a batch fashion. From the view point of Kalman filtering, the states (parameter solution) are directly observed (provided by a batch solution, either non-iterative or iterative). The Recursive Kalman filtering equations are used to properly weight the uncertainties in observations and states and update the uncertainty in the new states, but are not used to solve unknowns. In other words, by properly define states and observations, we convert the *nonlinear* Kalman filtering problem to a *linear* Kalman filtering problem, which is, with regard to the formulation to which the Kalman filter is applied, equivalent to a batch solution! In this way, advantages of sequential processing are kept and the performance of the algorithm is drastically improved from direct nonlinear iterated Kalman filtering.

Since the dimension of parameter space for each group is very large (including 3-D coordinates of all points as well as motion parameters), a direct search in such a large space is computationally prohibitive. Instead of using direct batch processing, we explore the structure of the problem, as we did before in space decomposition, so that the search space is reduced to motion parameters only. In order to reduce the number of iterations and prevent divergence, our closed-form matrix-weighted least-squares solution is used as an initial guess solution for batch optimization within each data group. (Simulations showed that initial guesses provided by the scalar-weighted least-squares solution generally leads to the same solution but require a larger number of iterations in optimization). It should be mentioned that our formulation is independent of the way the batch solution are computed (or states are "observed"). While iterative optimal solutions lead to more accurate (close to theoretical bound) results, a non-iterative batch solution (e.g., our matrix-weighted solution) often suffice, especially in case where the speed of the algorithm is critical.

We also investigate the representation of 3-D points in different coordinate systems: camera-centered and world centered. A camera-centered coordinate system is fixed with the stereo camera system and moves with it. The world coordinate system is fixed with the scene. In self-guided navigation, the use of a camera-centered reference system is preferable. However, if the navigation has to refer to a map or to construct a global 3-D map of the sensed world, a representation in a world-centered coordinate system is more appropriate. Parameter updating, error estimation and some important properties associated with these two representations are investigated.

7.8.1 Local and Global Coordinate Systems

Consider the case where a stereo camera system moves in a static environment. Let the world coordinate system coincide with the camera-centered coordinate system at time t_0. In that world coordinate system, the attitude (including position and orientation) of the stereo camera system, as a rigid entity, at time t_k can be represented by a rigid motion, from its original position at time t_0. The attitude of the camera system at time t_k can be reached by moving camera from its original attitude at time t_0 by a rotation about the origin (represented by a rotation matrix R_k) followed by a translation (represented by a vector $\mathbf{T}_k$). Letting $\mathbf{c}_k$ be any point on the camera system at time t_k, represented in world coordinate system, then $\mathbf{c}_k$ and $\mathbf{c}_0$ are related by the following transformation

$$\mathbf{c}_k = R_k \mathbf{c}_0 + \mathbf{T}_k . \tag{7.8.1}$$

The same transformation transforms a position vector $\mathbf{x}_k$ of a scene point, in the camera-centered coordinate system at t_k, back to the world coordinate system:

$$\mathbf{x}_0 = R_k \mathbf{x}_k + \mathbf{T}_k \tag{7.8.2}$$

where $\mathbf{x}_0$ is the position of the same point in the world coordinate system. Incrementing time index k by 1, (7.8.2) gives

$$\mathbf{x}_0 = R_{k+1} \mathbf{x}_{k+1} + \mathbf{T}_{k+1}. \tag{7.8.3}$$

Substituting $\mathbf{x}_0$ in (7.8.2) by the right-hand side of (7.8.3), we get

$$\mathbf{x}_{k+1} = R_{k+1}^{-1} R_k \mathbf{x}_k + R_{k+1}^{-1} (\mathbf{T}_k - \mathbf{T}_{k+1}). \tag{7.8.4}$$

If we define

$$R_{k+1,k} \triangleq R_{k+1}^{-1} R_k \tag{7.8.5}$$

and

$$\mathbf{T}_{k+1,k} \triangleq R_{k+1}^{-1} (\mathbf{T}_k - \mathbf{T}_{k+1}), \tag{7.8.6}$$

(7.8.4) gives

$$\mathbf{x}_{k+1} = R_{k+1,k} \mathbf{x}_k + \mathbf{T}_{k+1,k}. \tag{7.8.7}$$

Since $\mathbf{x}_k$ and $\mathbf{x}_{k+1}$ are represented in the camera-centered coordinate system at t_k and t_{k+1}, respectively, $R_{k+1,k}$ and $\mathbf{T}_{k+1,k}$ are the rotation matrix and the translation vector, of the interframe motion, from time t_k to time t_{k+1}, of a scene point seen by the camera. They correspond to the motion parameters estimated in Section 7.3 or Section 7.4. In this Section, we use double subscripts for the interframe motion parameters of the scene seen in the camera-centered coordinate system, and a single subscript for the global attitude of the camera as in (7.8.1). Equations (7.8.5) and (7.8.6) give the relations between these two types of parameters:

$$R_{k+1} = R_k R_{k+1,k}^{-1}, \tag{7.8.8}$$

$$\mathbf{T}_{k+1} = \mathbf{T}_k - R_{k+1} \mathbf{T}_{k+1,k} = \mathbf{T}_k - R_k R_{k+1,k}^{-1} \mathbf{T}_{k+1,k}. \tag{7.8.9}$$

Using the above two equations, we can update the global attitude of the camera system from its previous attitude and current interframe motion of the scene observed in the camera-centered coordinate system.

7.8.2 A Recursive-Batch Approach

With an image sequence, the parameters to be estimated include the 3-D positions of feature points of a scene, represented in some coordinate system and the interframe motion parameters between every pair of consecutive instants. The global attitude of the camera system can be determined from interframe motions based on (7.8.8) and (7.8.9).

According to the discussion in sections 2 to 4, we know that the estimation of structure and interframe motions are closely related. The accuracy of the estimated structure influences the accuracy of the estimated motion parameters, and vice versa. With a large number of image frames, the same part of a scene is observed many times and therefore the structure can be more accurately estimated after many image frames are processed. This implies, in particular, that the estimation of early interframe motions can benefit from the processing of later image frames. Due to this type of interaction between the structure and interframe motion, higher estimation accuracy can be obtained if all the image frames to be considered are processed in a batch fashion. In theory, the problem can still formulated as the estimation of a parameter vector (consisting of parameters for point structure and all interframe motions) from an observation vector (consisting of the measured image coordinates of all points in all frames). However, this is a computationally prohibitive task if the image sequence is virtually infinite.

In fact, each point will be visible only in certain time span. It will eventually go out of the field of view and disappear from the image sequence. This implies that in a long image sequence, the relation between two images is weak if the two images are separated by many image frames. For example, the accuracy of the estimated structure in a section of a scene will not affect very much that of a different section.

As we discussed in Section 7.1, Kalman filtering has some desirable advantages. For a linear problem, theoretically, the result of Kalman filtering is the same as that of a batch method. However, for a nonlinear problem, the result of Kalman filtering is not as good. As we discussed in Section 4.7, the key problem with Kalman filtering for a nonlinear problem is that the system Jacobian matrix for each old observation cannot use future observations. If all observations are processed in a batch fashion, the modification of parameters is much reliable and the system matrix of every observation is updated at each iteration. In other words, with a sequential algorithm, the contribution or influence of the later observations to the evaluation of the system matrices for the early observations are neglected (this is not a problem for a linear problem since the system Jacobian matrix is constant).

In order to achieve good performance without suffering from excessive computational cost, we need batch processing only for those data that have considerable interactions. The above observations have motivated our recursive-batch approach. In this approach, the observation sequence is processed in relatively small batches. For each batch of data, estimates are determined in a batch fashion from the previous estimates and the current batch of data. The approach is recursive because the newly estimated result depends on the previous result. From a global point of view, the data is processed sequentially by feeding through a processing algorithm that covers a certain length (batch size) of the sequence. This approach has the following advantages:

(1) It can process virtually infinitely long image sequences with a limited memory.

(2) Since old estimates can be used for new estimates, and only a relatively small amount of computation is required to update estimates, the algorithm is efficient.

(3) The algorithm out-performs a straight sequential method (e.g, Kalman filtering) because the data is processed in a batch fashion in which each overlapping batch sufficiently covers the interaction among data.

If the batch size of the recursive-batch approach is so large that the batch covers the whole image sequence, the recursive-batch approach degenerates into a pure batch approach. To the other extreme, if the batch size is equal to one, the recursive-batch approach becomes a sequential one. The choice of an appropriate batch size is important. For the problem to be studied in this chapter, a natural batch size corresponds to all observations in a single stereo pair. With each new stereo pair, a new interframe motion needs to be estimated and the structure of points is updated. The performance may be further improved if several stereo pairs are processed as a batch set (the batch will slide thorough the sequence so that batches are overlapping). However, this will significantly increase the computational cost.

At the level of groups, the recursive estimation is equivalent to Kalman filtering. However, with a batch solution of state vector directly as observation, the Kalman filtering structure at this level is *linear*. Therefore, the problems with nonlinear Kalman filtering do not apply here. To see this more clearly, we present the corresponding observation equation. Let the state vector (in our case, current attitude of the camera determined by R_k and $\mathbf{T}_k$) be $\mathbf{m}_k$. It is computed in a batch fashion from the previous attitude and the last interframe motion, with an estimated covariance matrix $\Gamma_{\mathbf{m}_k}$. The corresponding observation equation is

$$\hat{\mathbf{m}}_k = \mathbf{m}_k + \boldsymbol{\delta}_{\mathbf{m}_k}$$

where $\hat{\mathbf{m}}_k$ is the "observation", $\mathbf{m}_k$ is the true state vector and $\boldsymbol{\delta}_{\mathbf{m}_k}$ is a noise vector which has an estimated covariance matrix $\Gamma_{\mathbf{m}_k}$. As we know, with a linear observation equation, the result of Kalman filtering is equivalent to the batch minimum variance solution.

If the observation were directly defined as, e.g., the vector of image coordinates of points at time $\mathbf{u}_k$, one would have to deal with the non-linear observation equation

$$\hat{\mathbf{u}}_k = \mathbf{f}(\mathbf{m}_k) + \boldsymbol{\delta}_{u_k}$$

where $\mathbf{f}$ is a non-linear function. If the iterated non-linear Kalman filtering is used to solve this nonlinear problem, one will see an initial trajectory-divergence period which requires many image frames to "pull" the trajectory back [7.4], due to the reasons we discussed above.

7.8.3 Recursive-Batch Updating

The model represented by the objective function (7.4.7) is very suited for recursive-batch updating. With each new pair of stereo images, interframe motion is computed with a closed-form solution presented in Section 7.3, and then the parameters are further optimized by the method presented in Section 7.4. The structure is updated by the fusion of the estimated structure accumulated up to the previous time instant with the new stereo pair. In other words, the structure of points in the camera-centered coordinate system is the parameter vector (or called system state) that is updated through the image sequence, while interframe motions determine how the previous structure evolves to the current time instant. In order to provide the following recursive step with the updated structure and the associated uncertainty, we need to compute the structure in the coordinate system centered at the current camera position, as well as its error covariance matrix.

After the interframe motion parameter vector $\mathbf{m}_{k+1,k}$ is estimated based on points $\{\mathbf{x}_k\}$ before motion and $\{\mathbf{x}_{k+1}\}$ after motion, we compute the covariance matrix of the motion parameter vector $\Gamma_{\mathbf{m}_{k+1,k}}$ as discussed in Section 7.4. The structure at time t_{k+1} can be updated in a way similar to that for the structure at time t_k: Let $\hat{\mathbf{x}}_k$ be the estimated structure of $\mathbf{x}_k$ provided by processing all frames up to t_k, and $\Gamma_{\hat{\mathbf{x}}_k}$ be the associated covariance matrix. Moving the point $\hat{\mathbf{x}}_k$ to time t_{k+1} gives

$$\mathbf{p} \triangleq R_{k+1,k}\hat{\mathbf{x}}_k + \mathbf{T}_{k+1,k}$$

with the associated covariance matrix

$$\Gamma_{\mathbf{p}} = R_{k+1,k}\Gamma_{\hat{\mathbf{x}}_k}R_{k+1,k}^{-1} + \frac{\partial \mathbf{p}(\mathbf{m}_{k+1,k})}{\partial \mathbf{m}_{k+1,k}}\Gamma_{\mathbf{m}_{k+1,k}}\frac{\partial \mathbf{p}(\mathbf{m}_{k+1,k})^t}{\partial \mathbf{m}_{k+1,k}}.$$

From triangulation at time t_{k+1} we get the estimated position $\mathbf{q}$ of $\mathbf{x}_{k+1}$ and the associated covariance matrix $\Gamma_{\mathbf{q}}$. According to (4.5.3), the updated estimate of $\mathbf{x}_{k+1}$ is given by

$$\Gamma_{\mathbf{p}}(\Gamma_{\mathbf{q}} + \Gamma_{\mathbf{p}})^{-1}\mathbf{q} + \Gamma_{\mathbf{q}}(\Gamma_{\mathbf{q}} + \Gamma_{\mathbf{p}})^{-1}\mathbf{p} = \mathbf{q} + \Gamma_{\mathbf{q}}(\Gamma_{\mathbf{q}} + \Gamma_{\mathbf{p}})^{-1}(\mathbf{p} - \mathbf{q})$$

with a covariance matrix from (4.5.4)

$$\Gamma_{\mathbf{q}}(\Gamma_{\mathbf{q}} + \Gamma_{\mathbf{p}})^{-1}\Gamma_{\mathbf{p}} = \Gamma_{\mathbf{q}} - \Gamma_{\mathbf{q}}(\Gamma_{\mathbf{q}} + \Gamma_{\mathbf{p}})^{-1}\Gamma_{\mathbf{q}}.$$

The algorithm starts at time t_0 with the structure estimated through triangulation and the associated covariance matrix (Subsection 7.2.3). Then the recursive estimation proceeds by incrementing k in t_k by one to get the following stereo pair. The interframe motion is first estimated using the batch method together with the associated covariance matrix of the motion parameters (Sections 7.3 to 7.4). The structure after the interframe motion and its covariance matrix are estimated using the method presented in this subsection. These estimates are used for the next recursive step as the estimates for structure before the next interframe motion.

7.8.4 Local and Global Representations

As mentioned at the beginning of this section, two coordinate systems are used, a (local) camera-centered and a (global) world coordinate systems. The structure of the perceived world can be represented in either coordinate system. The representation in a camera-centered coordinate system is useful for path planning with respect to the motion of the navigation system on which the camera system is mounted. For example, it can be used directly to plan for the direction and distance of the next motion. The representation in a world coordinate system is useful for extended visual map generation, since what is perceived in a moving camera-centered coordinate system needs to be registed on a map which is represented in a fixed world coordinate system. The global attitude of the current camera system is also critical information for map guided navigation, where the navigation system constantly needs to keep track its current position on a map and to make decisions.

Equations (7.8.8) and (7.8.9) give the relation that can be used to update the current global attitude of the camera system (represented by a vector $\mathbf{m}_{k+1}(R_{k+1}, \mathbf{T}_{k+1})$ that consists of 6 independent motion parameters) from the previous attitude $\mathbf{m}_k(R_k, \mathbf{T}_k)$ and the current interframe motion $\mathbf{m}_{k+1,k}(\mathbf{R}_{k+1,k}, \mathbf{T}_{k+1,k})$. They define a vector updating function $\mathbf{h}$:

$$\mathbf{m}_{k+1} = \mathbf{h}(\mathbf{m}_k, \mathbf{m}_{k+1,k}). \tag{7.8.10}$$

Letting the difference between the estimated and true vector be denoted by $\boldsymbol{\delta}$ with the vector as a subscript, from (7.8.10) we have

$$\boldsymbol{\delta}_{\mathbf{m}_{k+1}} = \frac{\partial \mathbf{h}(\hat{\mathbf{m}}_k, \hat{\mathbf{m}}_{k+1,k})}{\partial \mathbf{m}_k} \boldsymbol{\delta}_{\mathbf{m}_k} + \frac{\partial \mathbf{h}(\hat{\mathbf{m}}_k, \hat{\mathbf{m}}_{k+1,k})}{\partial \mathbf{m}_{k+1,k}} \boldsymbol{\delta}_{\mathbf{m}_{k+1,k}}.$$

So the covariance matrix of the global attitude of the camera is updated by

$$\Gamma_{\mathbf{m}_{k+1}} = \frac{\partial \mathbf{h}(\hat{\mathbf{m}}_k, \hat{\mathbf{m}}_{k+1,k})}{\partial \mathbf{m}_k} \Gamma_{\mathbf{m}_k} \frac{\partial \mathbf{h}(\hat{\mathbf{m}}_k, \hat{\mathbf{m}}_{k+1,k})^t}{\partial \mathbf{m}_k}$$
$$+ \frac{\partial \mathbf{h}(\hat{\mathbf{m}}_k, \hat{\mathbf{m}}_{k+1,k})}{\partial \mathbf{m}_{k+1,k}} \Gamma_{\mathbf{m}_{k+1,k}} \frac{\partial \mathbf{h}(\hat{\mathbf{m}}_k, \hat{\mathbf{m}}_{k+1,k})^t}{\partial \mathbf{m}_{k+1,k}}$$

where $\Gamma_{\mathbf{m}_{k+1,k}}$ is the covariance matrix of the interframe motion parameter vector $\mathbf{m}_{k+1,k}$ estimated by the method in Section 7.4.

Similarly, the global position of the estimated structure can also be updated to take into account the new observations in the current stereo pair. Let $\mathbf{p} \triangleq \hat{\mathbf{x}}_0$ be the old estimate of the global position of a point. From the new estimate $\mathbf{x}_{k+1}$, represented in the camera-centered coordinate system, we can get the corresponding global position from (7.8.2):

$$\mathbf{q} \triangleq R_{k+1}\hat{\mathbf{x}}_{k+1} + \mathbf{T}_{k+1}$$

with an associated covariance matrix

$$\Gamma_{\mathbf{q}} = R_{k+1}\Gamma_{\hat{\mathbf{x}}_{k+1}}R_{k+1}^{-1} + \frac{\partial \mathbf{q}(\hat{\mathbf{m}}_{k+1})}{\mathbf{m}_{k+1}}\Gamma_{\mathbf{m}_{k+1}}\frac{\partial \mathbf{q}(\hat{\mathbf{m}}_{k+1})^t}{\mathbf{m}_{k+1}}.$$

According to (4.5.3), the updated global structure of the point is given by

$$\Gamma_{\mathbf{q}}(\Gamma_{\mathbf{p}} + \Gamma_{\mathbf{q}})^{-1}\mathbf{p} + \Gamma_{\mathbf{p}}(\Gamma_{\mathbf{p}} + \Gamma_{\mathbf{q}})^{-1}\mathbf{q} = \mathbf{p} + \Gamma_{\mathbf{p}}(\Gamma_{\mathbf{p}} + \Gamma_{\mathbf{q}})^{-1}(\mathbf{q} - \mathbf{p}).$$

Based on (4.5.4) the error covariance matrix of the global point is estimated by

$$\Gamma_{\mathbf{p}}(\Gamma_{\mathbf{p}} + \Gamma_{\mathbf{q}})^{-1}\Gamma_{\mathbf{q}} = \Gamma_{\mathbf{p}} - \Gamma_{\mathbf{p}}(\Gamma_{\mathbf{p}} + \Gamma_{\mathbf{q}})^{-1}\Gamma_{\mathbf{p}}.$$

The following is an outline of the recursive-batch algorithm:

Step (1): Let $k = 0$. Get stereo pair at time t_0 with stereo point correspondences. For every point, compute the estimated point position $\hat{\mathbf{x}}_0$, at time t_0, and the associated covariance matrix (Section 7.2). At time t_0, the camera-centered and the world coordinate systems coincide: $R_0 = I$ and $\mathbf{T}_0 = \mathbf{0}$.

Step (2): Get stereo pair at time t_{k+1} with stereo point correspondences. For every point, compute the estimated camera-centered point position $\hat{\mathbf{x}}_{k+1}$, at time t_{k+1}, and the associated covariance matrix (Section 7.2).

Step (3): Compute the closed-form matrix-weighted solution for the interframe motion parameter vector $\mathbf{m}_{k+1,k}$ using the point correspondences from t_k to t_{k+1} (Subsection 7.3.3).

Step (4): Further optimize the above solution for $\mathbf{m}_{k+1,k}$, through a few iterations, and compute the associated covariance matrix (Section 7.4).

Step (5): Update the camera-centered coordinates of the points at time t_{k+1} and the associated covariance matrices (Subsection 7.8.3).

Step (6): Update the global attitude of the camera system $\mathbf{m}_k$ (also R_k and $\mathbf{T}_k$), and the associated covariance matrices (Subsection 7.8.4). Update the global position of points and the associated covariance matrices (Subsection 7.8.4).

Step (7): If t_{k+1} is not the time for the last stereo image pair to be taken, let $k \leftarrow k+1$ and go to Step (2). Otherwise, the algorithm terminates.

7.8.5 Numerical Examples

The objectives of this group of simulations are to demonstrate the performance of the recursive-batch algorithms and investigate the relationships between error in the camera-centered coordinate system and that in the world coordinate system.

3-D points are generated between $z = 0$ to $z = 30$ with a uniform distribution. The simulated stereo cameras have a field of view of about 38° and a resolution of 512×512 pixels. The stereo setup has a baseline length of 0.2, and the optical axes of the two cameras form a convergence angle of 10°. The camera system navigates forward through the scene with slight rotations and lateral translations in navigation. To keep a relatively constant number of visible point in each image, only those points that lie in depth range [2, 10], in the camera-centered coordinate system, are used for motion analysis. This arrangement is designed to simulate situations where the stereo system navigates through a zigzag path and the points in the last view are totally different from those in the first. We obtain two sequences. The first one consists of 21 stereo frames, with each interframe motion traveling 1 meter in depth, and the second has a larger interframe motion in depth (20/7) and the whole sequence has 8 stereo pairs. In both sequences, the camera system has traveled a total of 20 meters in depth. In the first sequence with a smaller interframe motions, about 60 point correspondences are available between two consecutive stereo frames, while in the second sequence, only about 40 point correspondences are available because of larger interframe motions. The exact number of point correspondences available varies slightly from frame to frame. For both sequences, average errors of the algorithms are accumulated through a total of 50 random trials, each with a completely new set of scene points.

Fig. 7.16 shows the average errors in the camera global attitude (orientation and position) estimated by two algorithms respectively: unweighted least-squares and iterative optimization (with the solutions of unweighted least-squares as initial guesses). We can see that the error in the estimated global attitude of the camera system is accumulated through the navigation, which is an inherent nature of an extended navigation. However, the error after the iterative optimization is much smaller than that of the unweighted least-squares solutions. Our methods do not exhibit a long divergence period which is observed in the iterated nonlinear Kalman filtering. The result is accurate starting from the beginning (time t_1 when the second stereo frame is available). The error in the depth component of the camera is larger than that in the lateral components. This is contrary to the results for the interframe motions (in the camera centered coordinate system). This implies that global representation and local representation have different characteristics. Due to the error correlation, the composite effect of errors in interframe rotation and translation still makes the depth component of the global position less reliable than the corresponding lateral components. The method described in Section 7.8 here has also been tested with long real world image sequences [7.6, 7].

Fig. 7.17 is similar to Fig. 7.16 except that the cameras undergo a larger interframe motions with only 8 stereo image pairs covering the same distance. Interestingly, although fewer image frames are available, the global attitude of the camera system is more accurate than that from small interframe motions. This can be explained in the following way. The global attitude is accumulated through interframe motions. Each interframe motion introduces a certain amount of error. With more interframe motions, the accumulated error is larger. This seems to suggest that the result from a sparse sample is better than that from a dense sample. Obviously this is contrary to the common sense that with more information, the expected error in an estimator should not be worse. In fact, this again indicates the limit of sequential processing. We have defined a batch that covers all observations in a single stereo pair. Among different stereo pairs, the structure of the algorithm is still sequential. If the batch covers several stereo pairs, the results will be better. For example, if the batch for the dense image sequence covers the same distance as that between two consecutive pairs in the sparse sequence, then the result of dense sequence will be better than that of the sparse sequence. At the extreme, under a batch approach results from a dense image sequence will never be worse than those from a sparse sequence. But, as we mentioned, increasing batch size will significantly increase the computational cost.

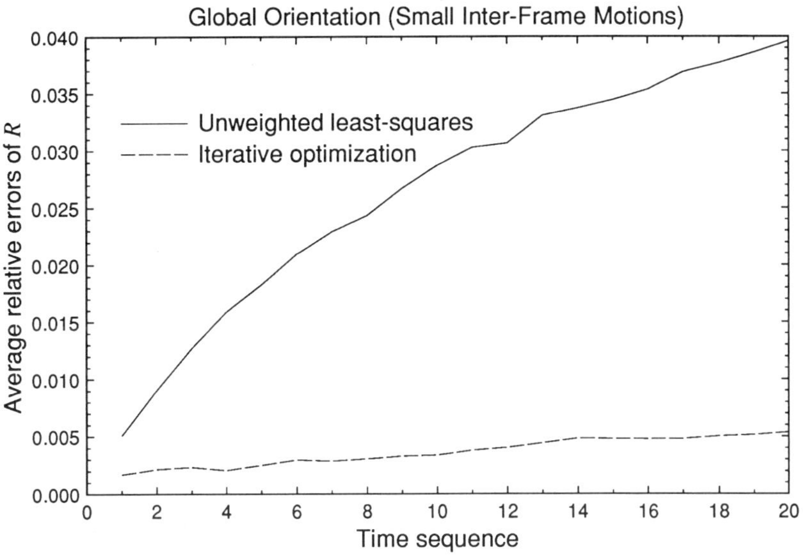

Fig. 7.16(a)

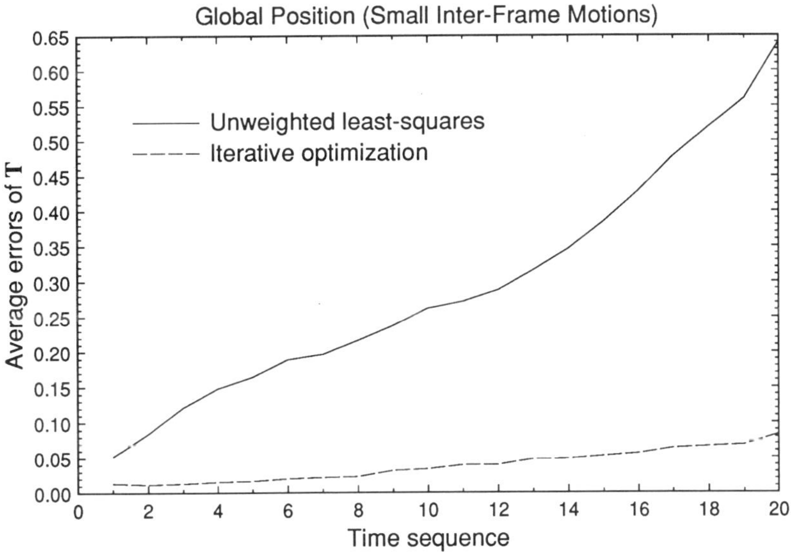

Fig. 7.16(b)

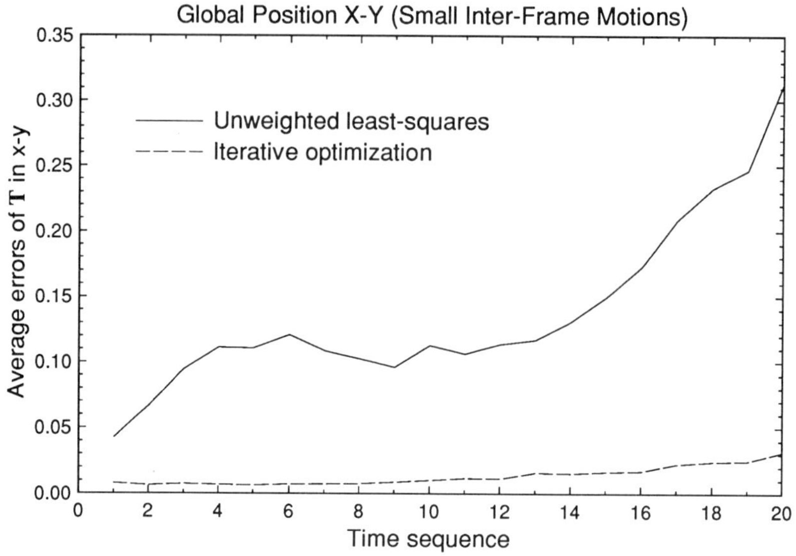

Fig. 7.16(c)

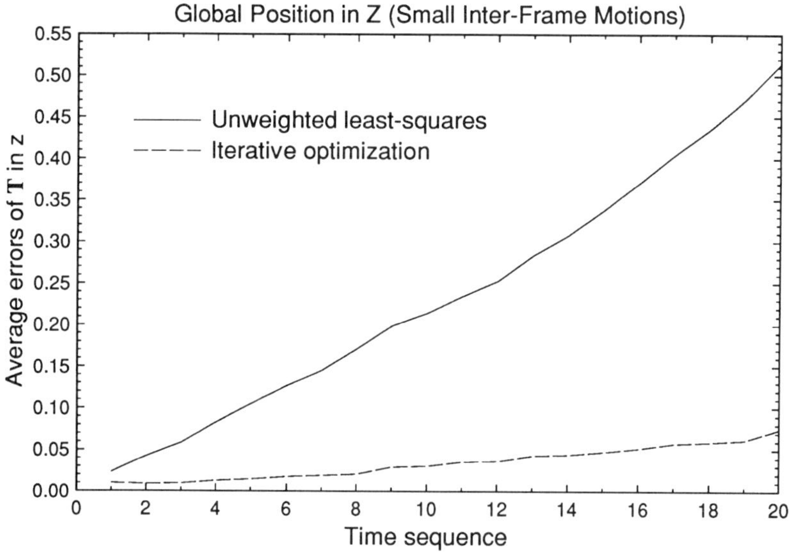

Fig. 7.16(d)

Fig. 7.16. Errors in the estimated global attitude of the camera system in navigation with small interframe motions. **(a)** Error of camera orientation represented by rotation matrix R; **(b)** error of camera position represented by translation vector **T**; **(c)** error of x-y component of **T**; **(d)** error of z component of **T**.

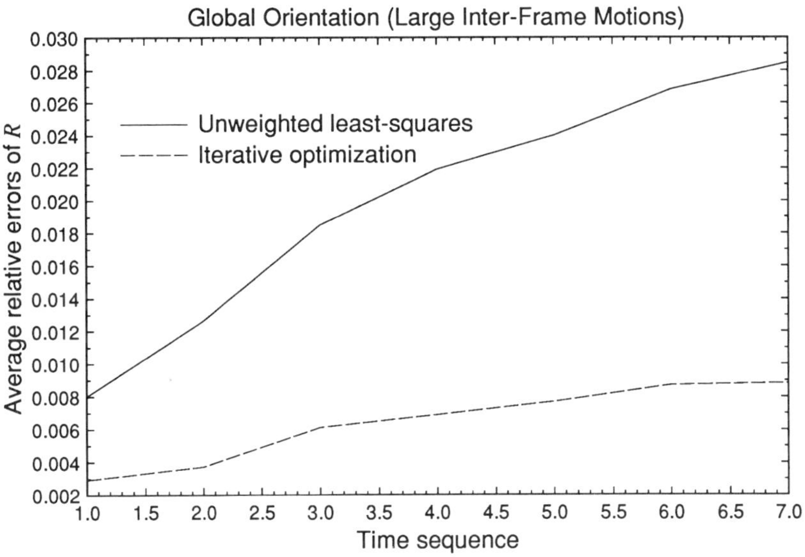

Fig. 7.17(a)

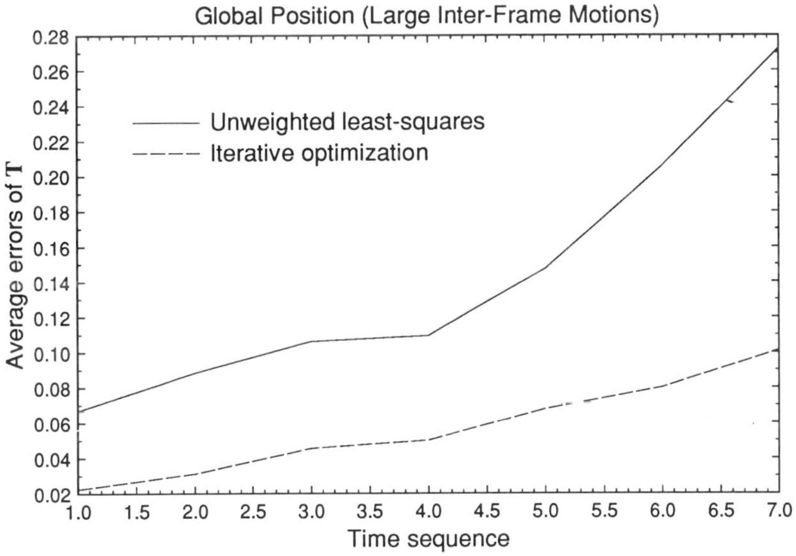

Fig. 7.17(b)

Fig. 7.17. Errors of the estimated global attitude of the camera system in navigation with large interframe motions. **(a)** Error of camera orientation represented by rotation matrix R; **(b)** error of camera position represented by translation vector $\mathbf{T}$.

7.9 Conclusions

In this chapter, we have presented a closed-form matrix-weighted solution to motion estimation from point correspondences in two pairs of stereo image pairs. The corresponding algorithm is non-iterative and fast. The simulation have demonstrated that the solutions from this algorithm are significantly more reliable than unweighted and scalar-weighted solutions if a sufficient number of points are available (over 6). The matrix-weighted least-squares solution can be used as a final solution if the speed requirement does not allow iterations.

An approach to optimal estimation of the motion parameters and the structure of the 3-D points has been introduced. In order to reduce the cost of computation and improve the numerical stability of the algorithm, the parameter space is decomposed in such a way that the structure of points is not included in the search space. Our frame work can be directly be extended to recursive estimation from long image sequences. The simulations have shown that, in the presence of Gaussian noise, the solution of this algorithm is the most accurate and it has essentially reached the theoretical bound. This method is useful in the presence of well-behaved noise and for applications where the best solution is needed and the speed is not so critical that iterations are allowed.

When input data are contaminated by gross errors, none of the above methods works very well. But the iterative optimization is especially vulnerable to gross errors. The robust method is appropriate for this type of measurement data. We have formulated the robust estimation as a least-squares problem with robust weights. To improve the detection of outliers, we determine the robust weights directly on image planes where measurements are taken. The closed-form solution, with the weighting matrices modified according to the robust weights, provides a good initial starting parameter vector for the computation of the robust estimates. Our simulations showed that the robust estimator remains stable in the presence of outliers in the data, while conventional least-squares methods fail.

Stereo vision from only temporal matching is possible. Complete motion and structure can be determined from stereo image sequences with only temporal matching. Under this scenario, the two cameras can take images of completely different parts of a rigid scene.

A recursive-batch approach has been introduced to efficiently compute motion and structure from long image sequences. The advantages of this recursive approach is that the algorithm is non-iterative, and the solution follows the true motion trajectory closely and thus avoids a long divergence period observed in conventional iterated nonlinear Kalman filtering.

7.10 Bibliographical Notes

Much attention has been drawn to the shape of the stereo uncertainty. Some publications are devoted exclusively to this issue, e.g., Blostein & Huang [7.3], Kamgar-Parsi & Kamgar-Parsi [7.13] and Rodriguez & Aggarwal [7.20].

Unweighted least-squares solutions to motion parameters from 3-D point correspondences have been derived by several researchers. The quaternion-based methods can be found in Shuster [7.21], Faugeras & Hebert [7.8], and Horn [7.12]. Arun, Huang & Blostein [7.1] presented a solution based on singular value decomposition. The results about unweighted and scalar weighted centroid-coincidence can be found in Huang, Blostein & Margerum [7.9], Horn [7.12], and Kiang, Chou & Aggarwal [7.14]. The result about matrix weighted centroid-coincidence was reported in [7.25].

Kalman filtering has been used widely for stereo motion estimation, including the studies by Broida & Chellappa [7.4], Ayache & Faugeras [7.2], and Matthies & Shafer [7.19]. Recently, recursive-batch methods have been proposed by Weng, Ahuja & Huang [7.23], Kumar, Tirumalai & Jain [7.15], and Weng & Cohen [7.24] to deal with early divergence and slow convergence in straight Kalman filtering for this nonlinear problem.

Huber's book [7.10], *Robust Statistics*, is a in-depth treatment of robust statistics. The book by Hoaglin, Mosteller & Tucky [7.11], *Understanding Robust and Exploratory Data Analysis*, gives a more practical treatment of the subject. The robust statistics has been applied to motion estimation, for example in Lee, Haralick & Zhuang [7.16], Weng & Cohen [7.25], and Tirumalai, Schunck & Jain [7.22].

Appendices

Appendix 7.A MWCC Theorem

MWCC Theorem. If R^* and $\mathbf{T}^*$ minimize (7.3.5) with the weighting matrix Γ_i^{-1} not depending on either R or $\mathbf{T}$, then the matrix-weighted centroids of $\{\hat{\mathbf{x}}_i'\}$ and $\{R^*\hat{\mathbf{x}}_i + \mathbf{T}^*\}$ must coincide:

$$\sum_{i=1}^{n} \Gamma_i^{-1}(R^*\hat{\mathbf{x}}_i + \mathbf{T}^*) = \sum_{i=1}^{n} \Gamma_i^{-1}\hat{\mathbf{x}}_i'. \tag{7.A.1}$$

Proof. Let

$$\hat{\mathbf{x}}_i'' \triangleq (\hat{x}_{i1}'', \hat{x}_{i2}'', \hat{x}_{i3}'')^t = R\hat{\mathbf{x}}_i + \mathbf{T}$$

for $i = 1, 2, \cdots, n$. Minimizing (7.3.5) is equivalent to the following: Given $\{\hat{\mathbf{x}}_i\}$ and $\{\hat{\mathbf{x}}_i'\}$, $i = 1, 2, \cdots, n$, determine $\{\hat{\mathbf{x}}_i''\}$ to minimize

$$\sum_{i=1}^{n}(\hat{\mathbf{x}}_i'' - \hat{\mathbf{x}}_i')^t\Gamma_i^{-1}(\hat{\mathbf{x}}_i'' - \hat{\mathbf{x}}_i'),$$

subject to the rigidity constraints

$$\|\hat{\mathbf{x}}_i'' - \hat{\mathbf{x}}_j''\|^2 = \|\hat{\mathbf{x}}_i - \hat{\mathbf{x}}_j\|^2$$

for $1 \le i \le n$, $1 \le j \le n$. As a necessary condition of this minimization problem with equality constraints, the partial derivatives of the corresponding Lagrangian

$$\mathrm{L} = \sum_{i=1}^{n}(\hat{\mathbf{x}}_i'' - \hat{\mathbf{x}}_i')^t\Gamma_i^{-1}(\hat{\mathbf{x}}_i'' - \hat{\mathbf{x}}_i')$$
$$+ \sum_{i=1}^{n}\sum_{j=1}^{n}\lambda_{ij}(\|\hat{\mathbf{x}}_i'' - \hat{\mathbf{x}}_j''\|^2 - \|\hat{\mathbf{x}}_i - \hat{\mathbf{x}}_j\|^2)$$

where $\lambda_{ij} = \lambda_{ji}$, must vanish. Differentiating L with respect to $\hat{x}_{kl}''$, $1 \le k \le n$ and $1 \le l \le 3$, yields,

$$0 = \frac{\partial \mathrm{L}}{\partial \hat{x}_{kl}''} = 2\Gamma_{kl}^{-1}(\hat{\mathbf{x}}_k'' - \hat{\mathbf{x}}_k') + 2\sum_{j=1}^{n}\lambda_{kj}2(\hat{x}_{kl}'' - \hat{x}_{jl}''), \tag{7.A.2}$$

where Γ_{kl}^{-1} denotes the l-th row of the matrix Γ_k^{-1}. Since Γ_k^{-1} in (7.3.10) does not depend on either R or $\mathbf{T}$, it does not depend on $\hat{\mathbf{x}}_k''$.

Summing up (7.A.2) for $k = 1$ to n gives

$$\sum_{k=1}^{n} \Gamma_k^{-1}(\hat{\mathbf{x}}_k'' - \hat{\mathbf{x}}_k') = 0 \tag{7.A.3}$$

since $\sum_{k=1}^{n}\sum_{j=1}^{n} \lambda_{kj}(\hat{x}_{kl}'' - \hat{x}_{jl}'') = 0$ for any integer l with $1 \le l \le 3$. According to the definition of $\hat{\mathbf{x}}_k''$, (7.A.1) directly follows from (7.A.3). This completes the proof. □

Appendix 7.B Least-Squares Matrix Fitting

Given two matrices, $C = [\mathbf{C}_1\ \mathbf{C}_2\ \cdots\ \mathbf{C}_n]$, $D = [\mathbf{D}_1\ \mathbf{D}_2\ \cdots\ \mathbf{D}_n]$, the solution of R in

$$\min_R \|RC - D\|, \quad \text{subject to: R is a rotation matrix} \tag{7.B.1}$$

is as follows.

Define a 4×4 matrix B by

$$B = \sum_{i=1}^{3} B_i^t B_i$$

where

$$B_i = \begin{bmatrix} 0 & (\mathbf{C}_i - \mathbf{D}_i)^t \\ \mathbf{D}_i - \mathbf{C}_i & [\mathbf{D}_i + \mathbf{C}_i]_\times \end{bmatrix}$$

where we define a mapping $[\cdot]_\times$ from a 3-dimensional vector to a 3×3 matrix:

$$[(x_1, x_2, x_3)^t]_\times = \begin{bmatrix} 0 & -x_3 & x_2 \\ x_3 & 0 & -x_1 \\ -x_2 & x_1 & 0 \end{bmatrix}.$$

Let $\mathbf{q} = (q_0, q_1, q_2, q_3)^t$ be a unit eigenvector of B associated with the smallest eigenvalue. The solution of rotation matrix R in (7.B.1) is

$$\begin{bmatrix} q_0^2+q_1^2-q_2^2-q_3^2 & 2(q_1q_2-q_0q_3) & 2(q_1q_3+q_0q_2) \\ 2(q_2q_1+q_0q_3) & q_0^2-q_1^2+q_2^2-q_3^2 & 2(q_2q_3-q_0q_1) \\ 2(q_3q_1-q_0q_2) & 2(q_3q_2+q_0q_1) & q_0^2-q_1^2-q_2^2+q_3^2 \end{bmatrix}$$

For proofs see Appendix 3.B.

8. Motion Modeling and Prediction

The trajectory of a moving object can be used to understand the motion pattern and predict future motion. In this chapter, we investigate the issues related to motion modeling, understanding and prediction. Based on dynamics, a locally constant angular momentum (LCAM) model is introduced. The model is local in the sense that it is applied to a limited number of image frames at a time. Specifically, the model constrains the motion, over a local frame subsequence, to be a superposition of precession and translation. Thus, the instantaneous rotation axis of the object is allowed to change with time. The trajectory of the rotation center is approximated by a vector polynomial. The parameters of the model evolve in time so that they can adapt to long term changes in motion characteristics. Based on the assumption that the motion is smooth, object positions and motion in the near future can be predicted, and short missing subsequences can be recovered.

8.1 Coherence of Motion

Perception of three-dimensional motion from images is an integral part of vision. It involves estimation of the nature and parameters of 3-D motion, and as a result, prediction of future positions of moving objects. Human vision is adept at using image sequences to understand and predict motion [8.11]. For example, after a football is kicked off, people can judge whether the football will pass through uprights long before it actually reaches there. In computer vision, cameras must be continuously reoriented to track a moving object for autonomous image acquisition. The motion of a robot arm or a vehicle may have to be estimated and predicted to plan safe motion trajectories. Retrieval and repair of satellites in space requires that the spacecraft rendezvous with the target, which, in turn, requires that the spin and the tumbling motion of the target be detected and estimated first. An understanding of the 3-D motion makes it possible to make predictions about future locations and configurations of the moving objects. Such prediction capability allows planning of manipulatory actions on moving objects, e.g., capturing a spacecraft.

We need to characterize quantitatively general 3-D motion from image sequences. The generality of the problem refers to the lack of knowledge about the structure of the objects undergoing motion as well as the type of motion they are undergoing. For example, it may not even be known if the objects are translating, rotating or precessing, much less the motion parameters. Our goal is to understand the motion with as little *a priori* knowledge as possible. The motion of an object is determined by underlying dynamics. By the analysis of the image sequence under a general dynamic model, the understanding and description of the motion can be derived. Furthermore, based on the motion parameters derived, we can make extrapolations and interpolations through image sequences to predict and recover part of the motion. Clearly, we do not in general know the forces acting on the object and the object structural response to the forces which would otherwise enable us to derive object's 3-D motion from the principles of dynamics. However, it is essential to impose a constraint on the object motion to make the inverse problem of 3-D inference solvable.

In general, the moving objects exhibit a smooth motion, i.e., the motion parameters between consecutive image pairs are correlated. From this assumption and given a sequence of images of a moving rigid object, we determine what kind of local motion the object is undergoing. A *l*ocally *c*onstant *a*ngular *m*omentum model, or LCAM model for short, is introduced. The model assumes short term conservation of angular momentum and a polynomial curve as the trajectory of rotation center. This constraint is the precise statement of what we mean by smoothness of motion. However, we allow the angular momentum, and hence, the motion characteristics of the object to change or evolve over long term. Thus, we do not constrain the object motion by some global model of allowed dynamics.

The kinematic rotational part of our model is a precessional motion derived from the conditions of constant angular momentum and symmetry of the object. However, such dynamic conditions are just the sufficient conditions. For motion analysis through vision, what matters is the kinematics of the object motion. Therefore, any dynamic conditions that cause the processional motion are applicable here. On the other hand, dynamic conditions are important for the kinematics of the motion model, since all the motion in the physical world is governed by dynamics. A kinematic motion model without reasonable dynamic background may rarely or never be applicable to the real world.

We will give a closed-form solution to motion parameters and structure from a sequence of images. As a result of the analysis presented in this chapter, some of the questions that we can answer are: whether there is precession or tumbling; what the precession is if it exists; how the rotation center of the object (which may be an invisible point!) moves in space; what the future motion would probably be; where a particular object point would be located in image frames or in 3-D at the next several time instants; where the object would be if it is missing from a image subsequence, and what the motion before the given sequence could be.

As a consequence of being able to predict future locations of feature points, only a neighborhood of the predicted position may need to be searched to obtain matching points in successive images.

The imposition of the local smoothness of motion constraint helps to combat the errors due to noise. One way to combat the effect of such noise would be to use a good estimator, which we have discused in the previous chapters. Another way would be to use the temporal smoothness of motion, which will be dealt with in this chapter.

Our approach is based on the two-view motion analysis of image sequences consisting of either monocular images, or stereo image pairs. (A view or an image frame refers to the image(s) obtained at one time instance.) Generally, two-view motion does not represent actual continuous motion undergone by the object between the two time instants. The physical location of the rotation axis is not determined by such two-view position transformation. Two-view motion estimation has been discussed in the previous chapters. With an arrangement of a single monocular image sensor, the 3-D translation and the range of the object can be determined up to a scale factor. If stereo images are used, we can determine the absolute translation velocities and the ranges of object points. If the optimal two-view motion solutions are used, the solution of model parameters will certainly be more accurate in the presence of noise.

The approach presented in this chapter is independent of the type of algorithms used to determine two-view motion parameters. To be specific, we use feature points for the discussion here. We assume that there is a single rigid object in motion, the correspondences of points between images are given, and the motion does not exhibit any discontinuities such as those caused by collisions.

8.2 The LCAM Model

We first need to study the laws of physics that govern the motion of a rigid body.

8.2.1 Motion of a Rigid Body in 3-D

All external forces acting on a body can be reduced to a total force **F** acting on a suitable point Q, and a total applied torque **N** about **Q**. For a body moving freely in space, the center of mass is to be taken as the point Q. If the body is constrained to rotate about a fixed point, then that point is to be taken as the point Q. That point may move with the supports. Letting m be the mass of the body, the motion of the center of mass is given by

$$\mathbf{F} = \frac{d}{dt}(m\,\mathbf{V}). \tag{8.2.1}$$

Let **L** be the angular momentum of the body. The torque **N** and the angular momentum **L** satisfy [8.6, 10]:

$$\mathbf{N} = \frac{d\mathbf{L}}{dt}. \tag{8.2.2}$$

The rotation is about the point Q, which will be referred to as the rotation center. In the remainder of this subsection, we concentrate on the rotation part of the motion. The motion of the rotation center Q will be discussed in the next subsection.

In matrix notation, the angular momentum **L** can be represented by

$$\mathbf{L} = G\boldsymbol{\omega}$$

or writing in components:

$$\begin{bmatrix} L_x \\ L_y \\ L_z \end{bmatrix} = \begin{bmatrix} g_{xx} & g_{yx} & g_{zx} \\ g_{xy} & g_{yy} & g_{zy} \\ g_{xz} & g_{yz} & g_{zz} \end{bmatrix} \begin{bmatrix} \omega_x \\ \omega_y \\ \omega_z \end{bmatrix}$$

where

$$g_{xx} = \int (y^2 + z^2) dm, \quad g_{yy} = \int (z^2 + x^2) dm, \quad g_{zz} = \int (x^2 + y^2) dm,$$

$$g_{zx} = g_{xz} = -\int xz \; dm, \quad g_{yx} = g_{xy} = -\int xy \; dm,$$

$$g_{zy} = g_{yz} = -\int zy \; dm.$$

The above integrals are over the mass of the body.

If the coordinate axes are the principal axes of the body [8.6, 10], the inertia tensor G takes the diagonal form:

$$G = \begin{bmatrix} g_{xx} & 0 & 0 \\ 0 & g_{yy} & 0 \\ 0 & 0 & g_{zz} \end{bmatrix}. \tag{8.2.3}$$

Referred to a coordinate system fixed on such a rotating body, (8.2.2) becomes

$$n_x = g_{xx} \dot{\omega}_x + \omega_y \omega_z (g_{zz} - g_{yy}),$$

$$n_y = g_{yy} \dot{\omega}_y + \omega_z \omega_x (g_{xx} - g_{zz}),$$

$$n_z = g_{zz} \dot{\omega}_z + \omega_x \omega_y (g_{yy} - g_{xx}),$$

where $(n_x, n_y, n_z) = \mathbf{N}$. These are known as Euler's equations for the motion of a rigid body. These equations are nonlinear and have generally no closed-form solutions. Numerical methods are generally needed to solve them.

Clearly the motion of a rigid body under external forces is complicated. In fact even under no external forces, the motion remains complex. Perspective projection adds further complexity to the motion as observed in the image. However, in a short time interval, realistic simplifications can be introduced. One simplification occurs if we ignore the impact of the external torque over short time intervals. If there is no external torque over a short time, there is no change in the angular momentum of the object. Thus, if we have a dense temporal sequence of images, we can perform motion analysis over a small number of successive frames under the assumption of locally constant angular momentum. Another simplification occurs if the body possesses an axis of symmetry. The symmetry here means that at least two of i_{xx}, i_{yy}, i_{zz} in (8.2.3) are equal. Cylinders and disks are such examples. Most satellites are also symmetrical or almost symmetrical in this sense.

Under the above two simplifications, Euler's equations are integrable [8.6, 10]. The motion is such that the body rotates about its axis of symmetry **m**, and at the same time the axis rotates about a spatially fixed axis **l**. The motion can be represented by a rotating cone that rolls along the surface of a fixed cone without slipping as shown in Fig. 8.1, where the body is fixed on the rolling cone, the axis of symmetry coincides with that of rolling cone, and the center of mass or the fixed point Q of the body coincides with the apices of the cones. Then, the motion of the rolling cone is the same as the motion of the body. Fig. 8.1 gives three possible configurations of the rolling cone and the fixed cone.

Let $\boldsymbol{\omega}_l$ be the angular velocity at which the rolling cone rotates about **l**, $\boldsymbol{\omega}_m$ be the angular velocity at which the rolling cone rotates about its own axis of symmetry **m**. Then the instantaneous angular velocity $\boldsymbol{\omega}$ is the vector sum of $\boldsymbol{\omega}_l$ and $\boldsymbol{\omega}_m$ as shown in Fig. 8.1. The magnitudes of $\boldsymbol{\omega}_m$ and $\boldsymbol{\omega}_l$ are constant. Thus, the magnitude of the

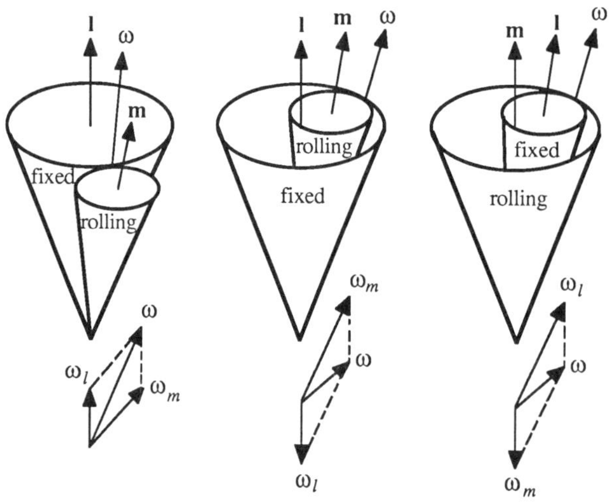

Fig. 8.1. The precessional motion of a symmetrical rigid body.

instantaneous angular velocity is also constant. This kind of motion about a point is called precession in the following sections and it represents the restriction imposed by our model on the allowed object rotation.

A special case occurs when **m** is parallel to **l**. Then $\boldsymbol{\omega}$ is also parallel to **l**. Therefore, the instantaneous rotation axis does not change its orientation in motion. This type of motion is called motion without precession.

8.2.2 Motion of Rotation Center

The location of rotation center $\mathbf{Q}(\mathbf{t})$ changes with time. Assume the trajectory of the rotation center is smooth, or specifically, it can be expanded into a Taylor series:

$$\mathbf{Q}(\mathbf{t}) = \mathbf{Q}(t_0) + \frac{1}{1!}\frac{d\mathbf{Q}(0)}{dt}(t - t_0) + \frac{1}{2!}\frac{d\mathbf{Q}(0)}{dt}(t - t_0)^2 + \cdots. \tag{8.2.4}$$

If the time intervals between image frames are short, we can estimate the trajectory by the first k terms. We get a polynomial of time t. The coefficients of the polynomial are three-dimensional vectors. Letting $\frac{1}{j!}\frac{d^j\mathbf{Q}(0)}{dt^j} = \mathbf{b}_{j+1}$, $j = 0, 1, 2, \cdots, k-1$, we have

$$\mathbf{Q}_i = \mathbf{b}_1 + \mathbf{b}_2(t_i - t_0) + \mathbf{b}_3(t_i - t_0)^2 + \cdots + \mathbf{b}_k(t_i - t_0)^{k-1}. \tag{8.2.5}$$

For simplicity, we assume the time intervals between image frames are constant c, i.e., $t_i = ci + t_0$. From (8.2.5) we get

$$\mathbf{Q}_i = \mathbf{b}_1 + c\,\mathbf{b}_2 i + c^2\mathbf{b}_3 i^2 + \cdots + c^{k-1}\mathbf{b}_k i^{k-1}. \tag{8.2.6}$$

Letting $\mathbf{a}_j = c^{j-1}\mathbf{b}_j$, $j = 1, 2, \cdots, k$, we get

$$\mathbf{Q}_i = \mathbf{a}_1 + \mathbf{a}_2 i + \mathbf{a}_3 i^2 + \cdots + \mathbf{a}_k i^{k-1}. \tag{8.2.7}$$

Equation (8.2.7) is the model for the motion of rotation center. The basic assumption we made is that the trajectory can be approximated by a polynomial. If the motion is smooth and the time interval covered by the model is relatively short, Equation (8.2.7) is a good approximation of the trajectory. In the sense of dynamics, (8.2.7) implies that the total force acting on the center of rotation has vanished high order temporal derivatives.

A polynomial trajectory of center of rotation in (8.2.7) together with the precession model presented in the previous subsection, we have the complete LCAM model. The model is characterized by locally constant angular momentum, i.e., the angular momentum of the moving object can be treated as constant over short time intervals.

A point should be mentioned here. Though we derive the model from the assumption of constant angular momentum and object symmetry, the condition leading to such motion is not unique. In other words, the motion model we derived applies to any moving objects whose rotation can be locally modeled by such motion: the rotation about a fixed-on-body axis that rotates about a spatially fixed axis, and whose translation can be locally modeled by a vector polynomial. It is important to motivate the kinematics from dynamic conditions. But in reality, many different dynamic conditions may result in the same type of motion.

Our goal here is to understand 3-D motion of an object over an extended time period using the two-view motion analysis of images taken at consecutive time instants. Thus we would first estimate the motion parameters of the moving object from the images taken at two time instants, using the approaches discussed in the previous chapters. Such motion parameters give the displacement between two time instants and do not describe the actual motion, since the object can move arbitrarily between the two time instants. The displacement can be represented by a rotation about an axis located at the origin of a world coordinate system, and a translation [8.1]. We have called this displacement two-view motion.

Let the column vector $\mathbf{p}_0$ be the 3-D coordinates of any object point at time t_0. $\mathbf{p}_1$ be that of the same point at time t_1, R_1 be the rotation matrix from time t_0 to t_1, and $\mathbf{T}_1$ be the corresponding translation vector. Then, $\mathbf{p}_0$ and $\mathbf{p}_1$ are related by

$$\mathbf{p}_1 = R_1 \mathbf{p}_0 + \mathbf{T}_1 \tag{8.2.8}$$

where R_1 represents a rotation about an axis through the origin.

Given a set of point correspondences, R_1 and $\mathbf{T}_1$ can be determined by two-view motion analysis. In the case of monocular vision, the translation vector can only be determined up to a positive scale factor, i.e., only the direction of $\mathbf{T}$, $\mathring{\mathbf{T}} = \mathbf{T}/\|\mathbf{T}\|$, can be determined from the perspective projection.

In equation (8.2.8) letting $\mathbf{p}_0$ be at the origin, it is clear that $\mathbf{T}_1$ is just the translation of the point at origin. For any point $\mathbf{Q}_0$, we can translate the rotation axis so that it goes through $\mathbf{Q}_0$ and rotate $\mathbf{p}_0$ about the axis at the new location. Mathematically, from (8.2.8) it follows that

$$\mathbf{p}_1 = R_1(\mathbf{p}_0 - \mathbf{Q}_0) + (R_1\mathbf{Q}_0 + \mathbf{T}_1). \tag{8.2.9}$$

Compared with (8.2.8), (8.2.9) tells us that the same motion can be represented by rotating $\mathbf{p}_0$ about $\mathbf{Q}_0$ by R_1, and then translating by $R_1\mathbf{Q}_0 + \mathbf{T}_1$. Because $\mathbf{Q}_0$ is arbitrarily chosen, there are infinitely many ways to select the location of the rotation axis. This is an ambiguity problem in motion understanding from image sequences. If we let the rotation axis always be located at origin, the trajectory described by R_i and $\mathbf{T}_i$, $i = 1, 2, 3 \cdots$ would be like what is showed in Fig. 8.2, which is very unnatural.

In Fig. 8.2 the real trajectory of the center of the body is the dashed line. However, neither the rotation nor the translation components show this trajectory. As we discussed in Subsection 8.2.1, the center of mass of a body in free motion satisfies Newton's equation of motion of a particle (8.2.1). Rotation is about the center of mass (or fixed point if it exists). Thus, motion should be expressed in two parts, the motion of the rotation center (the center of mass or the fixed point), and the rotation about the rotation center.

Let $\mathbf{Q}_i$ be the position vector of the rotation center at time t_i, R_i be the rotation matrix from t_{i-1} to t_i, $\mathbf{T}_i$ be the translation vector from t_{i-1} to t_i. From (8.2.8) it follows that

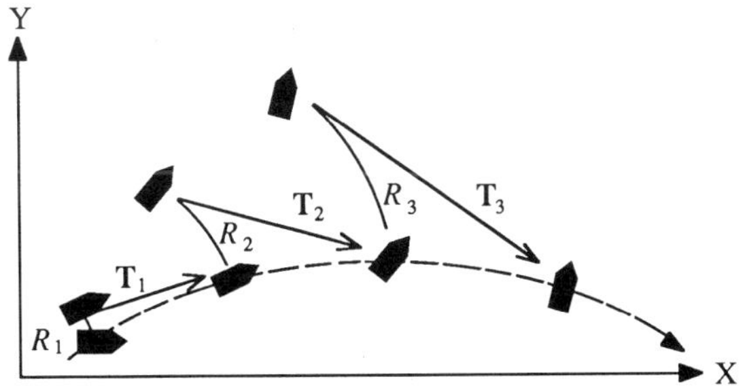

Fig. 8.2. Trajectory described by R_i and $\mathbf{T}_i$ if the rotation axis is always located at the origin.

$$\mathbf{Q}_1 = R_1\mathbf{Q}_0 + \mathbf{T}_1.$$

Or,

$$-R_1\mathbf{Q}_0 + \mathbf{Q}_1 = \mathbf{T}_1.$$

Similarly we get equations for the motion from t_{i-1} to t_i, $i = 1, 2, \cdots, f$:

$$\begin{aligned} -R_1\mathbf{Q}_0 + \mathbf{Q}_1 &= \mathbf{T}_1, \\ -R_2\mathbf{Q}_1 + \mathbf{Q}_2 &= \mathbf{T}_2, \\ &\cdots\cdots \\ -R_f\mathbf{Q}_{f-1} + \mathbf{Q}_f &= \mathbf{T}_f. \end{aligned} \tag{8.2.10}$$

Equations (8.2.10) give the relationship among the locations of the rotation center, the two-view rotation matrices and the two-view translation vectors.

Substituting (8.2.7) into (8.2.10), we get

$$\begin{aligned} &(I - R_1)\mathbf{a}_1 + \mathbf{a}_2 + \mathbf{a}_3 + \cdots + \mathbf{a}_k = \mathbf{T}_1, \\ &(I - R_2)\mathbf{a}_1 + (2I - R_2)\mathbf{a}_2 + (4I - R_2)\mathbf{a}_3 + \cdots + (2^{k-1}I - R_2)\mathbf{a}_k = \mathbf{T}_2, \\ &\cdots\cdots\cdots \\ &(I - R_f)\mathbf{a}_1 + (fI - (f-1)R_f)\mathbf{a}_2 + (f^2I - (f-1)^2R_f)\mathbf{a}_3 + \cdots \\ &\quad + (f^{k-1}I - (f-1)^{k-1}R_f)\mathbf{a}_k = \mathbf{T}_f. \end{aligned} \tag{8.2.11}$$

Vector equations (8.2.11) are referred to as the coefficient equations. Both sides of the equations are three-dimensional vectors. There are f equations in k unknown three-dimensional vectors. Let $\mathbf{A} = (a_1^t, a_2^t, \cdots, a_k^t)^t$, $\mathbf{T} = (\mathbf{T}_1^t, \mathbf{T}_2^t, \cdots, \mathbf{T}_f^t)^t$, D be the coefficient matrix of the unknowns in (8.2.11). Let the element of D at i-th row and j-th column be 3×3 matrix D_{ij}, i.e., $D = [D_{ij}]_{f\times k}$. We have

$$D_{ij} = i^{j-1}I - (i-1)^{j-1}R_i.$$

We can rewrite the coefficient equations (8.2.11) as

$$D\mathbf{A} = \mathbf{T}. \tag{8.2.12}$$

D and $\mathbf{T}$ are determined by two-view motion analysis. The problem here is to determine $\mathbf{A}$, the coefficients of the polynomial in (8.2.7).

8.2.3 Solutions of the Coefficient Equation

Let $f = k$ in (8.2.11). Then the matrix D is a square matrix. We wish to know whether the linear equations (8.2.12) have a solution. If a solution exists, is it unique? If it is not unique, what is the general solution?

The solution of the coefficient equations depends on the types of motion, or the rotation matrices R_i and the translation vectors $\mathbf{T}_i$. Let us first consider a simpler case, where $k = 2$. This means that the trajectory of the rotation center is locally approximated by a motion of constant velocity. Three frames are used in this case. The coefficient equations become

$$(I - R_1)\mathbf{a}_1 + \mathbf{a}_2 = \mathbf{T}_1, \tag{8.2.13}$$

$$(I - R_2)\mathbf{a}_1 + (2I - R_2)\mathbf{a}_2 = \mathbf{T}_2. \tag{8.2.14}$$

Solving for $\mathbf{a}_2$ in (8.2.13) and substituting it into (8.2.14), we get

$$(I - 2R_1 + R_2R_1)\mathbf{a}_1 = (2I - R_2)\mathbf{T}_1 - \mathbf{T}_2. \tag{8.2.15}$$

If $I - 2R_1 + R_2R_1$ is nonsingular, $\mathbf{a}_1$ can be uniquely determined from (8.2.15):

$$\mathbf{a}_1 = (I - 2R_1 + R_2R_1)^{-1}((2I - R_2)\mathbf{T}_1 - \mathbf{T}_2).$$

Then $\mathbf{a}_2$ is determined from (8.2.13):

$$\mathbf{a}_2 = \mathbf{T}_1 - (I - R_1)\mathbf{a}_1.$$

Appendix 8.A shows that $(I - 2R_1 + R_2R_1)$ is nonsingular if and only if the following two conditions are both satisfied: 1) the axes of rotations, represented by R_1 and R_2, respectively, are not parallel. 2) Neither rotation angle is zero. Condition 2) is usually satisfied if the motion is not pure translation. If condition 1) is not satisfied, the solution of equations (8.2.13) and (8.2.14) is not unique and has some structure. To show this, assume the rotation axes of R_1 and R_2 are parallel. Let $\mathbf{w}$ be any vector parallel to these axes. Because any point on the rotation axis remains unchanged after rotation, we have $R_1\mathbf{w} = \mathbf{w}$, $R_2\mathbf{w} = \mathbf{w}$. For any solution $\mathbf{a}_1$ and $\mathbf{a}_2$, $\mathbf{a}_1 + c\,\mathbf{w}$ and $\mathbf{a}_2$ is another solution, where c is an arbitrary real constant. Therefore, there exist infinitely many solutions.

The following theorem presents the results for the general case.

Theorem 8.1. In coefficient equations, let $f = k$. Define S_k^j to be a 3×3 matrix :

$$S_k^j = \sum_{l=0}^{k-j}(-1)^l \binom{k}{l} R_{k-l}R_{k-l-1}\cdots R_{j+1}I, \qquad j = 0, 1, 2, \cdots, k.$$

Define number $u_{i,j}$:

$$u_{i,j} = \sum_{m=1}^{i}(-1)^{i-m} m^{j-1}\binom{i}{m}, \qquad j = i+1, i+2, \cdots, k.$$

Then

$$S_k^0\mathbf{a}_1 = -\sum_{l=1}^{k} S_k^l \mathbf{T}_l,$$

$$\mathbf{a}_k = \frac{1}{(k-1)!}\Big(\sum_{m=1}^{k-1} S_{k-1}^m \mathbf{T}_m + S_{k-1}^0\,\mathbf{a}_1\Big),$$

$$\mathbf{a}_{k-1} = \frac{1}{(k-2)!}\Big(\sum_{m=1}^{k-2} S_{k-2}^m \mathbf{T}_m + S_{k-2}^0\,\mathbf{a}_1 - u_{k-2,k}\,\mathbf{a}_k\Big),$$

$$\mathbf{a}_{k-2} = \frac{1}{(k-3)!}\Big(\sum_{m=1}^{k-3} S_{k-3}^m \mathbf{T}_m + S_{k-3}^0\,\mathbf{a}_1 - u_{k-3,k}\,\mathbf{a}_k - u_{k-3,k-1}\mathbf{a}_{k-1}\Big),$$

....

$$\mathbf{a}_3 = \frac{1}{2!}\Big(\sum_{m=1}^{2} S_2^m \mathbf{T}_m + S_2^0\,\mathbf{a}_1 - u_{2,k}\,\mathbf{a}_k - u_{2,k-1}\mathbf{a}_{k-1} - \cdots - u_{2,4}\mathbf{a}_4\Big),$$

$$\mathbf{a}_2 = \frac{1}{1!}\Big(\sum_{m=1}^{1} S_1^1 \mathbf{T}_1 + S_1^0\,\mathbf{a}_1 - u_{1,k}\,\mathbf{a}_k - u_{1,k-1}\mathbf{a}_{k-1} - \cdots - u_{1,3}\mathbf{a}_3\Big).$$

Proof. See Appendix 8.A.

If S_k^0 is not singular, the first equation given by Theorem 8.1 uniquely determines $\mathbf{a}_1$. Then $\mathbf{a}_k$, $\mathbf{a}_{k-1}$, $\cdots$, $\mathbf{a}_2$ can be determined, sequentially, by the second, third, ..., and last equations in Theorem 8.1. Thus, if S_k^0 is not singular, solution is unique.

Theorem 8.2. In the case of rotation without precession, letting w be any column vector parallel to the rotation axes, then,

$$S_k^0 \mathbf{w} = 0 \tag{8.2.16}$$

and for any vector **a**

$$(S_k^0 \mathbf{a}) \cdot \mathbf{w} = 0. \tag{8.2.17}$$

Proof. See Appendix 8.B.

Using Theorem 8.1 gives

$$S_k^0 \mathbf{a}_1 = -\sum_{l=1}^{k} S_k^l \mathbf{T}_l . \tag{8.2.18}$$

In the case of rotation without precession, equation (8.2.16) implies S_k^0 is singular. From (8.2.17), the left-hand side of (8.2.18) is orthogonal to **w**. However if the real trajectory of the rotation center is not exactly a j-th degree polynomial with $j \le k-1$ in (8.2.7), the right-hand side of (8.2.18) can be any vector, which may not be orthogonal to **w**. This means that no solution exists for equation (8.2.18). If the real trajectory is a j-th degree polynomial with $j \le k-1$, then equation (8.2.18) has a solution by our derivation of (8.2.18). Since equation (8.2.7) is usually only an approximation of the real trajectory, a least-squares solution of (8.2.18) can serve our purpose. Let $\hat{a}_1$ be a least-squares solution of (8.2.18) which is solved by using independent columns of S_k^0. If the rank of S_k^0 is 2, which is generally true for motion without precession, the general solution is then $\mathbf{a}_1 = \hat{\mathbf{a}}_1 + c\,\mathbf{w}$, where c is any real number. All general solutions $\{\hat{\mathbf{a}}_1 + c\,\mathbf{w}\}$ form a straight line in 3-D space. From equation (8.2.7), this line gives the location and direction of the two-view rotation axis of the motion between time instants t_0 and t_1. From Theorem 8.2 it follows that

$$S_{k-1}^0 \mathbf{w} = 0, \;\; S_{k-2}^0 \mathbf{w} = 0, \;\; \cdots, \; S_1^0 \mathbf{w} = 0.$$

Then

$$S_{k-1}^0 \mathbf{a}_1 = S_{k-1}^0 \hat{\mathbf{a}}_1, \;\; S_{k-2}^0 \mathbf{a}_1 = S_{k-2}^0 \hat{\mathbf{a}}_1, \;\; \cdots, \;\; S_1^0 \mathbf{a}_1 = S_1^0 \hat{\mathbf{a}}_1.$$

Based on the equations given by Theorem 8.1, the unknowns $\mathbf{a}_k$, $\mathbf{a}_{k-1}$, $\cdots$, $\mathbf{a}_2$ are determined without knowing the undetermined number c.

If the motion is pure translation without rotation, all the rotation matrices R_i, $i = 1, 2, \cdots, k$, are unit matrix I. S_k^0 is zero matrix. The first three columns of D are zero. $\mathbf{a}_1$ can not be determined by coefficient equations. From Theorem 8.1, $\mathbf{a}_2$, $\mathbf{a}_3$, $\cdots$, $\mathbf{a}_k$, can still be determined by coefficient equations. Because no rotation exists, any point can be considered as a rotation center. Equation (8.2.7) can be used to approximate the trajectory of any object points.

Thus the solutions of the coefficient equations can be summarized as follows.

1. In the case of rotation with precession, the solution of the coefficient equations is generally unique. The trajectory of the rotation center is described by (8.2.7).

2. In the case of rotation without precession, the general solution of $\mathbf{a}_1$ gives the two-view rotation axis of the first two-view motion. All other coefficients $\mathbf{a}_2$, $\mathbf{a}_3$, $\cdots$, $\mathbf{a}_k$ are generally determined uniquely by Theorem 8.1. Thus, the two-view rotation axes of all two-view motions are determined by (8.2.7). Because no precession exists, any point on the rotation axis can be considered as the rotation center. This is the meaning of the general solution $\mathbf{a}_1$. Once a particular point on the rotation axis is chosen as the rotation center, its trajectory is described by equation (8.2.7). Fig. 8.3 shows the possible "parallel" trajectories of the rotation center depending on which point on the axis is chosen as the rotation center.

3. In the case of pure translation without rotation, $\mathbf{a}_2$, $\mathbf{a}_3$, $\cdots$, $\mathbf{a}_k$ can still be determined by coefficient equations. However $\mathbf{a}_1$ can not be determined by coefficient equations. $\mathbf{a}_1$ can be chosen to be the position of any object point at time t_0. Then equation (8.2.7) describes the trajectory of this point.

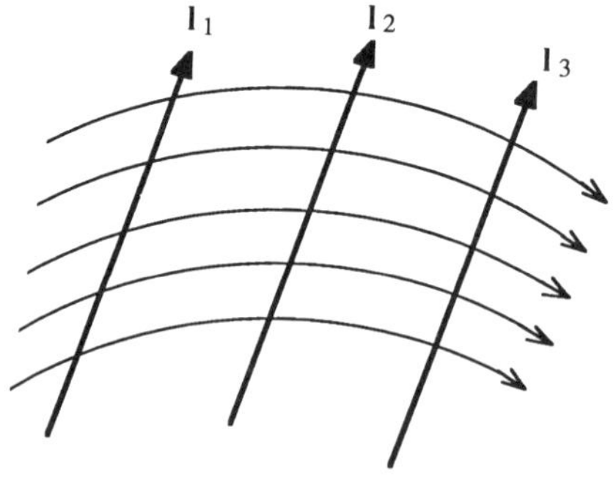

Fig. 8.3. The possible trajectories of rotation centers when rotation axes are parallel.

In the presence of noise, both a large number of point correspondences and a large number of image frames provide overdetermination. The algorithm presented in Chapter 7 can be used for the closed-form least-squares solution of two-view motion parameters. To use overdetermination based on a large number of frames, we let $f > k$ in the coefficient equations (8.2.11). In fact, the coefficient matrix S_k^0 is essentially a high order deference. This is shown in Lemma 8.1 of Appendix 8.A. S_k^0 tends to be ill-conditioned when k gets large. This means $f > k$ is more important when k is large. If $f > k$, equation (8.2.12) can be solved by a least-squares method. We find a solution $\mathbf{A}$ to minimize

$$\|D\,\mathbf{A} - \mathbf{T}\|.$$

In the case of motion with precession, all the columns of D are generally independent. The least-squares solution is

$$\mathbf{A} = (D^t D)^{-1} D^t \mathbf{T}.$$

In the case of motion without precession, the column vectors of D are linearly dependent. This can be shown by letting $\mathbf{a}_1$ in equation (8.2.11) be a non-zero vector parallel to the two-view rotation axes. Then the first three columns of D linearly combined by $\mathbf{a}_1$ is zero vector. To get the least-squares solution of the coefficient equations (8.2.11), the largest set of independent columns of D should be found or tolerance-based column pivoting should be made [8.9]. Theorem 8.1 solves $\mathbf{a}_2, \mathbf{a}_3, \cdots, \mathbf{a}_k$. This means the last $3k-3$ columns of D are always independent. In the presence of noise the columns of D are very unlikely to be exactly linearly dependent even in the case of motion without precession.

8.2.4 Continuous and Discrete Motions

The LCAM model we discussed is based on continuous precessional motion. We must find the relationship between continuous precession and two-view motion, before we can estimate the precessional parameters of our model based on discrete two-view motions.

As we discussed in Subsection 8.2.1, a precession can be considered as the motion of a rolling cone which rolls without slipping upon a fixed cone. The angular frequency at which the symmetrical axis of the rolling cone rotates about the fixed cone is constant.

Assuming at time t_1, an edge point A′ on the rolling cone touches an edge point A on the fixed cone as shown in Fig. 8.4. After a certain amount of rolling, the touching points become B′ on the rolling cone and B on the fixed cone at time t_2. Let θ be the central angle of points A′ and B′, and ϕ be that of A and B. Let r and r' be the radii of circles O and O′, respectively. The arc length between A and B is equal to that between A′ and B′. Thus, $\phi r = \theta r'$ or $\phi \sin\alpha = \theta \sin\beta$, where α and β are generating angles of the fixed cone and the rolling cone, respectively. We get

$$\frac{\theta}{\phi} = \frac{\sin\alpha}{\sin\beta}. \tag{8.2.19}$$

The precession consists of two rotational components. One is the rotation of the rolling cone about its own symmetrical axis. The other is the rotation of the rolling cone about the fixed cone. From Fig. 8.4 it can be readily seen that the relative position of the rolling cone and the fixed cone is uniquely determined if the touching points of the two cones are determined. Or alternatively, starting from the previous position, the new position of the rolling cone is determined if the two angles ϕ and θ are determined. Thus, no matter how we order these two rotational components, the final positions are identical as long as the angle ϕ and θ are kept unchanged. We can first rotate the rolling cone about its axis **m** and then rotate the rolling cone about the axis of the fixed cone, **l**, or vice versa.

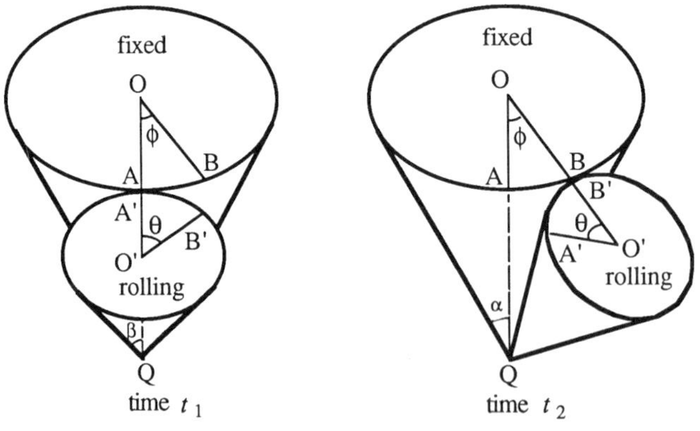

Fig. 8.4. The relation between rotation angles θ and ϕ.

We hope to find the equivalent two-view rotation axis of this continuous motion between two frames at time t_1 and time t_2, respectively in Fig. 8.4. If we can find two fixed points which stay in the same positions before and after the motion, then the two-view rotation axis must go through these points. One trivial fixed point is the apex Q of the cones. Another fixed point can be found as follows: In Fig. 8.5 let the midpoint of arc AB touch the rolling cone (at time $(t_1 + t_2)/2$). Extend line OB so that it intersects the plane containing Q, O′ and B′ at a point P_1. Extend line OA so that it intersects the plane containing Q, O′ and A′ at a point P_2. Draw a circle centered at O and passing through P_1 and P_2. Then the midpoint P of arc P_1P_2 is a fixed point. This can be seen by noting that the rolling cone can also reach its position at next time instant t_2 in an alternative manner as follows. First, rotate the rolling cone (slipping along the fixed cone) about **l** by angle $\phi/2$, thus rotating P to its new position at P_1, and axis **m** reaches the position shown in Fig. 8.5. Then rotate the rolling cone (slipping on the fixed cone) about its own axis **m** by angle θ. Point P now reaches position P_2. Finally, rotate the rolling cone (slipping along the fixed cone) about **l** again by angle $\phi/2$, taking the rolling cone to the position at time instant t_2. This takes the point P back to its starting position. Therefore, the two-view rotation axis ϕ found by two-view motion analysis from two image frames, goes through Q and P. Notice that the angular frequency, at which the

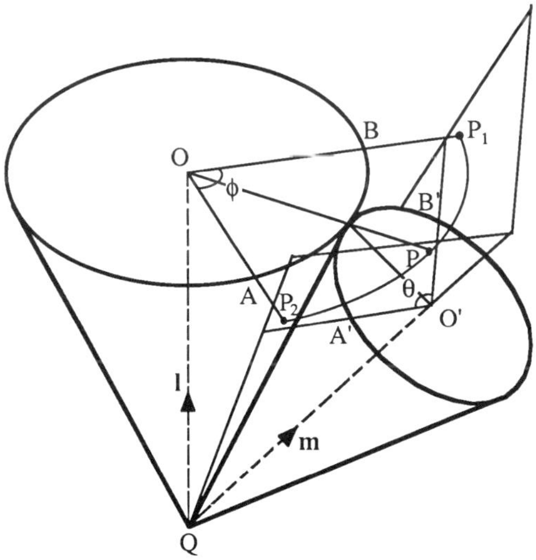

Fig. 8.5. Finding fixed points for two-view rotation.

symmetrical axis of the rolling cone rotates about the fixed cone is constant. From the way of finding P, it is clear that the two-view rotation axis also rotates about **l** by a constant angle between consecutive frames. We have the following theorem:

Theorem 8.3. If a rigid body undergoes a precessional motion of the LCAM model, the two-view rotation axis between constant time intervals changes by rotating about the precessional vector by a constant angle. □

Without loss of generality, we assume the time intervals between consecutive image frames are of unit length. We define the *precessional vector* to be a unit vector **l** parallel to the symmetrical axis of the fixed cone, define the *precessional angular frequency* ϕ to be the angular frequency at which the symmetrical axis of rolling cone rotates about the precessional axis, define the *i-th body vector* $\mathbf{m}_i$ to be a unit vector parallel to the symmetrical axis of the rolling cone at time t_i, and define the *body rotation angular frequency* θ to be the angular frequency at which the rolling cone rotates about its symmetrical axis (see Fig. 8.6).

From image sequences we find estimates of two-view motion parameters. They are the *i-th two-view rotation axis vector,* $\mathbf{n}_i$, a unit vector parallel to the two-view rotation axis between time instants t_{i-1} and t_i; the corresponding *i-th two-view rotation angle* ψ_i and *i-th two-view translation vector* $\mathbf{T}_i$. Fig. 8.6 shows the precession parameters of continuous motion and discrete two-view motion.

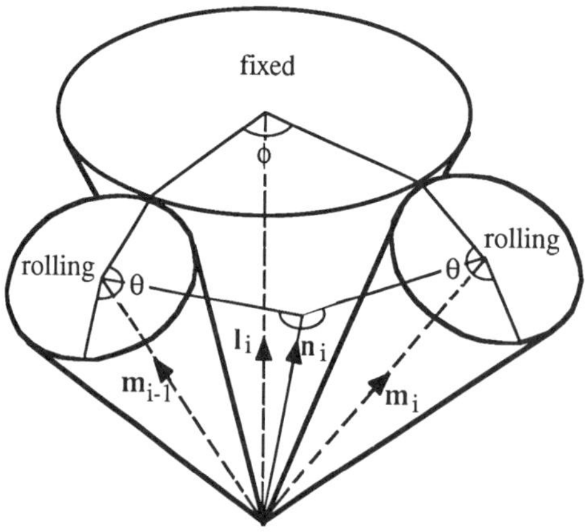

Fig. 8.6. Parameters of continuous precession and discrete two-view motion.

Letting $R(\mathbf{n},\theta) = [r_{ij}]$ denote the rotation matrix representing a rotation with axis unit vector $\mathbf{n} = (n_x, n_y, n_z)$ and rotation angle θ, then $R(\mathbf{n},\theta)$ is given by

$$\begin{bmatrix} (n_x^2-1)(1-\cos\theta)+1 & n_x n_y(1-\cos\theta)-n_z\sin\theta & n_x n_z(1-\cos\theta)+n_y\sin\theta \\ n_y n_x(1-\cos\theta)+n_z\sin\theta & (n_y^2-1)(1-\cos\theta)+1 & n_y n_z(1-\cos\theta)-n_x\sin\theta \\ n_z n_x(1-\cos\theta)-n_y\sin\theta & n_z n_y(1-\cos\theta)+n_x\sin\theta & (n_z^2-1)(1-\cos\theta)+1 \end{bmatrix}. \tag{8.2.20}$$

Theorem 8.4. The continuous precession parameters and discrete two-view motion parameters are related by

$$R(\mathbf{l},\phi)R(\mathbf{m}_{i-1},\theta) = R(\mathbf{n}_i,\psi_i), \tag{8.2.21}$$

$$R(\mathbf{m}_i,\theta)R(\mathbf{l},\phi) = R(\mathbf{n}_i,\psi_i). \tag{8.2.22}$$

Proof. From time t_{i-1}, to time t_i, the body moves from its previous position to a new position. From Fig. 8.6 the new position of the rolling cone (or the body) can be reached in the following way: First, the rolling cone rotates about its body vector $\mathbf{m}_{i-1}$ by angle θ. Then, the rolling cone rotates about the precessional vector $\mathbf{l}$ by angle ϕ. The two-view motion combines these two motions into one, which is the rotation about the two-view rotation axis vector $\mathbf{n}_i$ by angle ψ_i. We get equation (8.2.21). Similarly if we change the order of these two rotational components we get equation (8.2.22). □

From Theorem 8.3, the two-view rotation axis rotates about the precessional vector. Therefore, the precessional vector $\mathbf{l}$ is perpendicular to $\mathbf{n}_i - \mathbf{n}_{i-1}$ and $\mathbf{n}_{i-1} - \mathbf{n}_{i-2}$. The sign of $\mathbf{l}$ is arbitrary. Thus, $\mathbf{l}$ can be determined by

$$\mathbf{l} = \frac{(\mathbf{n}_i - \mathbf{n}_{i-1})\times(\mathbf{n}_{i-1} - \mathbf{n}_{i-2})}{\|(\mathbf{n}_i - \mathbf{n}_{i-1})\times(\mathbf{n}_{i-1} - \mathbf{n}_{i-2})\|}. \tag{8.2.23}$$

We will assume that precessional angular frequency, body rotation angular frequency and two-view rotation angle are not larger than half a turn between every two consecutive frames. This assumption is necessary in practice, since the rotation must be small enough for matching to be possible. The precessional angular frequency ϕ is equal to the the angle between $\mathbf{n}_i\times\mathbf{l}$ and $\mathbf{n}_{i-1}\times\mathbf{l}$:

$$|\phi| = \cos^{-1}\frac{(\mathbf{n}_i \times \mathbf{l})\cdot(\mathbf{n}_{i-1}\times \mathbf{l})}{\|\mathbf{n}_i\times\mathbf{l}\| \ \|\mathbf{n}_{i-1}\times\mathbf{l}\|}. \tag{8.2.24}$$

The sign of ϕ is the same as the sign

$$(\mathbf{n}_{i-1}\times\mathbf{n}_i)\cdot\mathbf{l}. \tag{8.2.25}$$

After $\mathbf{l}$ and ϕ are found by equations (8.2.23), (8.2.24) and (8.2.25), $R(\mathbf{l},\phi)$ can be calculated by (8.2.20). $R(\mathbf{m}_{i-1},\theta)$ and $R(\mathbf{m}_i,\theta)$ can be determined by (8.2.21) and (8.2.22):

$$R(\mathbf{m}_{i-1}, \theta) = R^{-1}(\mathbf{l}, \phi)R(\mathbf{n}_i, \psi_i), \tag{8.2.26}$$

$$R(\mathbf{m}_i, \theta) = R(\mathbf{n}_i, \psi_i)R^{-1}(\mathbf{l}, \phi). \tag{8.2.27}$$

We can determine $\mathbf{m}_{i-1}$, $\mathbf{m}_i$ and θ by (8.2.26) and (8.2.27), because $\mathbf{n}$ and θ can be determined from $R(\mathbf{n}, \theta)$ [8.1]:

$$\theta = \pm\cos^{-1}\left(\frac{r_{11}+r_{22}+r_{33}-1}{2}\right),$$

$$\mathbf{l} = \frac{[r_{32}-r_{23}, r_{13}-r_{31}, r_{21}-r_{12}]^t}{\|[r_{32}-r_{23}, r_{13}-r_{31}, r_{21}-r_{12}]\|}.$$

We get the following theorem.

Theorem 8.5. The precessional vector, precessional angular frequency, body axes and body rotation angular frequency which define the precession part of the LCAM model can all be determined from three consecutive two-view motions, or four consecutive image frames. □

In addition to these basic parameters which uniquely determine the motion of the model, some other parameters can also be determined from these basic parameters. For example, the generating angles α and β of the fixed cone and the rolling cone, respectively, in Fig. 8.4 can also be determined from $\mathbf{l}$, ϕ, $\mathbf{m}_i$, θ and equation (8.2.19).

8.3 Estimation and Prediction

The LCAM model is applied to subsequences of the images successively. The parameters of the model are estimated for every (overlapping) subsequence. The estimated model parameters can then be used to describe the current local motion. The following questions can be answered. Is there precession? If so, what are the precession parameters? What are the current or previous body vectors? What is the body rotation angular frequency? What is the probable motion for the next several time intervals? What are the probable locations of the feature points at the next several time instants? If the moving object is occluded in some of the previous image frames, what are the motion and the locations of these feature points during that time period?

The number of frames covered by a LCAM model can be made adaptive to the current motion. The number can be changed continuously to cover as many frames as possible so long as the constant angular momentum assumption is approximately true during the time period to be covered. The value of the number of frames chosen can be based on the accuracy with which the model describes the current set of consecutive frames. The residuals of least-squares solutions and the variances of the model parameter samples indicate the accuracy. The noise level also affects the residuals and the variances of parameter samples. However, the noise level is relatively constant or can be measured. The resolution of the cameras and the viewing angle covering the object generally determine the noise level. The noise can be smoothed by determining the best time intervals and the number of frames covered by the model, according to the current motion. Because the LCAM model is relatively general, the time interval a LCAM model can cover is expected to be relatively long in most cases.

The following part deals with the estimation of model parameters using overdetermination. Although one can derive formulation for minimum variance estimator here, the computation of the estimates requires iterations. We will discuss this optimal solution in Section 8.5. Here we give a closed-form solution that uses overdetermination. The solution can be directly computed without resorting to iterations.

After finding two-view rotation axis vectors $\mathbf{n}_1, \mathbf{n}_2, \cdots, \mathbf{n}_f$, precessional vector $\mathbf{l}$ should be orthogonal to $\mathbf{n}_2 - \mathbf{n}_1$, $\mathbf{n}_3 - \mathbf{n}_2$, $\cdots$, $\mathbf{n}_f - \mathbf{n}_{f-1}$. However, because of noise, this may not be true. Thus, we find $\mathbf{l}$ such that the sum of squares of the projections of $\mathbf{l}$ onto $\mathbf{n}_2 - \mathbf{n}_1, \mathbf{n}_3 - \mathbf{n}_2, \cdots, \mathbf{n}_f - \mathbf{n}_{f-1}$ is the smallest. Let

$$A = \begin{bmatrix} (\mathbf{n}_2 - \mathbf{n}_1)^t \\ (\mathbf{n}_3 - \mathbf{n}_2)^t \\ \cdots\cdots \\ (\mathbf{n}_f - \mathbf{n}_{f-1})^t \end{bmatrix}.$$

We are to find unit vector $\mathbf{l}$ in

$$\min_{\mathbf{l}} \|A\,\mathbf{l}\|, \quad \text{subject to: } \|\mathbf{l}\| = 1. \tag{8.3.1}$$

The solution $\mathbf{l}$ of (8.3.1) is the unit eigenvector corresponding to the smallest eigenvalue of A^tA (see for example, [8.2]).

Let the precessional angular frequency determined from (8.2.24) and (8.2.25) be ϕ_i. The precessional angular frequency of the model ϕ can be estimated by the mean

$$\phi = \frac{1}{f-1}\sum_{i=2}^{f}\phi_i .$$

Let the body rotation angular frequency determined by (8.2.27) be θ_i. Body rotation angular frequency of the model can be estimated by the mean

$$\theta = \frac{1}{f}\sum_{i=1}^{f}\theta_i .$$

Body vectors are estimated by averaging two consecutive two-view motion using (8.2.26) and (8.2.27), respectively.

Two-view angular frequency can also be estimated by mean

$$\psi = \frac{1}{f}\sum_{i=1}^{f}\psi_i .$$

According to the motion model, the $f+1$st two-view rotation axis vector $\mathbf{n}_{f+1}$ is $R(\mathbf{l}, (f-i+1)\phi)\mathbf{n}_i$ for any $1 \le i \le f$. In the presence of noise, we use the mean over all previous two-view motions to predict the next two-view rotation axis vector:

$$\mathbf{n}_{f+1} = f^{-1}\sum_{i=1}^{f} R(\mathbf{l}, (f-i+1)\phi)\mathbf{n}_i .$$

From (8.2.9) the next position of point $\mathbf{x}_f$ in frame f-th frame is predicted by

$$\mathbf{x}_{f+1} = R(\mathbf{n}_{f+1}, \psi)(\mathbf{x}_f - \mathbf{Q}_f) + \mathbf{Q}_{f+1}$$

where $\mathbf{Q}_f$ and $\mathbf{Q}_{f+1}$ are determined by (8.2.7). The prediction can be made for more than $p \geq 2$ frames by using the following equations successively.

$$\mathbf{n}_{f+p} = R(\mathbf{l}, \phi)\mathbf{n}_{f+p-1},$$

$$\mathbf{x}_{f+p} = R(\mathbf{n}_{f+p}, \psi)(\mathbf{x}_{f+p-1} - \mathbf{Q}_{f+p-1}) + \mathbf{Q}_{f+p}.$$

If the object was occluded in parts of image sequences, the positions and orientations of the object as well as the locations of the feature points on the object can be recovered by an interpolation similar to the prediction procedure discussed above. For the motion of the rotation center, occlusion just means that some rows in the coefficient equations are missing. The solution can still be found if we have enough rows. For the precession part of the motion, the interpolation can be made in a way similar to prediction or extrapolation. When making interpolation we use both the "history" and the "future" of the missing part. For prediction, only the "history" is available. Furthermore, we can also extrapolate backwards to find "history", i.e., to recall what has not been seen before. The essential assumption is that the motion is smooth.

8.4 Monocular Vision

For the monocular case, 3-D positions can be predicted only up to a global scale factor from two images. When more interframe motions are involved, how many scale factors that are unable to be determined? The answer is one, provided that at least one point is visible among every three consecutive image frames. In other words, from point correspondences over a sequence of monocular image sequences, the depth of any visible points and translation between any two image frames can be determined up to a scale factor c. Therefore, although the 3-D position of the points can only be determined up to a scale factor, the image coordinates of the points can be predicted without knowing the scale factor, since the scale factor c is canceled out in image coordinates. In the following, we derive those results.

Suppose a point P is located at $\mathbf{x}_i = (x_i, y_i, z_i)^t$ at time t_i, $i = 0, 1, 2, \cdots$. The image vectors are defined by $\mathbf{x}_i = z\mathbf{X}_i$. From image frame f_i to image frame f_{i+1} the motion, called i-th motion, is a rotation represented by R_i followed by a translation represented by $\mathbf{T}_i$.

$$\mathbf{x}_i = R_i \mathbf{x}_{i-1} + \mathbf{T}_i,$$

or

$$\frac{z_i}{\|\mathbf{T}_i\|}\mathbf{X}_i = R_i \frac{z_{i-1}}{\|\mathbf{T}_i\|}\mathbf{X}_{i-1} + \mathring{\mathbf{T}}_i,$$

$i = 1, 2, \cdots, n$. As discussed in Chapter 3, we can determine relative depths

$$\tilde{z}_i \triangleq \frac{z_i}{\|\mathbf{T}_i\|}, \tag{8.4.1}$$

and

$$\tilde{z}'_{i-1} \triangleq \frac{z_{i-1}}{\|\mathbf{T}_i\|},$$

for $i = 1, 2, \cdots, n$, assuming the point P is visible for t_0 to t_n. We have

$$\frac{\tilde{z}_i}{\tilde{z}'_i} = \frac{\|\mathbf{T}_{i+1}\|}{\|\mathbf{T}_i\|}. \tag{8.4.2}$$

Multiplying both sides for $i = 1$ to $i = k$ yields

$$\prod_{i=1}^{k} \frac{\tilde{z}_i}{\tilde{z}'_i} = \frac{\|\mathbf{T}_k\|}{\|\mathbf{T}_1\|}. \tag{8.4.3}$$

Letting $\|\mathbf{T}_1\| \triangleq s$ be the unknown scale factor, (8.4.3) gives

$$\|\mathbf{T}_k\| = s\prod_{i=1}^{k} \frac{\tilde{z}_i}{\tilde{z}'_i}. \tag{8.4.4}$$

That is, the norm of k-th translation, $k = 1, 2, \cdots, n$, is a product of the scale factor s and a number

$$\prod_{i=1}^{k} \frac{\tilde{z}_i}{\tilde{z}_i'} \tag{8.4.5}$$

which can be determined from the relative depths. Therefore, although many interframe motions are involved in a long monocular sequence, only one unknown scale factor exists, instead of many. If the norm of translation in any interframe motion is known, s is determined from (8.4.3) and then motion and structure in the entire sequence is determined completely. Similarly, if the absolute depth of a point at a time instance is known, the norm of translation vector is determined from (8.4.1), and then so does motion and structure in the entire sequence.

Now what happens if there exists no point that is visible through the entire sequence? We can see from (8.4.2) that (8.4.3) and (8.4.4) still hold true if each factor

$$\frac{\tilde{z}_i}{\tilde{z}_i'} \tag{8.4.6}$$

corresponds to a different point. For $\tilde{z}_i$ and $\tilde{z}_i'$ to be determined for a point P, the point P must be visible from t_{i-1} to t_{i+1} in three consecutive image frames. Therefore, as long as through any three consecutive frames there is at least one point visible, the norm of any translation vector can be completely determined from s as in (8.4.4). Now we have proved the following theorem.

Theorem 8.6. Suppose the rotation matrix and direction of the translation between every consecutive monocular image pairs can be determined. The magnitude of any interframe translation and depths of visible points can be determined up to the same scale factor s, provided that through every three consecutive images at least one point is visible. □

Since the depth of every point can be determined up to the same scale factor s, the 3-D position of the points can be predicted up to s as we discussed in the previous section. Since the image coordinates of a point cancel out this scale factor, the image coordinates of the point can be predicted without knowing the scale factor. For example, we can let $s = 1$ and compute the predicted 3-D position in the next frame based on the previous motion trajectory. Then by projecting the predicted 3-D point onto the image plane, we get the predicted image position of the point. This predicted position is the same for any positive s.

In the presence of noise, the error in a single point may significantly influence the accuracy of the product in (8.4.5). Therefore, the product should be determined based on many points instead of one. For example, the value of (8.4.6) can be averaged over many points before it is used to evaluate (8.4.5).

8.5 Optimization

The algorithm discussed above gives closed-form solutions to the model parameters. Simple least-squares solution is obtained when redundant data are available. Those solutions are generally good preliminary estimates. However, since the statistics of noise distribution is not employed, the solution is not optimal. If iterations are allowed, one can obtain higher accuracy through optimization.

The optimization for more than two image frames is very similar to the case of two frames (either monocular or stereo). We just need a natural extension. The two-step approach can be used. First, the algorithm discussed above provides initial guesses. Then the initial guesses are improved through iterative optimization. Similar to the notation in Section 7.4, let $\mathbf{m}$ denote the motion model parameter vector, $\mathbf{x} = (\mathbf{x}_1, \mathbf{x}_2, \cdots, \mathbf{x}_n)$ denote the structure of the scene at time t_0, $\mathbf{u}_{ijk}$ denote the observed i-th image point in the j-th camera (e.g., left: 1st, right: 2nd) at time t_k $(k = 0, 1, \cdots, f)$, and $\mathbf{h}_{ijk}(\mathbf{m}, \mathbf{x})$ denote the computed projection of the i-th point in the j-th camera at time t_k. If the additive noise in the coordinates of the image point is independently and identically distributed Gaussian noise,

$$\hat{\mathbf{u}}_{ijk} = \mathbf{h}_{ijk}(\mathbf{m}, \mathbf{x}) + \delta_{ijk}.$$

The maximum likelihood estimator of M and $\mathbf{x}$ is the one that minimize image plane errors:

$$\sum_{i=1}^{n}\sum_{j=1}^{2}\sum_{k=0}^{f} \|\hat{\mathbf{u}}_{ijk} - \mathbf{h}_{ijk}(\mathbf{m},\mathbf{x})\|^2.$$

Again, to make the dimension of the iterative search space low, only the independent model parameters should be included in $\mathbf{m}$. They include *precessional velocity* vector defined as $\phi\mathbf{l}$, 0-th *body velocity* vector defined as $\theta\mathbf{m}_0$, and coefficient vectors in the coefficient equation $\{\mathbf{a}_i\}$. Those parameters uniquely determine the two-view motion parameters through (8.2.21) and (8.2.22). Since the optimal 3-D position of a point $\mathbf{x}_i$ at time t_0 depends only on model parameters and the observations of the point in image sequences, the optimal $\mathbf{x}_i$ can be determined by minimizing

$$g_i(\mathbf{m}) \triangleq \min_{\mathbf{x}_i} \sum_{j=1}^{2} \sum_{k=0}^{f} \|\hat{\mathbf{u}}_{ijk} - \mathbf{h}_{ijk}(\mathbf{m},\mathbf{x}_i)\|^2. \tag{8.5.1}$$

The solution for $\mathbf{x}_i$ that minimizes (8.5.1) can be estimated in a closed-form as in Section 7.4. Then, using the space decomposition technique gives

$$\min_{\mathbf{m},\,\mathbf{x}}\{\sum_{i=1}^{n}\sum_{j=1}^{2}\sum_{k=0}^{1} \|\hat{\mathbf{u}}_{ijk} - \mathbf{h}_{ijk}(\mathbf{m},\mathbf{x})\|^2\} = \min_{\mathbf{m}} \sum_{i=1}^{n} g_i(\mathbf{m}).$$

This decomposition significantly reduces the dimension of search space. If the two-view motion parameters used to compute the model parameters are optimized, the initial guesses of $\mathbf{m}$ are usually very good. This will significantly reduce the number of iterations in computing the optimal $\mathbf{m}$. It is clear that, for monocular vision or any multi-ocular vision, we just need to change the upper limit of j accordingly.

Minimum variance estimator under a general noise distribution is the same as the above maximum likelihood estimator. Cramér-Rao lower error bound can also be evaluated in a way similar to that in Chapter 7.

If the image sequence is extremely long, it is computationally expensive to use batch nonlinear iterative optimization over entire image sequence. As discussed in Chapter 4, sequential processing is computationally less expensive but it performs poorly. The recursive-batch processing introduced in Section 4.7 can be used here. In fact, the LCAM model covers a certain number of frames only. This is especially suitable for recursive-batch processing. The recursive-batch process optimizes for the set of frames covered by the LCAM model at a time.

8.6 Experimental Examples

We present computer simulations as well as an experiment with a real image sequence. The results shown here were computed based on the closed-form solutions we discussed.

8.6.1 Computer Simulation

In the simulation, a sequence of 3-D coordinates of points on a moving cube were generated by a program. The motion of the rotation center is characterized by the coefficient vectors $\mathbf{a}_i$ in (8.2.7). By Theorem 8.3, the rotation of the body is about the two-view rotation axis $\mathbf{n}$ by angle θ between consecutive frames. The two-view rotation axis rotates about a fixed precessional vector $\mathbf{l}$ by angle ϕ. The object is assumed to be a transparent cube of side length 10cm. Simulated cameras are 100cm away from the object. The viewing lines from the two cameras to the object form an angle of 45°. The feature points used for point correspondences are the vertices of the cube. As the object is undergoing motion, these points generate a sequence of 3-D coordinates. These 3-D coordinates are digitized by simulated cameras of resolution from 64×64 up to 512×512. The cube covered about one fourth of the area of the whole image. Thus, the actual image resolution is reduced by a factor of 2, i.e., if the image resolution is denoted as 64×64 in the following figures, actually about 32×32 resolution is used.

The errors in the image coordinates of the feature points arise from several sources such as digitization noise, corner detector errors and lens distortion errors, etc. In a well calibrated data acquisition system, lens distortion is well compensated and thus could be ignored. The corner detector errors can be approximated as digitization noise. For example, an error range of plus or minus one pixel in the corner detector could be approximated by reducing the image resolution by a factor of two. The resolution of 64×64 approximates the case of a 512×512 image with the additional corner detector error of range ±4 pixels. For the simulations, the object covers only about one fourth of the images. Therefore, if the object covered the whole image, the errors shown in the simulation data should correspond to images at half the resolutions.

In the simulation the noise is represented by the choice of image resolution, i.e., noise is due to the spatial quantization of the image plane. An increase in noise can be represented by a decrease in image resolution. Because of digitization errors, the set of points ceases to satisfy the rigid body constraint. Therefore, to find the best two-view motion parameters, a least-squares solution is obtained using the algorithm presented in Section 7.3. No optimization has been performed for the results presented in this section.

For the data presented, the following motion parameters are used: the precession vector: $\mathbf{l} = (0, 0, 1)^t$; the precession angular frequency: $\phi = 0.4$; the two-view rotation axis of first two-view motion: $\mathbf{n}_1 \,//\, (1, 0, 4)^t$; the coefficient vectors:

$$\mathbf{a}_1 = (-2, -3, -1)^t,$$

$$\mathbf{a}_2 = (0.5, 0.5, 0.25)^t,$$

$$\mathbf{a}_3 = (0.005, 0.005, 0.0025)^t.$$

The results given are the mean values from 20 trials, each of which randomly chooses the original orientation of the cube.

Fig. 8.7 describes the estimation errors of $\mathbf{l}$ and ϕ. The errors of $\mathbf{l}$ are defined as the length of the difference vector of the estimated and the real unit vectors. The errors approach zero very fast as the number of two-view motions increases. Based on the estimated model parameters, the predictions of the 3-D coordinates of each point at the next several time instants are made. The predicted and the actual digitized positions are compared. Their distances are considered as prediction errors. The relative errors are defined as the prediction errors divided by the length of the diagonal of the cube. The results of predicting more frames are presented in Fig. 8.8. Fig. 8.9 shows the mean relative prediction errors at time t_5 versus image resolution for different numbers of point correspondences. Fig. 8.10 shows the mean relative prediction errors versus image resolution for different numbers of two-view motions covered. Fig. 8.11 gives the mean relative prediction errors versus the number of two-view motions covered for different numbers of point correspondences.

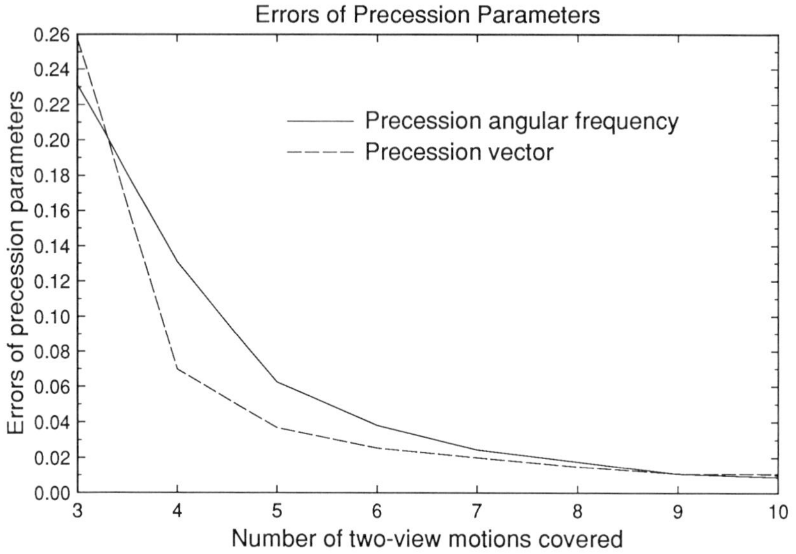

Fig. 8.7. Errors of precessional vector **l** and precessional angular angular frequency ϕ versus the number of two-view motions covered Number of point correspondences: 3; frame predicted: next; image resolution: 256×256.

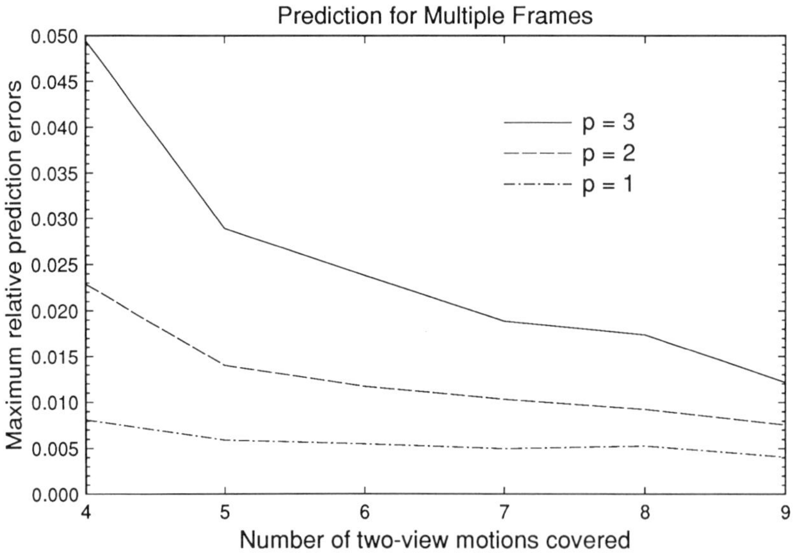

Fig. 8.8. Maximum relative prediction errors versus the number of two-view motions covered for different numbers of frames predicted (p). Frame predicted: next p-th frame; number of point correspondences: 3; image resolution: 512×512; degree of polynomial $k-1$ in equation (8.2.7): 2.

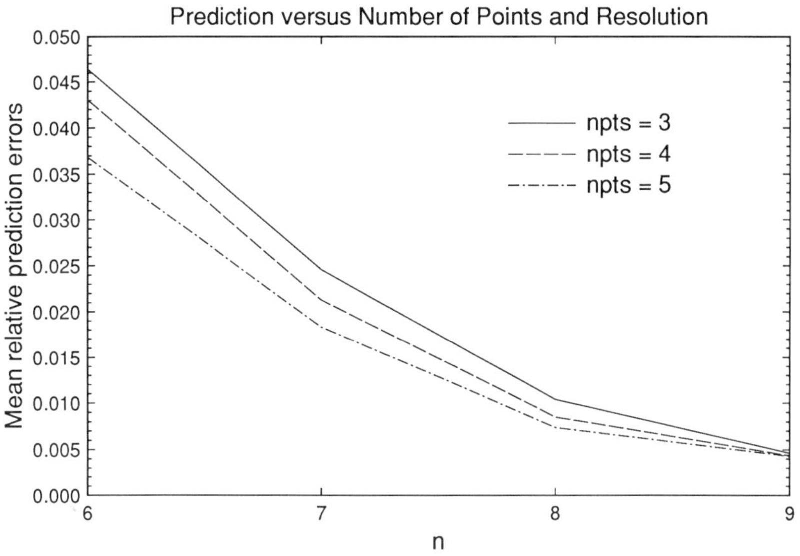

Fig. 8.9. Mean relative prediction errors versus the image resolution ($2^n \times 2^n$) for different numbers of point correspondences (npts). Frame predicted: next frame; number of two-view motions covered: 5; degree of polynomial $k-1$ in equation (8.2.7): 2.

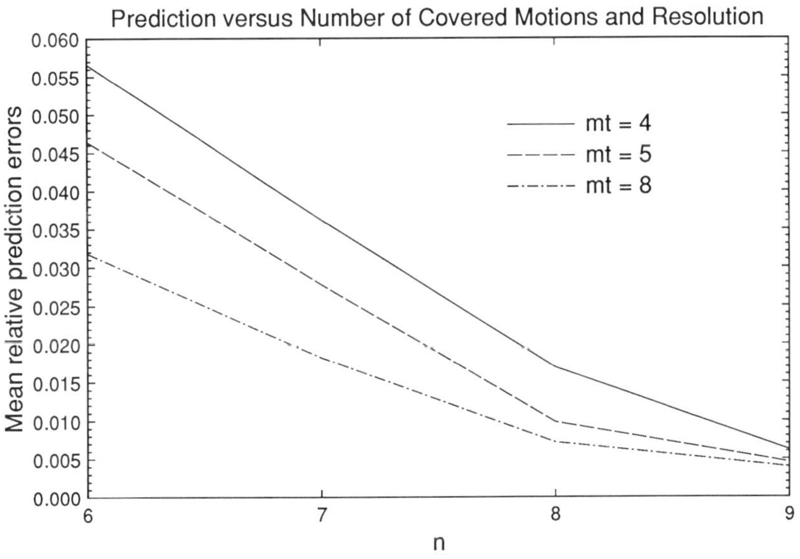

Fig. 8.10. Mean relative prediction errors versus image resolution ($2^n \times 2^n$) for different numbers of two-view motions covered (mt). Frame predicted: next frame; number of point correspondences: 3; degree of polynomial $k-1$ in equation (8.2.7): 2.

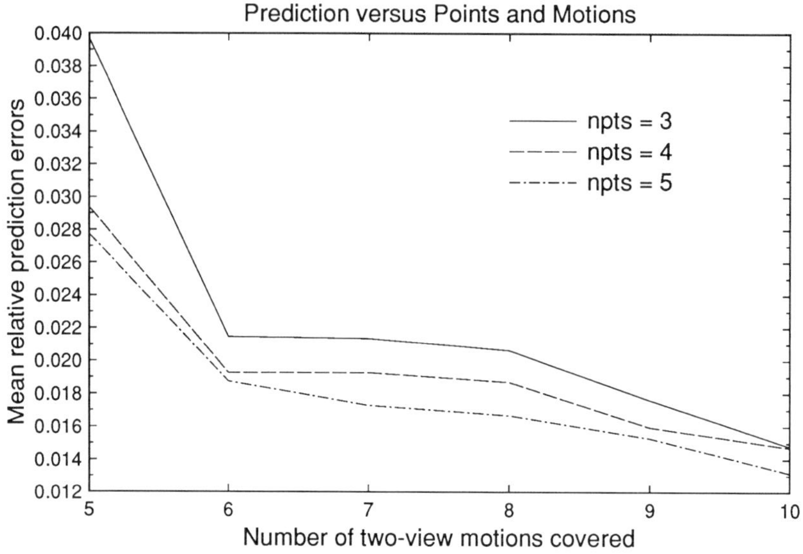

Fig. 8.11. Mean relative prediction errors versus the number of two-view motions covered for different numbers of point correspondences (npts). Frame predicted: next frame; image resolution: 128×128; degree of polynomial $k-1$ in equation (8.2.7): 3.

8.6.2 Experiment with Real Images

The data for the second experiments were taken from a model airplane. The setup of the real cameras is the same as that in the simulation experiment discussed above. Without translation, the model airplane rotates about a vertical line by about 15° per frame. Simultaneously it also rotates about its head-tail central line by about 8° per frame as shown in Fig. 8.12. The feature points used for point correspondences are at the tips of the wings, stabilizers and rudder of the airplane. Since not all of these 5 tips are visible in all images, 4 currently visible tips are used as feature point sets through the image sequence. The number of point correspondences is then 4. The feature points and their correspondences are determined manually based on the gray tone of the digitized images. The corner detector error here is then the error of manual detection based on the gray tone of the pixels. The camera resolution is 512×512. Starting from the fourth frame, the prediction is made for the 3-D coordinates of the feature points at the next time instant. The relative maximum prediction errors are shown in Fig. 8.12. The relative errors are defined by the error divided by the maximum distance between feature points. Compared with the synthetic data case, the additional error source is

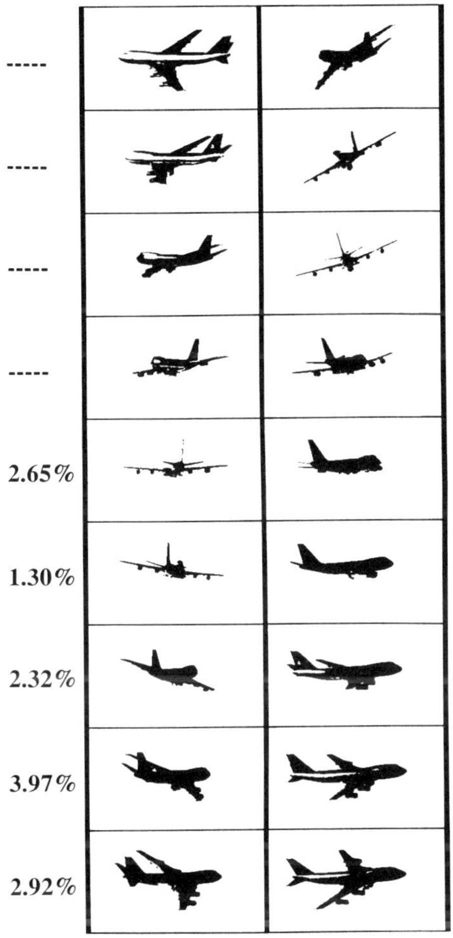

Fig. 8.12. Image frame sequence of a model airplane. Left column: left view; right column: right view. From the fifth frame, the maximum relative prediction errors are shown to the left of the corresponding frame pairs.

lens distortion. The relative errors in this experiment are close to the synthetic data case.

8.7 Summary

We have described an approach to modeling and estimating general 3-D motion of an object from image sequences. The dynamics of the moving object is modeled by two components. First, the rotation of the object is assumed to be a precession which can be modeled by such motion: the rotation about a fixed-on-body axis that rotates about a spatially fixed axis. One of the conditions of object dynamics leading to such motion is that the object under motion is symmetric and its angular momentum is constant. Second, the object is assumed to undergo a smooth translational motion. In particular, we assume that the location of the rotation center of the object can be represented by a vector polynomial in time. The motion of any points on the object can be expressed by the superposition of these two components. The problem of modeling motion then amounts to estimating the parameters of precession with respect to the rotation center, and the parameters of translation of the rotation center. This estimation can be performed from either point correspondences or optical flow over a sequence of image frames. We discuss the former. Using the technique of two-view motion analysis, estimates of two-view rotational and translational parameters can be derived. To reduce the sensitivity to noise, least-squares estimates are obtained from multiple features in two-view motion analysis. Based on the parameters of two-view motion, the parameters of LCAM model are estimated so as to understand the local motion and predict the future motion. Again, the least-squares method is used for model parameter estimation from multiple image frames to combat noise.

We have presented a linear algorithm that implements our approach. The experiments have been performed on image sequences obtained from simulated as well as actual moving objects. To test the accuracy of the model, the predictions of the locations of object points were obtained and the errors between the predicted and actual locations were measured. The errors of estimated model parameters have been presented for different numbers of image frames. The prediction errors have been shown for different image resolutions, different numbers of object points, different numbers of image frames covered, and different numbers of frames predicted.

8.8 Bibliographical Notes

The dynamics of the motion of a rigid object is a classical problem. In the field of astrodynamics, the motion estimation is addressed under the assumption that detailed data about the mass distribution of the moving objects are available. For instance, the principal moments of inertia and the structures of objects are required in [8.5, 8]. Because the Euler's equations of rigid body motion are nonlinear, numerical methods are necessary to solve the dynamics problem.

In the field of computer vision, constant motion has been widely assumed in studies that involve long image sequences. Broida & Chellappa [8.3] discuss the inference of 2-D motion from 1-D image sequences under the assumption that the object undergoes constant translational and rotational 2-D motion and the structure of the object is known. Yasumoto & Medioni [8.13] estimate motion parameters from long sequences of point matches, assuming that the motion is constant through the sequences.

Later, non-constant 3-D motion modeling was investigated. The LCAM model was originally proposed in Huang, Weng & Ahuja [8.7] and was later fully presented in Weng, Huang & Ahuja [8.12]. This investigation brought physics into computer understanding of visual motion. Broida & Chellappa extended the Kalman filtering approach in their early work [8.3] to 3-D motion in [8.4], where a smooth kinematic 3-D motion model was introduced under the assumption that the geometrical structure of the object is known, which in turn requires that the matching between the 3-D object and the model must be established first. The Kalman filtering method has been applied to the LCAM model in Young & Broida [8.14].

Appendices

Appendix 8.A Solution of Coefficient Equations

Lemma 8.1. Define

$$\tilde{S}_k^j = \sum_{l=0}^{k-j}(-1)^l \binom{k}{l} \prod_{m=j+2}^{k-l+1} R_m .$$

Let $\prod_{m=i}^{j} R_m = I$, if $i > j$. Then,

$$\tilde{S}_k^0 R_1 - S_k^0 = S_{k+1}^0.$$

Proof.

$$\tilde{S}_k^0 R_1 - S_k^0 = \sum_{l=0}^{k}(-1)^l \binom{k}{l} \prod_{m=1}^{k-l+1} R_m - \sum_{l=0}^{k}(-1)^l \binom{k}{l} \prod_{m=1}^{k-l} R_m$$

$$= \binom{k}{0} \prod_{m=1}^{k+1} R_m + \sum_{l=1}^{k}(-1)^l \binom{k}{l} \prod_{m=1}^{k-l+1} R_m - \sum_{l=0}^{k-1}(-1)^l \binom{k}{l} \prod_{m=1}^{k-l} R_m - (-1)^k \binom{k}{k} I$$

$$= \binom{k+1}{0}\prod_{m=1}^{k+l} R_m + \sum_{l=1}^{k}(-1)^l \binom{k}{l}\prod_{m=1}^{k-l+1} R_m \tag{8.A.1}$$

$$+ \sum_{j=1}^{k}(-1)^j \binom{k}{j-1}\prod_{m=1}^{k-j+1} R_m + (-1)^{k+1}\binom{k+1}{k+1} I.$$

Using the identity $\binom{k}{l} + \binom{k}{l-1} = \binom{k+1}{l}$, (8.A.1) becomes

$$\tilde{S}_k^0 R_1 - S_k^0 = \binom{k+1}{0}\prod_{m=1}^{k+1} R_m + \sum_{l=1}^{k}(-1)^l \binom{k+1}{l}\prod_{m=1}^{k-l+1} R_m + (-1)^{k+1}\binom{k+1}{k+1} I$$

$$= \sum_{l=0}^{k+1}(-1)^l \binom{k+1}{l}\prod_{m=1}^{k-l+1} R_m$$

$$= S_{k+1}^0.$$

□

Lemma 8.2. Let $k \geq 2$, $1 \leq m \leq k-1$. Then

$$\sum_{j=1}^{k}(-1)^{k-j} j^m \binom{k}{j} = 0, \tag{8.A.2}$$

and

$$\sum_{j=1}^{k}(-1)^{k-j} j^k \binom{k}{j} = k!. \tag{8.A.3}$$

Proof. From the binomial equation

$$(1 + x)^k = \sum_{j=0}^{k}\binom{k}{j} x^j.$$

Differentiating both sides and then multiplying by x give

$$(1 + x)^{k-1} kx = \sum_{j=1}^{k} j \binom{k}{j} x^j.$$

Differentiating both sides and then multiplying by x again, we get

$$(1 + x)^{k-2}(k(k-1)x^2 + k(1 + x)x) = \sum_{j=1}^{k} j^2 \binom{k}{j} x^j.$$

Generally, doing this m times, $1 \leq m \leq k$, gives

$$(1+x)^{k-m}[k(k-1)\cdots(k-m+1)x^m + (1+x)P(x)] = \sum_{j=1}^{k} j^m \binom{k}{j} x^j \tag{8.A.4}$$

where $P(x)$ is an $(m-1)$-th degree polynomial. This can be readily proved by induction on m. Let $x = -1$. For $1 \le m \le k-1$, (8.A.4) gives

$$0 = \sum_{j=1}^{k} j^m \binom{k}{j} (-1)^j = \sum_{j=1}^{k} (-1)^{-j} j^m \binom{k}{j}.$$

Multiplying both sides by $(-1)^m$ yields (8.A.2). For $m = k$, (8.A.4) gives

$$k!(-1)^k = \sum_{j=1}^{k} (-1)^j j^m \binom{k}{j},$$

Multiplying both sides by $(-1)^k$ yields (8.A.3). □

Theorem 8.1. Let $f = k$ in coefficient equations. Define S_k^j to be a 3×3 matrix :

$$S_k^j = \sum_{l=0}^{k-j} (-1)^l \binom{k}{l} \prod_{m=j+1}^{k-l} R_m \qquad j = 0, 1, 2, \cdots, k$$

where $\prod_{m=i}^{j} R_m = I$, if $i > j$. Define number u_{ij}:

$$u_{ij} = \sum_{m=1}^{i} (-1)^{i-m} m^{j-1} \binom{i}{m} \qquad j = i+1, i+2, \cdots, k.$$

Then

$$S_k^0 \mathbf{a}_1 = -\sum_{l=1}^{k} S_k^l \mathbf{T}_l,$$

$$\mathbf{a}_{i+1} = \frac{1}{i!}\left(\sum_{m=1}^{i} S_i^m \mathbf{T}_m + S_i^0 \mathbf{a}_1 - \sum_{j=i+2}^{k} u_{ij} \mathbf{a}_j\right), \quad i = k-1, k-2, \cdots, 2, 1.$$

Proof. We first prove

$$S_k^0 \mathbf{a}_1 = -\sum_{l=1}^{k} S_k^l \mathbf{T}_l. \tag{8.A.5}$$

Let $\mathbf{c}(t)$ be the position vector of the rotation center at time t, $\mathbf{c}_i$ be the position vector of the rotation center at time t_i. For conciseness, we define $\mathbf{T}_0 = \mathbf{c}_0$. (Note $\mathbf{c}_0 = \mathbf{a}_1$). We have

$$\mathbf{c}_0 = \mathbf{T}_0,$$

$$\mathbf{c}_1 = R_1\mathbf{c}_0 + \mathbf{T}_1 = R_1\mathbf{T}_0 + \mathbf{T}_1,$$

$$\mathbf{c}_2 = R_2\mathbf{c}_1 + \mathbf{T}_2 = R_2R_1\mathbf{T}_0 + R_2\mathbf{T}_1 + \mathbf{T}_2,$$

$$\cdots\cdots.$$

It follows that

$$\mathbf{c}_i = \sum_{l=0}^{i} \prod_{m=l+1}^{i} R_m \mathbf{T}_l, \qquad 0 \le i \le k. \tag{8.A.6}$$

The $(k-1)$-th degree vector polynomial that goes through $\mathbf{c}_{i,}$ $i = 1, 2, , \cdots, k$, is uniquely given by

$$\mathbf{c}(t) = \sum_{i=1}^{k} \frac{\prod\limits_{\substack{1 \le j \le k \\ j \ne i}} (t - t_j)}{\prod\limits_{\substack{1 \le j \le k \\ j \ne i}} (t_i - t_j)} \mathbf{c}_i .$$

Or, using (8.A.6)

$$\mathbf{c}(t) = \sum_{i=1}^{k} \sum_{l=0}^{i} \frac{\prod\limits_{\substack{1 \le j \le k \\ j \ne i}} (t - t_j)}{\prod\limits_{\substack{1 \le j \le k \\ j \ne i}} (t_i - t_j)} \prod_{m=l+1}^{i} R_m \mathbf{T}_l . \tag{8.A.7}$$

Since the time intervals are the same constant c, we have $t_i - t_j = c(i - j)$. Letting $t = t_0$ in (8.A.7), it follows that

$$\mathbf{c}_0 = \sum_{i=1}^{k}\sum_{l=0}^{i} \frac{(-1)^{k-1} k!/i}{(-1)^{k-i}(i-1)!(k-i)!} \prod_{m=l+1}^{i} R_m \mathbf{T}_l$$

$$= \sum_{i=1}^{k}\sum_{l=0}^{i} (-1)^{i-1} \binom{k}{i} \prod_{m=l+1}^{i} R_m \mathbf{T}_l . \tag{8.A.8}$$

Since $\mathbf{T}_0 = \mathbf{c}_0 = \mathbf{a}_1$, (8.A.8) gives

$$\mathbf{a}_1 = \sum_{i=1}^{k}\sum_{l=1}^{i}(-1)^{i-1}\binom{k}{i}\prod_{m=l+1}^{i} R_m \mathbf{T}_l + \sum_{i=1}^{k}(-1)^{i-1}\binom{k}{i}\prod_{m=1}^{i} R_m \mathbf{a}_1.$$

Moving the term of $\mathbf{a}_1$ to the left yields

$$\left(\sum_{i=0}^{k}(-1)^{i}\binom{k}{i}\prod_{m=1}^{i} R_m\right)\mathbf{a}_1 = -\sum_{i=1}^{k}\sum_{l=1}^{i}(-1)^{i}\binom{k}{i}\prod_{m=l+1}^{i} R_m \mathbf{T}_l.$$

Since $\sum_{i=1}^{k}\sum_{l=1}^{i}$ and $\sum_{l=1}^{k}\sum_{i=l}^{k}$ sum the same terms, we get

$$\left(\sum_{i=0}^{k}(-1)^{i}\binom{k}{i}\prod_{m=1}^{i} R_m\right)\mathbf{a}_1 = -\sum_{l=1}^{k}\sum_{i=l}^{k}(-1)^{i}\binom{k}{i}\prod_{m=l+1}^{i} R_m \mathbf{T}_l.$$

Multiplying both sides by $(-1)^k$ and letting $j = k-i$, it follows that

$$\left(\sum_{j=0}^{k}(-1)^{j}\binom{k}{j}\prod_{m=1}^{k-j} R_m\right)\mathbf{a}_1 = -\sum_{l=1}^{k}\sum_{j=0}^{k-l}(-1)^{j}\binom{k}{j}\prod_{m=l+1}^{k-j} R_m \mathbf{T}_l. \tag{8.A.9}$$

Using the definition of S_k^l (8.A.9) gives (8.A.5).

We now prove the rest of the equations. Let $2 \le i \le k-1$. Consider the first i equations in (8.2.11). Moving the terms of $\mathbf{a}_{i+1}, \cdots, \mathbf{a}_k$ to the right side, we get a set of i equations similar to (8.2.11) with the terms on the right side changed. Using the results in (8.A.5) for this set of equations to solve for $\mathbf{a}_1$ gives

$$\begin{aligned} S_i^0\mathbf{a}_1 &= -\sum_{m=1}^{i} S_i^m\left(\mathbf{T}_m - \sum_{j=i+1}^{k} D_{mj}\mathbf{a}_j\right) \\ &= -\sum_{m=1}^{i} S_i^m\mathbf{T}_m + \sum_{j=i+1}^{k}\left(\sum_{m=1}^{i} S_i^m D_{mj}\right)\mathbf{a}_j \end{aligned} \tag{8.A.10}$$

where

$$\begin{aligned} \sum_{m=1}^{i} S_i^m D_{mj} &= \sum_{m=1}^{i} S_i^m\left(m^{j-1}I - (m-1)^{j-1}R_m\right) \\ &= \sum_{m=1}^{i-1} m^{j-1}S_i^m + i^{j-1}I - \sum_{m=2}^{i}(m-1)^{j-1}S_i^m R_m \\ &= \sum_{m=1}^{i-1} m^{j-1}\sum_{l=0}^{i-m}(-1)^l\binom{i}{l}\prod_{n=m+1}^{i-l} R_n \end{aligned}$$

$$- \sum_{m=2}^{i} (m-1)^{j-1} \sum_{l=0}^{i-m} (-1)^l \binom{i}{l} \prod_{n=m}^{i-l} R_n + i^{j-1} I$$

$$= \sum_{m=1}^{i-1} m^{j-1} \sum_{l=0}^{i-m-1} (-1)^l \binom{i}{l} \prod_{n=m+1}^{i-l} R_n + \sum_{m=1}^{i-1} (-1)^{i-m} m^{j-1} \binom{i}{i-m} I$$

$$- \sum_{v=1}^{i-1} v^{j-1} \sum_{l=0}^{i-v-1} (-1)^l \binom{i}{l} \prod_{n=v+1}^{i-l} R_n + i^{j-1} I$$

$$= \left(\sum_{m=1}^{i-1} (-1)^{i-m} m^{j-1} \binom{i}{m} + i^{j-1} \right) I$$

$$= \left(\sum_{m=1}^{i} (-1)^{i-m} m^{j-1} \binom{i}{m} \right) I$$

$$= u_{ij} I.$$

So, (8.A.10) yields

$$S_i^0 \mathbf{a}_1 = - \sum_{m=1}^{i} S_i^m \mathbf{T}_m + \sum_{j=i+1}^{k} u_{ij} \mathbf{a}_j .$$

Or,

$$u_{i\ (i+1)} \mathbf{a}_{i+1} = \sum_{m=1}^{i} S_i^m \mathbf{T}_m + S_i^0 \mathbf{a}_1 - \sum_{j=i+2}^{k} u_{ij} \mathbf{a}_j .$$

By Lemma 8.2, $u_{i\ (i+1)} = i\,!$, and we have

$$\mathbf{a}_{i+1} = \frac{1}{i\,!} \left(\sum_{m=1}^{i} S_i^m \mathbf{T}_m + S_i^0 \mathbf{a}_1 - \sum_{j=i+2}^{k} u_{ij} \mathbf{a}_j \right).$$

Letting $i = k-1, k-2, \cdots, 2, 1$, we get the equations of Theorem 8.1 presented in Subsection 8.2.3. □

Appendix 8.B Rotation without Precession

Theorem 8.2. In the case of rotation without precession, letting $\mathbf{w}$ be any vector parallel to the rotation axes, then,

$$S_k^0 \mathbf{w} = 0 \tag{8.B.1}$$

and for any vector $\mathbf{a}$

$$(S_k^0\mathbf{a}) \cdot \mathbf{w} = 0. \tag{8.B.2}$$

Proof. We just prove (8.B.2). The proof of equation (8.B.1) is very similar. We prove another equation at the same time: Under the same condition, for any vector **b** the following equation holds (see Lemma 8.1 in Appendix 8.A)

$$(\tilde{S}_k^0\mathbf{b}) \cdot \mathbf{w} = 0. \tag{8.B.3}$$

We use induction on k.

Let k=1. $(S_1^0\mathbf{a}) \cdot \mathbf{w} = ((I - R_1)\mathbf{a}) \cdot \mathbf{w} = (\mathbf{a} - R_1\mathbf{a}) \cdot \mathbf{w}$. R_1 is a rotation matrix with rotation axis **w**. Thus, the difference of the original vector **a** and the rotated vector $R_1\mathbf{a}$ is orthogonal to the rotation axis. i.e., $(\mathbf{a} - R_1\mathbf{a}) \cdot \mathbf{w} = 0$. Similarly $(\tilde{S}_1^0\mathbf{b}) \cdot \mathbf{w} = 0$.

Assume $(S_k^0\mathbf{a}) \cdot \mathbf{w} = 0$ and $(\tilde{S}_k^0\mathbf{b}) \cdot \mathbf{w} = 0$ for $k \le n$. Let $k = n + 1$. From Lemma 8.1, we get $S_{n+1}^0 = \tilde{S}_n^0 R_1 - S_n^0$. Then

$$(S_{n+1}^0\mathbf{a}) \cdot \mathbf{w} = (\tilde{S}_n^0 R_1\mathbf{a} - S_n^0\mathbf{a}) \cdot \mathbf{w}$$

$$= (\tilde{S}_n^0 R_1\mathbf{a}) \cdot \mathbf{w} - (S_n^0\mathbf{a}) \cdot \mathbf{w} = 0 - 0 = 0.$$

Similarly (8.B.3) holds for $k = n+1$. □

Appendix 8.C Singularity of the Matrix

Let R_1 and R_2 be the rotation matrices with rotation axes n_1 and $\mathbf{n}_2$, rotation angle θ_1 and θ_2, respectively. $A = R_2R_1 - 2R_1 + I$ is singular if and only if $\mathbf{n}_1$ and $\mathbf{n}_2$ are parallel or $\theta_1 = 0$, or $\theta_2 = 0$.

Proof. Without loss of generality, let $\mathbf{n}_1$ and $\mathbf{n}_2$ be unit vectors. Since the points on the rotation axis do not change the positions after motion, we have $R_i n_i = \mathbf{n}_i$, $i = 1,2$.

If $\theta_1 = 0$, then $R_1 = I$ and we have $A\mathbf{n}_2 = R_2\mathbf{n}_2 - 2\mathbf{n}_2 + \mathbf{n}_2 = 0$. Thus, A is singular if $\theta_1 = 0$. Similarly if $\theta_2 = 0$, A is singular. If $\mathbf{n}_1$ and $\mathbf{n}_2$ are parallel, we have $R_2\mathbf{n}_1 = \mathbf{n}_1 = R_1\mathbf{n}_1$. Thus

$$A\mathbf{n}_1 = R_2\mathbf{n}_1 - 2R_1\mathbf{n}_1 + \mathbf{n}_1 = 0,$$

and therefore, A is singular.

Conversely, assume A is singular. There is a unit vector **x** such that $A\mathbf{x} = (R_2R_1 - 2R_1 + I)\mathbf{x} = 0$. Or,

$$R_2R_1\mathbf{x} = 2R_1\mathbf{x} - \mathbf{x}. \tag{8.B.4}$$

There exists a unit vector $\mathbf{y}$ which is perpendicular to $\mathbf{x}$ such that $R_1\mathbf{x}$ can be represented by $R_1\mathbf{x} = a\mathbf{x} + b\mathbf{y}$, where a and b are real numbers. Since $R_1\mathbf{x}$ is a unit vector, $a^2 + b^2 = 1$. From (8.B.4) we get

$$1 = \|\mathbf{x}\|^2 = \|R_2R_1\mathbf{x}\|^2 = \|2R_1\mathbf{x} - \mathbf{x}\|^2 = \|(2a - 1)\mathbf{x} + 2b\mathbf{y}\|^2$$
$$= (2a - 1)^2 + (2b)^2 = 4a^2 - 4a + 1 + 4b^2 = 5 - 4a.$$

Thus $a = 1$. Thus, $b = 0$. We have

$$R_1\mathbf{x} = \mathbf{x}. \tag{8.B.5}$$

Using (8.B.4) and (8.B.5) yields

$$R_2\mathbf{x} = R_2R_1\mathbf{x} = 2R_1\mathbf{x} - \mathbf{x} = 2\mathbf{x} - \mathbf{x} = \mathbf{x}. \tag{8.B.6}$$

Unless $\theta_1 = 0$ or $\theta_2 = 0$, (8.B.5) and (8.B.6) imply that $\mathbf{x}$ is parallel to the rotation axes of R_1 and R_2. Therefore, $\mathbf{n}_1$ and $\mathbf{n}_2$ are parallel unless $\theta_1 = 0$ or $\theta_2 = 0$. □

References

Chapter 1

1.1 J. K. Aggarwal, I. K. Nandhakumar: On the computation of motion from sequences of images - a review. Proc. IEEE, **76**, 917-935 (1988)

1.2 J. K. Aggarwal, N. I. Badler (eds.): *Abstracts for the Workshop on Computer Analysis of Time-Varying Imagery* (Univ. Pennsylvania Press, Philadelphia, PA 1979)

1.3 V. Cappallini (ed.): *Time-Varying Image Processing and Moving Object Recognition* (Elsevier, Amsterdam 1987)

1.4 V. Cappallini (ed.): *Time-Varying Image Processing and Moving Object Recognition, 2* (Elsevier, Amsterdam 1990)

1.5 E. C. Hildreth: *The Measurement of Visual Motion* (MIT Press, Cambridge, MA 1983)

1.6 T. S. Huang (ed.): *Image Sequence Processing and Dynamic Scene Analysis* (Springer, New York 1975)

1.7 T. S. Huang (ed.): *Image Sequence Processing and Dynamic Scene Analysis* (Springer, New York 1983)

1.8 T. S. Huang (ed.): *Advances in Computer Vision and Image Processing, Vol. 3: Time-Varying Imagery Analysis* (JAI Press, Greenwich, Connecticut 1988)

1.9 W. N. Martin, J. K. Aggarwal: Dynamic scene analysis: a survey. Computer Graphics and Image Processing **7**, 356-374 (1978)

1.10 W. N. Martin, J. K. Aggarwal (eds.): *Motion Understanding: Robot and Human Motion* (Kluwer, Boston, MA 1988)

1.11 H. -H. Nagel: Analysis techniques for image sequences, in Proc. Int'l Joint Conf. on Pattern Recognition (IEEE Computer Soc. Press, Washington D.C. 1978) pp. 186-211

1.12 H. -H. Nagel: Overview on image sequence analysis, in Image Sequence Processing and Dynamic Scene Analysis, ed. by T. S. Huang (Springer, New York, 1983) pp. 2-39

1.13 H. -H. Nagel, Image sequences - ten (octal) years - from phenomenology towards a theoretical foundation, in Proc. the Eighth International Conference on Pattern Recognition (IEEE Computer Soc. Press, Washington D.C. 1986) pp. 1174-1185

1.14 S. Ullman, *The Interpretation of Visual Motion* (MIT Press, Cambridge, MA 1979)

Chapter 2

2.1 E. H. Adelson and J. R. Bergen: Spatiotemporal energy models for the perception of motion. J. Opt. Soc. Amer., A, **2**, 284-299 (1985)

2.2 P. Anandan and R. Weiss: Introducing a smoothness constraint in a matching approach for the computation of optical flow fields, in Proc. Workshop on Computer Vision: Representation and Control (IEEE Computer Soc. Press, Washington D.C. 1985) pp. 186-194

2.3 N. Ayache and B. Faverjon: Efficient registration of stereo images by matching graph descriptions of edge segments. Int'l J. Computer Vision, **1**, 107-131 (1987)

2.4 S. T. Barnard, W. B. Thompson: Disparity analysis of images. IEEE Trans. Pattern Anal. Machine Intell., **2**, 333-340 (1980)

2.5 H. H. Baker, T. O. Binford: Depth from edges and intensity based stereo, in Proc. Int'l Joint Conf. Artif. Intell. (William Kaufmann, Los Angeles, LA 1981) pp. 631-636

2.6 H. H. Baker, R. C. Bolles: Generalizing epipolar-plane image analysis on the spatiotemporal surface, in Proc. IEEE Conf. Computer Vision Image Processing (IEEE Computer Soc. Press, Washington D.C. 1988) pp. 2-9

2.7 M. Born, E. Wolf: *Principles of Optics: Electromagnetic Theory of Propagation, Interference and Diffraction of Light* (Pergamon Press, Oxford, England, 1975)

2.8 O. J. Braddick: A short-range process in apparent motion. Vision Research, **14**, 519-527 (1974)

2.9 O. J. Braddick: Low-level and high-level processes in apparent motion. Philosophical Transactions of the Royal Society of London, B, **290**, 137-151 (1980)

2.10 L. Dreschler, H. -H. Nagel: Volumetric model and 3-D trajectory of a moving car derived from monocular TV frame sequences of a street scene. Computer Graphics and Image Processing, **20**, 199-228 (1982)

2.11 C. L. Fennema, W. R. Thompson: Velocity determination in scenes containing several moving objects. Computer Graphics and Image Processing, **9**, 301-315 (1979)

2.12 S. V. Fogel: A nonlinear approach to the motion correspondence problem, in Proc. 2nd Int'l Conf. Computer Vision (IEEE Computer Soc. Press, Washington D.C. 1988) pp. 619-628

2.13 F. Glazer, G. Reynolds, P. Anandan: Scene matching by hierarchical correlation, in Proc. IEEE Conf. Computer Vision Pattern Recognition (IEEE Computer Soc. Press, Washington D.C. 1983) pp. 432-441

2.14 W. E. L. Grimson: *From Images to Surfaces: A Computational Study of the Human Early Visual Systems* (MIT Press, Cambridge, MA, 1981)

2.15 W. E. L. Grimson: Computational experiments with a feature based stereo algorithm. IEEE Trans. Pattern Analysis and Machine Intelligence, **7**, 17-34 (1985)

2.16 W. K. Gu, J. Y. Yang, T. S. Huang: Matching perspective views of a polyhedron using circuits. IEEE Trans. Pattern Anal. Machine Intell., **9**, 390-400 (1987)

2.17 D. J. Heeger: Optical flow using spatiotemporal filters. Int'l J. Computer Vision, **2**, 279-302 (1987)

2.18 E. C. Hildreth: *The Measurement of Visual Motion* (MIT Press, Cambridge, MA 1983)

2.19 W. Hoff, N. Ahuja: Surface from stereo: integrating feature matching, disparity estimation and contour detection. IEEE Trans. Pattern Anal. Machine Intell., **11**, 121-136 (1989)

2.20 B. K. P. Horn, B. G. Schunck: Determining optical flow. Artificial Intelligence, **17**, 185-203 (1981)

2.21 B. K. Horn: Motion fields are hardly ever ambiguous. Int'l J. Computer Vision, **1**, 259-273 (1987)

2.22 J. J. Hwang, E. L. Hall: Matching of featured objects using relational tables from stereo images. Computer Graphics and Image Processing, **20**, 22-42 (1982)

2.23 J. K. Kearney, W. B. Thompson, D. L. Boley: Optical flow estimation: an error analysis of gradient-based methods with local optimization. IEEE Trans. Pattern Anal. Machine Intell., **9**, 229-244 (1987)

2.24 R. Kingslake: *Lens Design Fundamentals* (Academic Press, New York, 1978)

2.25 L. Kitchen, A. Rosenfeld: Gray-level corner detection. Pattern Recognition Letters, **1**, 95-102 (1982)

2.26 H. S. Lim, T. O. Binford: Stereo correspondence: a hierarchical approach, in Proc. Image Understanding Workshop (Science Applications Corp. Mclean, Virginia 1987)

2.27 J. O. Limb, J. A. Murphy: Estimating the velocity of moving images in television signals. Computer Graphics and Image Processing, **4**, 311-327 (1975)

2.28 D. Marr, T. Poggio: A theory of human stereo vision. Proc. Royal Society of London, B **204**, 301-328 (1979)

2.29 D. Marr: *Vision: A Computational Investigation into the Human Representation and Processing of Visual Information* (Freeman, San Francisco, CA, 1982)

2.30 J. E. W. Mayhew, J. P. Frisby: Psychophysical and computational studies towards a theory of human stereopsis. Artificial Intelligence, **17**, 349-385 (1981)

2.31 G. Medioni, R. Nevatia: Segment-based stereo matching. Computer Vision, Graphics and Image Processing, **31** (1985)

2.32 H. P. Moravec: Towards automatic visual obstacle avoidance, in Proc. 5th Int'l Joint Conf. Artif. Intell. (William Kaufmann, Los Angeles, LA 1977)

2.33 H. -H. Nagel, W. Enkelmann: An investigation of smoothness constraints for the estimation of displacement vector fields from image sequences. IEEE Trans. Pattern Anal. Machine Intell., **8**, 565-593 (1986)

2.34 A. N. Netravali, J. D. Robbins: Motion-compensated television coding: part I. Bell System Technical J., **58**, 631-670 (1979)

2.35 Y. Ohta, T. Kanade: Stereo by intra- and inter-scanline search using dynamic programming. IEEE Trans. Pattern Anal. Machine Intell., **7**, 139-154 (1985)

2.36 S. Ullman: *The Interpretation of Visual Motion* (MIT Press, Cambridge, MA, 1979)

2.37 A. Verri, T. Poggio: Motion field and optical flow: qualitative properties. IEEE Trans. Pattern Anal. Machine Intell., **11**, 490-498 (1989)

2.38 A. M. Waxman: An image flow paradigm, in Proc. Workshop on Computer Vision: Representation and Control (IEEE Computer Soc. Press, Washington D.C. 1984) pp. 49-57

2.39 J. Weng, N. Ahuja, T. S. Huang: Two-view matching, in Proc. 2nd Int'l Conf. Computer Vision (IEEE Computer Soc. Press, Washington D.C. 1988) pp. 64-73. Also ref.: Mathing two perspective views, accepted and to appear in IEEE Trans. Pattern Anal. Machine Intell., **14** (1992)

2.40 J. Weng, T. S. Huang, N. Ahuja: Motion from images: image matching, parameter estimation and intrinsic stability, in Proc. IEEE Workshop on Visual Motion (IEEE Computer Soc. Press, Washington D.C. 1989) pp. 359-366

2.41 A. Witkin, D. Terzopoulos, M. Kass: Signal matching through scale space. Int'l J. Computer Vision, **1**, 133-144 (1987)

2.42 O. A. Zuniga, R. M. Haralick: Corner detection using the facet model, in Proc. IEEE Conf. Computer Vision and Pattern Recognition (IEEE Computer Soc. Press, Washington D.C. 1983) pp. 30-37

Chapter 3

3.1 G. Adiv: Determining three-dimensional motion and structure from optical flow generated by several moving objects. IEEE Trans. Pattern Anal. Machine Intell., **7**, 348-401 (1985)

3.2 O. Bottema, B. Roth: *Theoretical Kinematics* (North-Holland, New York, 1979)

3.3 A. R. Bruss, B. K. Horn: Passive navigation. Computer Vision, Graphics and Image Processing, **21**, 3-20 (1983)

3.4 J. Q. Fang, T. S. Huang: Some experiments on estimating the 3-D motion parameters of a rigid body from two consecutive image frames. IEEE Trans. Pattern Anal. Machine Intell., **6**, 547-554 (1984)

3.5 O. D. Faugeras, M. Hebert: A 3-D recognition and positioning algorithm using geometrical matching between primitive surfaces, in Proc. 8th Int'l Joint Conf. Artif. Intell. (William Kaufmann, Los Angeles, LA 1983) pp. 996-1002

3.6 O. D. Faugeras, F. Lustman, G. Toscani: Motion and structure from point and line matches, in Proc. Int'l Conf. Computer Vision (IEEE Computer Soc. Press, Washington D.C. 1987)

3.7 O. D. Faugeras, S. Maybank: Motion from point matches: multiplicity of solutions, in Proc. Workshop on Visual Motion (IEEE Computer Soc. Press, Washington D.C. 1989) pp. 248-255

3.8 W. R. Hamilton: *Elements of Quaternions*, 3rd ed. (Chelsea, New York, 1969)

3.9 R. Jain, H. -H. Nagel: On the analysis of accumulative difference pictures from image sequences of real world scenes. IEEE Trans. Pattern Anal. Machine Intell., **1**, 206-214 (1979)

3.10 H. C. Longuet-Higgins: A computer program for reconstructing a scene from two projections. Nature, **293**, 133-135 (1981)

3.11 H. C. Longuet-Higgins: The reconstruction of a scene from two projections — configurations that defeat the 8-point algorithm, in Proc. IEEE 1st Conf. on Artif. Intell. Applications (IEEE Computer Soc. Press, Washington D.C. 1984) pp. 395-397

3.12 H. C. Longuet-Higgins: Multiple interpretations of a pair of images of a surface. Proc. Royal Society London, A, **418**, 1-15 (1988)

3.13 S. J. Maybank: The angular velocity associated with the optical flow field arising from motion through a rigid environment. Proc. Royal Society London, A, **401**, 317-326 (1985)

3.14 A. Mitiche, J. K. Aggarwal: A computational analysis of time-varying images, in *Handbook of Pattern Recognition and Image Processing*, ed. by T. Y. Young and K. S. Fu (Academic Press, New York, 1986)

3.15 S. Negahdaripour: Multiple interpretations of the shape and motion of objects from two perspective images. IEEE Trans. Pattern Anal. Machine Intell., **12**, 1025-1039 (1990)

3.16 A. N. Netravali, T. S. Huang, A. S. Krishnakumar, R. J. Holt: Algebraic methods in 3D motion estimation from two-view point correspondences. Int'l J. Imaging Systems and Technology, **1**, 78-99 (1989)

3.17 J. Philip: Estimation of three-dimensional motion of rigid objects from noisy observations. IEEE Trans. Pattern Anal. Machine Intell., **13**, 61-66 (1991)

3.18 J. W. Roach, J. K. Aggarwal: Determining the movement of objects from a sequence of images. IEEE Trans. Pattern Anal. Machine Intell., **2**, 554-562 (1980)

3.19 M. D. Shuster: Approximate algorithms for fast optimal attitude computation, in Proc. AIAA Guidance and Control Specialist Conference (Palo Alto, 1978) pp. 88-95

3.20 M. E. Spetsakis, J. Aloimonos: Optimal motion estimation, in Proc. Workshop on Visual Motion (IEEE Computer Soc. Press, Washington D.C. 1989) pp. 229-237

3.21 R. Y. Tsai, T. S. Huang: Uniqueness and estimation of 3-D motion parameters of rigid bodies with curved surfaces. IEEE Trans. Pattern Anal. Machine Intell., **6**, 13-27 (1984)

3.22 A. M. Waxman, B. Kamgar-Parsi, M. Subbarao: Closed-form solutions to image flow equations for 3-D structure and motion, Int'l J. Computer Vision, **1**, 239-258 (1987)

3.23 J. Weng, T. S. Huang, N. Ahuja: Error analysis of motion parameter determination from image sequences, in Proc. 1st Int'l Conf. Computer Vision (IEEE Computer Soc. Press, Washington D.C. 1987) pp. 703-707

3.24 J. Weng, T. S. Huang, N. Ahuja: Motion and structure from two perspective views: algorithms, error analysis and error estimation. IEEE Trans. Pattern Anal. Machine Intell., **11**, 451-476 (1989)

3.25 J. H. Wilkinson, *The Algebraic Eigenvalue Problem* (Oxford University Press, England, 1965)

3.26 B. L. Yen, T. S. Huang: Determining 3-D motion and structure of a rigid body using the spherical projection. Computer Vision, Graphics and Image Processing, **21**, 21-32 (1983)

3.27 X. Zhuang, R. M. Haralick: Rigid body motion and the optic flow image, in Proc. IEEE 1st Conf. Artif. Intell. Applications (IEEE Computer Soc. Press, Washington D.C. 1984) pp. 366-375

3.28 X. Zhuang, T. S. Huang, R. M. Haralick: Two-view motion Analysis: a unified algorithm. J. Opt. Soc. Amer., A, **3**, 1492-1500 (1986)

3.29 X. Zhuang, T. S. Huang, N. Ahuja, R. M. Haralick: A simplified linear optic flow-motion algorithm. Computer Vision, Graphics and Image Processing, **42**, 334-344 (1988)

Chapter 4

4.1 G. Adiv: Determining three-dimensional motion and structure from optical flow generated by several moving objects. IEEE Trans. Pattern Anal. Machine Intell., **7**, 348-401 (1985)

4.2 G. Adiv: Inherent ambiguities in recovering 3-D motion and structure from a noisy flow field. IEEE Trans. Pattern Anal. Machine Intell., **11**, 477-489 (1989)

4.3 J. Aisbett: An iterated estimation of the motion parameters of a rigid body from noisy displacement vectors. IEEE Trans. Pattern Anal. Machine Intell., **12**, 1092-1098 (1990)

4.4 B. D. Anderson, J. B. Moore: *Optimal Filtering* (Prentice-Hall, New Jersey, 1979)

4.5 A. Ben-Israe, T. N. Greville: *Generalized Inverses: Theory and Applications* (Wiley, New York, 1974)

4.6 T. J. Broida, R. Chellappa: Estimation of object motion parameters from noisy images. IEEE Trans. Pattern Anal. Machine Intell., **8**, 90-99 (1986)

4.7 K. M. Brown, J. E. Dennis: Derivative free analogues of the Levenberg-Marquardt and Gauss algorithms for nonlinear least squares approximation. Numeriche Mathematik, **18**, 289-297 (1972)

4.8 A. R. Bruss, B. K. Horn: Passive Navigation. Computer Vision, Graphics and Image Processing, **21**, 3-20 (1983)

4.9 H. Cramér: *Mathematical Methods of Statistics* (Princeton Univ. Press, New Jersey, 1946)

4.10 O. D. Faugeras, F. Lustman, G. Toscani: Motion and structure from point and line matches, in Proc. Int'l Conf. Computer Vision (IEEE Computer Soc. Press, Washington D.C. 1987)

4.11 R. J. Fitzgerald: Divergence of the Kalman Filter. IEEE Trans. Automatic Control, **16**, 736-747 (1971)

4.12 A. Gelb (ed): *Applied Optimal Estimation* (MIT Press, Cambridge, MA, 1974)

4.13 A. A. Giordano, F. M. Hsu: *Least Squares Estimation with Applications to Digital Signal Processing* (Wiley, New York, 1985)

4.14 R. E. Kalman: A new approach to linear filtering and prediction problems. J. Basic Eng., Series 82D, 35-45 (1960)

4.15 C. L. Lawson, R. J. Hanson: *Solving Least Squares Problems* (Prentice-Hall, New Jersey, 1974)

4.16 K. Levenberg: A method for the solution of certain nonlinear problems in least squares. Quart. Appl. Math., **2**, 164-168 (1944)

4.17 H. C. Longuet-Higgins: A computer program for reconstructing a scene from two projections. Nature, **293**, 133-135 (1981)

4.18 D. G. Luenberger: *Optimization by Vector Space Methods* (Wiley, New York, 1969)

4.19 D. G. Luenberger: *Linear and nonlinear programming*, 2nd ed. (Addison-Wesley, Massachusetts, 1982)

4.20 D. W. Marquardt: An algorithm for least squares estimation of nonlinear parameters. SIAM J. Appl. Math., **11**, 431-441 (1963)

4.21 P. S. Maybeck: *Stochastic Models, Estimation, and Control*, Vol. 1 (Academic Press, New York, 1979)

4.22 P. S. Maybeck: *Stochastic Models, Estimation, and Control*, Vol. 2 (Academic Press, New York, 1982)

4.23 A. Mitiche, J. K. Aggarwal: A computational analysis of time-varying images, in *Handbook of Pattern Recognition and Image Processing*, ed. by Y. Young and K. S. Fu (Academic Press, New York, 1986)

4.24 E. H. Moore: *General Analysis*, Part I Memoirs (Amer. Philosophical Soc. 1, 1935)

4.25 J. M. Ortega, W. C. Rheinboldt: *Iterative solution of nonlinear equations in several variables* (Academic Press, New York, 1970)

4.26 J. M. Ortega: *Matrix Theory* (Plenum, New York, 1987)

4.27 A. Papoulis: *Probability, Random Variables, and Stochastic Processes* (McGraw-Hill, New York, 1984)

4.28 R. Penrose: A generalized inverse for matrices. Cambridge Philosophical Soc., **51**, 406-413 (1955)

4.29 R. Penrose: On best approximate solutions of linear matrix equations. Cambridge Philosophical Soc., **52**, 17-19 (1956)

4.30 C. R. Rao: *Linear Statistical Inference and Its Applications*, 2nd ed. (Wiley, New York, 1973)

4.31 J. W. Roach, J. K. Aggarwal: Determining the movement of objects from a sequence of images. IEEE Trans. Pattern Anal. Machine Intell., **2**, 554-562 (1980)

4.32 F. H. Schlee, C. J. Standish, N. F. Tota: Divergence in the Kalman Filter. AIAA J., **5**, 1114-1120 (1967)

4.33 H. W. Sorenson: *Parameter Estimation: Principles and Problems* (Marcel Dekker, New Yor, 1980)

4.34 H. W. Sorenson (ed): *Kalman Filtering: Theory and Application* (IEEE Press, New York, 1985)

4.35 M. E. Spetsakis, J. Aloimonos: Optimal motion estimation, in Proc. IEEE Workshop on Visual Motion (IEEE Computer Soc. Press, Washington D.C. 1989) pp. 229-237

4.36 G. Toscani, O. D. Faugeras, Structure from motion using the reconstruction & reprojection technique, in Proc. IEEE Workshop Computer Vision (IEEE Computer Soc. Press, Washington D.C. 1987) pp. 345-348

4.37 H. L. Van Trees: *Detection, Estimation, and Modulation Theory*, Vol. 1 (Wiley, New York, 1969)

4.38 R. Y. Tsai, T. S. Huang: Uniqueness and estimation of 3-D motion parameters of rigid bodies with curved surfaces. IEEE Trans. Pattern Anal. Machine Intell., **6**, 13-27 (1984)

4.39 A. M. Waxman, B. Kamgar-Parsi, M. Subbarao: Closed-form solutions to image flow equations, in Proc. 1st Int'l Conf. Computer Vision (IEEE Computer Soc. Press, Washington D.C. 1987) pp. 12-24

4.40 J. Weng, N. Ahuja, T. S. Huang: Error analysis of motion parameters estimation from image sequences, in Proc. 1st Int'l Conf. Computer Vision (IEEE Computer Soc. Press, Washington D.C. 1987) pp. 703-707

4.41 J. Weng, T. S. Huang, N. Ahuja: A two-step approach to optimal motion and structure estimation, in Proc. IEEE Workshop Computer Vision (IEEE Computer Soc. Press, Washington D.C. 1987) pp. 355-357

4.42 J. Weng, N. Ahuja: Octree of objects in arbitrary motion: representation and efficiency. Computer Vision, Graphics, and Image Processing, **39**, 167-185 (1987)

4.43 J. Weng, N. Ahuja, T. S. Huang: Closed-form solution + maximum likelihood: a robust approach to motion and structure estimation, in Proc. IEEE Conf. Computer Vision and Pattern Recognition (IEEE Computer Soc. Press, Washington D.C. 1988) pp. 381-386

4.44 J. Weng, N. Ahuja, T. S. Huang: Optimal motion and structure estimation, in Proc. IEEE Conf. Computer Vision and Pattern Recognition (IEEE Computer Soc. Press, Washington D.C. 1989) pp. 144-152

4.45 J. Weng, T. S. Huang, N. Ahuja: Motion from images: image matching, parameter estimation and intrinsic stability, in Proc. IEEE Workshop on Visual Motion (IEEE Computer Soc. Press, Washington D.C. 1989) pp. 359-366

4.46 S. S. Wilks: *Mathematical Statistics* (Wiley, New York, 1962)

4.47 Y. Yasumoto, G. Medioni: Robust estimation of three-dimensional motion parameters from sequence of image frames using regularization. IEEE Trans. Pattern Anal. Machine Intell., **8**, 464-471 (1986)

4.48 S. Zacks: *The Theory of Statistical Inference* (Wiley, New York, 1971)

4.49 X. Zhuang, T. S. Huang, R. M. Haralick: Two-view motion analysis: a unified algorithm. J. Opt. Soc. Amer., A, **3**, 1492-1500 (1986)

4.50 X. Zhuang, T. S. Huang, N. Ahuja, R. Haralick: A simplified linear optic flow-motion algorithm. Computer Vision, Graphics and Image Processing, **42**, 334-344 (1988)

Chapter 5

5.1 O. D. Faugeras, S. Maybank: Motion from point matches: multiplicity of solutions, in Proc. Workshop on Visual Motion (IEEE Computer Soc. Press, Washington D.C. 1989) pp. 248-255

5.2 C. -H. Lee: Structure and motion from two perspective views via planar patch, in Proc. 2nd Int'l Conf. Computer Vision (IEEE Computer Soc. Press, Washington D.C. 1988) pp. 158-164

5.3 H. C. Longuet-Higgins, The visual ambiguity of a moving plane, in Proc. Roy. Soc. London, B, **223**, 165-175 (1984)

5.4 H. C. Longuet-Higgins: The reconstruction of a scene from two projections — configurations that defeat the 8-point algorithm, in Proc. 1st Conf. Artif. Intell. Applications (IEEE Computer Soc. Press, Washington D.C. 1984) pp. 395-397

5.5 S. Negahdaripour, B. K. P. Horn: Direct passive navigation. IEEE Trans. Pattern Anal. Machine Intell., **9**, 168-176 (1987)

5.6 A. N. Netravali, T. S. Huang, A. S. Krishnakumar, R. J. Holt: Algebraic methods in 3D motion estimation from two-view point correspondences. Int'l J. Imaging systems and Technology, **1**, 78-99 (1989)

5.7 R. Y. Tsai, T. S. Huang: Estimating three-dimensional motion parameters of a rigid planar patch, II: singular value decomposition. IEEE Trans. Acoustics, Speech, and Signal Process., **30**, 525-534 (1983)

5.8 R. Y. Tsai, T. S. Huang: Estimating three-dimensional motion parameters of a rigid planar patch, III: finite point correspondences and the three-view problem. IEEE Trans. Acoustics, Speech, Signal Processing, **32**, 213-220 (1984)

5.9 A. M. Waxman, K. Wohn: Contour evolution, neighborhood deformation, and global image flow: planar surfaces in motion. Int'l J. Robotics Research, **4**, 95-108 (1985)

5.10 J. Weng, N. Ahuja, T. S. Huang: Motion and structure from point correspondences: a robust algorithm for planar case with error estimation, in Proc. Int'l Conf. Pattern Recognition (IEEE Computer Soc. Press, Washington D.C. 1988) pp. 247-251

5.11 J. Weng, N. Ahuja, T. S. Huang: Motion and structure from point correspondences with error estimation: planar surfaces. IEEE Trans. Acoustics, Speech, and Signal Processing, **39**, 2691-2717 (1991)

Chapter 6

6.1 N. Ayache, B. Faverjon: Efficient registration of stereo images by matching graph descriptions of edge segments. Int'l J. Computer Vision, **1**, 107-131 (1987)

6.2 J. Burns, A. Hanson, E. Riseman: Extracting straight lines. IEEE Trans. Pattern Anal. Machine Intell., **8**, 425-455 (1986)

6.3 O. D. Faugeras, F. Lustman, G. Toscani, Motion and structure from point and line matches, in Proc. Int'l Conf. Computer Vision (IEEE Computer Soc. Press, Washington D.C. 1987)

6.4 Y. Liu, T. S. Huang: Estimation of rigid body motion using straight line correspondences, further results, in Proc. Int'l Conf. Pattern Recognition (IEEE Computer Soc. Press, Washington D.C. 1986) pp. 306-307

6.5 Y. Liu, T. S. Huang: A linear algorithm for determining motion and structure from line correspondences. Computer Vision, Graphics and Image Processing, **44**, 35-57 (1988)

6.6 J. H. McIntosh, K. M. Mutch: Matching straight lines. Computer Vision, Graphics and Image Processing, **43**, 386-408 (1988)

6.7 A. Mitiche, S. Seida, J. K. Aggarwal: Interpretation of structure and motion using straight line correspondences, in Proc. Int'l Conf. Pattern Recognition (IEEE Computer Soc. Press, Washington D.C. 1986) pp. 1110-1112

6.8 R. Nevatia, K. R. Babu: Linear feature extraction and description. Computer Graphics and Image Processing, **13**, 257-269 (1980)

6.9 M. Spetsakis, J. Aloimonos: Closed form solution to the structure from motion problem from line correspondences, in Proc. 6th AAAI National Conf. on Artif. Intell. (Amer. Assoc. Artif. Intell., 1987) pp. 738-743

6.10 J. Weng, Y. C. Liu, T. S. Huang, N. Ahuja: Estimating motion/structure from line correspondences: a robust linear algorithm and uniqueness theorems, in Proc. IEEE Conf. Computer Vision and Pattern Recognition (IEEE Computer Soc. Press, Washington D.C. 1988) pp. 387-392

6.11 J. Weng, T. S. Huang, N. Ahuja: Motion and structure estimation from line matches: performance obtained and beyond, in Proc. 10th Int'l Conf. Pattern Recognition (IEEE Computer Soc. Press, Washington D.C. 1990) pp. 168-172. Also ref.: Motion and structure from line correspondences: closed-form solution, uniqueness, and optimization, accepted and to appear in IEEE Trans. Pattern Anal. Machine Intell., **14** (1992)

6.12 B. L. Yen, T. S. Huang: Determining 3-D motion and structure of a rigid body using straight line correspondences, in *Image Sequence Processing and Dynamic Scene Analysis* (Springer, New York, 1983)

Chapter 7

7.1 K. S. Arun, T. S. Huang, S. D. Blostein: Least-squares fitting of two 3-D point sets. IEEE Trans. Pattern Anal. Machine Intell., **9**, 698-700 (1987)

7.2 N. Ayache, O. Faugeras: Building, registration, and fusing noisy visual maps, in Proc. 1st Int'l Conf. Computer Vision (IEEE Computer Soc. Press, Washington D.C. 1987) pp. 73-82

7.3 S. D. Blostein, T. S. Huang: Error analysis in stereo determination of 3-D point positions. IEEE Trans. Pattern Anal. Machine Intell., **9**, 752-765 (1987)

7.4 T. J. Broida, R. Chellappa: Estimation of object motion parameters from noisy images. IEEE Trans. Pattern Anal. Machine Intell., **8**, 90-99 (1986)

7.5 K. M . Brown, J. E. Dennis: Derivative free analogues of the Levenberg-Marquardt and Gauss algorithms for nonlinear least squares approximation. Numeriche Mathematik, **18**, 289-297 (1972)

7.6 N. Cui, J. Weng, P. Cohen: Extended structure and motion analysis from monocular image sequences, in Proc. 3rd Int'l Conf. Computer Vision (IEEE Computer Soc. Press, Washington D.C. 1990) pp. 222-229

7.7 N. Cui, J. Weng, P. Cohen: Motion and structure from long stereo image sequences, in Proc. IEEE Workshop on Visual Motion (IEEE Computer Soc. Press, Washington D.C. 1991) pp. 75-80

7.8 O. D. Faugeras, M. Hebert: A 3-D recognition and positioning algorithm using geometrical matching between primitive surfaces, in Proc. 8th Int'l Joint Conf. Artif. Intell. (William Kaufmann, Los Angeles, LA 1983) pp. 996-1002

7.9 T. S. Huang, S. D. Blostein, E. A. Margerum: Least-squares estimation of motion parameters from 3-D point correspondences, in Proc. IEEE Conf. Computer Vision and Pattern Recognition (IEEE Computer Soc. Press, Washington D.C. 1986) pp. 198-201

7.10 P. J. Huber: *Robust Statistics* (Wiley, New York, 1981)

7.11 D. C. Hoaglin, F. Mosteller, J. W. Tucky: *Understanding Robust and Exploratory Data Analysis* (Wiley, New York, 1983)

7.12 B. K. Horn: Closed-form solution of absolute orientation using unit quaternions. J. Opt. Soc. Amer., A, **4**, 629-642 (1987)

7.13 B. Kamgar-Parsi, B. Kamgar-Parsi: Evaluation of quantization error in computer vision. IEEE Trans. Pattern Anal. Machine Intell., **11**, 929-940 (1989)

7.14 S. M. Kiang, R. J. Chou, J. K. Aggarwal: Triangulation errors in stereo algorithms, in Proc. IEEE Workshop on Computer Vision (IEEE Computer Soc. Press, Washington D.C. 1987) pp. 72-78

7.15 R. V. Kumar, A. Tirumalai, R. C. Jain: A non-linear optimization algorithm for the estimation of structure and motion parameters, in Proc. IEEE Conf. Computer Vision and Pattern Recognition (IEEE Computer Soc. Press, Washington D.C. 1989) pp. 136-143

7.16 C. Lee, R. M. Haralick, X. Zhuang: Recovering 3-D motion parameters from image sequences with gross errors, in Proc. Workshop on Visual Motion (IEEE Computer Soc. Press, Washington D.C. 1989) pp. 46-53

7.17 K. Levenberg: A method for the solution of certain nonlinear problems in least squares. Quart. Appl. Math., **2**, 164-168 (1944)

7.18 D. W. Marquardt: An algorithm for least squares estimation of nonlinear parameters. SIAM J. Appl. Math., **11**, 431-441 (1963)

7.19 L. Matthies, S. Shafer: Error Modeling in Stereo Navigation. IEEE J. Robotics and Automation, **3**, 239-248 (1987)

7.20 J. J. Rodriguez, J. K. Aggarwal: Stochastic analysis of stereo quantization error. IEEE Trans. Pattern Anal. Machine Intell., **12**, 467-470 (1990)

7.21 M. D. Shuster: Approximate algorithms for fast optimal attitude computation, in Proc. AIAA Guidance and Control Specialist Conf. (Palo Alto, 1978) pp. 88-95

7.22 A. Tirumalai, B. Schunck, R. C. Jain: Robust dynamic stereo for incremental disparity map refinement (University of Washington, Seattle, WA 1990)

7.23 J. Weng, N. Ahuja, T. S. Huang: Optimal motion and structure estimation, in Proc. IEEE Conf. Computer Vision and Pattern Recognition (IEEE Computer Soc. Press, Washington D.C. 1989) pp. 144-152

7.24 J. Weng, P. Cohen, N. Rebibo: Fusion of stereo views in long image sequences, in Proc. First International Workshop on Multi-Sensor Fusion and Environment Modeling (Int'l Advanced Robotics Programme, Toulouse, France 1989). Also ref.: Motion and structure estimation from stereo image sequences, accepted and to appear in IEEE Trans. Robotics and Automation (1992)

7.25 J. Weng, P. Cohen: Robust motion and structure estimation using stereo vision, in Proc. Int'l Workshop Robust Computer Vision (University of Washington, Seattle, WA 1990)

7.26 J. Weng, P. Cohen, M. Herniou: Stereo camera calibration with nonlinear corrections, in Proc. 10th Int'l Conf. Pattern Recognition (IEEE Computer Soc. Press, Washington D.C. 1990) pp. 246-253. Also ref.: Camera calibration with distortion models and accuracy evaluation, accepted and to appear in IEEE Trans. Pattern Anal. Machine Intell., **14** (1992)

Chapter 8

8.1 O. Bottema, B. Roth: *Theoretical Kinematics* (North-Holland, New York, 1979)

8.2 W. L. Brogan: *Modern Control Theory* (Prentice-Hall, New Jersey, 1982)

8.3 T. J. Broida, R. Chellappa: Estimation of object motion parameters from noisy images. IEEE Trans. Pattern Anal. Machine Intell., **8**, 90-99 (1986)

8.4 T. J. Broida, R. Chellappa: Kinematics and structure of a rigid object from a sequence of noisy images, in Proc. IEEE Workshop on Motion: Representation and Analysis (IEEE Computer Soc. Press, Washington D.C. 1986) pp. 95-100

8.5 B. A. Conway, J. E. Tulinglowski, P. D. Webber: Dynamics of remote orbital capture, in Proc. AAS/AIAA Astrodynamics Specialist Conference (1983)

8.6 G. R. Fowles, *Analytical Mechanics*, 3rd ed. (Holt, Rinehart and Winston, Now York, 1977)

8.7 T. S. Huang, J. Weng, N. Ahuja: 3-D motion from image sequences: modeling, understanding and prediction, in Proc. IEEE Workshop on Motion: Representation and Analysis (IEEE Computer Soc. Press, Washington D.C. 1986)

8.8 M. H. Kaplan, A. A. Nadkarni: Control and stability problems of remote orbital capture. Mechanism and Machine Theory, **12** 57-64 (1977)

8.9 C. L. Lawson, R. J. Hanson: *Solving Least Squares Problems* (Prentice-Hall, New Jersey, 1974)

8.10 W. D. Macmillan: *Dynamics of Rigid Bodies* (Mcgraw-Hill, New Jersey, 1936)

8.11 S. Ullman: *The Interpretation of Visual Motion* (MIT press, Cambridge, MA, 1979)

8.12 J. Weng, T. S. Huang, N. Ahuja: 3-D motion estimation, understanding and prediction from noisy image sequences. IEEE Trans. Pattern Anal. Machine Intell., **9**, 370-389 (1987)

8.13 Y. Yasumoto, G. Medioni, Robust estimation of three-dimensional motion parameters from a sequence of image frames using regularization. IEEE Trans. Pattern Anal. Machine Intell., **8**, 464-471 (1986)

8.14 G. S. Young, R. Chellappa, 3-D motion estimation using a sequence of noisy stereo images: models, estimation, and uniqueness results. IEEE Trans. Pattern Anal. Machine Intell., **12**, 735-759 (1990)

Subject Index